# Numerical Methods in HEAT TRANSFER

*Edited by*
R. W. Lewis
K. Morgan
O. C. Zienkiewicz
*Department of Civil Engineering*
*University of Wales, Swansea*

*A Wiley–Interscience Publication*

**JOHN WILEY & SONS**
Chichester · New York · Brisbane · Toronto

Copyright © 1981 by John Wiley & Sons Ltd.

All rights reserved.

**British Library Cataloguing in Publication Data:**

Numerical methods in heat transfer.—(Wiley
  series in numerical methods in engineering).
  1. Heat—Transmission—Mathematics—Congresses
  2. Numerical calculations—Congresses
  I. Lewis, Roland W. II. Morgan, Kenneth
  III. Zienkiewicz, Olegierd Cecil
  536′.2′05117    QC320.2     80-49973

  ISBN 0 471 27803 3

Printed in Great Britain by J. W. Arrowsmith Ltd., Bristol

# Numerical Methods
## in
# HEAT TRANSFER

## WILEY SERIES IN
## NUMERICAL METHODS IN ENGINEERING

Consulting Editors
**R. H. Gallagher,** College of Engineering,
University of Arizona
and
**O. C. Zienkiewicz,** Department of Civil Engineering,
University College Swansea

---

# Contributing Authors

M. A. ALADJEM     *Applied Mathematics Centre, P.O. Box 384, Sofia 1000, Bulgaria*

M. D. ALMOND     *Intern, Nuclear Medicine, Walter Reed General Hospital, Washington, D.C. 20014, U.S.A.*

M. BORSETTO     *ISMES, Bergamo, Italy*

C. A. BREBBIA     *Department of Civil Engineering, University of Southampton, Southampton SO9 5NH, U.K.*

G. CARRADORI     *ISMES, Bergamo, Italy*

A. CHAUDOUET     *CETIM, B.P.67, 60304 Senlis Cedex, France*

J. CRANK     *Brunel University, Uxbridge UB8 3PH, Middlesex, U.K.*

D. DUTOYA     *Office National d'Etudes et de Recherches Aérospatiales, 92320 Châtillon, France*

W. D. L. FINN     *Department of Civil Engineering, The University of British Columbia, Vancouver, B.C., Canada V6T 1W5*

F. K. FONG     *Department of Chemical Engineering, The Ohio State University, Columbus, Ohio 43210, U.S.A.*

M. A. HOGGE     *Laboratoire D'Aéronautique, Université de Liège, Belgium*

R. J. HOPKIRK     *Electrowatt Engineering Services Ltd, P.O. Box 8022, Zurich, Switzerland*

L. IMRE     *Institute of Heat and Systems Engineering, Technical University of Budapest, 1111, Budapest, Goldman Gy. tér 3, Hungary*

E. E. KHALIL — Department of Mechanical Engineering, Cairo University, Cairo, Egypt

R. A. KNAPP — Engineering and Research Staff, Ford Motor Company, Dearborn, Michigan 48121, U.S.A.

R. M. KRUDENER — Consultant, Owner Therm-Search-Associates, P.O. Box 10, Jackson, Wyoming 83025, U.S.A.

D. W. LARSON — Sandia Laboratories, Albuquerque, New Mexico 87115, U.S.A.

E. C. LEMMON — Intermountain Technologies Inc., P.O. Box 1604, Idaho Falls, Idaho 83401, U.S.A.

E. LEONARDI — University of New South Wales, Kensington, 2033, Australia

G. LOUBIGNAC — CETIM, B.P.67, 60304 Senlis Cedex, France

D. R. LYNCH — Thayer School of Engineering, Dartmouth College, Hanover, NH, U.S.A.

I. L. MACLAINE-CROSS — Department of Mining and Mineral Sciences, University of New South Wales, Broken Hill, 2880, Australia

P. MICHARD — Office National d'Etudes et de Recherches Aérospatiales, 92320 Châtillon, France

M. D. MIKHAILOV — Applied Mathematics Centre, P.O. Box 384, Sofia 1000, Bulgaria

H. MOEN — Institute for Energy Technology, Box 40, N-2007 Kjeller, Norway

T. V. NGUYEN — School of Mechanical and Industrial Engineering, University of New South Wales, Kensington, 2033, Australia

K. O'NEILL — U.S. Army Cold Regions Research and Engineering Laboratory, Hanover, NH, U.S.A.

P.-J. PRALONG — Electrowatt Engineering Services Ltd, P.O. Box 8022, Zurich, Switzerland

M. PREDELEANU — Laboratoire de Mécanique et Technologie, Université Paris VI—Enset, 61 Avenue du Président Wilson, 94230 Cachan, France

J. A. REIZES — *University of New South Wales, Kensington, 2033, Australia*

R. RIBACCHI — *Institute of Mining, University of Rome, Rome, Italy*

M. A. SERAG-ELDIN — *Heat Transfer Section, Mechanical Engineering Department, Imperial College, London SW7, U.K.*

D. SHARMA — *Dames & Moore, Denver, Colorado, U.S.A.*

D. R. SKIDMORE — *Department of Chemical Engineering, The Ohio State University, Columbus, Ohio 43210, U.S.A.*

D. B. SPALDING — *Heat Transfer Section, Mechanical Engineering Department, Imperial College, London SW7, U.K.*

E. A. THORNTON — *Old Dominion University, Norfolk, Virginia 23508, U.S.A.*

G. deVAHL DAVIS — *School of Mechanical and Industrial Engineering, University of New South Wales, Kensington, 2033, Australia*

E. VAROĞLU — *Faculty of Graduate Studies, The University of British Columbia, Vancouver, B.C., Canada V6T 1W5*

A. R. WEITING — *NASA, Langley Research Centre, Hampton, Virginia 23665, U.S.A.*

L. D. WILLS — *Instructor, Mechanical Engineering Department, University of Arkansas, Fayetteville, Arkansas 72701, U.S.A.*

H. WOLF — *Raymond F. Giffels Distinguished Professor, Mechanical Engineering Department, University of Arkansas, Fayetteville, Arkansas 72701, U.S.A.*

L. C. WROBEL — *Department of Civil Engineering, University of Southampton, Southampton SO9 5NH, U.K.*

H. WU — *Engineering and Research Staff, Ford Motor Company, Dearborn, Michigan 48121, U.S.A.*

O. C. ZIENKIEWICZ — *Civil Engineering Department, University of Wales, Swansea, U.K.*

# Preface

The increasingly important role of numerical methods in the analysis of heat transfer was the major factor in presenting an international conference on this topic in Swansea during July, 1979. The authors present at this conference were drawn from universities, research institutions, and an industrial environment. It therefore appeared that the best overall picture of the 'state of art' would best be served by an organized multi-author presentation from these three environments. With this in mind the editors invited a series of contributions from some of the authors present at the Swansea conference.

Twenty-four contributions in all are included with the opening paper by Zienkiewicz on 'Finite Elements in Thermal Problems'. This discusses the history of numerical methods in thermal problems with particular reference to the finite element method. The background of the generalized finite element method to an ever-increasing range of thermal situations is discussed and particular reference is made to topics such as thermal stress analysis, non-linearity in steady-state and transient situations, convective–diffusive heat transport and, finally, coupled thermal flow problems.

The second chapter, by Mikhailov and Aladjem, discusses the development of software systems for automatic solution of ordinary and partial differential equations with particular emphasis on a new interactive system in which the boundary conditions, order of approximation, and special instructions are specified by the user. The versatility of the new package is demonstrated via the solution of some specific thermal problems. The succeeding chapter by Imre discusses the importance of not only thermal but also hydrodynamical, electrical, and mechanical interaction in composite devices built from components of various materials and having different geometries. An example is given of the solution of thermohydrodynamical and thermomechanical structural systems.

A review of the many algorithms available for the numerical integration of the evolution equations of non-linear heat condition is presented by Hogge in Chapter 4. From this study a recommendation is made with regard to the two- and three-level schemes discussed in the chapter. The next chapter, by Wrobel and Brebbia, introduces the boundary element method in the context of thermal problems. The technique is presented for both steady-state and transient heat conduction problems with the base equations being deduced by using weighted residuals. Several applications are presented to show some of the potentialities of the method in the solution of such problems. This chapter is followed by Chaudouet and Loubignac's adaptation of the governing equations for elasticity and temperature distribution into a boundary integral equation formulation. Three problems are discussed: a hollow cylinder of incompressible material, a typical industrial example, and the analysis of a roller in the continuous casting of steel slabs. The advantages of their technique compared with finite difference or finite element formulations are discussed, leading to the conclusions that computer savings and greater accuracy are possible. Another French contribution by Predeleanu also discusses a boundary solution approach for the dynamic problem of thermoviscoelasticity. The basic integral representations of displacement and temperature fields are deduced for a general class of viscoelastic bodies defined by Riemann–Stieltjes integral convolutions and two applications of the technique are given. The first is concerned with the boundary integral equation method and the second with an 'interface' condition for the coupling of the boundary approach with other numerical and analytical methods.

Chapter 8, by Wolf *et al.*, describes the application of finite difference methods to the delineation of temperature histories in composite nuclear fuel rods of the type used in fast oxide reactors. A detailed exposition is given of the numerical techniques employed as well as the relevant material properties and boundary conditions. Results are quoted for the particular case of the adiabatic centre void.

The extremely complicated problem of moving boundaries in thermal problems is discussed by Crank in Chapter 9. Two examples of such problems are given, namely melting ice and the oxygen diffusion and consumption problem. Analytic solutions, both exact and approximate, are given for the two particular situations, along with finite difference solutions involving fixed grids, variable space grids and with fitted curvilinear coordinates.

The chapters by Lemmon and by O'Neill and Lynch, deal with phase change problems. Lemmon's chapter describes a technique to account for convective heat transfer at the moving phase-change interface. A novel feature of the method is that the mesh is not changed even though the phase-change interface moves through it. Comparisons are made between the technique and other solutions, either approximate or exact, and it is concluded that favourable answers are obtained. The chapter by O'Neill and Lynch presents a general method for the finite element simulation of problems involving phase change or unusually high heat flux. The technique is based upon the deformation and/or translation of a single initially specified mesh. An example is given of a one-dimensional problem involving heat conduction with and without phase change.

There then follow two chapters, which discuss the important role of geothermal energy and some associated problems. The first, by Borsetto, Carradori and Ribacchi, presents a mathematical model for the coupled problem of heat and mass transfer and stress fields under transient conditions. Specific attention is paid to jointed or microfractured rocks and some typical geothermal problems are presented. The next, by Hopkirk, Sharma and Pralong, investigates the geothermal reservoir initiated by a hydraulic fracture. The particular problem of rock shrinkage during operation of the geothermal reservoir is investigated and a simplified method of treating this is demonstrated via an example.

Chapter 14 by Moen presents a survey of simulation models for industrial processes with particular reference to metallurgical problems. A finite difference form of the dynamic heat conduction equation is used for analysis of typical problems such as casting of aluminium ingots as a semi-continuous process. Skidmore and Fong present a heat flow model in underground coal liquefaction. This process involves the introduction of a solvent or comminution reagent into a coal seam and converting monolithic coal particles in slurry or into coal-derived oils. A new technique, called the method of alternating variables, is used. This combines the ADI iteration method for the parabolic equation and the implicit central difference method for the hyperbolic equation by solving for alternate variables at successive time intervals. Chapter 16, by Nguyen, Maclaine-cross and de Vahl Davis, investigates the simultaneous development of the velocity and temperature distributions in a fluid flowing through a cascade of parallel horizontal plates. Results are quoted for different values of Reynolds and Prandtl numbers and for various values of plate spacings. Contour plots of streamlines and isotherms show the effects of free convection and of the channel entrance on the flow pattern and temperature distribution.

In Chapter 17, a characteristics based finite element method is presented for heat transport problems involving convection. Finn and Varoğlu give examples of their novel technique for heat transport problems in rivers, and indicate the utility and accuracy of the method. Leonardi and Reizes then discuss a compressible vector potential–vorticity formulation for the particular case of a Newtonian fluid with variable properties. They show that the conventional parameters, the Rayleigh and Prandtl numbers, aspect ratio, and angle of inclination do not fully specify a problem and that additional parameters are necessary. Dutoya and Michard in Chapter 19 present a program for calculating the development of a boundary layer along a turbine blade profile, starting from the stagnation point, and through to its laminar, transitional, and turbulent states. Examples are given for the particular cases of boundary layer development along an adiabatic flat plate, and along the pressure and suction sides of a cooled turbine inducer blade.

In Chapter 20, Wieting and Thornton discuss the relationship between the thermal environment and the structural design for the case of a convectively-cooled aircraft. Recent developments and applications in finite element methods for such problems are presented, and they conclude that these techniques are competitive to the well established finite difference lumped-parameter method. Serag-Eldin and Spalding in the next chapter describe a numerical prediction procedure for the computation of three-dimensional recirculating and swirling flows inside can combustion chambers. They also compare with measurements some predictions for the turbulent non-reacting flow in a model of a gas turbine combustion chamber.

Chapters 22 and 23 deal respectively with conditions induced by a fire and the numerical computation of turbulent reacting combustor flows. The chapter by Larson develops a mathematical model of a fire in a room size enclosure and comes to the conclusion that radiative heat transfer is predominant and must be included, along with turbulent flow, to achieve realistic modelling of enclosed fires. In Chapter 23, Khalil solves the governing conservation equations of mass, momentum, and energy using a finite difference simulator for problems such as the design of furnaces and combustion chambers.

The final chapter by Wu and Knapp presents an analytical method for studying the thermal conditions in automotive internal combustion gasoline engines. Calculated temperature distributions and heat rejection rates in sections of three different types of engines are presented as examples to illustrate the procedure and potential use of the technique.

## Acknowledgements

The editors wish to thank the *International Journal of Numerical Methods in Engineering* for its cooperation, and each of the authors for giving their time in the preparation and reviewing of the manuscripts.

R. W. LEWIS
K. MORGAN
O. C. ZIENKIEWICZ

# Contents

## Contents

*Numerical Methods in Heat Transfer*
Edited by R. W. Lewis, K. Morgan, and O. C. Zienkiewicz
© 1981 John Wiley & Sons Ltd

Chapter 1

# Finite Element Methods in Thermal Problems

*O. C. Zienkiewicz*

## 1.1 INTRODUCTION

The heat diffusion equation

$$\nabla^{\mathrm{T}}\mathbf{k}\nabla T + Q - c\frac{\partial T}{\partial t} = 0 \tag{1.1}$$

is surely the most widely studied and classical example of a continuum mathematical problem to the solution of which numerical techniques have been widely applied. Finite differences, starting with the well known work of Richardson[1] at the turn of the century, aided by the relaxation methods of Southwell[2] in the forties have today reached an extremely high level of efficiency in their solution both for steady and transient states. The more recent finite element method, which first made its impact in structural engineering in the late fifties, was first applied to the problem in the sixties by Zienkiewicz and Cheung,[3] Visser,[4] Wilson and Nickell[5] and others.

It soon became evident that all the features of the finite difference method and indeed of other numerical procedures are but particular examples of the generalized trial function, finite element, method.[6] This must therefore contain all the merits of other techniques and possibly offer additional advantages.

Why then at the present time the widespread interest in numerical methods for thermal problems? Obviously the answer lies both in the continuing development of the methodology and in the need for development of new procedures for an ever increasing range of thermal situations encountered. Thus, in this first chapter, we shall try to indicate both the background of the generalized finite element method as well as some of the special problems encountered. These will include:

(a) problems of thermal stress analysis;
(b) treatment of non-linearity in steady-state and transient situations of heat diffusion;
(c) convective–diffusive heat transport; and
(d) coupled thermal flow problems.

As these and indeed many other areas of thermal problem analysis will be dealt with in detail in succeeding chapters, the treatment will of necessity be brief. Indeed the elaboration of the finite element technique itself will have to be very general leaving many details to such texts as Zienkiewicz.[6]

## 1.2  THE ESSENTIALS OF THE FINITE ELEMENT (GENERALIZED) METHOD

The numerical 'discretization' of a continuum problem, such as that defined by Equation (1.1) and its associated boundary conditions, e.g.

$$T - \bar{T} = 0 \quad \text{on } \Gamma_1$$
$$\mathbf{n}^T \mathbf{k} \nabla T - q = 0 \quad \text{on } \Gamma_2 \tag{1.2}$$

is the essence of all numerical solutions. (In the above $\Gamma_1$ and $\Gamma_2$ define parts of the boundary on which temperatures $\bar{T}$ or fluxes $q$ are given and $\mathbf{n}$ represents the vector of unit magnitude normal to the boundary.)

The finite element method is defined in its generality as any process in which:

(1)  The continuous function $T$ is approximated by a series of parameters $a_i$ and specified trial functions $N_i(x, y, z)$ in the problem domain $\Omega$ as

$$T \approx \hat{T} = \sum N_i a_i \quad i = 1 - n \tag{1.3}$$

(2)  If the differential equation (such as Equation (1.1)) and its boundary conditions (such as Equation (1.2)) are written in a general form as (Figure 1.1)

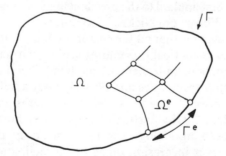

**Figure 1.1**  Problem domain ($\Omega$), boundary ($\Gamma$), and elements

$$A(T) = 0 \quad \text{in } \Omega$$
$$B(T) = 0 \quad \text{in } \Gamma \tag{1.4}$$

then the approximating equations from which solution is to be obtained are written as a set

$$\int_\Omega W_j A(\hat{T})\, d\Omega + \int_\Gamma \bar{W}_j B(\hat{T})\, d\Gamma = 0 \quad j = 1 - n \tag{1.5}$$

where $W_j$ and $\bar{W}_j$ are suitably defined weighting functions which ensure that as $n \to \infty$

$$A(\hat{T}) \to 0 \quad \text{in } \Omega$$

and

$$B(\hat{T}) \to 0 \quad \text{in } \Gamma \tag{1.6}$$

i.e. that at all points the approximation tends to the exact solution.

The reader will recognize the above as an expression of the *weighted residual method*[7,8] which appears to have its origins early in this century in the works of Galerkin and his associates.[9] This is indeed the case, but an important property allowing a local, narrowly based definition of shape functions appears to have been missed in the early work. This simple property is based on the fact that a definite integral of the type occurring in Equation (1.5) is simply a sum of such integrals occurring on 'subdomains' into which the whole 'domain' is divided (Figure 1.1). Thus if

$$\Omega = \sum_{e=1}^m \Omega^e \quad \text{and} \quad \Gamma = \sum_{e=1}^m \Gamma^e \tag{1.7}$$

where $\Omega^e$ and $\Gamma^e$ are associated with 'elements' into which we divide the problem, then for all finite functions ( ) we have

$$\int_\Omega (\ )\, d\Omega \equiv \sum_{e=1}^m \int_{\Omega^e} (\ )\, d\Omega$$

$$\int_\Gamma (\ )\, d\Gamma \equiv \sum_{e=1}^m \int_{\Gamma^e} (\ )\, d\Gamma \tag{1.8}$$

This allows the whole region to be divided into standard type of subregions in which the parameters $a_i$ are usually the nodal values of the independent function $T$ and in which the trial functions are defined in a local manner. The integrals can then be evaluated element by element and the approximating equations (such as Equation (1.5)) obtained by a simple addition of element contributions.

The approximating equation (Equation 1.5)) will be a simple algebraic equation system if $a_i$ are constants and the shape functions $N_i$ are given as functions of all the independent variables in the problem domain. In such a case if the basic equations are linear we arrive at an algebraic system of the form

$$\mathbf{Ha} + \mathbf{f} = \mathbf{0} \qquad \mathbf{a} = [a_1, a_2, \ldots, a_n]^T \tag{1.9}$$

from which $\mathbf{a}$ and hence the approximation $\hat{T}$ can be found by computation.

If we define $a_i$ as a function of one of the independent variables and use trial functions defined in terms of the remaining independent variables, then the process is termed '*semi-discrete*' and results in a system of ordinary differential equations which can often be solved analytically (or to which the numerical approximation can once again be applied).

To make matters more specific we shall illustrate the discretization for the example of Equation (1.1) with boundary condition (1.2).

Now we shall pursue the semi-discretization process and write

$$T(x, y, z, t) \approx \hat{T} = \sum N_i(x, y, z)a_i(t)$$

with

$$W_j = W_j(x, y, z) \qquad \bar{W}_j = \bar{W}_j(x, y, z) \tag{1.10}$$

We can now write explicitly for a space domain $\Omega$

$$\int_\Omega W_j\left(\nabla^{\mathrm{T}} \mathbf{k}\nabla\hat{T} + Q - c\frac{\partial\hat{T}}{\partial t}\right) \mathrm{d}x\,\mathrm{d}y\,\mathrm{d}z$$

$$+ \int_{\Gamma_1} \bar{W}_j(\hat{T} - \bar{T})\,\mathrm{d}\Gamma + \int_{\Gamma_2} \bar{\bar{W}}_j(\mathbf{n}^{\mathrm{T}}\mathbf{k}\nabla\hat{T} - q)\,\mathrm{d}\Gamma = 0 \tag{1.11}$$

and immediately we have an ordinary differential equation set which can be written as

$$\mathbf{Ha} + \mathbf{C}\frac{\mathrm{d}}{\mathrm{d}t}\mathbf{a} + \mathbf{f} = 0 \tag{1.12}$$

where

$$H_{ji} = \int_\Omega W_j\nabla^{\mathrm{T}}\mathbf{k}\nabla N_i\,\mathrm{d}x\,\mathrm{d}y\,\mathrm{d}z + \int_{\Gamma_1} \bar{W}_jN_i\,\mathrm{d}\Gamma + \int_{\Gamma_2} \bar{\bar{W}}_j\mathbf{n}^{\mathrm{T}}\mathbf{k}\nabla N_i\,\mathrm{d}\Gamma$$

$$C_{ji} = \int_\Omega W_jcN_i\,\mathrm{d}x\,\mathrm{d}y\,\mathrm{d}z \tag{1.13}$$

$$f_j = \int_\Omega W_jQ\,\mathrm{d}x\,\mathrm{d}y\,\mathrm{d}z - \int_{\Gamma_1} \bar{W}_j\bar{T}\,\mathrm{d}\Gamma - \int_{\Gamma_2} \bar{\bar{W}}_jq\,\mathrm{d}\Gamma$$

In above, as we have already stated, each integral can be found as a sum of the contribution of the 'elements'.

In the steady-state heat flow problem (e.g. posed in Equation (1.12) if $C = 0$), the discretization is reduced to an ordinary set of algebraic equations of the form

$$\mathbf{Ha} + \mathbf{f} = 0 \tag{1.14}$$

The particular (or general) form of the approximation means that we can immediately show that all well known and evidently different numerical techniques are embraced in the above definitions. Thus if we take

(1)  $W_j = \bar{W}_i = \delta_j(\mathbf{x}_j)$, i.e. the Dirac function, we obtain a general class of *point collocation procedures*. Indeed if the trial functions $N_i$ are locally defined by polynomial expansions all features of the *finite difference method are recovered*.

(2)  If the problem is linear and if the trial functions $N_i$ are so chosen that

$$A(N_i) \equiv 0 \tag{1.15}$$

then the approximating equation (Equation (1.5)) contains only the integrals of the boundary contributions, i.e.

$$\int_\Gamma \bar{W}B(\hat{T})\,\mathrm{d}\Gamma = 0 \tag{1.16}$$

and *boundary solution* or Trefftz solution procedures[10,11,12] are obtained.

In most finite element applications we use the Galerkin–Bubnov approximation, i.e. take

(3)  $$W_j = N_j \tag{1.17}$$

and in addition specify locally defined polynomial functions. It is convenient in many cases to apply an integration by parts to the matrix terms of such equations as (1.13) to obtain an equal order of differentiation occurring in the $W_i$ and $N_j$ terms. Thus for instance we note that in Equation (1.13) we can rewrite $H_{ji}$ by such integration by parts (or Green's theorem) as

$$H_{ji} = -\int \nabla^{\mathrm{T}} W_j \mathbf{k} \nabla N_i \,\mathrm{d}x\,\mathrm{d}y\,\mathrm{d}z + \int_\Gamma W_j \mathbf{n}^{\mathrm{T}} \mathbf{k} \nabla N_i \,\mathrm{d}\Gamma$$

$$+ \int_\Gamma W_j N_i \,\mathrm{d}\Gamma + \int_{\Gamma_2} \bar{\bar{W}}_j \mathbf{n}^{\mathrm{T}} \mathbf{k} \nabla N_i \,\mathrm{d}\Gamma \tag{1.18}$$

Using the Galerkin–Bubnov trial function with

$$W_j = N_j \quad \bar{W} = N_j \quad \bar{\bar{W}}_j = -N_j \tag{1.19}$$

we can write simply

$$H_{ji} = -\int_\Omega \nabla N_j \mathbf{k} \nabla N_i + \int_{\Gamma_1} N_j N_i \,\mathrm{d}\Gamma \tag{1.20}$$

This form is symmetric, showing in this case the variational origin of the differential equation; it is simpler to compute too.

Each of the processes outlined above has its merits and in recent years much effort has been made to combine these in single computer programs.[13,14]

In what follows we shall restrict our discussion to the standard finite element processes.

Many alternative choices exist for both weighting ($W_j$) and trial (or shape) ($N_j$) functions and we have not yet discussed any limitations as to their choice. The most essential of these limitations are:

(a) *Linear independence*, ensuring that contributions of each term $N_i a_i$ cannot reproduce combined contributions of other terms.

(b) *Integrability*, which ensures that at each point of the domain the values of the integrand in Equation (1.15) remain finite. This imposes certain restrictions on the continuity of $W_j$ and $N_j$. For instance the definition (1.20) of $H_{ji}$ requires that $N_j$ be $C^0$ continuous (i.e. that the function itself be simply continuous).

(c) *Completeness*, or in other words the possibility of satisfying the requirement (1.6) as $n \to \infty$. This is probably the most difficult matter to ensure generally; but with (piecewise) polynomials used in the finite element expansions it is simply assured if the *complete* polynomial of order $p$ is such that

$$p \geqslant m \tag{1.21}$$

where $m$ is the highest order of derivative in the integrands occurring in the problem formulation.

For a full discussion of mathematical requirements in the general approximation the reader is referred to such tests as[6] or more mathematical exposition such as[15] and[16]. However, it is (almost) obvious that the error occurring in a finite element approximation of $T$ will be of order

$$O(h^{p+1}) \tag{1.22}$$

where $h$ is the element size. This follows from consideration of the properties of a local Taylor series expansion.

Similarly it is evident that the $m$th derivatives of $T$ will be approximated to the order

$$O(h^{p-m+1}) \tag{1.23}$$

Thus for instance the first derivative will have an approximation of the order $O(h^p)$ only—and in many cases such low approximation order is undesirable as the first derivative (corresponding to flux in thermal calculations or to strain in problems of solid mechanics) is of primary interest. In such cases fortunately we have recourse to a theorem which shows that for all elements (in a problem which is self-adjoint and has therefore a variational basis) there exist points at which the *first derivatives have one order higher approximation*.[17,18] Such points often coincide with certain Gaussian quadrature points which are used for numerical integration of the element integrals and in Figure 1.2 we show an approximation to a heat conduction problem of a one-dimensional steady-state kind with linear elements. It can be shown that in such a case the approximation

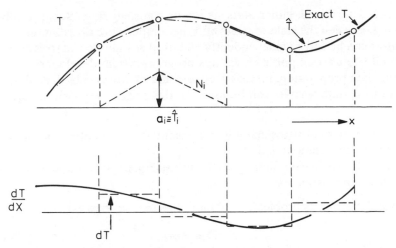

**Figure 1.2** Optimal sampling points for first derivatives in a one-dimensional approximation with linear elements ($p = 1$)

$\hat{T}$ and exact $T$ solutions coincide at nodes[19]—but what is more important to note here we observe that the approximation of the first derivative is best (and of one order higher convergence) at the element mid-point which is here the *optimal sampling point.*

In Figure 1.3 we show some such optimal sampling points for several commonly used finite elements in two-dimensional domains.

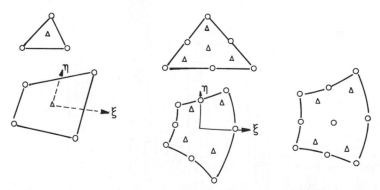

**Figure 1.3** Optimal sampling points for the first derivatives in $C^0$ continuous commonly used elements in two-dimensional problems

## 1.3  THERMAL STRESS ANALYSIS

One of the most common thermal problems is that of determining stress distributions in machine components or other solids for which the thermal field has been determined, either in its steady or transient state. As generally the coupling between the state of stress and the temperature distribution is weak, the latter can be solved independently and used as input to the stress analysis. Much effort has been put in by various program system developers to ensure that the data preparation, meshes used, etc. are identical in both analyses and that a direct data transfer can be used. In this section we wish to make two points:

(a)  that in thermal stress analysis it is essential to use optimal sampling for stress computation; and

(b)  that often a coarser mesh analysis suffices for stress rather than temperature computation.

In solid mechanics problems we copute thermal stresses as

$$\boldsymbol{\sigma} = \mathbf{D}(\boldsymbol{\varepsilon} - \boldsymbol{\varepsilon}_T) \tag{1.24}$$

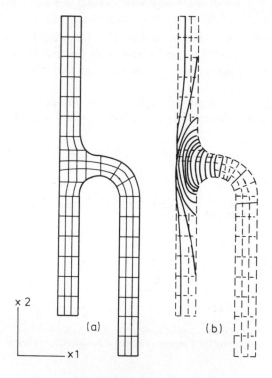

**Figure 1.4**  An axisymmetric machine component: (a) mesh of eight-node isoparametrics; (b) temperature contours

where $\sigma$ and $\varepsilon$ are the stress and strain tensors respectively, $\mathbf{D}$ is the elasticity matrix and $\varepsilon_T$ is the, essentially volumetric, initial, thermal strain:

$$\varepsilon_T = \mathbf{m}\alpha T \qquad m = [1, 1, 1, 0, 0, 0]^T \qquad (1.25)$$

It is well known that in elastic problems where near incompressibility occurs it is essential not only to sample the stresses at optimal points but to use 'reduced' integration rules to avoid overconstrained results (viz. reference (6), Chapter 11 and reference (18)). Similar precautions appear necessary in thermal stress analysis where severe volumetric constraints are imposed by thermal action. The sampling is particularly important as we illustrate in the example of Figures 1.4 and 1.5 (kindly supplied to us by Dr Beer[20]). In these figures we show an axisymmetric machine component in which a temperature distribution shown has been determined by (transient) finite element analysis (Figure 1.4) and the detail of stresses computed by a direct calculation from the displacement function at surface nodes is contrasted with a computation in which the values found at optimal sampling points are extrapolated to the surface by a bilinear fit (Figure 1.5). The results conclusively show the need for such an extrapolation—direct computation leading to values of normal stress comparable in magnitude to the maximum values. Clearly the accuracy gain is considerable.

Examination of Equations (1.24) and (1.25) indicates another phenomenon. While $\varepsilon$ is computed with a 'finite element' approximation accuracy, the temperature $T$ may be known precisely (or at least with an accuracy independent of the mesh used in the finite element stress analysis). Thus the second term of Equation (1.25) is not subject to the errors involved in the stress analysis. In stress analysis we shall find therefore that if the thermal 'forces', i.e.

$$\int_\Omega \mathbf{B}^T \mathbf{D} \mathbf{m} \alpha T \, d\Omega \qquad (1.26)$$

(using the notation of Zienkiewicz[6]) are correctly evaluated—the stress computation can be performed on a coarser mesh incapable of reproducing the full strain variation.

In Figure 1.6 we show how a quite complex thermal stress distribution (in a long bar subject to an edge intensive temperature field) is obtained using a very crude mesh of finite elements of a linear type. In practical problems we are quite often concerned with 'thermal shock' situations where a very sharp gradient of temperature exists near the surface. In such cases it is often possible to perform the stress analysis on a quite crude mesh and yet obtain high accuracy by correctly evaluating the temperature contributions.

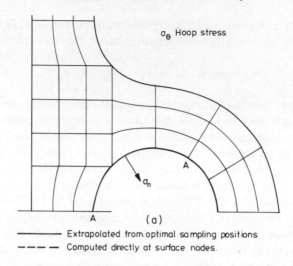

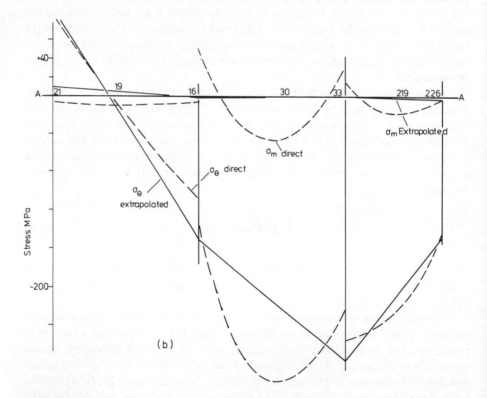

**Figure 1.5** Problem of Figure 1.4. (a) Mesh detail; (b) distribution of $\sigma_n$ (normal) and $\sigma_\theta$ (hoop) stresses along free surface AA: full curves, extrapolated from optimal sampling positions; broken curves, computed directly at surface nodes

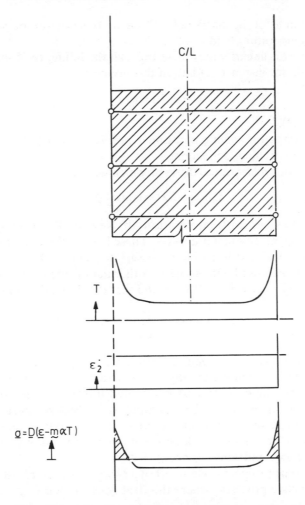

**Figure 1.6** A long bar with linear elements: the capability of
a coarse stress analysis mesh reproducing thermal stresses
with a high precision

## 1.4 NON-LINEAR PROBLEMS

In the discretization processes outlined in Section 1.2 we have not *a priori*
assumed linear behaviour. Thus if the conductivity **k**, specific heat $c$ and heat
generation term $Q$ are functions of temperature we shall simply arrive at
problems characterized by Equations (1.12) (or Equation (1.14)) for the steady
state, i.e.

$$\mathbf{H}\mathbf{a}+\mathbf{C}\frac{\mathrm{d}}{\mathrm{d}t}\mathbf{a}+\mathbf{f}=\mathbf{0} \tag{1.27}$$

in which some or all of the matrices $\mathbf{H}$, $\mathbf{C}$ or $\mathbf{f}$ are functions of the parameters $\mathbf{a}$ defining the temperature field.

Indeed such non-linearity may arise through the influence of the boundary conditions with, for instance, radiation flux giving

$$q = q(T) \approx q(\mathbf{a}) \tag{1.28}$$

in Equation (1.2).

Many techniques for solving non-linear problems exist today, the optimal ones often being problem dependent.

*For steady-state problems*, i.e. those in which we solve

$$\mathbf{H}(\mathbf{a})\mathbf{a} + \mathbf{f}(\mathbf{a}) \equiv \boldsymbol{\psi}(\mathbf{a}) = \mathbf{0} \tag{1.29}$$

Three basic techniques exist: (a) iteration; (b) Newton–Raphson (or modified NR) methods; (c) incremental processes. These basic techniques are described fully in Zienkiewicz[6] and it is only necessary to mention here that all three processes have been used with success in thermal problems. If the Newton–Raphson technique is used it will be found that the derivative matrix

$$\mathbf{H}_{\mathrm{T}} \equiv \frac{d\mathbf{H}}{d\mathbf{a}} \tag{1.30}$$

is in general non-symmetric even if Galerkin processes are adopted in discretization. This is computationally inconvenient and most frequently a modified Newton–Raphson process is adopted in which some approximate, symmetric, form of $\bar{\mathbf{H}}_{\mathrm{T}}$ is used in computing successive corrections. The practical interest in such non-linear steady-state solutions is generally associated with, mild, dependence of $\mathbf{k}$ on the temperature and in such situations the solutions show rapid convergence.

In radiation type problems (where often a power law is involved, $q \propto T^n$) or in chemically active processes where the heat generation is of type

$$Q = Q_0\, e^{\beta T} \tag{1.31}$$

the convergence may be slow or indeed non-existent as occasionally no solution is available. One such problem is associated with the storage of explosive material and when the parameter $Q_0$ reaches a certain 'critical value' solutions cease to exist. Such a problem has been studied extensively by Anderson and Zienkiewicz[21] and in Figure 1.7 we show how two non-linear solutions for $Q_0 < Q_{\mathrm{crit}}$ coalesce at $Q_0 = Q_{\mathrm{crit}}$ yielding no solution for $Q_0 > Q_{\mathrm{crit}}$.

For some steady-state, non-linear problems it is possible to linearize the solution by an artifice well documented in mathematical literature but not widely known. The process introduces no approximation within the limits of its validity and we shall describe this below.

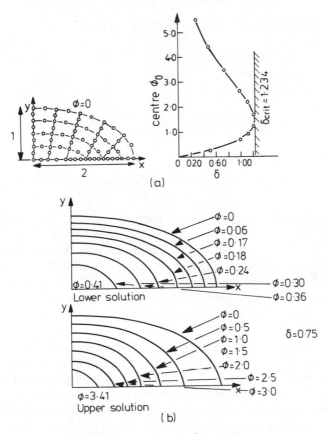

**Figure 1.7** Steady-state solution of $\nabla^2\phi + \delta\,e^\phi = 0$. Spontaneous ignition problem (above $\delta = \delta_{crit}$ no solutions exist)

If we are faced with steady-state heat conduction in which the material is homogeneous and isotropic the governing equation is

$$\nabla^{T}k\nabla T + Q \equiv k\nabla^{2}T + Q = 0 \tag{1.32}$$

where $k$ is now a position-independent function of $T$.

It is possible in this case to establish a unique, scalar, independent, variable $\phi$ defined as

$$\phi = \int_{0}^{T} k(\mathscr{T})\,d\mathscr{T} \qquad \frac{\partial\phi}{\partial T} = k(T) \tag{1.33}$$

As

$$\frac{\partial\phi}{\partial x} = \frac{\partial\phi}{\partial T}\frac{\partial T}{\partial x} = k\frac{\partial T}{\partial x} \tag{1.34}$$

and so on, Equation (1.32) can be rewritten simply as

$$\nabla^2\phi + Q = 0 \tag{1.35}$$

which is a linear situation.

Boundary conditions of the type

$$T - \bar{T} = 0 \quad \text{and} \quad k\frac{\partial T}{\partial n} - q = 0 \tag{1.36}$$

can now be simply translated to

$$\phi - \bar{\phi} = 0 \qquad \frac{\partial \phi}{\partial n} - q = 0 \tag{1.37}$$

and for such cases the substitution is extremely useful—and many practical problems can be solved without any iterative computation.

Unfortunately the substitution will simply transfer the non-linearity to the boundary if the radiation boundary condition $(q = q(T))$ is encountered. Further, if the transient equation (Equation (1.2)) is considered the above linearization is not applicable. Despite these shortcomings the use of the substitute variable can be of considerable practical significance in many problems.

*For transient problems* the technique of dealing with non-linearities is very dependent on the nature of the time stepping algorithm adopted. In Zienkiewicz,[6] Chapter 21 we show that there is a complete equivalence between the use of finite difference time approximations to Equation (1.27) and the use of 'finite elements in time' involving a weighted residual formulation of this differential equation using some shape function applicable to the time domain.

Thus for instance if we start with Equation (1.27) and assume a linear expansion (trial function approximation) in a time domain $\Delta t$ (from step $n$ to $n + 1$) we can write

$$\mathbf{a}(t) \approx \hat{\mathbf{a}}(t) = \mathbf{a}_n N_n + \mathbf{a}_{n+1} N_{n+1} \tag{1.38}$$

with

$$\begin{aligned} N_n &= (\Delta t - t)/\Delta t \quad (t = 0 \text{ at } n) \\ N_{n+1} &= t/\Delta t \qquad\qquad (t = \Delta t \text{ at } n + 1) \end{aligned} \tag{1.39}$$

and then the weighted residual approximation to (1.27) can be written as

$$\int_0^{\Delta t} W(\mathbf{H}\hat{\mathbf{a}}(t) + \mathbf{C}\dot{\hat{\mathbf{a}}}(t) + \mathbf{f}(t))\, \mathrm{d}t = 0 \tag{1.40}$$

For linear problems a well known algorithm corresponding to a finite difference approximation results:

$$\mathbf{H}(\mathbf{a}_n(1-\theta) + \mathbf{a}_{n+1}\theta) + \mathbf{C}(\mathbf{a}_{n+1} - \mathbf{a}_n)/\Delta t + \bar{\mathbf{f}} = \mathbf{0} \tag{1.41}$$

where

$$\theta = \frac{\int_0^{\Delta t} Wt \, dt}{(\int W \, dt)\Delta t} \qquad \bar{\mathbf{f}} = \frac{\int_0^{\Delta t} W\mathbf{f} \, dt}{\Delta t} \qquad (1.42)$$

and $\bar{\mathbf{f}}$ is an averaged forcing vector.

The above time stepping process allows $\mathbf{a}_{n+1}$ to be computed simply from known $\mathbf{a}_n$ values and is frequently used in practice. With $\theta \geqslant \frac{1}{2}$ unconditional stability exists; the lower limit being the well known Crank–Nicholson scheme.

In non-linear situations Equation (1.40) remains valid but Equation (1.41) is no longer generally correct. Clearly some interpolation needs to be applied to the $\mathbf{H}$ (and $\mathbf{C}$ or $\mathbf{f}$) values if these are functions of $\mathbf{a}(t)$. If these matrices change only monotonically during a time interval $\Delta t$, the integral of Equation (1.41), by well known properties of integration, can be written as a valid and convergent approximation with

$$\mathbf{H} = \mathbf{H}_{n+\bar{\theta}} \qquad \mathbf{C} = \mathbf{C}_{n+\bar{\theta}} \qquad \mathbf{f} = \mathbf{f}_{n+\bar{\theta}} \qquad (1.43)$$

where $0 \leqslant \bar{\theta} \leqslant 1$ represents some point within the interval. If $\bar{\theta} \neq 0$ then at each step of computation a non-linear problem needs to be solved. If however we take $\bar{\theta} = 0$, i.e.

$$\mathbf{H} = \mathbf{H}_n \qquad \mathbf{C} = \mathbf{C}_n \qquad \mathbf{f} = \mathbf{f}_n \qquad (1.44)$$

the computation at each step is not iterative. For many practical problems it is convenient to use the latter approximation (and possibly a smaller time step) than to iterate.

Particularly convincing in the above context of solving non-linear problems are two-step algorithms which quite generally (see Zienkiewicz,[6] (Chapter 21)) can be written for a linear case as

$$[\gamma \mathbf{C} + \beta \Delta t \mathbf{K}]\mathbf{a}_{n+1} + [(1-2\gamma)\mathbf{C} + (\tfrac{1}{2} - 2\beta + \gamma)\Delta t \mathbf{K}]\mathbf{a}_n$$
$$+ [-(1-\gamma)\mathbf{C} + (\tfrac{1}{2} + \beta - \gamma)\Delta t \mathbf{K}]\mathbf{a}_{n+1} + \bar{\mathbf{f}} = 0 \qquad (1.45)$$

Here a weighted residual form is applied with parabolic shape functions over a time interval $2\Delta t$ and $\mathbf{a}_{n+1}$ is computed from known values $\mathbf{a}_n$ and $\mathbf{a}_{n-1}$. Here once again for non-linear problems an intermediate value of the matrices $\mathbf{H}$, etc. can be taken, and a convenient and accurate approximation is to put $\mathbf{H} = \mathbf{H}_n$ and so on, which now represents a mid-interval value. This approximation has been used successfully with the algorithm of Equation (1.45) taking

$$\gamma = \tfrac{1}{2} \qquad \beta = \tfrac{1}{6} \qquad (1.46)$$

and thus reducing to the Lees[22] form. In this context problems of phase change and non-linear radiation have been effectively solved by Comini *et al.*[23]

It is of interest to note that other similar algorithms with different $\gamma$ and $\beta$ values could again be so employed with increased stability.

Much research in the area of transient non-linear analysis is now in progress and clearly new findings will be available with time with applications ranging from 'freezing' problems to metal solidification.

## 1.5 CONVECTIVE–DIFFUSIVE HEAT TRANSPORT

When a velocity field **u** exists in a medium through which heat is conducted the basic heat balance equation (Equation (1.1)) is augmented by a convective term and becomes

$$\nabla^T \mathbf{k} \nabla T + Q - \mathbf{u}^T \nabla T - c \frac{\partial T}{\partial t} = 0 \tag{1.47}$$

Even if the velocity field is known '*a priori*' and the problem is linear, difficulties arise when the standard processes of discretization of Section 1.2 are applied. If the Galerkin–Bubnov process is applied then, even in the steady-state case $(\partial T/\partial t = 0)$ an oscillatory solution will be found with the local Péclet number

$$Pe = \frac{\bar{u}h}{k} \tag{1.48}$$

exceeding a value of 2.

In the above the mesh size is indicated by $h$, $\bar{u}$ is the magnitude of the velocity vector and $k$ a typical conductivity. Such an oscillation is indicated in Figure 1.8 for a typical one-dimensional problem at two values of $Pe$.

It turns out that if a non-symmetric weighting function is used in place of the standard shape function (as illustrated for linear, one-dimensional elements in Figure 1.9(a)) then it is possible not only to avoid oscillation but improve the accuracy of the solution. For the one-dimensional problem it is indeed possible with such weighting functions to achieve exact solutions of nodes.[24,25] We shall not go into the details of such 'upwinded' weighting functions which are fully described elsewhere[24,25,26] (Figure 1.9(b)), and have been extended to two- and even three-dimensional problems.[24–27] It is necessary however to remark that their directionality presents some computational difficulties although various simplifications have been suggested.[28] To overcome these difficulties it is possible to proceed in an alternative manner. In this Equation (1.47) is modified by the introduction of an artificial diffusivity **k**′ such that

$$\mathbf{k}' \to 0 \quad h \to 0 \tag{1.49}$$

to

$$\nabla^T (\mathbf{k} + \mathbf{k}') \nabla T + Q - \mathbf{u}^T \nabla T - c \frac{\partial T}{\partial t} = 0 \tag{1.50}$$

Clearly the problem is convergent to the original one as the size of the mesh is decreased. If in a one-dimensional, steady-state problem (with linear elements)

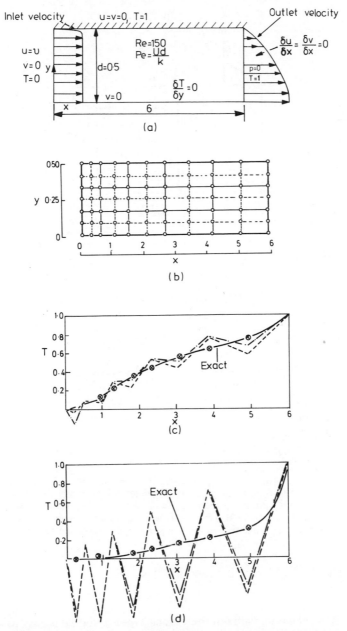

**Figure 1.8**  Thermal convection–diffusion in entry flow. (a) Problem statement (velocity distribution determined independently); (b) mesh detail; (c) computations for $Pe = 50$; (d) computations for $Pe = 150$. In (c) and (d) 'Exact' represents a solution with a very fine eight-node element mesh

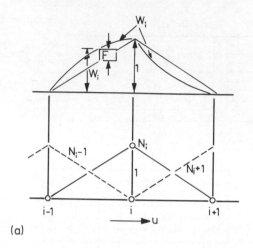

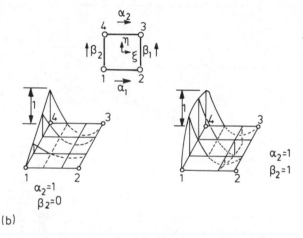

Typical weighting functions for a bi-linear two-dimensional element (parent-coordinates). Velocity sign convention

One dimensional problem. Shape function (N) and weighting function (W). Constant velocity u.

**Figure 1.9** Linear shape and weighting functions for: (a) one-dimensional problem, shape function $N$, weighting function $W$, constant velocity $u$; (b) a bi-linear two-dimensional element (parent coordinates), velocity sign convention—typical weighting functions only are shown

we use a standard Galerkin process and if we take

$$k' = \alpha |\bar{u}| h \qquad (1.51)$$

where $|\bar{u}|$ is the absolute value of the velocity vector, then it is easy to show that for a certain optimal value of $\alpha$, i.e.

$$\alpha = \coth\left(\tfrac{1}{2}Pe\right) - 2Pe \qquad (1.52)$$

an exact nodal solution can be obtained.

The concept of such a generalized artificial diffusion can be extended to two- or three-dimensional solutions and indeed to transient problems where it increases both accuracy and stability.[29] Much further research clearly needs to be undertaken in this direction.

An alternative to the above processes is presented if we realize that the addition of convective terms changes the basic problem from a 'parabolic' to a 'hyperbolic' one.[30] Thus if we introduce a change of variables from $t$, $x$ and $y$ to $t'$, $x'$ and $y'$ such that

$$dt' = dt \quad dx' = dx - \frac{u}{c}dt \quad dy' = dy - \frac{v}{c}dt \tag{1.53}$$

where $u$ and $v$ are the $x$- and $y$-components of velocity, then the convective transport equation (Equation (1.47)) returns simply to the original, diffusive form (1.1). This modification, while simplifying the basic equation, introduces (in the $t'$, $x'$ and $y'$) domain a problem with time-varying boundaries.

To avoid such difficulties Finn and Varoglu[31] use, very effectively, special combined space–time element discretization. If the sides of the elements are oriented along $x'$- and $y'$-(characteristic) directions it is found that excellent accuracy can be obtained with a standard Galerkin–Bubnov process, albeit at the expense of somewhat more complicated algorithms.

## 1.6 COUPLED PROBLEMS AND CONCLUDING REMARKS

Despite the apparent simplicity of thermal conduction problems we have identified diverse areas where at the present moment of time difficulties remain and towards which much research effort is directed. Many other practical problems arise in which even more intense difficulties are encountered and which today are still not solved by routine computation. One such problem is posed for instance by coupled flow–heat transfer equations. In such cases we are faced with the solution of a flow problem characterized by the Navier–Stokes equations

$$\rho\left(\frac{\partial u}{\partial t} + u\frac{\partial u}{\partial x} + v\frac{\partial u}{\partial y} + w\frac{\partial u}{\partial z}\right) = -\frac{\partial p}{\partial x} + \rho g_x + 2\frac{\partial}{\partial x}\mu\frac{\partial u}{\partial x} + \frac{\partial}{\partial y}\mu\left(\frac{\partial u}{\partial y} + \frac{\partial v}{\partial x}\right)$$

$$+ \frac{\partial}{\partial z}\mu\left(\frac{\partial u}{\partial z} + \frac{\partial w}{\partial x}\right) \tag{1.54}$$

with similar equations for $y$- and $z$-directions and an auxilliary incompressibility condition

$$\frac{\partial u}{\partial x} + \frac{\partial v}{\partial y} + \frac{\partial w}{\partial z} = 0 \tag{1.55}$$

where $u$, $v$ and $w$ are the components of the velocity vector $\mathbf{u}$. As the density $\rho$

and the viscosity $\mu$ may be significantly affected by the temperature $T$ and further, as the heat generation term $Q$ includes the work dissipation due to viscosity, Equations (1.47), (1.54) and (1.55) have to be solved simultaneously in their discrete forms. It is evident that the problem is of considerable complexity but at the same time includes many important practical cases necessitating solution, such as thermally induced circulation and flow of highly viscous fluids. We shall not go into the details of computation but for the interested reader in Figures 1.10 and 1.11 we show two interesting solutions recently obtained. In the first a thermally induced circulation is studied in a

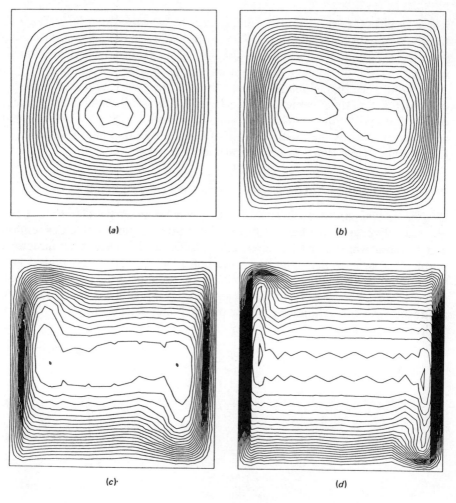

(a)

(b)

(c)

(d)

**Figure 1.10**   Stream function contours: (a) $Ra = 10^4$, (b) $Ra = 10^5$, (c) $Ra = 10^6$, and (d) $Ra = 10^7$

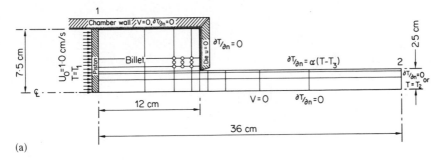

(a)

**Figure 1.11** Metal flow in an extrusion situation. (a) Steady-state plane extrusion, full slip boundary: reduction 0.6; $u/p = 39$ elements, $K = \sigma Y/\sqrt{3} = 1000$ kg cm$^{-2}$, $\rho = 1$ kg cm$^{-3}$, $C_p = 1$ cal cm$^{-3}$ °C$^{-1}$; (b) computed velocity fields for various entry velocities; (c) velocity profiles along AA ($x = 11.5$ cm); (d) isotherms for various entry velocities

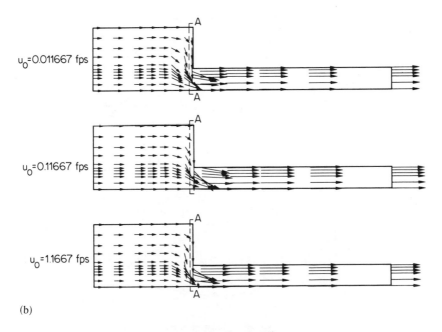

(b)

**Figure 1.11** Continued

simple geometric form.[32,33] In the second an equally complex problem of metal flow in an extrusion situation is shown to affect substantially the temperature field.[34]

Both these examples are indicative of problems to which many chapters of this book are devoted and the areas in which present-day research is intensive.

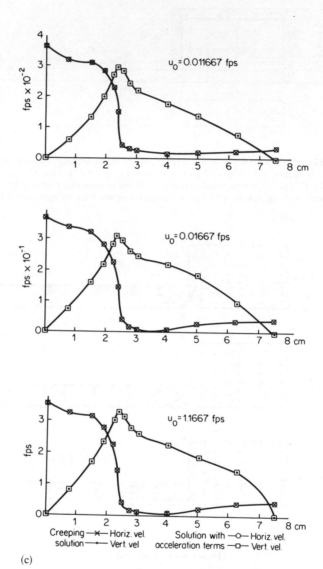

(c)

**Figure 1.11** Continued

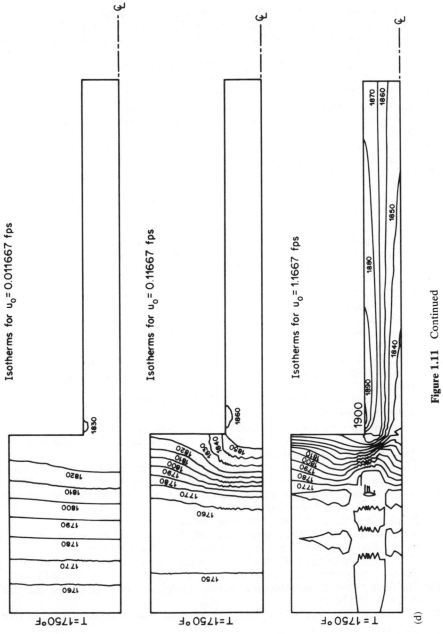

**Figure 1.11** Continued

## REFERENCES

1. L. F. Richardson (1910). 'The approximate arithmetical solution by finite differences of physical problems', *Trans. R. Soc. (Lond.) A*, **210**, 307–357.
2. R. V. Southwell (1946). *Relaxation Methods in Theoretical Physics*, Clarendon Press. Oxford.
3. O. C. Zienkiewicz and Y. K. Cheung (1965). 'Finite elements in the solution of field problems', *The Engineer*, 507–510, Sept.
4. W. Visser (1965). 'A finite element method for the determination of non-stationary temperature distribution and thermal deformations', *Proc. Conf. on Matrix Methods in Structural Mechanics, Air Force Inst. Techn. Wright-Patterson AF Base, Ohio, 1965.*
5. E. L. Wilson and R. E. Nickell (1966). 'Application of finite element method to heat conduction analysis', *Nucl. Engng Des.*, **4**, 1–11.
6. O. C. Zienkiewicz (1977). *The Finite Element Method*, 3rd edn, McGraw-Hill, New York.
7. S. H. Crandall (1956). *Engineering Analysis*, McGraw-Hill, New York.
8. B. A. Finlayson (1972). *The Method of Weighted Residuals and Variational Principles*, Academic Press, New York.
9. B. G. Galerkin (1915). 'Series solution of some problems of elastic equilibrium of rods and plates', *Vestn. Inzh. Tech.*, **19**, 897–908 (in Russian).
10. E. Trefftz (1926). 'Gegenstück zom Ritz'schen Verfahren', *Proc. Sec. Int. Congr. Applied Mechanics, Zurich, 1926.*
11. S. C. Mikhlin (1964). *Variational Methods in Mathematical Physics*, Pergamon Press, Oxford.
12. T. A. Cruse and F. J. Rizzo (Eds) (1975). 'Boundary-integral equation method: computational applications in applied mechanics', *Proc. Am. Soc. Mech. Engrs, Spec. Publ.* 11.
13. O. C. Zienkiewicz, D. W. Kelly and P. Bettess (1977). 'The coupling of the finite element method and boundary solution procedures', *Int. J. Num. Meth. Engng*, **11**, 355–376.
14. O. C. Zienkiewicz, D. W. Kelly and P. Bettess (1979). 'Marriage *à la mode*—the best of both worlds (finite elements and boundary integrals)', *Energy Methods in Finite Element Analysis*, Ch. 5, John Wiley, Chichester.
15. G. Strang and G. J. Fix (1973). *An Analysis of the Finite Element Method*, p. 106, Prentice-Hall, Englewood Cliffs, NJ.
16. J. T. Oden (1969). 'A general theory of finite elements—I: Topological considerations', and 'II: Applications', *Int. J. Num. Meth. Engng*, **1**, 295–321, 247–260.
17. (a) T. Moan (1973). 'On the local distribution of errors by the finite element approximation', in *Theory and Practice in Finite Element Standard Analysis*, Eds Y. Yamada and R. H. Gallager, University of Tokyo Press, Tokyo.
    (b) T. Moan (1974). 'Orthogonal polynomials and "best" numerical integration formulas on a triangle', *Zeit. angew. Math. Mech.*, **54**, 501–580.
18. O. C. Zienkiewicz and E. Hinton (1976). 'Reduced integration, function smoothing and non-conformity in finite element analysis', *J. Franklin Inst.*, **302**, 443–461.
19. Pin Tong (1969). 'Exact solution of certain problems by the finite element method'. *A.I.A.A.* **7**, 178–180.
20. G. Beer (1978). Technische Datenverarbeitung, Graz, Austria (private communication).
21. C. A. Anderson and O. C. Zienkiewicz (1974). 'Spontaneous ignition: finite element solutions for steady and transient conditions', *Am. Soc. Mech. Engrs, J. Heat Transfer*, 398–404, August.

22. M. Lees (1966). 'A linear three level difference scheme for quasilinear parabolic equations', *Math. Comp.*, **20**, 516–622.
23. (a) G. Comini, S. del Guidice, R. W. Lewis and O. C. Zienkiewicz (1974). 'Finite element solution of non-linear heat conduction problems with special reference to phase change', *Int. J. Num. Meth. Engng*, **8**, 613–624.
    (b) K. Morgan, R. W. Lewis and O. C. Zienkiewicz (1978). 'An improved algorithm for heat conduction problems with phase change', *Int. J. Num. Meth. Engng*, **12**, 1191–1195.
24. I. Christie, D. F. Griffiths, A. R. Mitchell and O. C. Zienkiewicz (1976). 'Finite element methods for second order differential equations with significant first derivatives', *Int. J. Num. Meth. Engng*, **10**, 1389–1396.
25. O. C. Zienkiewicz, J. C. Heinrich, P. S. Huyakorn and A. R. Mitchell (1977). 'An upwind finite element scheme for two dimensional convective transport equations', *Int. J. Num. Meth., Engng*, **11**, 131–144.
26. J. C. Heinrich and O. C. Zienkiewicz (1977). 'Quadratic finite element schemes for two dimensional convective-transport problems', *Int. J. Num. Meth. Engng*, **11**, 1831–1844.
27. D. A. Dillard and R. L. Davies (1979). 'A quadratic finite element for the three dimensional convective transport equation'. *Joint ASME/A.I.Ch.E. National Heat Transfer Conf., San Diego, California, August 1979*.
28. T. J. R. Hughes (1978). 'A simple scheme for developing "upwind" finite elements', *Int. J. Num. Meth. Engng*, **12**, 1359–1365.
29. S. Nakazawa, D. W. Kelly, O. C. Zienkiewicz, I. Christie and M. Kawahara (1980). 'An analysis of explicit finite element approximations for the shallow water equation', *Proc. 3rd Int. Conf. on Finite Elements in Flow Problems, Banff, Canada 1980*.
30. E. Varoğlu and W. D. L. Finn (1978). 'A finite element method for the diffusion-convection equation with constant coefficients', *Adv. Water Resources*, **1**, 337–343.
31. W. D. L. Finn and E. Varoğlu (1979). 'An efficient solution for heat transfer problems', *Proc. Conf. on Num. Meth. in Thermal Problems, University College of Swansea, Swansea, 1979*.
32. O. C. Zienkiewicz, R. H. Gallagher and P. Hood (1977). 'Newtonian and non-Newtonian viscous incompressible flow. Temperature induced flows. Finite element solutions', in *The Mathematics of Finite Elements and Applications II*, Ed. J. Whiteman, Academic Press, New York.
33. R. S. Marshall, J. C. Heinrich and O. C. Zienkiewicz (1978). 'Natural convection in a square enclosure by a finite element penalty function method using primitive variables', *Num. Heat Transfer*, **1**, 315–330.
34. O. C. Zienkiewicz, E. Oñate and J. C. Heinrich (1979). 'Plastic flow in metal forming, 1. Coupled thermal, 2. Thin sheet forming', in *Applications of Numerical Methods for Forming Processes*, Eds H. Armen and R. F. Jones, Am. Soc. Mech. Engrs.

*Numerical Methods in Heat Transfer*
Edited by R. W. Lewis, K. Morgan, and O. C. Zienkiewicz
© 1981 John Wiley & Sons Ltd

*Chapter 2*

# Automatic Solution of Thermal Problems

*M. D. Mikhailov and M. A. Aladjem*

## SUMMARY

The purpose of this chapter is to give a brief account of the development of software systems for automatic solution of ordinary and partial differential equations (ODEs and PDEs) and to describe the new interactive system for the solution of non-linear PDEs—ARIEL. Equations, boundary conditions, order of approximation, and special instructions are specified by the user. ARIEL automatically generates, compiles, loads, and executes a new program according to specifications. The system's application is illustrated by the solution of some thermal problems.

## 2.1 INTRODUCTION

The study of ordinary and partial differential equations (ODEs and PDEs) has many important applications to a great variety of thermal problems. With the development of digital computers an ever increasing number of numerical methods have been developed for their solution. The choice of an appropriate numerical technique to solve a given problem is governed by the number of independent variables in PDEs, its order, type, and boundary conditions. An immense number of algorithms are described in the literature, each having its limits of application. The engineer or scientist is interested in obtaining results with a minimum personal effort and without the necessity of studying the numerical methods and without writing computer programs. For this reason, there has been considerable interest in developing computer systems for automatic solution of PDEs allowing also for the solution of ODEs.

The first attempts for the automatic solution of PDEs started with the creation of digital computers. They comprise the special programs for their solution written in assembly code.

Later, with the second generation hardware, PDE packages for the solution of a number of PDE types and coordinate systems were prepared. They are characterized by the presence of a drive program which forms a complete numerical procedure from several subroutines. These PDE packages have been defined as subroutine collections rather than as complete programs.

The third generation hardware led to the development of software for the solution of PDEs. Two different approaches were formed: (a) generalized PDE packages, and (b) PDE languages. All available generalized packages and languages for PDEs are evolving towards facilities for a wider range of realistic problems, and more towards enabling the user to define 'what to do' and shifting the 'how to do it' to the computer.

The development of PDE software systems is a straightforward task for two reasons: (a) there is a tendency for the real problem to involve particular PDEs which are defined by a large number of factors and have different mathematical structure, and (b) a great variety of numerical methods for the solution of PDEs exist, none of which suffices to solve all PDEs.

To overcome these difficulties a number of authors have developed software for the numerical solution of PDEs as listed below. For more details concerning software packages and special purpose languages for ODEs and PDEs the reader is referred to excellent reviews.[1]

The state of the art of software for PDEs implies that there is a real need for development of new systems. The authors of the present chapter have developed a new iterative system for the automatic solution of non-linear PDEs—ARIEL. Equations, boundary conditions, order of approximation, and special instructions are specified by the user. ARIEL automatically generates, compiles, loads, and executes a new program according to specifications. The purpose of this chapter is to give a brief review of software systems for automatic solution of PDEs and to describe the new interactive system ARIEL. The system's application is illustrated by the solution of some thermal problem.

## 2.2  SOFTWARE FOR PDEs

One of the earliest reported packages was the SPADE project of the Rand Corporation,[2] which developed a series of generalized subroutines for elliptic and parabolic PDEs. SPADE was limited by the available second generation hardware and its use of assembly language.

The subroutine-oriented SALEM package[3] was developed at Lehigh University for automatic solving of one-dimensional parabolic and hyperbolic equations, and two dimensional elliptic and parabolic equations under various conditions, but was limited to simple regular geometries. It was a predecessor of the DSS and LEANS system discussed later.

At the University of Paris and at CNRS a set of ALGOL-60 subroutines were developed.[4-6] They are specifically designed to solve general one- and two-dimensional elliptic equations for irregular geometries and any type of boundary conditions (Dirichlet, Neumann or mixed). The references[4-6] do not mention details on how easily to specify the problem to the package.

The DSS (*Distributed System Simulator*)[7] is a FORTRAN program for the solution of PDEs of up to two dimensions. This system is predicated on the assumption that the user has a basic knowledge of FORTRAN, including subroutines. PDEs, initial and boundary conditions are defined in a number of subroutines following carefully prescribed rules. The user describes his problem as a special case of the 'master PDE'

$$\frac{1}{y^b}\frac{\partial}{\partial y}\left(y^b A_{6j}\frac{\partial u_j}{\partial y}\right)+\frac{1}{x^a}\frac{\partial}{\partial x}\left(x^a A_{1j}\frac{\partial u_j}{\partial x}\right)$$

$$-\frac{\partial}{\partial x}(A_{2j}u_j)-A_{3j}\frac{\partial u_j}{\partial t}-A_{4j}u_j-A_{5j}=0 \tag{2.1a}$$

and the 'master boundary conditions'

$$B_{1j}\frac{\partial u_j}{\partial x}+B_{2j}u_j-B_{3j}=0 \tag{2.1b}$$

by defining $A_{ij}$ ($i = 1, 6$) and $B_{ij}$ ($i = 1, 3$), i.e. writing a subroutine for each of the coefficients. DSS solves the problems defined by equations (2.1) using the alternating direction Peaceman–Rachfort method for two-dimensional cases ($A_{6j} \neq 0$) and implicit stable techniques for one-dimensional cases.

The LEANS-III (*Lehigh Analog Simulator, Version III*),[8,9] like DSS, is subroutine oriented, and the user describes his problem in relation to master equations

$$A_{2j}\frac{\partial^2 u_j}{\partial t^2}+A_{1j}\frac{\partial u_j}{\partial t}=\frac{1}{x^a}\frac{\partial}{\partial x}\left(x^a A_{4j}\frac{\partial u_j}{\partial x}\right)+\frac{\partial}{\partial x}(A_{3j}u_j)$$

$$+A_{5j}u_j+A_{6j}+\frac{1}{y^b}\frac{\partial}{\partial y}\left(y^b A_{7j}\frac{\partial u_j}{\partial y}\right)+\frac{\partial}{\partial y}(A_{8j}u_j)$$

$$+\frac{1}{z^c}\frac{\partial}{\partial z}\left(z^c A_{9j}\frac{\partial u_j}{\partial z}\right)+\frac{\partial}{\partial z}(A_{10j}u_j) \tag{2.2a}$$

and a general boundary condition

$$B_{1j}\frac{\partial u_j}{\partial w}+B_{2j}u_j=B_{3j} \tag{2.2b}$$

where $w = x$, $y$ or $z$ and $a, b, c$ are programable coordinate factors with values of 0, 1, and 2 for Cartesian, cylindrical, and spherical coordinates respectively. $A_{1j}$ to $A_{10j}$ and $B_{1j}$ to $B_{5j}$ are programed by the user in a FORTRAN-IV subroutine and may be functions of the independent and dependent variables. LEANS-III solves the problem (2.2) using the method of lines.

The generalized FORTRAN package FORSIM[10] can solve problems characterized by the equation

$$A_1 \frac{\partial^2 u}{\partial t^2} + A_2 \frac{\partial u}{\partial t} = \frac{\partial}{\partial x}(A_3 u) + \frac{1}{x^c}\frac{\partial}{\partial x}\left(x^c A_4 \frac{\partial u}{\partial x}\right) + A_5 u + A_6 \qquad (2.3a)$$

with boundary conditions of the type

$$\frac{\partial}{\partial x}(B_1 u) + B_2 u = B_3 \qquad (2.3b)$$

where the coefficient $A_i$ $(i = 1, 6)$ and $B_i$ $(i = 1, 3)$ are specified by the user and may be constants or functions of $x$ and/or $t$ and the exponent $c$ defines the coordinate system. Either a three point or a five point difference approximation to the spatial derivatives may be used. FORSIME assumes that the user is familiar with FORTRAN subroutine usage, as in the case of DSS and LEANS. FORSIM, like LEANS, was originally an ODE simulation package and was converted for PDE applications through the methods of lines and master equations (2.3).

It is obvious that master equations (2.1) to (2.3) can describe a great number of PDE types and coordinate systems. However, there are many other PDEs which cannot be forced into equations such as (2.1) to (2.3), for example the conservation equations of fluid flow. For this reason DSS and FORSIM were considerably generalized to solve a broad class of PDEs.

DSS/2[11] is a modern package allowing for the PDEs to be defined in the following manner

$$\frac{\partial u_j}{\partial t} = F_j\left(x, t, u_1, u_2, \ldots u_n, \frac{\partial u_1}{\partial x}, \frac{\partial u_2}{\partial x}, \ldots \frac{\partial u_n}{\partial x}, \frac{\partial^2 u_1}{\partial x^2}, \ldots \frac{\partial^2 u_n}{\partial x^2}\right) \qquad (2.4a)$$

subject to boundary conditions

$$G_j\left(x, t, u_1, u_2, \ldots u_n, \frac{\partial u_1}{\partial t}, \frac{\partial u_2}{\partial t}, \ldots \frac{\partial u_n}{\partial t}, \frac{\partial u_1}{\partial x}, \frac{\partial u_2}{\partial x}, \ldots \frac{\partial u_n}{\partial x}\right) = 0 \qquad (2.4b)$$

and initial conditions

$$u_j(t_0) = u_{j0}(x) \qquad (2.4c)$$

DSS/2 contains a rather extensive series of subroutines to assist the user in the numerical methods of lines integration of ODEs and PDEs.

FORSIM,[12] like DSS/2, now solves PDEs defined as follows:

$$\left[\frac{\partial u_i}{\partial t} = \phi(x, t, u_i, u_j, \ldots \delta u_i, \delta u_j)\right]_{\substack{i=1,\text{NPOINT} \\ j=1,\text{NEQN}}} \qquad (2.5)$$

where $\phi$ is any function and $\delta$ is any spatial derivative. FORSIM has a sparse matrix implementation of the Gear method of integration.

Both DSS/2 and FORSIM incorporate high-order approximation of the spatial derivatives. They are designed for PDEs in one dimension, but two and three dimensional cases have been solved.

All the packages reviewed herein require the user to define his problem generally by appropriately specifying input parameters to subroutines. Hence, the user must have some familiarity with subroutine utilization, the organization of the package, and the way to activate it. This considerably restricts the number of users who, as a rule, are not interested in studying the package and need results with a minimum of personal effort.

Development of software systems for automatic solution of PDEs is due to the user's need to define the problem and let the computer choose the way to solve it. Existing systems can conditionally be divided into two classes: (i) systems with a limited spectrum of performance, and (ii) systems with a broad spectrum of performance.

Limited spectrum systems are realized on the second generation hardware and solve particular types of PDE. Such a system is the program generator FIELD.[13] It comprises an archive of ALGOL-60 modules, syntax analyser, administrative system, etc. FIELD generates programs automatically and solves in two dimensional regions the second order PDE

$$\Delta u = V \quad \text{or} \quad \Delta u - cu = V \qquad (2.6a, b)$$

the fourth order PDE

$$\Delta \Delta u = V \quad \text{or} \quad \Delta \Delta u - cu = V \qquad (2.6c, d)$$

and eigenvalue problems of the form

$$\Delta u + ku = 0 \quad \text{or} \quad \Delta \Delta u - ku = 0 \qquad (2.6e, f)$$

The boundary of the regions where problems are solved is formed by arcs of circles and straight lines. In the case of Dirichlet boundary conditions, the latter can be formed by means of arcs of second order curves. The user can choose the method of solution (Ritz, Galerkin, variational-difference and finite difference) and use several systems of coordinate functions. The problem is defined by means of a special system of symbols for the type of equation, boundary conditions, form of the region, method of solution, and the form of presentation of the results. On the basis of this information FIELD automatically synthesizes a program from the modules present in the archive. The FIELD system realizes a fully automatic programing process, but solves quite a limited class of problems.

Another limited spectrum system is KSI–BESM.[14] It solves stationary problems describing the motion of charged particles in an electromagnetic field and optimization problems of electronic–optical systems. From a mathematical standpoint these problems reduce to the solution of Laplace or Poisson equations in three dimensional regions of complex geometry and Dirichlet, Neumann or mixed boundary conditions, as well as to the minimization of given functional depending on a prescribed number of parameters subject to stated restrictions. The system KSI–BESM comprises a bank of modules, which realize a number of numerical methods and a complex of service and control devices (a compiler for the input language, devices for development and modification of the system, blocks controlling the functioning of the system, etc.). The automatic KSM–BESM system appears to be one of the most successful attempts to realize an automated solution of a class of problems.

Another system solving the Laplace and Poisson equations in multi-connected regions is the ASPP.[15] It utilizes the method of the P-transformations.

A number of useful ideas concerning the organization of problem-oriented systems have been developed by Kabulov and have been realized by his collaborators.[16]

The broad spectrum systems utilize more general numerical methods and, consequently, their efficiency decreases. This drawback is balanced by the use of powerful and fast computers.

The first problem-oriented language for the solution of PDEs is PDEL (*Partial Differential Equation Language*).[17,18] It is oriented somewhat towards a user who has no programing experience. The problem is defined in terms of natural, problem-oriented, programming statements. PDEL system is a formal translator–compiler with a high gain (i.e. the ratio of the size of the PL/1 program generated to the size of the source program is at least 10 to 1, and usually much higher). PDEL is oriented to the solution of linear and quasi-linear systems of PDEs of hyperbolic, elliptic, and parabolic types defined in complex regions and Dirichlet or Neumann boundary conditions. The language has special operators for the definition of region's geometry and the location of the boundary conditions. It renders possibilities for controlling the type of the results obtained. As a matter of fact from the rich variety of PDEs which could formally be described by the realized subset of PDEL, only five types of PDEs have been solved. Besides this, there are several drawbacks in the description of the language, namely: it is not possible to choose the numerical method following the user's instructions; in the formal description of the language the boundary conditions refer to one equation, etc. Never-theless, the PDEL plays an important role in the future development of systems for the automated solution of PDEs, because of the following new ideas: (a) a macroprocessor for the generation of the program has been used; (b) a language for the automatic solution of PDE has been described, and (c) a subset of this language has been realized.

Other broad spectrum systems are the SAI and SAII.[19] They are vector-oriented and permit the user to solve PDEs within the framework of an ALGOL 60 program by defining operators for CURL, DIV, CROSS, etc.

These systems have been successfully realized on the IBM 360, CDS STAR-100.

The analysis of the generalized packages and software systems for the automatic solution of PDEs realized up to the present moment shows that regardless of great progress, we are still very far from the ideal system, which, according to us, should satisfy the following requirements:

(1)  Solve the broadest spectrum of ODEs and PDEs.
(2)  The boundary conditions have to be of arbitrary type.
(3)  The solution's region has to be complex in a one, two, or three dimensional space.
(4)  The numerical methods used have to be numerous.
(5)  The solution has to be obtained with a prescribed accuracy and in a form most convenient for the user.
(6)  The most appropriate numerical method has to be automatically chosen.
(7)  The inverse problem has also to be solved.
(8)  A dialogue or a problem-oriented language has to be used.
(9)  The program generated by the system should be of high quality and allow for execution on different computers.
(10) The system should be easily expanded.

As a first step towards the creation of a system, comprising some of the qualities mentioned before, the system ARIEL was developed.

## 2.3  DESCRIPTION OF ARIEL

ARIEL operates on the NOVA-840 minicomputer (64 K core) and the ECLIPSE C-300. ARIEL's central part is the module generator. Its purpose is to generate a group of modules using available information which form the program to be executed. Besides, the module generator prepares a job containing all statements necessary for the translation, linkage, and cataloguing of the user's program. The module generator required the development of an original MACROALGOL language—a macroextension of the extended ALGOL 60.

The aim of the MACROALGOL language is the description of algorithms for synthesis of modules and is similar to the PL/1 macroprocessor, but is not oriented towards a particular basic language. This means that it could be used to generate modules in all algorithmic languages and assembler.

ARIEL is developed to solve automatically PDEs of the following kind

$$\frac{\partial u_i}{\partial t} = L_i(x, y, z, t, u_1, \ldots u_n, \delta u_1, \ldots \delta u_n) \tag{2.7a}$$

subject to initial conditions

$$u_i = f_i(x, y, z) \tag{2.7b}$$

and boundary conditions

$$B_i(x, y, z, t, u_1, \ldots u_n, \delta u_1, \ldots \delta u_n) = 0 \tag{2.7c}$$

where: $1 \leqslant i \leqslant 31$; $L_i$ are non-linear or linear operators; $B_i$ are linear operators; $t$ is an initial value independent variable; $x$, $y$, $z$ are boundary value independent variables; $u_i$ are dependent variables; $\delta u_i$ are space derivatives of the type

$$\delta = \frac{\partial^{n_1 + n_2 + n_3}}{\partial x^{n_1} \partial y^{n_2} \partial z^{n_3}} \tag{2.7d}$$

Practically there is no restriction for the orders of the derivatives ($n_1$, $n_2$, and $n_3$) since ARIEL automatically produces the finite difference patterns needed in the discretization process.

ARIEL can solve PDEs defined in 1, 2, or 3 dimensional complex space regions. A variable step mesh covers the solution region, but the boundaries must be a series of horizontal and vertical lines. The maximum number of nodes is 2000.

ARIEL has prompted entry of input of information. The problem to be solved is defined by means of a dialogue between system and user utilizing a tree hierarchy structure of four levels. To illustrate this Figure 2.1 shows the tree of the following problem:

$$\frac{\partial u}{\partial t} = A(x, y) \frac{\partial^2 u}{\partial x^2} + B(x, y) \frac{\partial^2 v}{\partial y^2} \tag{2.8a}$$

$$\frac{\partial y}{\partial t} = \frac{1}{1 + t^2} uv \tag{2.8b}$$

with three types of boundary conditions

$$\frac{\partial u}{\partial x} - 0.01 y = 0 \tag{2.8c}$$

$$u + \frac{1}{1 + t} 5 = 0 \tag{2.8d}$$

$$v - u + 1 = 0 \tag{2.8e}$$

The first level of the tree defines the number of differential equations and the number of boundary condition types. In the case under consideration, there are two PDEs and three types of boundary conditions. ARIEL can solve up to 31 equations with up to 31 types of boundary conditions.

The second level defines the number of 'members' in each of the equations. A 'member' means any dependent variable or any of its derivatives. It is

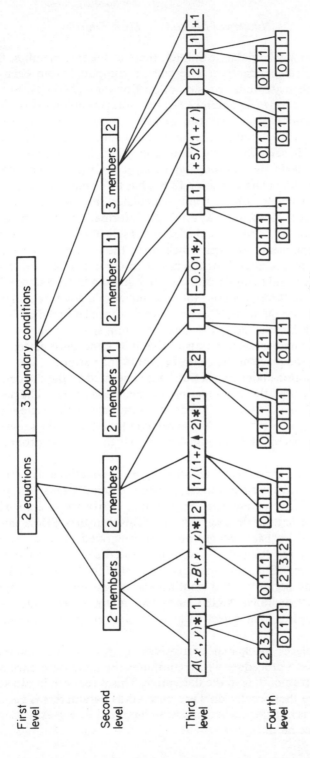

**Figure 2.1** The tree of problem (2.8)

assumed that in any boundary condition there is one free member. One has to prescribe which dependent variable will be computed from each boundary condition. In the case under consideration Equation (2.8a) has two members ($\partial^2 u/\partial x^2$ and $\partial^2 v/\partial y^2$). Equation (2.8b) has also two members ($u$ and $v$). The boundary condition (2.8c) has two members ($\partial u/\partial x$ and the free member $-0.01y$) and it will be used for the computation of the first dependent variable (as such $u$ is chosen). Similarly, Equation (2.8d) has two members ($u$ and $1/(1+t)$) and again the first dependent variable will be calculated from it. Eventually, the boundary condition (2.8e) has three members ($v$, $u$ and 1) and the second dependent variable will be calculated from it (that is $v$).

The third level comprises: (a) symbolic information concerning the arithmetic expressions to be found in the left hand side of each member, and (b) the number of dependent variables which form the member. For example, the arithmetic expression preceding the first member in Equation (2.8b) is $1/(1+t2)$, while the second member of the same equation is preceded only by $*$.

The last, fourth level, contains for each member a few three digit numbers, which are equal to the number of space variables. The first digit indicates the order of the derivative along the corresponding space variable. The second digit gives the number of the points of the finite-difference pattern. The third digit indicates the place of the central point in the pattern.

The problem tree under consideration is generated by the dialogue between ARIEL and the user, the interrogating side being the system. The problem tree is automatically entered in the information module of the system where it is stored until the user destroys it. When repeated solution of the same problem is required the system uses the same tree to generate the program for the solution.

ARIEL uses the method of lines for all types of equations. This method, also called a method of discrete ordinates, has been studied extensively, particularly in the USSR.[20] In this method, each spatial derivative is approximated by an appropriate finite difference pattern. This converts PDEs into a set of ODEs, coupled in space, which are then integrated in time. In most early applications the resulting ODEs were integrated continuously in time on an analog computer.[20] Still later, the development of sophisticated ODE integration algorithms permitted this continuous integration to be approximated sufficiently well digitally. ARIEL uses two methods for ODE integration: Euler and Leapfrog. The selection of numerical integration methods will be widened.

ARIEL replaces all spatial derivatives by appropriate finite difference patterns. It uses a procedure which computes the finite difference expression coefficients corresponding to the derivative. This is realized in two stages: (i) a program where these coefficients are entered as parameters is generated, (ii) the actual values of the coefficients are computed in correspondence to the finite-difference mesh.

The ordinary derivatives are approximated by the formula:

$$\frac{d^k u(x_p)}{dx^k} = \sum_{i=1}^{n} C_i^{(k)} u(x_i) \tag{2.9}$$

In this case the procedure 'dicol' proposed by Giammo can be used.[21] ARIEL utilizes an improved version of 'dicol' which we called DICOM. A complete listing of DICOM is presented in Figure 2.2. As input parameters, this routine has: $k$—the order of the derivative; $n$—number of nodes in the pattern; XP—the abscissa of the central point, i.e. the point at which the $k$th derivative is to be evaluated; XTAB—an $N$-element array, containing the abscissa of the nodes in the pattern. As results DICOM produces the coefficients for the $n$-ordinates, corresponding to the given abscissa, which are stored in the CDFF array.

The DICOM is convenient for calculation of a separate pattern. ARIEL utilizes the D1LAG procedure to compute the coefficients $C_i^{(k)}$ at each node of the fixed finite-difference mesh separately. A complete listing of D1LAG is presented in Figure 2.3. The input parameters are: KGL $[1:3]$—a three-element array which indicates the order of the derivative, the number of nodes in the pattern, and the position of the central node, that is, the node at which the

```
PROCEDURE DICOM (K,N,XP, XTAB,COEF);
VALUE K,N; INTEGER K,N; REAL XP;
ARRAY XTAB,COEF;
BEGIN INTEGER ARRAY XUSE [0:N-1];
         REAL FACTK, SUM,DENOM, PART;
         INTEGER I,TERMS,J, M,HIGH;
           FACTK:= 1.0; FOR I:= 2 STEP 1 UNTIL K DO FACTK := I*FACTK;
           TERMS := N-K-1; IF TERMS <0 THEN GO TO Z;
           FOR J:= 1 STEP 1 UNTIL N DO
    BEGIN
         DENOM :=1.0;
                   FOR I:= 1 STEP 1 UNTIL N DO
                   IF I<> J THEN DENOM := DENOM *(XTAB[J]- XTAB[I]);
         LOOP : SUM := 0; PART :=1.0;
                   IF TERMS=0 THEN GO TO Y;
                   M:=1; HIGH :=1 ;
           A: IF (HIGH= J) OR (XTAB[HIGH] =XP) THEN
              A1 : BEGIN HIGH := HIGH +1; GO TO A; END A1;
              IF HIGH >N THEN A2:BEGIN M:=M-1;
              IF M>0 THEN
              A3 : BEGIN HIGH := XUSE[M]+1; GO TO A;END A3;
              GO TO X; END;
              XUSE [M]:= HIGH; M:=M+1;
              IF M=<TERMS THEN BEGIN HIGH := HIGH +1; GO TO A END;
              FOR I := 1 STEP 1 UNTIL TERMS DO
                  PART := PART *(XP-XTAB[XUSE[I]]);
              SUM := SUM + PART; M := TERMS; PART :=1.0;
              HIGH := XUSE[TERMS] +1; GO TO A;
           Y:SUM:= 1.0;
           X:COEF[J] := SUM*FACTK/DENOM END LOOP;
              GO TO EXIT;
           Z: FOR I:= 1 STEP 1 UNTIL N DO COEF[I] :=0;
         EXIT: END DICOM;
```

**Figure 2.2**  A complete listing of DICOM

```
PROCEDURE D1LAG (KGL,GEOM,IGR,DCOEF);
  VALUE KGL,IGR;
  REAL ARRAY GEOM,DCOEF;
  INTEGER ARRAY KGL; INTEGER IGR;
    BEGIN  INTEGER I;
           EXTERNAL PROCEDURE DICOM;
           REAL ARRAY XTAB [1:KGL[2]];
           FOR I:= 1 STEP 1 UNTIL KGL[2] DO
              XTAB[I] := GEOM[IGR+I-KGL[3]];
           DICOM (KGL[1],KGL[2],GEOM[IGR],XTAB,DCOEF);
    END  D1LAG ;
```

**Figure 2.3**   A complete listing of D1LAG

derivative has to be evaluated; GEOM $[1:L]$—an $L$-element array containing all coordinates of the mesh; IGR—an integer indicating the node at which the derivative is computed. The D1LAG produces the coefficients of the finite-difference pattern which are stored in the DCOEF $[1:KGL(2)]$ array.

The partial derivatives in two-dimensional meshes are approximated according to

$$\frac{\partial^{k_1+k_2}u}{\partial x^{k_1}\partial y^{k_2}} = \sum_{i=1}^{n_1}\sum_{j=1}^{n_2} C_{ij}^{(k_1,k_2)}u(x_i, y_j) \qquad (2.10\text{a})$$

where

$$C_{ij}^{(k_1,k_2)} = C_i^{(k_1)}C_j^{(k_2)} \qquad (2.10\text{b})$$

and $C_i^{(k_1)}$, $C_j^{(k_2)}$ are computed by means of DICOM. To achieve this ARIEL utilizes the D2LAG, whose listing is presented in Figure 2.4. The input parameters are: KGL $[1:2, 1:3]$—an array whose elements indicate the orders

```
PROCEDURE D2LAG (KGL,GEOM,IGR,DCOEF);
  ARRAY GEOM,DCOEF; INTEGER ARRAY IGR,KGL;
  BEGIN INTEGER I1, I2, N, L, IND;
        EXTERNAL PROCEDURE DICOM;
    N := IF KGL[1,2] > KGL[2,2] THEN KGL[1,2] ELSE KGL[2,2];
      BEGIN ARRAY HELP[1:2,1:N], XTAB      , COEF[1:N+2];
        FOR L:=1,2 DO
          BEGIN
            FOR I1 := 1 STEP 1 UNTIL KGL[L,2] DO
               XTAB[I1] := GEOM[L,IGR[L] + I1 - KGL[L, 3]];
            DICOM(KGL[L,1],KGL[L,2], GEOM[L,IGR[L] ],  XTAB,COEF);
            FOR I1 := 1 STEP 1 UNTIL KGL[L,2] DO
               HELP[L,I1] := COEF[I1] ;
          END L;
          IND := 1;
          FOR I1 := 1 STEP 1 UNTIL KGL[1,2] DO
          FOR I2 := 1 STEP 1 UNTIL KGL[2,2] DO
            BEGIN DCOEF[IND] := HELP[1,I1]*HELP[2,I2];
              IND := IND + 1;
            END I2,I1;
      END;
  END D2LAG;
```

**Figure 2.4**   A complete listing of D2LAG

of derivatives, the numbers of nodes for approximation, and the indices of the central points; GEOM $[1:2, 1:1]$, where the coordinates of nodes along $x$ and $y$, respectively, are stored; IGR $[1:2]$—an integer array where the indices of the node at which the derivatives are evaluated are stored.

As an example, for the derivative $\partial^4 u / \partial x^2 \partial y^2$ for an uniform mesh and a three point pattern along each coordinate, that is

$$KGL = \begin{vmatrix} 2 & 3 & 2 \\ 2 & 3 & 2 \end{vmatrix}$$

one will obtain

$$C_{ij} = \begin{vmatrix} 1 & -2 & 1 \\ -2 & 4 & -2 \\ 1 & -1 & 1 \end{vmatrix}$$

The partial derivatives in three dimensional meshes are approximated according to

$$\frac{\partial^{k_1+k_2+k_3}}{\partial x^{k_1} \partial y^{k_2} \partial z^{k_3}} = \sum_{i=1}^{n_1} \sum_{j=1}^{n_2} \sum_{k=1}^{n_3} C_{ijk}^{(k_1,k_2,k_3)} u(x_i, y_j, z_k) \qquad (2.11a)$$

where

$$C_{ijk}^{(k_1,k_2,k_3)} = C_i^{(k_1)} C_j^{(k_2)} C_k^{(k_3)} \qquad (2.11b)$$

Here the D3LAG procedure, the listing of which is presented in Figure 2.5, is used. The input parameters differ from those of D2LAG in that the first

```
PROCEDURE D3LAG(KGL,GEOM,IGR,DCOEF);
 ARRAY GEOM,DCOEF; INTEGER ARRAY IGR,KGL;
  BEGIN INTEGER I1, I2, I3, N, L, IND;
      EXTERNAL PROCEDURE DICOM;
   N := IF KGL[1,2] > KGL[2,2] THEN KGL[1,2] ELSE KGL[2,2];
   N := IF     N    > KGL[3,2] THEN    N    ELSE KGL[3,2];
   BEGIN ARRAY HELP[1:3,1:N], XTAB    , COEF[1:N+2];
       FOR L:=1,2,3 DO
        BEGIN
           FOR I1 := 1 STEP 1 UNTIL KGL[L,2] DO
              XTAB[I1] := GEOM[L,IGR[L] + I1 - KGL[L, 3]];
           DICOM(KGL[L,1],KGL[L,2], GEOM[L,IGR[L] ],  XTAB,COEF);
           FOR I1 := 1 STEP 1 UNTIL KGL[L,2] DO
              HELP[L,I1] := COEF[I1] ;
        END L;
      IND := 1;
      FOR I1 := 1 STEP 1 UNTIL KGL[1,2] DO
      FOR I2 := 1 STEP 1 UNTIL KGL[2,2] DO
      FOR I3 := 1 STEP 1 UNTIL KGL[3,2] DO
      BEGIN DCOEF[IND] := HELP[1,I1]*HELP[2,I2]*HELP[3,I3];
           IND := IND + 1;
         END I2,I1,I3;
      END;
  END;
 END D3LAG;
```

**Figure 2.5** A complete listing of D3LAG

indices of the arrays vary from 1 to 3 to account for the third space variable $z$; that is KGL $[1:3, 1:3]$, GEOM $[1:3, 1:3]$, IGR $[1:3]$.

The ARIEL utilizes improved versions of the procedures D1LAG, D2LAG and D3LAG, just considered, where checks are made for the pattern approximation to the boundary, and indication for user's logical errors is yielded. In this way ARIEL allows for: (a) the solution of the problems with mixed derivatives; (b) the usage of non-uniform meshes; (c) the usage of finite difference patterns with a larger number of nodes, and (d) places no restrictions for the order of derivatives.

## 2.4 HOW TO WORK WITH ARIEL

ARIEL is an interactive system. The statement ARIEL, initializes the system and when it's ready, the 'ENTER COMMAND' appears. ARIEL recognizes the following nine commands:

ENTER—enters all the information necessary to define a new initial boundary value problem. For this purpose the command starts that module of the system, which secures prompt entry of information. The latter is saved in the ARIEL information module. The user has to define the finite-difference approximations. ARIEL imposes no restrictions on the number of nodes and their location.

GENERATE—generates, compiles, and loads the computer program which solves the previously defined problem. It initiates a dialogue, in the course of which the user has to indicate the name of his program and choose one of the methods proposed by the ARIEL for the solution of his problem.

SOLVE ⟨name⟩—starts execution of the program with the name indicated.

PRINT—prints a current dialogue record on the corresponding device.

LIST—lists the information module of all previously solved problems.

SEARCH ⟨name⟩—prints all information on the stated problem.

DELETE ⟨name⟩—deletes the stated problem from the information module.

CLEAR—denotes the current record of the dialogue.

HELP—answers user questions in English. The procedure is borrowed from Data General Corporation.[22] The user may learn the kinds of commands, get information on different circuits, the numerical methods that may be used, the structures of the ARIEL systems, etc.

As a whole, after the appearance of the 'ENTER COMMAND' the user may enter each of the previously mentioned eight commands in English. The entered one is analysed by the system and the command is recognized by its first four letters. That is why the following statements are equivalent:

PLEASE BE SO KIND TO GENERATE MY NEW TASK, GENE, or GENERATE NEW PROBLEM.

All statements directed to ARIEL are checked and errors are indicated through announcements. Such errors could be: unknown statement, incorrectly entered information, formal errors concerning the definition of finite-difference pattern, etc.

The problem solution with ARIEL requires the following sequence:

*Define the PDEs.* The problem has to be written in the form given by Equations (2.7). The user must define the finite-difference approximation for every space derivative.

*Define the type of boundary conditions.* The boundary conditions have to be written in the form given in Equation (2.7c).

*Choose the integration method for ODEs.* ARIEL uses only two methods: Euler and Leapfrog.

*Generate the computer program.* ARIEL automatically generates, compiles, links, and catalogues the problem solving program.

*Execution of the program.* The mesh type, the boundary conditions, location as well as the initial conditions must be defined. A variable step mesh may also be used. Result may be displayed or printed. The experiment points out whether the solution is stable. In case it is not, the user has no difficulty in changing the mesh or the time step of integration. If this does not help either, the user can quickly generate a new program with other finite-difference approximations.

Consider now two examples. As a first example consider the following equation

$$U_t = U_{xx} + U_{yy} \tag{2.12a}$$

in a two-dimensional region and first kind boundary condition

$$U - x^2 + y^2 = 0 \tag{2.12b}$$

A five-point difference approximation will be used. To start, use the statement:

ARIEL
%WELCOME!THE SYSTEM FOR AUTOMATIC SOLUTION OF PARTIAL AND
%ORDINARY DIFFERENTIAL EQUATIONS WAITS FOR YOUR REQUEST!
%ENTER COMMAND:

Each line of system information starts with '%'. To enter information defining the problem (2.12), we have to use the ENTER command.

%ENTER COMMAND: ENTER MY NEW TASK

As we have already noted each of the nine commands may be given in English. That's why ARIEL answers how it has understood the user

%YOU WILL ENTER A NEW PROBLEM

and asks if the information will be entered from the card reader ($CDR) or the terminal ($TTI)

%SELECT DIVICE (0-$CDR/$LPT, 1-$TTI/$TT0):   1
%CHOOSE NAME OF YOUR PROBLEM: FOURIER

Our answer chooses Fourier as a name of the program which will be generated by the system

%NUMBER OF EQUATIONS:   1
%NUMBER OF SPACE VARIABLES:   2
%MAX.NUMBER OF MEMBERS IN EQUATIONS:   2
%TOTAL NUMBER OF BOUNDARY CONDITIONS:   1
%HAVE YOU ANY ERROR (YES-1,NO-0):   0

We replied that the problem considered has 1 equation, 2 space variables—$x$ and $y$, the right-hand side of the equation contains 2 members and 1 type of boundary conditions exists.

%NUMBER OF MEMBERS IN EQUATION 1:   2
%EXPRESSION:
%DERIV(X):   2
%NUMBER OF POINTS OF APPROXIMATION:   3
%PLACE OF CENTRAL POINT:   2
%DERIV(Y):   0
%HAVE YOU ANY ERROR(YES-1,NO-0):   0

Information about $U_{xx}$ is entered through the last 6 questions '%EXPRES-SION:' was not answered because there is no expression before $U_{xx}$. The next questions refer to $+U_{yy}$.

%EXPRESSION:   +
%DERIV(X):   0
%DERIV(Y):   2
%NUMBER OF POINTS OF APPROXIMATION:   3
%PLACE OF CENTRAL POINT:   2
%HAVE YOU ANY ERROR?(YES-1,NO-0):   0

If the answer to the last question is 1, ARIEL repeats the last sentence of questions.

%NUMBER OF MEMBERS IN BOUNDARY CONDITION 1:   1
%EXPRESSION:
%DERIV(X):   0
%DERIV(Y):   0
%EXPRESSION:   $-X{\uparrow}2 + Y{\uparrow}2$
%HAVE YOU ANY ERROR?(YES-1,NO-0):   0

The answer states that boundary conditions have 1 member, namely $U$. The question before the last one requires information about the known function $-x^2 + y^2$ in the boundary condition.

Arithmetic expressions, which could contain user's procedures are entered after 'EXPRESSION:'. That is why a question follows

%ENTER NAMES OF EXTERNAL REAL PROCEDURES:

In the present case there are no procedures. If such are present, their names are entered, separated by a comma.

All of the previously entered information is automatically saved in ARIEL's information module. It is sufficient for the generation, linkage, and cataloguing of the user program through the GENERATE command.

%ENTER COMMAND:   GENERATE
%WHAT IS THE NAME OF YOUR PROBLEM:   FOURIER
%CHOOSE METHODS (EULER-1,LEAPFROG-2):   1
%THE PROGRAM WAS GENERATED SUCCESSFULLY!
%ENTER COMMANDS:   SOLVE FOURIER

The last command starts the FOURIER program. The dialogue which follows, is

%USER'S PROGRAM 'FOURIER'
%INPUT FILE:   $CDR
%OUTPUT FILE:   $LPT

This means, that the input is the card reader, and the output unit is the line printer.

%MAX.NUMBER OF X-POINTS:   11
%MAX.NUMBER OF Y-POINTS:   11
%ENTER GRID SIZE AND DISPOSITION FROM $CDR

The real coordinates are entered through the $CDR (card reader), and they are along $x$ $(0, 0.1, 0.2, \ldots, 0.9, 1)$ and along $y$ $(0, 0.1, 0.2, \ldots, 0.9, 1)$. A variable step mesh may also be used. We have chosen an array of grid points with $11 \times 11 = 121$ nodes. This uniform finite difference mesh covers the solution region, but the boundaries must be a series of horizontal and vertical lines. The region and the type of boundary condition are entered through the card reader as

```
1, 1, 1, −1, −1, −1, −1, 1, 1, 1, 1,
1, 0, 1, −1, −1, −1, −1, 1, 0, 0, 1,
1, 0, 1,  1, −1, −1, −1, 1, 0, 0, 1,
1, 0, 0,  1, −1, −1, −1, 1, 0, 0, 1,
1, 0, 0,  1, −1, −1, −1, 1, 0, 0, 1,
```

```
1, 0, 0,   1,   1,   1,   1, 1, 0, 0, 1,
1, 0, 0,   0,   0,   0,   0, 0, 0, 0, 1,
1, 0, 0,   0,   0,   0,   0, 0, 0, 0, 1,
1, 0, 0,   0,   0,   0,   0, 0, 0, 0, 1,
1, 0, 0,   0,   0,   0,   0, 0, 0, 0, 1,
1, 1, 1,   1,   1,   1,   1, 1, 1, 1, 1,
```

where the codes 0 and −1 mean that the point is in or out of the solution region, respectively. At the boundary the code 1 means that this point is computed by the first boundary condition.

%HAVE YOU INITIAL CONDITIONS(YES-1,NO-0):   0

In this case ARIEL assigns the value 1 to all possible grid points. If the answer is 1, initial values must be entered from card reader.

%ENTER DELTA T:   0.002
%ENTER TIME FOR PRINTING THE SOLUTION:   0.1, 0.8, 1, 1.5, 2

We chose the time integration step 0.002 and we entered the values of time for which we must obtain the printed results.

The difference between the exact steady state solution and the numerical solution obtained by ARIEL is less than 0.001.

As a second example consider the solution of laminar free convection from a vertical plate. The problem may be described by PDEs. Using similarity variables the problem can be transformed into ODEs subject to appropriate boundary conditions. The final problem and the boundary conditions as given by Schlichting[23] are

$$U_{xxx} + 3 \times U \times U_{xx} - 2 \times U_x^2 + V = 0 \qquad (2.13a)$$

$$V_{xx} + 3 \times Pr \times U \times V_x = 0 \qquad (2.13b)$$

with boundary conditions

$$U = 0, \quad U_x = 0 \quad \text{and} \quad V - 1 = 0 \qquad \text{at } x = 0 \quad (2.13c, d, e)$$

$$U_x = 0 \quad \text{and} \quad V = 0 \qquad \text{at } x = 5 \qquad (2.13f, g)$$

In order to place the problem (2.13) within the class that may be solved by ARIEL, it must be written as

$$U_t = U_{xxx} + 3 \times U \times U_{xx} - 2 \times (U_x)^2 + V \qquad (2.14a)$$

$$V_t = V_{xx} + 3 \times Pr \times U \times V_x \qquad (2.14b)$$

When reaching steady state $U_t = V_t = 0$ and the problem (2.14) is equivalent to problem (2.13).

To solve the problem by means of ARIEL the following dialogue must take place:

ARIEL
%WELCOME!THE SYSTEM FOR AUTOMATIC SOLUTION OF PARTIAL AND
%ORDINARY DIFFERENTIAL EQUATIONS WAITS FOR YOUR REQUEST!
%ENTER COMMAND:   ENTER
% YOU WILL ENTER A NEW PROBLEM
%SELECT DIVICE(0-$CDR/$LPT,1-$TTI/$TT0):   1
%CHOOSE NAME OF YOUR PROBLEM:   SCHLICHTING
%NUMBER OF EQUATIONS:   2
%NUMBER OF SPACE VARIABLES:   1
%MAX.NUMBER OF MEMBER IN EQUATIONS:   5
%TOTAL NUMBER OF BOUNDARY CONDITIONS:   4
%HAVE YOU ANY ERROR?(YES-1,NO-0):   0

The reply states that the problem has 2 equations, 1 space variable $x$, the right-hand side of Equation (2.14a) has 5 members and we have 4 different types of boundary conditions.

%NUMBER OF MEMBERS IN EQUATION 1:   5
%EXPRESSION:
%WHICH FUNCTION DO YOU DIFFERENTIATE:   1
%DERIV(X):   3
%NUMBER OF POINTS OF APPROXIMATION:   4
%PLACE OF CENTRAL POINT:   3

In this way we entered the first member $U_{xxx}$. The second member $+3 \times U$ follows:

%EXPRESSION:   +3*
%WHICH FUNCTION DO YOU DIFFERENTIATE:   1
%DERIV(X):   0

Then the third member $\times U_{xx}$ is entered:

%EXPRESSION:   *
%WHICH FUNCTION DO YOU DIFFERENTIATE:   1
%DERIV(X):   2
%NUMBER OF POINTS OF APPROXIMATION:   4
%PLACE OF CENTRAL POINT:   3

Then the fourth member $-2 \times (U_x$ is entered:

%EXPRESSION:   −2*(
%WHICH FUNCTION DO YOU DIFFERENTIATE:   1
%DERIV(X):   1
%NUMBER OF POINTS OF APPROXIMATION:   3
%PLACE OF CENTRAL POINT:   2

The description of the fifth member $)^2 + V$ follows:

%EXPRESSION:   )↑2+
%WHICH FUNCTION DO YOU DIFFERENTIATE:   2
%DERIV(X):   0
%HAVE YOU ANY ERROR?(YES-1,NO-0):   0

This concludes the entering of information for Equation (2.14a). The dialogue for Equation (2.14b) follows:

%NUMBER OF MEMBERS IN EQUATION 2:   3
%EXPRESSION:
%WHICH FUNCTION DO YOU DIFFERENTIATE:   2
%DERIV(X):   2
%NUMBER OF POINTS OF APPROXIMATION:   3
%PLACE OF CENTRAL POINT:   2

The description of first member of Equation (2.14b), $V_{xx}$, ends and the description of the second member, $+3 \times Pr \times U$, begins:

%EXPRESSION:   +3*PR*
%WHICH FUNCTION DO YOU DIFFERENTIATE:   1
%DERIV(X):   0

The description of the last member, $\times V_x$, follows:

%EXPRESSION:   *
%WHICH FUNCTION DO YOU DIFFERENTIATE:   3
%****FUNCTION NUMBER ERROR
%WHICH FUNCTION DO YOU DIFFERENTIATE:   2
%DERIV(X):   1
%NUMBER OF POINTS OF APPROXIMATION:   3
%PLACE OF CENTRAL POINT:   2
%HAVE YOU ANY ERROR?(YES-1,NO-0):   0

This concludes the input information of Equation (2.14b) and initiates the dialogue for the information of the types of boundary conditions.

%NUMBER OF MEMBERS IN BOUNDARY CONDITION 1:   1
%EXPRESSION:
%WHICH FUNCTION DO YOU DIFFERENTIATE:   1
%DERIV(X):   0
%EXPRESSION:   +0
%WHICH FUNCTION WILL YOU DEFINE FROM THIS BOUNDARY CONDITION:   1
%HAVE YOU ANY ERROR?(YES-1,NO-0):   0

In this way we entered information for the first boundary condition type $U + 0 = 0$, which will be used at $x = 0$.

%NUMBER OF MEMBERS IN BOUNDARY CONDITIONS 2:   1
%EXPRESSION:
%WHICH FUNCTION DO YOU DIFFERENTIATE:
%DERIV(X):   1
%NUMBER OF POINTS OF APPROXIMATION:   2
%PLACE OF CENTRAL POINT:   2
%EXPRESSION:   +0
%WHICH FUNCTION WILL YOU DEFINE FROM THIS BOUNDARY
CONDITION:   1
%HAVE YOU ANY ERROR?(YES-1,NO-0):   0

We entered information for the second type boundary conditions, $U_x + 0 = 0$, which will be used later when $x = 0$ and $x = 5$.

%NUMBER OF MEMBERS IN BOUNDARY CONDITION 3:   1
%EXPRESSION:
%WHICH FUNCTION DO YOU DIFFERENTIATE:   2
%DERIV(X):   0
%EXPRESSION:   −1
%WHICH FUNCTION WILL YOU DEFINE FROM THIS BOUNDARY
CONDITION:   2
%HAVE YOU ANY ERROR?(YES-1,NO-0):   0

We entered information for the third type boundary condition $V - 1 = 0$, which will be used at $x = 0$. Now the dialogue for the last type boundary condition begins:

%NUMBER OF MEMBERS IN BOUNDARY CONDITION 4:   1
%EXPRESSION:
%WHICH FUNCTION DO YOU DIFFERENTIATE:   2
%DERIV(X):   0
%EXPRESSION:   +0
%WHICH FUNCTION WILL YOU DEFINE FROM THIS BOUNDARY
CONDITION:   2
%HAVE YOU ANY ERROR?(YES-1,NO-0):   0

This concludes the boundary condition description.

%ENTER NAMES OF EXTERNAL REAL PROCEDURES:   PR
%ENTER COMMAND:   GENERATE
%WHAT IS THE NAME OF YOUR PROBLEM:   SCHLICHTING
%CHOOSE METHODS (EULER-1,LEAPFROG-2):   1
%THE PROBLEM WAS GENERATED SUCCESSFULLY
%ENTER COMMAND:   SOLVE

The execution of the SCHLICHTING program starts

%USER'S PROGRAM 'SCHLICHTING'
%INPUT FILE:    $TTI
%OUTPUT FILE:    $LPT
%MAX.NUMBER OF X POINT:    11
%ENTER GRID SIZE AND DISPOSITION FROM $TTI
    0, 0.5, 1, 1.5, 2, 2.5, 3, 3.5, 4, 4.5, 5
    1,   2, 0,   0, 0,   0, 0,   0, 0,   0, 2
    3,   0, 0,   0, 0,   0, 0,   0, 0,   0, 4

The real coordinates are $x = 0, 0.5, \ldots, 5$. The region and the type of boundary conditions for the first function $U$ and the second function $V$ are 1, 2, 0, 0, 0, 0, 0, 0, 0, 0, 2 and 3, 0, 0, 0, 0, 0, 0, 0, 0, 0, 4, respectively. The codes 1, 2, 3, and 4 mean that the corresponding point is computed by the first, second, third, and fourth boundary condition, respectively. The code 0 shows that the point is in the solution region.

%HAVE YOU INITIAL CONDITION(YES-1,NO-0):    1
2, 2, 2, 2, 2, 2, 2, 2, 2, 2, 2,
2, 2, 2, 2, 2, 2, 2, 2, 2, 2, 2
%ENTER DELTA T:    0.01
%ENTER TIME FOR PRINTING THE SOLUTION:
0.1, 1, 2, 3.5, 5

The numerical results obtained by ARIEL fully correspond to the figure given by Schlichting.[23]

## 2.5  FUTURE DEVELOPMENT

The two examples presented illustrate how ARIEL works. Systems like ARIEL will support the cause of numerical methods in engineering practice.

Currently work is being done on a problem-oriented language for ARIEL. The selection of numerical integration methods will also be widened. Non-linear boundary conditions will also be included.

## REFERENCES

1. M. B. Carver (1975). 'Simulation packages for the solution of partial differential equation systems', *Summer Computer Simulation Conference*, 57–64.
2. D. M. Young and M. D. Juncosa (1959). 'SPADE-A set of subroutines for solving elliptic and parabolic partial differential equations', *Rand Corporation Report No. P-1709*, Santa Monica, California, May 21.
3. S. M. Morris and W. E. Schiesser (1968). 'SALEM: A programming system for the simulation of systems described by PDEs', *Proceedings, 1968 Fall Joint Computer Conference*, Vol. 33, Part I, 353–357, Thompson Book Co, Washington, D.C.

4. J. Cea, B. Nivelet, L. Schmidt and G. Terrine (1966). 'Techniques numeriques de l'approximation variationnelle des problems elliptiques', Vol. 1, Centre National de la Recherche Scientifique, *Publication No. MMC/10.12.5/AI*, Paris, April.

5. J. Cea, B. Nivelet, L. Schmidt and G. Terrine (1967). 'Techniques numeriques de l'approximation variationnelle des problems elliptiques', Vol. 3, Centre National de la Recherche Scientifique, *Publication NO. FT/6.3.7/AI*, Paris, March.

6. J. Cea, B. Nivelet and G. Terrine (1968). 'Techniques numeriques de l'approximation variationnelle des problems elliptiques', Vol. 2, Centre National de la Recherche Scientifique, *Publication No. MMC/10.5.8/AI*, Paris, November.

7. M. G. Zellner (1970). 'DSS-Distributed system simulator', *Ph.D. Dissertation*, Lehigh University, June.

8. W. E. Schiesser (1971). 'A digital simulation system for mixed ordinary/partial differential equation modes', *Proceedings, IFAC Symposium on Digital Simulation of Continuous Systems*, 2, p. S2–1 to S2–9, Gyor Hungary, September.

9. W. E. Schiesser (1972). 'A digital simulation system for higher-dimensional partial differential equations', *Proceedings of the 1972 Summer Computer Simulation Conference*, pp. 62–72, San Diego, June 14–16.

10. Carver, M. B. (1973). 'A FORTRAN oriented simulation system for the general solution of partial differential equations', *Proceedings 1973 Summer Computer Simulation Conference*, Montreal, Canada, July.

11. W. E. Schiesser (1978). 'DSS/2 (Differential systems simulator, version 2)', *Introductory Programming Manual*, Third printing, January.

12. M. B. Carver (1974). 'FORSIM: A FORTRAN package for the automated solution of coupled partial and/or ordinary differential equation systems—User's manual', *Atomic Energy of Canada Limited Report AECL-4844*, November.

13. *Boundary Value Problems in Regions of Complex Boundary*, Kiev, Inst. of Cybernetics, AS USSR, 1974.

14. V. P. Ilin (1974). *Numerical methods for the solution of electrooptics problems*, Novosibirsk, Science Press.

15. I. I. Liashko, I. V. Sergienko, G. E. Mistetskii and V. V. Skopetskii (1977). *Computer-automated solution of filtration problems*, Kiev, Naukova dumka.

16. V. K. Kabulov (1979). *Algorithmisation in Continuum Mechanics*, FAN Press, Usbekian SSR, Tashkent.

17. A. F. Cardenas and W. J. Karplus (1970). 'PDEL: A language for partial differential equations', *Comm. of the ACM*, **13**, 3, March, 184–191.

18. A. F. Cardenas and W. J. Karplus (1970). 'Design and organization of a translator for a partial differential equation language', *Proc. AFIPS 1970*, Vol. 36, Pt. 1, AFIPS Press, Montvale, N.J., 513–523.

19. K. V. Roberts (1971). 'The solution of partial differential equations using a symbolic style of ALGOL', *J. Computational Physics*, **8**, 83–105.

20. J. H. Giese (1971). A Bibliography for the numerical solution of partial differential equations, AD730662 US Army Aberdeeb R & D Centre, Maryland.

21. T. P. Giammo (1962). 'Difference expression coefficients algorithm', 79, *Comm. ACM Collected Algorithms*.

22. Data General Corporation (1970). *Extended Algol-User's Manual*.

23. H. Schlichting (1958). *Grenzschicht-Theorie*, Verlag G. Braun.

*Numerical Methods in Heat Transfer*
Edited by R. W. Lewis, K. Morgan, and O. C. Zienkiewicz
© 1981 John Wiley & Sons Ltd

## Chapter 3

# Heat Transfer Simulation of Composite Devices

*L. Imre*

### SUMMARY

Composite devices are built from components of varying materials, geometry and function.

In structural systems different transport processes often take place simultaneously. In such cases the structural elements may be not only in the thermal, but also in hydrodynamical, electrical and mechanical interaction.

Processes taking place simultaneously and being in interaction with the heat transport can be simulated by means of simultaneous models.

Simultaneous models are built from part models pertaining to the basic processes. The part models are coupled by dual- or multifunction elements that take part functionally in at least two part models.

The part models serving to determine the fields can be derived by decomposition into discrete parts and built up as networks. The state functions of some parts handled as elements in the network model have to be derived on the basis of a separate model, on a lower hierarchy level. These types of elements are termed 'variators'.

A mathematical description of network models can be achieved by the network equation system. The complete describing equation system can be solved economically mostly by numerical schemes. A description of mechanical part models is possible by different methods (e.g. force method, displacement method, method with mixed variables).

Repeated solution of the complete describing equation system gives the elements of the sensitivity matrix, which can be used with advantage for estimating the effect of each parameter. The simultaneous modelling is illustrated by examples of thermohydrodynamical and thermomechanical structural systems.

### 3.1  SIMULTANEOUS MODELS

#### 3.1.1  Simultaneous processes in composite devices

Composite devices are built from elements of various sizes, shapes, materials and functions. This chapter will consider those in which, during their normal operation, heat is produced which has to be dissipated in some suitable way. For this purpose also some cooling material can be made to flow through the system by means of a pump or by the thermosyphon effect.

In some composite devices (e.g. in thermomechanical actuating systems) the displacement of the elements is produced by the temperature variation of the components. Their temperature field may also be affected by the flow conditions of some cooling media. In certain cases also the displacement of components exerts an effect upon the flow and heat transfer conditions.

### 3.1.2  Physical mathematical modelling conception

The model serving for the description of simultaneous processes can be built from the models of the individual 'basic' processes. Often, these part models themselves are complex too. In cases when the basic processes are in interaction with each other, the models of these interactions are indispensable to couple the part models (Figure 3.1).

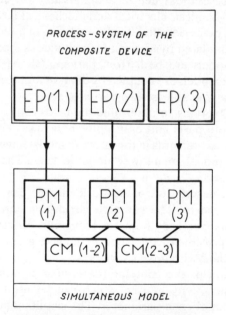

**Figure 3.1**  Structure of the simultaneous model: EP, elementary processes; PM, part models; CM, coupling elements

There are various model concepts that can be applied advantageously in the construction of the simultaneous model and the part models. At any rate, it appears practical to choose a model concept permitting increase in accuracy by refining the model, which is clear-cut and can be checked easily, which brings out the physical essence of the processes modelled and makes it perceptible. In addition, efforts should be made to accomplish a harmony of the goal and the means used to reach it, as well as to consider the economy of the investigation.[1]

### 3.1.3 Network models

For modelling the transport processes of devices consisting of many components, it is network modelling that meets the requirements mentioned above. The principle is to lump the system, from the viewpoint of the transport process, into parts, which are then substituted by network elements modelling the functions of each part and connected with each other by network elements expressing their interconnections.[2,3,4,5] Such functional network elements are current sources and sinks, voltage sources, capacitors and resistors. Those elements that cannot be substituted for by the network elements enumerated will be termed variators. A separate network model can be used for modelling each transport process. The network part models are coupled by network elements occurring in two networks at least. These are the dual- or multi-function elements.

### 3.1.4 Hierarchic and mixed models

Sometimes, parts that are subsystems in reality are treated as elements (e.g. as variators). To produce the state functions of these, further models—sometimes including details—are required (e.g. to produce the state function of a heat exchanger).

In other cases it may be desirable to define the temperature field of some parts in particular detail and with high accuracy. This, in fact, is sometimes the purpose of the examination.

The calculation of subsystems constructed within the simultaneous model can be ranged into a lower hierarchy level (hierarchic models). The models of lower hierarchy level are not necessarily network models. Also the model structure based on the finite difference or finite elements methods can be employed (mixed models). In such cases, the model of higher hierarchy level serves to define the boundary conditions that develop as an interaction between the elements of the system as a whole and which are needed in modelling the subsystems or a single component.

## 3.2 SIMULATION OF THERMOHYDRODYNAMICAL SYSTEMS

Complex devices in which both heat transfer and macroscopic mass transfer occur in interaction with each other are considered to be thermohydrodynamical systems. Simultaneous heat and mass flow network models can be employed in many cases to model them efficiently and economically.

### 3.2.1 The heat flow network part model

The heat flow network (HFN) part model is constructed by coupling properly the network elements modelling functionally the discrete parts of the device: heat sources, heat capacitances, conductive, convective and radiative resistances, as well as the thermal voltage sources determining the environmental temperatures. If the HFN also contains convective heat flow paths (which is the situation in the case of thermohydrodynamical systems), then those dual function elements and variators (e.g. heat exchanger, mixer) similarly appear in the HFN model.

For the description of heat transfer processes of thermal systems, energy balance equations and heat transfer equations are applied. The equation system describing the heat flow networks are in fact, a form of these equations arranged advantageously for numerical solution.[1]

The nodel equation system of the HFN expresses the heat flow balance. With the reduced incidence matrix $\mathbf{A}_h$ and reduced cutset matrix $\mathbf{B}_h$ of the graph, the nodal equation system (Kirchhoff I) will be:

$$\mathbf{A}_h(\mathbf{Q}_d + \mathbf{Q}_s) = \mathbf{0} \tag{3.1}$$

and

$$\mathbf{B}_h(\mathbf{Q}_d + \mathbf{Q}_s) = \mathbf{0} \tag{3.2}$$

respectively.

In the equations $\mathbf{Q}$ is the heat flow rate matrix, while the indices mean heat ($h$), the number of branches containing heat transfer resistance or heat capacitance ($d$) and the heat source ($s$).

With the loop matrix $\mathbf{H}_k$ of the graph of the heat flow network, on the basis of the temperature differences $\Delta T_d$ between the nodes bounding the branches and of the thermal voltage source $\Delta T_s$ connected into the loop (e.g. in modelling the boundary condition of the first kind), the loop equation (Kirchhof II) will be

$$\mathbf{H}_h(\Delta \mathbf{T}_d + \Delta \mathbf{T}_s) = \mathbf{0} \tag{3.3}$$

The loop equation system (Fourier) for the branches containing the heat transfer resistance is

$$\Delta \mathbf{T}_d = \mathbf{R}_h \mathbf{Q}_{dk} \tag{3.4}$$

where $\mathbf{R}_h$ is the diagonal matrix consisting of the heat transfer resistances of the branches. For the $i$th branch containing the heat capacitance $C_i$:

$$C_i \frac{\mathrm{d}(\Delta T_i)}{\mathrm{d}t} = Q_{ic} \tag{3.5a}$$

where $t$ is time, and for all the branches

$$\mathbf{C}\dot{\mathbf{T}} = \mathbf{Q}_{dc} \tag{3.5b}$$

where

$$\mathbf{Q}_d = \mathbf{Q}_{dh} + \mathbf{Q}_{dc} \tag{3.5c}$$

The state equation of the variators of the heat flow network furnishes the relationship of the input ($T_{\text{in}}$) and output ($T_{\text{out}}$) temperatures. For example, for the $j$th variator:

$$(T_{\text{out}})_j = f_j(T_{\text{in}})_j \tag{3.6}$$

### 3.2.2 The mass flow network part model

The mass flow network (MFN) part model is built by coupling the source and sink (supply and removal) of the mass, the pressure source (e.g. a pump), the elements of flow resistance and the storing elements, all of which model discrete parts. Inasmuch as the mass flow is not isothermal, the dual-function elements of the convective flow parts (e.g. the pressure sources deriving from the temperature difference) and the variator elements too, appear in the MFN part model.

To describe the flows, the equations of mass balance and impulse balance are applied. The equation system of the MFN is also a form of those equations arranged advantageously for numerical solution.[1,2]

With the reduced incidence matrices $\mathbf{A}_M$ and the reduced cutset matrix $\mathbf{B}_M$ of the graph, the nodal equation system of the MFN model will be:

$$\mathbf{A}_M(\mathbf{M}_d + \mathbf{M}_s) = \mathbf{0} \tag{3.7}$$

and

$$\mathbf{B}_M(\mathbf{M}_d + \mathbf{M}_s) = \mathbf{0} \tag{3.8}$$

respectively, where $\mathbf{M}$ is the mass flow rate matrix. The loop equation system of the mass flow network, with loop matrix $\mathbf{H}_M$ and the pressure differences of the branches $\Delta\mathbf{P}$, is:

$$\mathbf{H}_M(\Delta\mathbf{P}_d + \Delta\mathbf{P}_s) = \mathbf{0} \tag{3.9}$$

The branch equation system of the MFN for the pressure drop caused by the flow resistance is given by

$$\Delta\mathbf{P}_{dk} = \mathbf{R}_M(\mathbf{M}_d, \Delta\mathbf{T}_d) \cdot \mathbf{M}_d \tag{3.10}$$

where $\mathbf{R}_M$ is the diagonal matrix containing the flow resistances of the branches, which is in many cases dependent on the mass flow rate and the temperature. The pressure difference in the case of non-steady flow is a function of the mass flow rate time derivative ($\dot{\mathbf{M}}$):

$$\mathbf{P}_{da} = F(\dot{\mathbf{M}}_d) \tag{3.11}$$

### 3.2.3 Dual-function elements

A dual-function (DF) element is the convective thermal source:[1]

$$\Delta T_{sk} = \frac{Q_d}{\Phi_C} = -R_{dk}Q_d \tag{3.12}$$

where $\Phi_C = Mc_p$ is the heat capacity flow rate of the flowing medium,

$$R_{dk} = -\frac{1}{\Phi_C} \tag{3.13}$$

is the formal resistance and

$$Q_d = \frac{T_m - T_w}{R_{mw}} \tag{3.14}$$

is the heat flow getting into the flowing medium through the coupling resistance $R_{mw}$ between nodes $m$ and $w$ of HFN.

A DF is the temperature-dependent pressure source deriving from the density variation of the flowing medium. For example, for gas flow

$$\Delta P_{ST} = \frac{R}{(c_p)_{av}}\left[\rho_{av}\frac{Q_d}{M_d} - \frac{M_d^2}{2A^2}\left(\frac{1}{\rho_m} - \frac{1}{\rho_w}\right)\right] \tag{3.15}$$

where $(c_p)_{av}$ is the average specific heat capacitance at constant pressure, $\rho_{av}$ is the average density, $A$ is the cross section of the flow, and $R$ is the gas constant.

## 3.3 APPLICATION OF SIMULTANEOUS HEAT AND MASS FLOW NETWORK MODELS

### 3.3.1 Purpose of the examination

The application of the model conception discussed in the foregoing will be illustrated by the example of the steady-state warming of large oil transformers.[7,8,9,10]

The designing engineer examines the warming of oil transformers to find the answer to two questions:

(a)  How high is the 'hot-spot' temperature of the windings?
(b)  What is the influence exerted on the 'hot-spot' by the parameters that must be decided or taken into consideration in the design work?

Figures 3.2(a) and (b) show the operation scheme of transformers with natural and forced cooling systems, respectively.

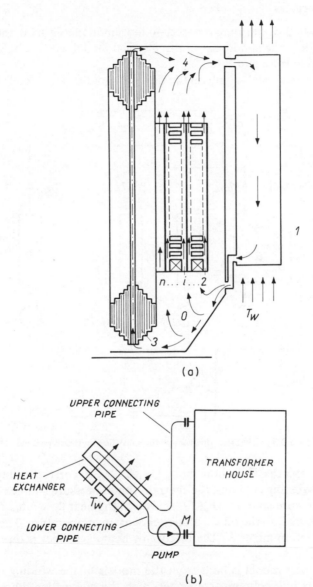

(a)

(b)

**Figure 3.2** (a) Simplified scheme of an oil transformer (ONAN) with thermosyphon cooling—the arrows indicate the flow of oil; (b) Schematic arrangement of an oil transformer (OFAF) with forced convection cooling

### 3.3.2  The simultaneous heat and mass flow network model of oil transformers

The examination of warming is based on the simultaneous heat and mass flow network model. The conditions of the physical model will not be discussed in detail here (see Imre and Bitai[10]).

The MFN part model is built on the flow paths of the cooling oil (Figure 3.3).

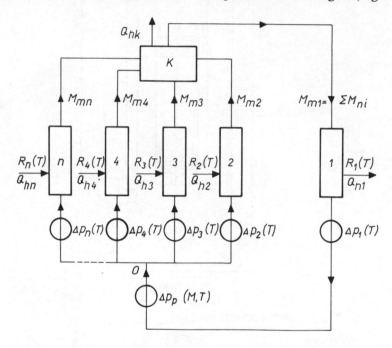

**Figure 3.3**  Schematic diagram of the mass flow network part model

The flow resistances $R_i$ in branches $i = 1, 2, \ldots, n$, corresponding to the oil ducts of the winding columns, the thermosyphon pressure sources $\Delta P_i(T)$ and the pump pressure sources $\Delta P_p(T)$, as well as the heat flow rates $Q_i$ from the coil and the core into the oil are all dependent on temperature. The branch oil flows are mixed in mixer $K$. The re-cooling heat exchanger is represented by branch 1.

The HFN part model is built from the models of the winding discs of the individual winding columns, connected with the convective thermal voltage sources according to Equation (3.12). Figure 3.4 shows the HFN model of a single disc (the thin lines indicate the turns of the disc). The part networks are connected by the DF elements $\Delta T_w$ and $\Delta P$ into a simultaneous heat and mass flow network (Figure 3.5). The squares along the branches represent the HFN model of a winding disc (marked with a number) each.

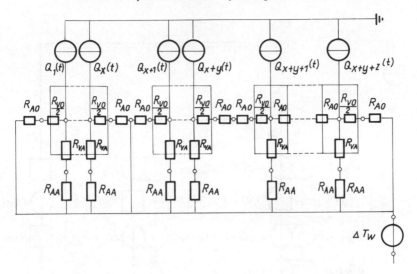

**Figure 3.4** Heat flow network model of a winding disc of divided construction. Lumping by turns. Number of the turns on each disc part: $x$, $y - x$, $z - y$

### 3.3.3 The equation system describing the heat flow network

The equation system of the HFN will be written for the part network according to Figure 3.4. For the steady state, after transcribing the combination of equation systems (3.1) and (3.4) into the nodal form, the valid HFN form of Poisson's equation will be obtained:[9]

$$\mathbf{KT} = \mathbf{Q}_s + \mathbf{VT}_w \tag{3.16}$$

where $\mathbf{K}$ is the corrected heat conduction hypermatrix of the order corresponding to the node numbers of the winding parts; it consists of quadratic matrices $\mathbf{K}_{1,x}$, $\mathbf{K}_{x,y}$, $\mathbf{K}_{y,z}$:

$$\mathbf{K} = \begin{bmatrix} \mathbf{K}_{1,x} & \mathbf{0} & \mathbf{0} \\ \mathbf{0} & \mathbf{K}_{x,y} & \mathbf{0} \\ \mathbf{0} & \mathbf{0} & \mathbf{K}_{y,z} \end{bmatrix} \tag{3.17}$$

$\mathbf{T}$ and $\mathbf{Q}$ are hypermatrices consisting of the column matrices of the nodal temperatures and heat sources, and $\mathbf{V}$ is the conduction hypermatrix consisting of the environmental coupling conductions:

$$\mathbf{T} = \begin{bmatrix} \mathbf{T}_{1,x} \\ \mathbf{T}_{x,y} \\ \mathbf{T}_{y,z} \end{bmatrix} \quad \mathbf{Q} = \begin{bmatrix} \mathbf{Q}_{1,x} \\ \mathbf{Q}_{x,y} \\ \mathbf{Q}_{y,z} \end{bmatrix} \quad \mathbf{V} = \begin{bmatrix} \mathbf{V}_{1,x} \\ \mathbf{V}_{x,y} \\ \mathbf{V}_{y,z} \end{bmatrix} \tag{3.18}$$

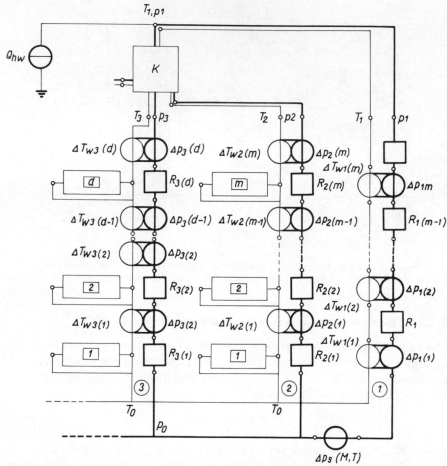

**Figure 3.5**  Simultaneous heat and mass flow network model of an oil transformer: thin line: heat flow network model; connected circles, dual-function elements

### 3.3.4  Equation system describing the mass flow network

With the notations of Figures 3.3 and 3.5 and based on Equations (3.7)–(3.11) one has:

(a)  The nodal equations system,

$$\left(M_1 - \sum_{b=2}^{n} M_b\right)_c = 0 \quad (b = 2, \ldots, n; c = 0, K) \tag{3.19}$$

(b)  The loop equation system for $l = 1, 2, \ldots, n-1$ independent loops:

$$\sum_{e} \sum_{z_e} (\Delta P_{ze} + \Delta P_{Rze})_l = 0 \tag{3.20}$$

where $e$ is the number of branches in the loop, and $z_e$ the number of network elements in the branches.

(c)  Branch equations corresponding to Equation (3.10) (see also Figure 3.5):

$$\Delta P_{Rze} \equiv \Delta P_{Rb(v)} = R_{b(v)}(M_b, T_{wb(v)}) \cdot M_b \qquad (3.21)$$

where $(v)$ is the number of the discrete parts in branch $b$.

(d)  The state functions of the dual-function elements will be given as follows.
Hydrostatic pressure source:

$$\Delta P_{ze} \equiv \Delta P_{b(v)} = g \int_{h_{b(i-1)}}^{h_{b(i)}} \rho_i[T_{wb(v)}(h)] \, dh \qquad (3.22)$$

where $h_{b(i)} - h_{b(i-1)}$ is the difference in height between the nodes in branch $b$
Convective thermal voltage source (Equation (3.12)):

$$\Delta T_{sk} = \Delta T_{wb(v)} = \frac{Q_{b(v)}}{\Phi_{Cw}} \qquad (3.23)$$

(e)  The state functions of the variators (see Equation (3.6)) are as follows.
Heat exchanger (on the air side $T_w \cong$ constant):

$$T_{out} = T_w + (T_{in} - T_w) \exp\left[-kA_H\left(\frac{1}{\Phi_{Cl}} + \frac{1}{\Phi_{Cw}}\right)\right] \qquad (3.24)$$

where $k$ is the overall heat conductance, $A_H$ the decisive heat transfer surface, and $\Phi_{Cw}$ the heat capacitance flow of the external air;
Mixer (its heat balance in Figure 3.5)

$$\sum_b M_b c_{pb}(T_b)_{out} - Q_{h,k} = 0 \qquad (3.25)$$

in which $Q_{h,k}$ is the heat loss of the transformer house.

## 3.3.5  Strategy for the solution

As was discussed in detail in Section 3.3.1 one of the aims of the solution is to determine the temperature of the 'hot-spot'. The highest temperatures develop in the uppermost winding discs of the winding columns. To determine them, it is necessary to find the solution of the equation system describing the simultaneous heat and mass flow network.

To find the proper strategy for the solution, a second goal also has to be taken into account: to examine the effect intensity of the different parameters applied in the design work. Of course, the influence on the 'hot-spot' temperature is of primary interest, but the influence of each parameter on other variables may also furnish important information for the designer.

The parameter sensitivity of systems is expressed properly by the sensitivity matrix, the elements of which are the relative sensitivity coefficients.[6,9]

For the purpose of this investigation we consider arbitrarily as input parameters those whose influence it is desired to examine (e.g. the external decisive temperature, $T_{Wj}$; the heat transfer surface of the heat exchanger, $A_H$; the overall conductance of the heat exchanger, $k$; the heat transfer coefficient on the winding surface, $h$).

The influence of a given input parameter $I_i$ upon a given output parameter $\Theta_j$ is interpreted in the sufficiently narrow vicinity of the working point ($W$). The relative sensitivity coefficient[6] is then:

$$A_{ie} = \left(\frac{\partial \Theta_j}{\partial I_i}\right)_W \left(\frac{I_i}{\Theta_j}\right)_W \tag{3.26}$$

The $i$th row of the sensitivity matrix contains the sensitivity coefficients of the output characteristics, pertaining to the $i$th input parameter. The relative sensitivity matrix (for $x$ input and $y$ output parameters) will be:

$$\mathbf{A} = \begin{vmatrix} A_{1,1} & \cdots & A_{1,j} & \cdots & A_{1,y} \\ \vdots & & \vdots & & \vdots \\ A_{i,1} & \cdots & A_{i,j} & \cdots & A_{i,y} \\ \vdots & & \vdots & & \vdots \\ A_{x,1} & \cdots & A_{x,y} & \cdots & A_{x,y} \end{vmatrix}_W \tag{3.27}$$

In the vicinity of the working point the individual elements of the relative sensitivity matrix will be found in the numerical way, by substituting the difference quotient for the differential quotient in Equation (3.26):

$$A_{i,j} \simeq \left(\frac{\Delta \Theta_j}{\Delta I_i}\right)_W \left(\frac{I_i}{\Theta_j}\right)_W \tag{3.28}$$

Using the original value[1] of $I_i$, then its value changed to a sufficient extent,[2] the solutions of $\Theta_{j(1)}$ and $\Theta_{j(2)}$ will be found, and $\Delta\Theta_j$ produced as their difference:

$$\Delta\Theta = \Theta_{j(2)} - \Theta_{j(1)} \tag{3.29}$$

The main steps in the numerical solution are (for more detail see (Reference 10)):

(1) Input data and control information.
(2) Selecting the program in compliance with the type and the cooling system of the transformer.
(3) Generation of the approximating values $M_b^{(0)}$ and $T_0^{(0)}$.
(4) Progressing in the direction of the oil flow, according to Figure 3.5, generation of the DF elements (solving at the same time the equation systems of the HFN models of the discs), and producing the approximating solution of the coupled simultaneous equation system.

(5) Generation of the system–level function relationships

$$P_b = P_b(M_b, T_0)_k$$

and
$$T_b = T_b(M_b, T_0)_k$$ 
(3.30)

and determining the mass flow rates in the branches (e.g. by the Newton–Raphson method).

(6) After having produced sufficiently exact values of the branch mass flow rates $M_b$, repeated solution of the HFN equation system and production of the steady-state temperature field.

(7) Changing the input parameter under examination to a sufficiently small extent and producing the solution repeatedly, determination of the elements present in the corresponding line of the relative sensitivity matrix, according to Equations (3.28) and (3.29).

### 3.3.6  Results

The results obtained for an ONAN cooled transformer are summarized in Table 3.1, which is essentially a tabulated form of the sensitivity matrix. The first column of the table contains the input parameters and their given values, while its first line indicates the output parameters and their values and the working point $W$ (see Table 3.1).

Table 3.1  Relative sensitivity matrix of an oil transformer

| $I_i$ | $\Theta_j$ | $M_1$ (kg s$^{-1}$) 21.67 | $M_2$ (kg s$^{-1}$) 9.549 | $M_3$ (kg s$^{-1}$) 6.746 | $M_4$ (kg s$^{-1}$) 5.378 | $T_{hp2}$ (°C) 68.38 | $T_{hp3}$ (°C) 71.06 |
|---|---|---|---|---|---|---|---|
| Heat transfer coefficient $h_2$ (W m$^{-2}$ °C$^{-1}$) | 60.0 | −0.55 | −1.05 | −0.30 | 0.00 | −11.70 | 0.00 |
| Overall conductance $k_1^{(ov)}$ (W m$^{-2}$ °C$^{-1}$) | 5.0 | −31.65 | −31.65 | −32.62 | −30.12 | −38.9 | −36.59 |
| Outer temperature $T_w$ (°C) | 25.0 | 31.65 | 32.05 | 32.61 | 29.75 | 37.73 | 35.74 |
| Hydraulic radius $r_2$ (mm) | 4.12 | 36.64 | 109.12 | −12.75 | −31.24 | −1.75 | 11.82 |

The other rows of the table contain the relative sensitivity coefficients of the output parameters pertaining to the corresponding input parameters. Since the system is not linear, the sensitivity matrix can be applied only to examining the changes occurring in the close vicinity of the working point.

Among the data obtained, there are some surprising results (see, for example, the influence on the 'hot-spot' temperatures of the ambient temperature and of the heat transfer coefficient arising in the winding gap; further details are given by Imre, Bitai and Csényi[11]).

## 3.4  SIMULATION OF THERMOMECHANICAL SYSTEMS

Compound devices in which the heat transfer processes and the movement of the components of the structure are connected or interact are considered to be thermomechanical systems.

### 3.4.1  The modelling of thermomechanical systems

Thermomechanical systems can be described by simultaneous models consisting of thermal and mechanical part models (Figure 3.1).

As was the case with the thermal part model, it appears economical and advantageous in this case also to apply the heat flow network model, which is non-linear in most cases.

The construction and method of description of the mechanical part model are decisively influenced by the characteristic features of the design of the object. Simple mechanisms, for example, can be approximated by models built from rigid bodies, linear and non-linear springs. To the description of the mechanical part model, the method of force, the method of displacement,[12] the method based on the transfer matrix and the method with mixed variables[20] can be applied.

Here, too, the thermal and mechanical part models are coupled by the DF elements model.

### 3.4.2  Application of simultaneous thermomechanical models

In engineering one uses a large number of thermomechanical actuating systems. The feature common to such systems is that one of their components undergoes a suitable deformation as a result of a temperature change; this causes the movement of a mechanism connected with it structurally and in some limiting position of the displacement it triggers the action of an actuator.

Among such thermomechanical actuators, thermoswitches, constructed to prevent overheating, play an important role.

The application of simultaneous thermomechanical models is illustrated in connection with the thermoswitch sketched in Figure 3.6, which has been developed to protect electrical rotating machines against overheating. In the closed space drawn with dash–dot lines are placed phase bimetals which are warmed up by the heating shields to an extent proportional to the motor current. These phase bimetals, deformed in a way depending on their temperature field, bend upward and by displacing the stirrups, the bridges and the arms $Z$ and $M$ into their limiting position they actuate the switch $S_W$. The

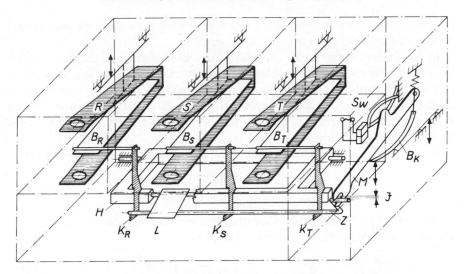

**Figure 3.6** Schematic arrangement of a thermoswitch: $R$, $S$, $T$, heating shields; $B_R$, $B_S$, $B_T$, phase bimetals; $H$, bridge-like frame; $K_R$, $K_S$, $K_T$, stirrups; $L$, connecting plate; $Z$, arm; $M$, switch-off arm; $S_W$, switch; $B_K$, compensating bimetal; $J$, clearance

influence of the variation of the ambient temperature $T_W$ is compensated for by the bimetal $B_k$. In the case of a phase drop-out the mechanism similarly produces a cut-off situation.

For the design engineer the simulation of the thermoswitch is important in order to predetermine, in the design stage, the behaviour to be expected of the construction under various operational conditions.

### 3.4.3 Construction and solution of the simultaneous thermomechanical model of the thermoswitch

Proceeding according to Figure 3.1, the simultaneous thermomechanical model will be built from thermal and mechanical part models; the part models will then be coupled by the coupling model of the DF elements.

The thermal part model serves to determine the temperature field of the thermoswitch. The DF elements of the thermoswitch are the bimetallic strips. By their temperature-dependent deformation, the mechanical part model assigns a position field to the temperature field.

### (a)  *A HFN model can be used as thermal part model*

Taking into account the properties of the structural elements and the purpose of the examination, the thermoswitch will be divided into discrete parts. For the decisive elements (such as the bimetallic strips and the heating shields) the spacings have to be decreased. In this way a heat flow network part model having 104 degrees of freedom was constructed.

In the thermoswitch under consideration—as in thermomechanical actuating systems in general—all three forms of heat transfer (conduction, convection and radiation) play important roles and, accordingly, the heat transfer resistances $R_{ki}$ present in the branch equation (Equation (3.4)) may consist of three parts. Let the total heat conductance between the $i$th and $j$th nodes of the heat flow network be written $K_{i,j} = 1/R_{i,j}$ (see (9)).

Based on the heat flow rate transmitted by conduction (with $k$ = constant):

$$K_{i,j}^{k} = \frac{Q_{i,j}^{k}}{T_i - T_j} \tag{3.31}$$

Based on $Q_{i,j}^{c}$ transmitted by natural convection (with $Nu = C(Gr\,Pr)^{0.25}$), the heat conductance of natural convection is

$$K_{i,j}^{c} = \frac{Q_{i,j}^{c}}{T_i - T_j} = b_{i,j}^{c}(T_i - T_j)^{0.25} \tag{3.32}$$

Based on the heat flow rate transferred by radiation $Q_{i,j}^{r}$ (with $Q_{i,j}^{r} = \varepsilon_r\sigma(T_i^4 - T_j^4)$), the radiation conductance will be formally

$$K_{i,j}^{r} = \frac{Q_{i,j}^{r}}{T_i - T_j} = b_{i,j}^{r}(T_i^2 + T_j^2)(T_i + T_j) \tag{3.33}$$

The conductances $K_{i,w}$ between the $i$th node and the environment at temperature $T_w$ can be written in a similar manner (in agreement with the boundary conditions). The total conductance will be

$$K_{i,j} = K_{i,j}^{k} + K_{i,j}^{c} + K_{i,j}^{r}, \tag{3.34a}$$

and

$$K_{i,w} = K_{i,w}^{k} + K_{i,w}^{c} + K_{i,w}^{r} \tag{3.34b}$$

As a combination of Equations (3.1), (3.4) and (3.5) let the heat flow balance relative to the $i$th node be written:

$$C_i\dot{T}_i + \sum_{\substack{j=1 \\ i \neq j}}^{n} K_{ij}(T_i - T_j) + K_{i,w}(T_i - T_w) = Q_{si} \tag{3.35}$$

Combine the heat flows passing through the environmental couplings (sinks) and the source heat flow rates:

$$Q_i = Q_{si} + K_{i,w}(T_i - T_w) \tag{3.36}$$

With this, the non-linear equation system describing the HFN model is

$$\mathbf{C\dot{T}} + \mathbf{K}(T_{i,j})\mathbf{T} = \mathbf{Q}(\mathbf{T}_i, \mathbf{T}_w) \qquad i = 1, 2, \ldots, n \quad j = 1, 2, \ldots, n \quad i \neq j \tag{3.37}$$

Various difference schemes can be applied to the solution of the non-linear equation system obtained. Let a three-time-level difference scheme be chosen:[15,16,17]

$$\mathbf{C}\frac{\mathbf{T}_2-\mathbf{T}_0}{2\Delta t}+\mathbf{K(T)}\frac{\mathbf{T}_2+\mathbf{T}_1+\mathbf{T}_0}{3}=\mathbf{Q}_s \tag{3.38}$$

where $\Delta t$ is the time step selected

$$\mathbf{T}_0=\mathbf{T}(t-\Delta t)$$
$$\mathbf{T}_1=\mathbf{T}(t) \tag{3.39}$$
$$\mathbf{T}_2=\mathbf{T}(t-\Delta t)$$

To obtain a linearized form the heat conduction matrix $\mathbf{K(T)}$ will be replaced by its time average:

$$\hat{\mathbf{K}}=\frac{1}{2\Delta t}\int_{t-\Delta t}^{t+\Delta t}\mathbf{K}[\mathbf{T}(t)]\,\mathrm{d}t \tag{3.40}$$

To calculate the time average it is assumed that the temperature function can be approximated by a linear variation within the time period:

$$\frac{\mathbf{T}_2-\mathbf{T}_1}{\Delta t}=\frac{\mathbf{T}_1-\mathbf{T}_0}{\Delta t} \tag{3.41}$$

Within the time step $(-\Delta t\leqslant t\leqslant+\Delta t)$, introduce the variable $\xi=t/\Delta t(-1\leqslant\xi\leqslant+1)$, and so write the temperature of the $i$th and $j$th nodes:

$$T_i=T_{i,1}+\xi(T_{i,1}-T_{i,0})=T_{i,j}+\xi\Delta T_i \tag{3.42a}$$

and

$$T_j=T_{j,1}+\xi(T_{j,1}-T_{j,0})=T_{j,1}+\xi\Delta T_j \tag{3.42b}$$

Now, derive the time average of radiation conductance according to Equation (3.33):

$$\hat{K}_{i,j}^{\mathrm{r}}=\frac{b_{i,j}^{\mathrm{r}}}{2\Delta t}\int_{t-\Delta t}^{t+\Delta t}(T_i^2+T_j^2)(T_i+T_j)\,\mathrm{d}t \tag{3.43}$$

With substitutions

$$\hat{K}_{i,j}^{\mathrm{r}}\cong\frac{b_{i,j}^{\mathrm{r}}}{2\Delta t}\int_{\xi=-1}^{\xi=1}[(T_{i,1}^2+T_{j,1}^2)+2\xi(\Delta T_iT_{i,1}+\Delta T_jT_{j,1})$$

$$+\xi^2(\Delta T_i^2+\Delta T_j^2)][(T_{i,1}+T_{j,1})+\xi(\Delta T_i+\Delta T_j)]\Delta t\,\mathrm{d}\xi \tag{3.44}$$

After integration:

$$\hat{K}_{i,j}^{\mathrm{r}}\cong b_{i,j}^{\mathrm{r}}[(T_{i,1}^2+T_{j,1}^2)(T_{i,1}+T_{j,1})$$

$$+\tfrac{1}{3}(\Delta T_i^2+\Delta T_j^2)(T_{i,1}+T_{j,1})+\tfrac{2}{3}(\Delta T_iT_{i,1}+\Delta T_jT_{j,1})](\Delta T_i+\Delta T_j) \tag{3.45}$$

The linearization can also be performed with the use of other relationships.[16,17,18,19]

In a similar way one can determine the time average of the radiation conductance $\hat{K}^{r}_{i,w}$ connected with the environment, and thus

$$\hat{\mathbf{Q}} = \mathbf{Q}_s + \hat{\mathbf{K}}_w \mathbf{T}_w \tag{3.46}$$

The natural convective conductance is calculated on the basis of the temperatures at time $t$:

$$\hat{K}^{c}_{i,j} = b^{c}_{i,j}(T_i - T_{j,1})^{0.25} \tag{3.47}$$

With the use of the time average $\hat{\mathbf{K}}$ of the total conductance, Equation (3.38) can be written in linearized form

$$\mathbf{C} = \frac{\mathbf{T}_2 - \mathbf{T}_0}{2\Delta t} + \hat{\mathbf{K}}\frac{\mathbf{T}_2 + \mathbf{T}_1 + \mathbf{T}_0}{3} = \mathbf{Q} \tag{3.48}$$

It appears practical to solve Equation (3.48) with varying time steps.[13]

As solutions of the equation system (3.48) the time functions of the nodal temperatures of the heat flow network will be obtained. From the viewpoint of the solution, the temperature conditions of the bimetallic strips as DF elements are of decisive importance. Figure 3.7 shows the temperature characteristic of the phase 'S' bimetal root as an example.

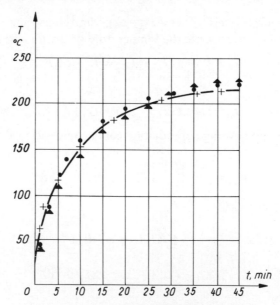

**Figure 3.7** Temperature characteristic of the 'S' phase bimetal root, $I_{\text{rated}} = 36$ A: ▲, with calculated network elements; ●, with identified values of some parameters; +, by measurement

## (b) *A model built from rigid bodies and springs*

Figure 3.8 can be applied as a mechanical part model.[12,13,14] Considering the very small and slow movement of the components, the variation of the kinetic energy will be neglected and the accuracy of the first-order theory accepted for the description of the elastic deformations. Contact between components that

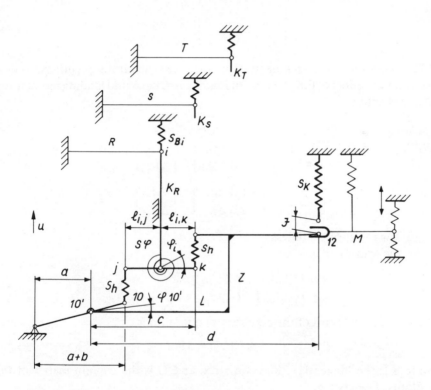

**Figure 3.8** Scheme of the mechanical part model of a thermoswitch: $s_{Bi}$, substituting springs modelling the coupling of rigid bodies; $s_k$, equivalent linear spring modelling the effect of omitted parts

can be modelled as rigid bodies will be modelled by means of springs loadable only for pressure. The elements of the thermoswitch examined are unloaded in the horizontal plane, and therefore only deformation in vertical direction will be considered.

The displacement method[12,14] was found to be particularly applicable[13] to describe the model in question. The equation system describing the mechanical part model will be written with the notations of Figure 3.8.

The transformational equation system of the displacements of the rigid bodies can be written:

$$\begin{bmatrix} \Delta u_k \\ \Delta u_j \end{bmatrix} = \begin{bmatrix} 1 & l_{i,k} \\ 1 & -l_{i,j} \end{bmatrix} \begin{bmatrix} \Delta u_i \\ \Delta \phi_i \end{bmatrix} \tag{3.49}$$

$$u_{10} = \frac{a+b}{a} \Delta u_{10'} \tag{3.50}$$

$$\begin{bmatrix} u_{10'} \\ u_{11} \end{bmatrix} = \begin{bmatrix} 1 & 0 \\ 1 & c \end{bmatrix} \begin{bmatrix} u_{10'} \\ \Delta \phi_{10'} \end{bmatrix} \tag{3.51}$$

The equilibrium equation system for the substituting springs, which describes the effect of the variation of loading force $\Delta F$ and loading momentum $\Delta M$, is given by

$$s_i \Delta u_i = \Delta F_i \tag{3.52}$$

$$s_h \begin{bmatrix} 1 & -1 \\ -1 & 1 \end{bmatrix} \begin{bmatrix} \Delta u_j \\ \Delta u_{10} \end{bmatrix} = \begin{bmatrix} \Delta F_j \\ \Delta F_{10} \end{bmatrix} \tag{3.53}$$

$$s_h \begin{bmatrix} 1 & -1 \\ -1 & 1 \end{bmatrix} \begin{bmatrix} \Delta u_k \\ \Delta u_{11} \end{bmatrix} = \begin{bmatrix} \Delta F_k \\ \Delta F_{11} \end{bmatrix} \tag{3.54}$$

where $s$ is the spring stiffness.

For the arm $M$ (if $J = 0$):

$$\begin{bmatrix} s_k & s_k d \\ s_k d & s_k d^2 \end{bmatrix} \begin{bmatrix} \Delta u_{10'} \\ \Delta \phi_{10'} \end{bmatrix} = \begin{bmatrix} \Delta F_{10'} \\ \Delta M_{10'} \end{bmatrix} \tag{3.55}$$

The incremental form of the equilibrium equation is

$$\mathbf{S(U)\Delta U = \Delta F} \tag{3.56}$$

where $\mathbf{S}$ is the tangential stiffness matrix, and $\mathbf{U}$ is the column matrix of the positions. The change

$$\Delta \mathbf{U} = \mathbf{U}(t + \Delta t) - \mathbf{U}(t) \tag{3.57}$$

of the position field takes place as a result of a change

$$\Delta \mathbf{F} = \mathbf{F}(t + \Delta t) - \mathbf{F}(t) \tag{3.58}$$

of the loading force and the loading momentum $\Delta M$. With the notations of Figure 3.8:

$$\Delta \mathbf{U} = [\Delta u_i \; \Delta u_j \; \Delta \phi_i \; \Delta \phi_{10} \; \Delta u_{10}] \tag{3.59}$$

The initial condition is:

$$\mathbf{U}(t = 0) = \mathbf{U}_0 \tag{3.60}$$

To describe the change of relationship, let the difference between positions $u_1$ and $u_2$ of the spring ends 1 and 2 coupled with each other be written as $u_{12} = u_1 - u_2$. The criterion for the change of connection is given by:

$$\operatorname{sgn}\left[u_{12}(t)\right] + \operatorname{sgn}\left[u_{12}(t + \Delta t)\right] = 0 \qquad (3.61)$$

A connection arises when $u_{12}(t) < 0$ and ceases when $u_{12}(t) > 0$ (as it was before the change). If $u_{12}$ is positive (i.e. the spring is extended), the spring will be eliminated.

Within the time step, the position field belonging to time $t^*$ $(t \leqslant t^* \leqslant t + \Delta t)$ relating to the first change of relationship is

$$\mathbf{U}(t^*) = \mathbf{U}(t) + \lambda^* \Delta \mathbf{U} \qquad (3.62)$$

where

$$\lambda^* = \min\left[\operatorname{abs}\left(\frac{u_{12}(t)}{u_{12}(t) - u_{12}(t + \Delta t)}\right)\right] \qquad (3.63)$$

The change of the next connection and of the following ones, respectively, within the time step is calculated on the basis of the state of the actual time $t^*$ (as 'initial' state), in the manner discussed above. The 'state at time $t^*$' means the position field, the unbalanced load distribution $(\Delta F(t^*) = \Delta F(t)(1 - \lambda^*))$ and the model valid at the time of the change of connection.

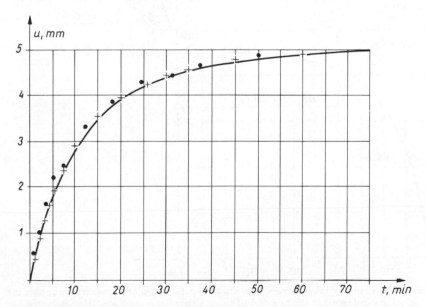

**Figure 3.9** Characteristic curve of the free deformation of an 'S' phase bimetal, for the case of $I_{\text{rated}} = 36$ A: ●, calculated values; +, measured values

(c)  *The role of the coupling model*

This is fulfilled by the DF model of the bimetallic strips (signal shaping function: temperature change; function of control action: deformation). It is assumed that owing to the very small displacement of the components, the modification of the thermal model can be neglected. To model the operation of the bimetallic strips, those values of the external concentrated forces acting on their ends are determined that produce exactly the free deformations $(x)$ relating to the temperature changes of the bimetals.[13] From this general form of the equation of coupling (with $i = 1, 2, 3$) can be written

$$F_i = s_i[T(t, x)]\Delta u_i[T(t, x)] \tag{3.64}$$

(d)  *The equation system describing the simultaneous thermomechanical model*

This system can be solved through the following main steps:

(1)   selection of the time step;

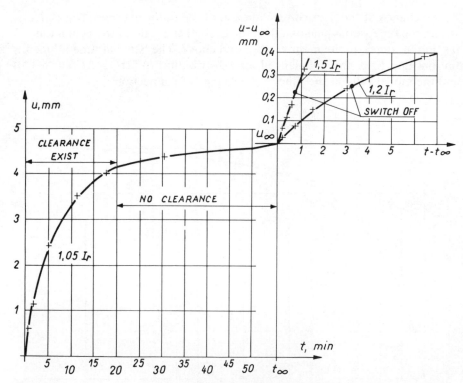

**Figure 3.10**   Position-time function of the end of switch-off arm $M$. Lower figure: approximation of $u_\infty$ in steady state in the case of $1.05I_{\text{rated}}$. Upper figure: position diagram for the case of changing from the steady state of $1.05I_{\text{rated}}$ to $1.2I_{\text{rated}}$ and $1.5I_{\text{rated}}$ load respectively; ●, switch-off

(2) generation of the HFN;
(3) solution of the HFN equation system;
(4) with known $T_2$, calculation of the next time step;
(5) based on $T_2$, solution of coupling Equation (3.64);
(6) (a) generation and solution of the equation system (3.49)–(3.56) describing the mechanical part model; (b) when a change of the connection occurs according to (3.61), then Equations (3.62)–(3.64) have to be solved, the model will be modified and the computation continued with step 6(a); after this one has to return to step 2.

### 3.4.4 Results

Figures 3.9 and 3.10 are presented to illustrate the results. In the latter the cut-off times with various overloads are also indicated; on the basis of these it is possible to judge whether normal operation can be expected.

REFERENCES

1. L. Imre (1981). *Heat Transfer in Composite Devices* (in Hungarian), Akadémiai Kiadó, Budapest.
2. W. J. Karplus and W. W. Soroka (1969). *Analog Methods: Computation and Simulation*, McGraw-Hill, New York.
3. H. Schenk (1958). *Heat Transfer Engineering*, Prentice Hall, London.
4. *Représentation Analogique et Homologiques dans les Techniques de la Chaleur* (1965). Dunod, Paris.
5. A. D. Kraus (1965). *Cooling Electronic Equipment*, Prentice Hall, London.
6. I. Szabó and L. Imre (1974). 'Determination of expected operational characteristics of driers', in *Handbook of Drying*, Ed. L. Imre, Ch. 18 (in Hungarian), Müszaki Könyvkiadó, Budapest.
7. L. Imre (1976). 'Determination of the steady-state temperature distribution of transformer windings by the heat flux network method', *Periodica Polytechn., El. Engng*, **20**, 461.
8. L. Imre, I. Szabó and A. Bitai (1978). 'Determination of the steady-state temperature field in naturally oil-cooled disc-type transformers, EC-20', *Proc. 6th Int. Heat Transfer Conf., Toronto, Canada, 1978*.
9. L. Imre, A. Bitai and P. Csényi (1979). 'Parameter sensitivity investigation of the warming of oil-transformers', *1st Int. Conf. on Numerical Methods in Thermal Problems, Swansea, 1979*.
10. L. Imre and A. Bitai (1979). 'A conception of simultaneous and hierarchic network modelling for computing the steady-state warming of naturally oil-cooled transformers', *Periodica Polytech., Mech. Engng*, **23**, 265.
11. L. Imre, A. Bitai and P. Csényi (1980). 'Warming analysis of oil-transformers by the sensitivity matrix', *Periodica Polytech., El. Engng*, **23**, 72.
12. I. Szabó and B. Roller (1971). *Theory and Calculation of Bar Structures* (in Hungarian), Müszaki Könyvkiadó, Budapest.
13. J. Barcza and L. Imre (1979). 'Model-concept for a transient analysis of non-linear thermo-mechanical systems', *1st Int. Conf. on Numerical Methods in Thermal Problems, Swansea, 1979*.

14. R. Kersten (1962). *Das Reduktionsverfahren der Baustatik*, Springer-Verlag, Berlin.
15. M. Lees (1966). 'A linear three-level difference scheme for quasilinear parabolic equations', *Math. Comp.*, **20**, 516.
16. C. Bonacina and G. Comini (1973). 'On the solution of the non-linear heat conduction equations by numerical methods', *Int. J. Heat Mass Transfer*, **16**, 581.
17. G. Comini, S. Del Guidice, R. W. Lewis and O. C. Zienkiewicz (1974). 'Finite element solution of non-linear heat conduction problems with special reference to phase-change', *Int. J. Num. Meth.*, **8**, 613.
18. W. E. Mason Jr (1979). 'Finite element analysis of coupled heat conduction and enclosure radiation', *1st Int. Conf. on Numerical Methods in Thermal Problems, Swansea, 1979*.
19. R. D. Karam (1979). 'Optimum solution of linearized radiation equation', *1st Int. Conf. on Numerical Methods in Thermal Problems, Swansea, 1979*.
20. N. E. Wiberg (1974). 'Matrix structural analysis with mixed variables', *Int. J. Num. Meth. Engng*, **8**, 167.

*Numerical Methods in Heat Transfer*
Edited by R. W. Lewis, K. Morgan, and O. C. Zienkiewicz
© 1981 John Wiley & Sons Ltd

*Chapter 4*

# A Comparison of Two- and Three-Level Integration Schemes for Non-Linear Heat Conduction

*M. A. Hogge*

## SUMMARY

The chapter discusses various algorithms that have been proposed for the numerical integration of the evolution equations of non-linear heat conduction after spatial (finite element) discretization. The algorithms reviewed comprise one-step schemes of order one and higher-order one-step schemes based on Hermitian polynomials, a set of linearized two-step schemes and a completely explicit two-sweep scheme.

The study is an attempt to define criteria for an optimum choice among such algorithms, where emphasis is given to the accuracy achievable. To that end, the different methods are used on a model problem for which an error analysis is performed. The stability properties of the schemes are also recalled or investigated.

From this study particular choices among two- and three-level schemes can be recommended and should be confirmed by using them for multidimensional problems of transient heat conduction.

## 4.1 INTRODUCTION

Among the numerous numerical techniques used in the field of non-linear parabolic problems such as non-linear transient heat transfer a widespread method follows a two-fold discretization procedure:[1]

(1) A semi-discretization in space is realized (e.g. by finite elements) yielding a set of first-order non-linear differential equations such as

$$\mathbf{K}(\mathbf{x})\mathbf{x}(t) + \mathbf{C}(\mathbf{x})\dot{\mathbf{x}}(t) = \mathbf{g}(\mathbf{x}, t)$$

$$\mathbf{x}(0) = \mathbf{x}_0$$

$$(4.1)$$

where $\mathbf{x}(t)$ is the vector of the local temperatures (with a superscript dot denoting its time derivative). $\mathbf{K}$ is the symmetrical and positive semi-definite conductivity matrix, $\mathbf{C}$ is the symmetrical and positive definite capacity matrix and $\mathbf{g}$ is a vector of thermal loads corresponding to $\mathbf{x}$; $\mathbf{x}_0$ is a vector of given initial temperatures.

(2) The time integration of these equations is accomplished by assuming some difference scheme leading to a step-by-step procedure.

The final non-linear algebraic system remains then to be solved at each time step either iteratively by application of some Newton's method or directly by using some linearization technique.

In this chapter we shall consider and compare some of the methods currently used or proposed for the second stage of this solution procedure, i.e. the time integration of system (4.1). Thus we leave completely apart the problem of the spatial discretization of such field problems (see e.g. references 1, 2) as well as the iterative solution of the resulting set of non-linear algebraic equations (see e.g. reference 3).

In order to undertake such a review and to facilitate the comparison with computational techniques applied to typical first-order differential equations,[4] we rewrite system (4.1) in the form

$$\dot{\mathbf{x}}(t) = -\mathbf{A}(\mathbf{x})\mathbf{x}(t) + \mathbf{d}(\mathbf{x}, t) = \mathbf{f}(\mathbf{x}, t)$$
$$\mathbf{x}(0) = \mathbf{x}_0 \tag{4.2}$$

where $\mathbf{A} = \mathbf{C}^{-1}\mathbf{K}$ has real positive eigenvalues and $\mathbf{d} = \mathbf{C}^{-1}\mathbf{g}$ is a new forcing vector. Such a system is a stiff one, i.e. composed of exponentially decaying components with widely spread time constants which, in addition, change continually in the non-linear case.

In current heat conduction applications (aside from thermal shock phenomena) one deals with situations where the slowly decaying components dominate the response; hence attention will be focused on time integration techniques that treat adequately the long-time components of the response while retaining numerical stability with respect to the fast varying excitations.

## 4.2 TWO-LEVEL INTEGRATION SCHEMES

### 4.2.1 $A_0$-Stable linear one-step methods[2,4-7]

Many numerical techniques available for the solution of the initial value problem (4.2) are linear $k$-step methods.[2,4] Their general form is

$$\sum_{j=0}^{k} \alpha_j \mathbf{x}_{n+j} = h \sum_{j=0}^{k} \beta_j \mathbf{f}_{n+j} \quad n = 0, 1, \ldots \tag{4.3}$$

where $\mathbf{x}_n$ is the approximation to $\mathbf{x}(nh)$, $h$ being the time step, and $\mathbf{f}_n$ stands for $\mathbf{f}(\mathbf{x}_n, nh)$. The constants $\alpha_j$ and $\beta_j$ are arbitrary to the extent of a constant multiplier; they can be normalized by assuming

$$\alpha_k = 1 \quad \text{or} \quad \sum_{j=0}^{k} \beta_j = 1$$

The simplest kind of such methods is the class of consistent one-step schemes ($k = 1$) in which a two-level difference approximation is chosen for $\dot{x}$ and a linear variation of $\mathbf{f}$ is assumed over the interval $[t_n, t_{n+1}]$, thus yielding

$$\mathbf{x}_{n+1} - \mathbf{x}_n = h[(1 - \theta)\mathbf{f}_n + \theta\mathbf{f}_{n+1}] \quad 0 \leqslant \theta \leqslant 1 \tag{4.4}$$

These methods will be referred to as the generalized trapezoidal schemes (GTS).

A slightly different form of one-step schemes is obtained from the same two-level difference approximation for $\dot{x}$ and from an assumed linear behaviour of $\mathbf{x}$ over $[t_n, t_{n+1}]$:

$$\mathbf{x}_{n+1} - \mathbf{x}_n = h\mathbf{f}_\theta = h\mathbf{f}(\mathbf{x}_\theta, t_\theta) \tag{4.5}$$

in which

$$\left. \begin{aligned} \mathbf{x}_\theta &= (1 - \theta)\mathbf{x}_n + \theta\mathbf{x}_{n+1} \\ t_\theta &= (1 - \theta)t_n + \theta t_{n+1} = t_n + \theta h \end{aligned} \right\} \quad 0 \leqslant \theta \leqslant 1 \tag{4.6}$$

They will be referred to as the generalized mid-point schemes (GMS).

The two families reduce to one if $\mathbf{f}$ is a linear function of the unknown (e.g. linear heat conduction): the so-called '$\theta$-method'. Particular cases are well known:

$$\theta = \begin{cases} 0 & \text{Euler explicit scheme;} \\ \frac{1}{2} & \text{either trapezoidal or Crank–Nicolson} \\ & \text{(mid-point rule) scheme;} \\ \frac{2}{3} & \text{Galerkin scheme;} \\ 1 & \text{fully implicit scheme.} \end{cases}$$

All the schemes are consistent ones of the first order, except for $\theta = \frac{1}{2}$, in which case the two families are second-order accurate in the time step size.

Stability and accuracy properties of those schemes have been intensively described in linear situations (e.g. references 1, 2, 8) and recently in non-linear cases.[5-7] In particular, only the second family has been shown to retain exactly the stability properties of linear situations (i.e. zero-stability[4] and $A_0$-stability[2] when $\theta \geqslant \frac{1}{2}$). Let us note that this latter family is not strictly of the form (4.3) if $\theta \neq \frac{1}{2}$. Applied to system (4.2), the schemes (4.4) and (4.5) yield the following set of non-linear algebraic equations, with $\mathbf{I}$ denoting the identity matrix:

$$(\mathbf{I}+\theta h \mathbf{A}_{n+1})\mathbf{x}_{n+1} = [\mathbf{I}-(1-\theta)h \mathbf{A}_n]\mathbf{x}_n$$

$$+(1-\theta)h\mathbf{d}_n + \theta h\mathbf{d}_{n+1} \quad \text{by GTS} \tag{4.7}$$

$$(\mathbf{I}+\theta h \mathbf{A}_\theta)\mathbf{x}_{n+1} = [\mathbf{I}-(1-\theta)h \mathbf{A}_\theta]\mathbf{x}_n$$

$$+h\mathbf{d}_\theta \quad \text{by GMS} \tag{4.8}$$

Thus every step involves the construction and solution of a new set of equations; therefore iterative solution techniques (except for $\theta = 0$) are required in which tangent or secant approximations may be used.[3] The following linearization techniques have been proposed, however, in order to avoid an iterative solution technique:[9,10]

(a)   A straight linearization over a time step by assuming

$$\mathbf{A}_{n+1} \simeq \mathbf{A}_n, \quad \mathbf{d}_{n+1} \simeq \mathbf{d}(\mathbf{x}_n, t_{n+1}) \quad \text{for GTS} \tag{4.9}$$

or

$$\mathbf{A}_\theta \simeq \mathbf{A}_n, \quad \mathbf{d}_\theta \simeq \mathbf{d}(\mathbf{x}_n, t_\theta) \qquad \text{for GMS} \tag{4.10}$$

This corresponds obviously to a secant or 'initial load' method of solution for the non-linear systems (4.7) or (4.8).

(b)   An extrapolation technique for the GMS family by taking

$$\mathbf{x}_\theta = (1+\theta)\mathbf{x}_n - \theta\mathbf{x}_{n-1} \tag{4.11}$$

(obviously not valid for the first time step).

(c)   A predictor–corrector technique for the GMS family in which a predicted value $\mathbf{x}_{n+1}^*$ stems from (4.8) by assuming (4.10); a corrected value is then obtained by solving again (4.8) with

$$\mathbf{x}_\theta^* = (1-\theta)\mathbf{x}_n + \theta\mathbf{x}_{n+1}^*$$

The procedure thus requires the solution of two linear systems of equations per time step.

## 4.2.2   Higher-order one-step schemes[11]

Following the finite element concept of polynomial approximation in space, we can introduce higher-order polynomial expansions for the time dependence of the unknown. One possibility among many others is to choose Hermitian interpolations for $\mathbf{x}(t)$ within a time step. The time derivatives $\dot{\mathbf{x}}, \ddot{\mathbf{x}}, \ldots$ involved in such higher-order expansions at the bounds of the time interval are back-substituted for the corresponding value of $\mathbf{x}$ through the equations to be solved.

   The families (4.7) and (4.8) are just the simplest case of these Hermitian operators: they are the result of a linear expansion of $\mathbf{x}$ within a time step.

A cubic polynomial, corresponding to a Hermitian expansion of order two, is written for the interval $[t_n, t_{n+1}]$:

$$\mathbf{x}(\xi) = (1 + 3\xi^2 + 2\xi^3)\mathbf{x}_n + h(\xi - 2\xi^2 + \xi^3)\dot{\mathbf{x}}_n$$

$$+ (3\xi^2 - 2\xi^3)\mathbf{x}_{n+1} + h(-\xi^2 + \xi^3)\dot{\mathbf{x}}_{n+1} \quad 0 \leqslant \xi \leqslant 1 \quad (4.12)$$

Differentiation and substitution into the homogeneous counterpart of system (4.2) after elimination of the nodal derivatives yields the higher-order one-step scheme

$$\left(\mathbf{I} + \frac{1+\xi}{3} h \mathbf{A}_{n+1} + \frac{\xi}{6} h^2 \mathbf{A}_{n+1}^2\right)\mathbf{x}_{n+1} = \left(\mathbf{I} + \frac{\xi-2}{3} h \mathbf{A}_n + \frac{1-\xi}{6} h^2 \mathbf{A}_n^2\right)\mathbf{x}_n \quad (4.13)$$

The same procedure can be used for a quintic polynomial associated with a Hermitian expansion of order three yielding thus a scheme involving up to the third power of the matrix $\mathbf{A}$. The practical usefulness of such schemes has not yet been demonstrated.

### 4.3 THREE-LEVEL INTEGRATION SCHEMES

#### 4.3.1 $A_0$-Stable linear two-step methods[2,4,12–14]

Mathematical studies show that two-step schemes ($k = 2$) are consistent to the second order iff we choose in (4.3)

$$\alpha_1 = 1 - 2\alpha_2 \qquad \alpha_0 = -1 + \alpha_2$$
$$\beta_1 = \tfrac{1}{2} + \alpha_2 - 2\beta_2 \qquad \beta_0 = \tfrac{1}{2} - \alpha_2 + \beta_2 \qquad (4.14)$$

This yields two-parameter $(\alpha_2, \beta_2)$ schemes for (4.2) of the form

$$\sum_{j=0}^{2} \alpha_j \mathbf{x}_{n+j} + h \sum_{j=0}^{2} \beta_j \mathbf{A}_{n+j} \mathbf{x}_{n+j} = h \sum_{j=0}^{2} \beta_j \mathbf{d}_{n+j} \quad (4.15)$$

$A_0$-stability is implied iff

$$\alpha_2 \geqslant \tfrac{1}{2} \qquad \beta_2 > \tfrac{1}{2}\alpha_2 \quad (4.16)$$

but the associated non-linear algebraic systems require some linearization technique for practical computations. Such a technique is due to Zlamal[2,12] and embodies the majority of two-step schemes in use. It rests on an extrapolation formula generalizing the one by Dupont *et al.*[13] and requiring that for any time-dependent function $w(t)$, if we compute

$$w_* = C_1 w_{n+1} + C_0 w_n \quad (4.17)$$

it satisfies

$$\sum_{j=0}^{2} \beta_j w_{n+j} = w_* + O(h^2)$$

$$w_* = w(t_*) + O(h^2)$$

(4.18)

The results of Zlamal lead to a one-parameter extrapolation formula by taking

$$C_1 = 2\beta_2 + \beta_1 = \tfrac{1}{2} + \alpha_2 = 1 + \theta$$

$$C_0 = \beta_0 - \beta_2 = \tfrac{1}{2} - \alpha_2 = -\theta$$

(4.19)

$$t_* = t_n + C_1 h = t_{n+1} + \theta h$$

in which $\theta = \alpha_2 - \tfrac{1}{2}$ ($\theta \geqslant 0$). Replacing in the non-linear terms of system (4.15) the time-dependent terms by their extrapolated value according to (4.17) and (4.19) yields the final linear algebraic system

$$(\theta + \tfrac{1}{2})\mathbf{x}_{n+2} - 2\theta\mathbf{x}_{n+1} + (\theta + \tfrac{1}{2})\mathbf{x}_n + h\mathbf{A}_*[\beta_2\mathbf{x}_{n+2} + (1 + \theta - 2\beta_2)\mathbf{x}_{n+1}$$

$$+ (\beta_2 - \theta)\mathbf{x}_n] = h\mathbf{d}_*$$

(4.20)

Clearly every step implies the construction and solution of a new system of linear equations with the positive definite system matrix $[(\tfrac{1}{2} + \theta)\mathbf{I} + \beta_2 h\mathbf{A}_*]$. A different starting procedure is needed since the first application of (4.20) requires knowledge of $\mathbf{x}_1$. One of the previous one-step schemes can be used for this purpose but the problems associated with that shift (e.g. step length change) will not be considered here.

Table 4.1 shows how most of the three-level schemes that have been presented in the literature are derived from (4.20) by particular choices of $\theta$ and $\beta_2$.

Table 4.1  Linearized three-level schemes

| $\theta$ | $\beta_2$ | $\mathbf{x}_*$ | $t_*$ | Algorithm name | Integration scheme |
|---|---|---|---|---|---|
| 0 | $\tfrac{1}{3}$ | $\mathbf{x}_{n+1}$ | $t_{n+1}$ | Lees[14] | $(\tfrac{1}{2}\mathbf{I} + \tfrac{1}{3}h\mathbf{A}_{n+1})\mathbf{x}_{n+2} = -\tfrac{1}{3}h\mathbf{A}_{n+1}\mathbf{x}_{n+1}$ $+ (\tfrac{1}{2}\mathbf{I} - \tfrac{1}{3}h\mathbf{A}_{n+1})\mathbf{x}_n + h\mathbf{d}_{n+1}$ (4.21) |
| 0 | $\alpha$ $(\alpha > \tfrac{1}{4})$ | $\mathbf{x}_{n+1}$ | $t_{n+1}$ | Dupont I[13] | $(\tfrac{1}{2}\mathbf{I} + \alpha h\mathbf{A}_{n+1})\mathbf{x}_{n+2} = -(1 - 2\alpha)h\mathbf{A}_{n+1}\mathbf{x}_{n+1}$ $+ (\tfrac{1}{2}\mathbf{I} - \alpha h\mathbf{A}_{n+1})\mathbf{x}_n + h\mathbf{d}_{n+1}$ (4.22) |
| $\tfrac{1}{2}$ | $\tfrac{1}{2} + \alpha$ $(\alpha > 0)$ | $\tfrac{3}{2}\mathbf{x}_{n+1} - \tfrac{1}{2}\mathbf{x}_n$ | $t_{n+\frac{3}{2}}$ | Dupont II[13] | $[\mathbf{I} + (\tfrac{1}{2} + \alpha)h\mathbf{A}_*]\mathbf{x}_{n+2} = [\mathbf{I} - (\tfrac{1}{2} - 2\alpha)h\mathbf{A}_*]\mathbf{x}_{n+1}$ $- \alpha h\mathbf{A}_*\mathbf{x}_n + h\mathbf{d}_*$ (4.23) |
| 1 | 1 | $2\mathbf{x}_{n+1} - \mathbf{x}_n$ | $t_{n+2}$ | Linearized fully implicit[15] | $(\tfrac{3}{2}\mathbf{I} + h\mathbf{A}_*)\mathbf{x}_{n+2} = 2\mathbf{x}_{n+1} + \tfrac{1}{2}\mathbf{x}_n + h\mathbf{d}_*$ (4.24) |

It should be noted that some of these schemes are equivalent to particular one-step schemes: this is the case for the Dupont I scheme (4.22) when $\alpha = \tfrac{1}{2}$ and the Dupont II scheme (4.23) when $\alpha = 0$ that restore both the Crank-

Nicolson algorithm (scheme (4.8) with $\alpha = \frac{1}{2}$). In addition a linearization of the unknown with $[t_n, t_{n+2}]$, i.e. the assumption that

$$\mathbf{x}_{n+1} = (\mathbf{x}_{n+2} + \mathbf{x}_n)/2 \tag{4.25}$$

restores also particular one-step schemes, for instance:

(a)  the Crank–Nicolson scheme for (4.21) and (4.22) whatever the value of $\alpha$;
(b)  the GMS scheme (4.8) with $\theta = \frac{3}{4}$ for (4.23) whatever the value of $\alpha$;
(c)  the fully implicit GMS scheme ($\theta = 1$) for (4.24).

### 4.3.2  A special two-step technique[16]

The method is based first on the splitting of the matrix $\mathbf{A}$ into strictly lower and upper triangular matrices, i.e.

$$\mathbf{A} = \mathbf{L} + \mathbf{U} \tag{4.26}$$

and second on an alternating approximation to the linearized trapezoidal or Crank–Nicolson scheme, i.e. schemes (4.7) or (4.8) with $\theta = \frac{1}{2}$ and with the assumptions (4.9) and (4.10):

$$(\mathbf{I} + \tfrac{1}{2}h\mathbf{A}_n)\mathbf{x}_{n+1} = (\mathbf{I} - \tfrac{1}{2}h\mathbf{A}_n)\mathbf{x}_n + h\mathbf{d}_n \tag{4.27}$$

The two-sweep technique is thus

$$(\mathbf{I} + h\mathbf{L}_n)\mathbf{x}_{n+1} = (\mathbf{I} - h\mathbf{U}_n)\mathbf{x}_n + h\mathbf{d}_n$$
$$(\mathbf{I} + h\mathbf{U}_n)\mathbf{x}_{n+2} = (\mathbf{I} - h\mathbf{L}_n)\mathbf{x}_{n+1} + h\mathbf{d}_n \tag{4.28}$$

and yields an explicit evaluation of the unknown without solving a system of equations. Eliminating $\mathbf{x}_{n+1}$ between the two equations leads to

$$(\mathbf{I} + h\mathbf{A}_n + h^2\mathbf{L}_n\mathbf{U}_n)\mathbf{x}_{n+2} = (\mathbf{I} - h\mathbf{A}_n + h^2\mathbf{L}_n\mathbf{U}_n)\mathbf{x}_n + 2h\mathbf{d}_n \tag{4.29}$$

which can be interpreted as a higher-order linearized one-step scheme by taking $h' = 2h$.

## 4.4  APPLICATION TO A MODEL EQUATION

In order to qualify the overall accuracy of the proposed schemes, we shall examine their application to the homogeneous model equation

$$\dot{x}(t) + (\lambda_0 + \lambda_1 x)x(t) = 0$$
$$x(0) = x_0 \tag{4.30}$$

corresponding to system (4.2) for a single-degree-of-freedom homogeneous situation.

Numerical arguments related to the stiff character of the differential system are thus discarded in this study and the exact solution to which the results are to be compared is readily:

$$x_E(t) = x_0\lambda_0 e^{-\lambda_0 t}/[\lambda_0 + \lambda_1 x_0(1 - e^{-\lambda_0 t})] \tag{4.31}$$

All the proposed schemes will be reviewed except (4.13) and (4.28): the latter because in one-dimensional cases this two-sweep scheme is exactly equivalent to two successive applications of the linearized Crank–Nicolson scheme, and the former because these higher-order schemes involve tremendous computational effort in solving the (also) higher-order final non-linear algebraic system.

Particularisation of the reviewed schemes to Equation (4.30) yields respectively

*GTS schemes* (Equation (4.7)):

$$x_{n+1} = \frac{1 - (1 - \theta)h\lambda_n}{1 + \theta h\lambda_{n+1}} x_n \quad 0 \leq \theta \leq 1 \tag{4.32}$$

where $\lambda_n = \lambda_0 + \lambda_1 x_n$

*GMS schemes* (Equation (4.8)):

$$x_{n+1} = \frac{1 - (1 - \theta)h\lambda_\theta}{1 + \theta h\lambda_\theta} x_n \quad 0 \leq \theta \leq 1 \tag{4.33}$$

where

$$\lambda_\theta = \lambda_0 + \lambda_1 x_\theta \qquad x_\theta = (1 - \theta)x_n + \theta x_{n+1}$$

*Dupont I schemes* (Equation (4.22)) *and Lee's* ($\alpha = \frac{1}{3}$) *algorithm* (Equation (4.21)):

$$x_{n+2} = \frac{-2(1 - 2\alpha)h\lambda_{n+1}}{1 + 2\alpha h\lambda_{n+1}} x_{n+1} + \frac{1 - 2\alpha h\lambda_{n+1}}{1 + 2\alpha h\lambda_{n+1}} x_n \quad \alpha > \frac{1}{4} \tag{4.34}$$

*Dupont II schemes* (Equation (4.23)):

$$x_{n+2} = \frac{1 - (\frac{1}{2} - 2\alpha)h\lambda_{n+\frac{3}{2}}}{1 + (\frac{1}{2} + \alpha)h\lambda_{n+\frac{3}{2}}} x_{n+1} \frac{\alpha h\lambda_{n+\frac{3}{2}}}{1 + (\frac{1}{2} + \alpha)h\lambda_{n+\frac{3}{2}}} x_n \quad \alpha \geq 0 \tag{4.35}$$

where

$$\lambda_{n+\frac{3}{2}} = \lambda_0 + \lambda_1 x_{n+\frac{3}{2}} \qquad x_{n+\frac{3}{2}} = \frac{3}{2}x_{n+1} - \frac{1}{2}x_n$$

*Fully implicit scheme* (Equation (4.24) before extrapolation):

$$x_{n+2} = \frac{4x_{n+1} + x_n}{3(1 + \frac{2}{3}h\lambda_{n+2})} \tag{4.36}$$

where $\lambda_{n+2} = \lambda_0 + \lambda_1 x_{n+2}$

For all these schemes an analytical expression of the converged solution is thus available in the form

$$x_{n+1} = f(x_n, h, \theta) \qquad \text{for one-step schemes}$$

or

$$x_{n+2} = f(x_{n+1}, x_n, h, \alpha) \qquad \text{for two-step schemes}$$

In the latter case, the schemes are self-started by taking the best possible value of $x_1 = x_E(h)$.

These solutions will now be compared to the exact one (Equation (4.31)) for different members of the families and different time step sizes in the case of mildly and highly non-linear situations ($\lambda_0 = 1.00$, $\lambda_1 = 0.01$ or $0.03$).

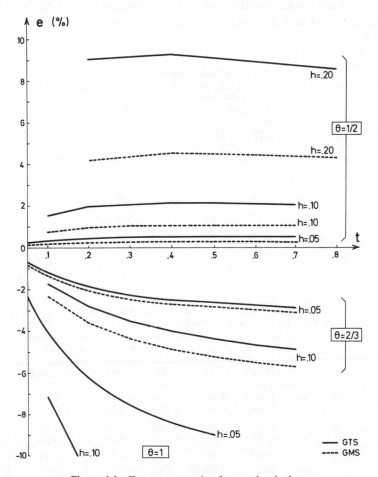

**Figure 4.1** Error propagation for two-level schemes

### 4.4.1 Two-level schemes (Figure 4.1 and Tables 4.2 and 4.3)

The results may be commented on as follows:

(a) for a given time step and a fixed level of non-linearity, the lowest error is encountered using $\theta = \frac{1}{2}$ in each family; in this case, the GMS is slightly more accurate;

(b) if the time step magnitude is increased with all other parameters fixed, the accuracy is decreasing as expected, but all the preceding remarks still hold; in particular deterioration of the accuracy is slower for $\theta = \frac{1}{2}$;

(c) if the level of non-linearity is raised, the solution suffers a loss of accuracy, the time step size being fixed. The alteration is greater when $h$ is greater.

This study of the model equation confirms the choice $\theta = \frac{1}{2}$ and the GMS family as being the most accurate one-step consistent scheme.

Table 4.2　Errors at the first step for two-level schemes
$e = 100(x_E - x)/x_E$　　　$e_1 = e$　for $\lambda_1 = 0.01$　　　$e_2 = e$　for $\lambda_1 = 0.03$

| $h$ | | $\theta = \frac{1}{2}$ | | $\theta = \frac{2}{3}$ | | $\theta = 1$ |
|---|---|---|---|---|---|---|
| | | GTS | GMS | GTS | GMS | GTS or GMS |
| 0.05 | $e_1$ | 0.024 | 0.013 | −0.20 | −0.21 | −0.63 |
| | $e_2$ | 0.23 | 0.12 | −0.70 | −0.79 | −2.35 |
| 0.10 | $e_1$ | 0.18 | 0.09 | −0.67 | −0.74 | −2.19 |
| | $e_2$ | 1.51 | 0.74 | −1.76 | −2.31 | −7.12 |
| 0.20 | $e_1$ | 1.22 | 0.65 | −1.87 | −2.28 | −7.01 |
| | $e_2$ | 9.10 | 4.20 | −2.39 | −5.49 | −19.36 |

Table 4.3　Errors at $t = 1.00$ for two-level schemes
$e = 100(x_E - x)/x_E$　　　$e_1 = e$　for $\lambda_1 = 0.01$　　　$e_2 = e$　for $\lambda_1 = 0.03$

| $h$ | | $\theta = \frac{1}{2}$ | | $\theta = \frac{2}{3}$ | | $\theta = 1$ |
|---|---|---|---|---|---|---|
| | | GTS | GMS | GTS | GMS | GTS or GMS |
| 0.05 | $e_1$ | 0.14 | 0.09 | −1.88 | −1.93 | −5.92 |
| | $e_2$ | 0.48 | 0.26 | −3.16 | −3.36 | −10.60 |
| 0.10 | $e_1$ | 0.58 | 0.34 | −3.48 | −3.69 | −11.58 |
| | $e_2$ | 1.96 | 1.03 | −5.44 | −6.25 | −20.86 |
| 0.20 | $e_1$ | 2.35 | 1.37 | −5.90 | −6.73 | −22.18 |
| | $e_2$ | 8.32 | 4.22 | −7.40 | −10.75 | −40.46 |

### 4.2.2　Three-level schemes: short-time response (Table 4.4)

Table 4.4　Error for three-level schemes ($\lambda_1 = 0.03$)
$e = 100(x_E - x)/x_E$　　$e_1 = e$　at first step　　$e_2 = e$　at $t = 1.00$

| | | Dupont I | | | Dupont II | | | Fully implicit |
|---|---|---|---|---|---|---|---|---|
| $h$ | | $\alpha = \frac{1}{3}$ (Lees) | $\alpha = \frac{1}{2}$ | $\alpha = 1$ | $\alpha = \frac{1}{4}$ | $\alpha = \frac{1}{2}$ | $\alpha = 1$ | |
| 0.05 | $e_1$ | −0.18 | 0.14 | 0.94 | −0.02 | 0.20 | 0.60 | 0.46 |
| | $e_2$ | −0.60 | 0.39 | 1.46 | 0.15 | 0.75 | 1.97 | 1.93 |
| 0.10 | $e_1$ | −0.97 | 0.94 | 5.19 | −0.12 | 1.10 | 3.20 | 2.48 |
| | $e_2$ | −2.71 | 1.80 | 5.57 | 0.51 | 2.86 | 7.65 | 7.87 |
| 0.20 | $e_1$ | −3.95 | 5.83 | 25.21 | −0.84 | 4.83 | 14.13 | 11.46 |
| | $e_2$ | 6.22 | 0.10 | 18.79 | 1.32 | 10.14 | 29.31 | 31.10 |

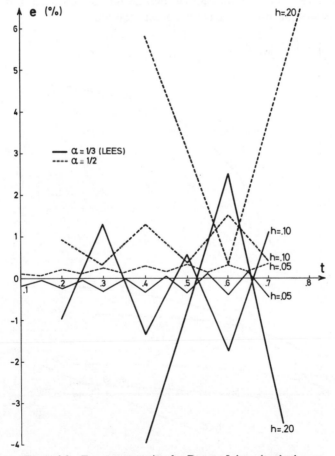

**Figure 4.2**　Error propagation for Dupont I three-level scheme

(a)  *Lees' algorithm* (Figure 4.2)

The short-time response exhibits a strong oscillatory behaviour, stronger if the
time step is greater. Accuracy is comparable to one-step solutions (Table 4.4).

(b)  *Dupont I scheme with* $\alpha = \frac{1}{2}$ (Figure 4.2)

The oscillatory behaviour remains for this other member of the Dupont I
family. For this particular choice of $\alpha$, we have shown that the scheme is
equivalent to the GMS scheme $\alpha = \frac{1}{2}$, but its implementation into a three-level
scheme introduces a strong oscillatory behaviour and a loss of accuracy.

(c)  *Dupont I scheme with* $\alpha = 1$ (Table 4.4)

No improvement in the oscillatory behaviour is achieved but accuracy
deteriorates so that this choice is not recommended.

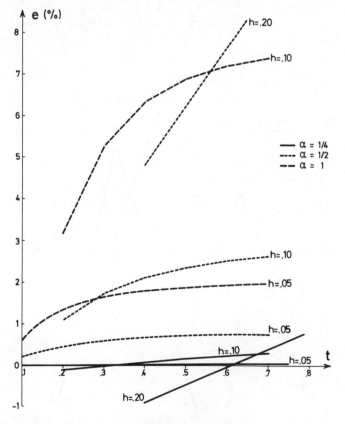

**Figure 4.3**  Error propagation for Dupont II three-level scheme

(d)  *Dupont II scheme with* $\alpha = \frac{1}{4}$ (Table 4.4, Figure 4.3)

This particular choice of the parameter eliminates the intermediate level at the second member of Equation (4.23). The results reported in Table 4.4 and Figure 4.3 show that this scheme is excellent for the problem at hand: accuracy is high even for large time steps and no oscillations are exhibited by the solutions.

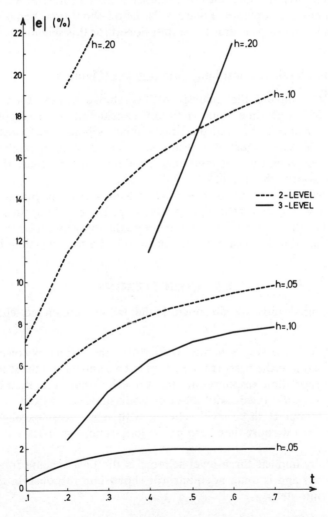

**Figure 4.4**  Error propagation for two- and three-level fully implicit schemes

(e)  *Other Dupont II schemes* ($\alpha = \frac{1}{2}$, $\alpha = 1$) (Table 4.4, Figure 4.3)

Other members of the family are less interesting from an accuracy point of view. Still they exhibit oscillation-free solutions and should in any case be preferred to the corresponding member in the Dupont I family.

(f)  *Fully implicit scheme* (Table 4.4, Figure 4.4)

This furnishes other oscillation-free solutions but accuracy deteriorates quickly when the time step size increases. Advantage over the corresponding one-step schemes is however obvious (Figure 4.4). Good short-time behaviour in the case of small time steps is thus to be mentioned for this scheme.

### 4.4.3  Three-level schemes: long-time reponse (Figure 4.5)

In order to complete the analysis of two-step schemes, their long-time behaviour for large time steps is finally considered. Lees' algorithm, the Dupont II scheme ($\alpha = \frac{1}{4}$) and the fully implicit scheme are investigated for $h = 1$, 2 and 5. As for short-time responses, Lees' algorithm exhibits a strong oscillatory behaviour that becomes uncontrolled for $h = 5$. Hence this scheme, as other Dupont I schemes, has to be discarded.

The Dupont II scheme ($\alpha = \frac{1}{4}$) quickly damps out the initial error and converges to zero even for the highest time step. The best long-time behaviour is exhibited by the fully implicit scheme where the small initial error disappears in a few steps; thus it constitutes the safest scheme for long-time behaviour.

### 4.5  CONCLUSIONS

From this limited study, we can conclude that the recommended schemes to be used are:

(a)  The GMS one-step schemes with $\theta \geqslant \frac{1}{2}$; the Crank–Nicolson scheme ($\theta = \frac{1}{2}$) exhibits the best error constant but we know from the study of cases where short-time responses are not smooth[2,7] that other members of the family (e.g. $\theta = \frac{2}{3}$) can exhibit better (short-time) behaviour.
(b)  The Dupont II three-level scheme with $\alpha = \frac{1}{4}$ gives excellent results, especially for short-time responses; long-time responses are also quite accurate.
(c)  The fully implicit three-level scheme is the best scheme for long-time responses and is quite adequate for short-time response in the case of small time steps.

Problems left open are: their behaviour in the case of multidimensional problems; their use in conjunction with iterative solution techniques if no

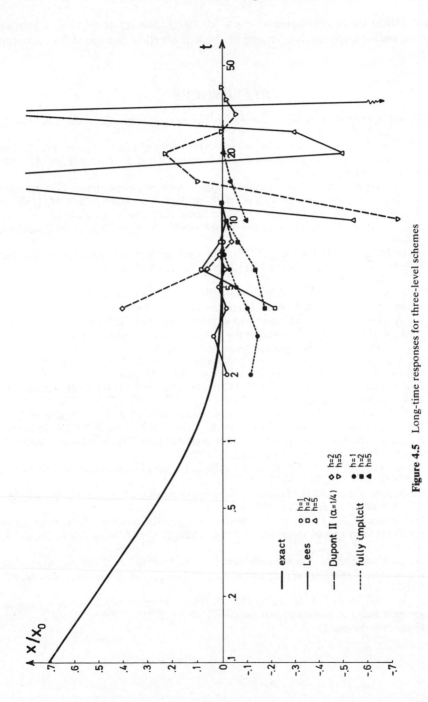

**Figure 4.5** Long-time responses for three-level schemes

linearization or extrapolation is used for the resulting non-linear algebraic system; and the appropriate choice of the self-starting technique for two-step methods.

## REFERENCES

1. O. C. Zienkiewicz (1977). *The Finite Element Method*, 3rd edn, Ch. 20–21, McGraw-Hill, London.
2. M. Zlamal (1976). 'Finite element methods in heat conduction problems', in *The Mathematics of Finite Elements and Applications II*, Ed. J. Whiteman, pp. 85–104, Academic Press, London.
3. M. A. Hogge (1980). 'Secant versus tangent methods in nonlinear heat transfer analysis', *2nd Int. Conf. on Comp. Meth. in Nonlinear Mech., TICOM, Univ. of Texas, Austin, March 1979, Int. J. Num. Meth. Engng* to be published.
4. J. D. Lambert (1973). *Computational Methods in Ordinary Differential Equations*, Wiley, London.
5. G. Dahlquist and B. Lindberg (1973). 'On some implicit one-step methods for stiff differential equations', *Report* TRITA-NA-7302, *Dept. of Information Processing, KTH, Stockholm*.
6. T. J. R. Hughes (1977). 'Unconditionally stable algorithms for nonlinear heat conduction', *Comp. Meth. Appl. Mech. Engng*, **10**, 135–139.
7. M. A. Hogge (1978). 'Accuracy and cost of integration techniques for nonlinear heat transfer', in *Finite Elements in the Commercial Environment*, Ed. J. Robinson, Vol. 1, pp. 133–154, Robinson & Ass., Bournemouth.
8. M. A. Hogge (1977). 'Integration operators for first-order linear matrix differential equations', *Comp. Meth. Appl. Mech. Engng*, **11**, 281–294.
9. J. Douglas and T. Dupont (1970). 'Galerkin methods for parabolic equations', *SIAM J. Num. Analysis*, **7**, 575–626.
10. R. E. Nickell *et al.* (1979). 'Spectral decomposition in advection-diffusion analysis by finite element methods', *Comp. Meth. Appl. Mech. Engng*, **17/18**, 561–580.
11. J. H. Argyris *et al.* (1977). 'Higher order methods for transient diffusion analysis', *Comp. Meth. Appl. Mech. Engng*, **12**, 243–278.
12. M. Zlamal (1977). 'Finite element methods for non-linear parabolic equations', *RAIRO Num. Analysis*, **11**, 93–107.
13. T. Dupont *et al.* (1974). 'Three-level Galerkin methods for parabolic equations', *SIAM J. Num. Analysis*, **11**, 392–410.
14. G. Comini *et al.* (1974). 'Finite element solution of nonlinear heat conduction problems with special reference to phase change', *Int. J. Num. Meth. Engng*, **8**, 613–623.
15. W. L. Wood (1978). 'On the Zienkiewicz 3- and 4-time-level schemes applied to the numerical integration for parabolic equations', *Int. J. Num. Meth. Engng*, **12**, 1717–1726.
16. D. M. Trujillo and H. R. Busby (1977). 'Finite element nonlinear heat transfer analysis using a stable explicit method', *Paper B2/12, Trans. SMIRT-4, San Francisco, August 1977.*

*Numerical Methods in Heat Transfer*
Edited by R. W. Lewis, K. Morgan, and O. C. Zienkiewicz
© 1981 John Wiley & Sons Ltd

*Chapter 5*

# Boundary Elements in Thermal Problems

*L. C. Wrobel and C. A. Brebbia*

## SUMMARY

In this chapter, the boundary element method is applied to steady-state and transient heat conduction problems. The direct formulation of the method, i.e. in terms of temperatures and fluxes as unknowns, is presented and the basic equations deduced by using weighted residuals. It is shown that boundary conditions of the convection or 'radiation' type can be easily included in the formulation.

Since two-dimensional and axisymmetric problems are independent of one coordinate, their fundamental solutions can be obtained from the three-dimensional ones by integrating with relation to this coordinate. For axisymmetric problems, it is also pointed out how the fundamental solution can be integrated analytically by using Legendre functions.

Transient problems are dealt with by employing fundamental solutions which are time, as well as space, dependent. Although it is necessary to use time intervals in order to obtain good numerical accuracy, this approach does not require the use of time stepping techniques of the finite difference kind.

Finally, several applications are presented to show some of the potentialities of the method in the solution of thermal problems.

## 5.1 INTRODUCTION

The numerical techniques most used by engineers for the analysis of thermal problems are finite differences and finite elements. Both methods divide the domain using a mesh or grid and defining a series of nodal points which can be internal or on the surface of the body. The boundary element method instead, is a new technique based on the combination of classical integral equations and finite elements concepts. In the method, nodes are only defined on the external surface and internal unknowns are not required as the problem mathematically has been reduced into a boundary solution. This reduction is possible by applying Green's theorem and using fundamental solutions which satisfy the governing equations or part of them. Satisfaction of boundary conditions produces a system of equations which can be solved to find the (boundary) unknowns. Once all values on the boundary are known, one can calculate any internal variables as functions of the boundary values.

The main advantages of the technique are the reduction in the number of unknowns governing the problem and the simplicity of the input data required to run it. In addition the method generally gives better results than domain type techniques, especially for regions with high gradients.

The boundary element method is described in detail in this chapter for steady and unsteady heat conduction problems. The steady-state case includes mixed boundary conditions of the convection or 'radiation' type. It is shown that inclusion of these conditions can be as easily done for boundary elements as for finite elements. The fundamental solutions corresponding to three- and two-dimensional problems are given, including the one for axisymmetric problems. The analytical integration of the fundamental solution for the axisymmetric case is explained in detail.

Transient heat conduction requires time as well as space integration to make the problem amenable to a boundary solution. The time integrals can be handled in a similar manner to the space integrals, i.e. using Green's theorem. This procedure produces a boundary integral relationship depending on time. Finding the corresponding time-dependent fundamental solution one can in principle, integrate these equations without having to use a time stepping technique. In practice however, it is necessary to use some time steps to obtain accurate results, but they can be much larger than those required by finite difference or finite element methods. In addition, it is important to point out that although the inclusion of initial conditions produces a domain integral, the nodal unknowns are still only on the boundaries, i.e. the problem continues to be a boundary problem.

The chapter presents several applications to illustrate the potentialities of the new technique. Wherever possible results are compared against analytical and finite element solutions, to show the advantages of boundary elements.

## 5.2  STEADY-STATE HEAT CONDUCTION

The governing equation for steady-state heat conduction in a homogeneous, isotropic solid body occupying a region $\Omega$ in space is,

$$k\nabla^2 T = p \quad \text{in } \Omega \tag{5.1}$$

where $T$ is temperature, $k$ is the thermal conductivity of the material and $p = -Q(x, y, z)$, $Q$ is the rate of heat generated per unit volume at the internal point $(x, y, z)$. The boundary conditions corresponding to this problem are of two types,

$$T = \bar{T} \quad \text{on } \Gamma_1 \tag{5.2a}$$

$$k\frac{\partial T}{\partial n} + hT = \bar{q} \quad \text{on } \Gamma_2 \tag{5.2b}$$

where $h$ is the heat transfer coefficient. The total boundary is $\Gamma = \Gamma_1 + \Gamma_2$. When $h$ is zero and $\bar{q}$ is the negative of a given heat flux, Equation (5.2b) corresponds to the heat input boundary condition. When $\bar{q} = hT_0$, where $T_0$ is the temperature of the surrounding medium, Equation (5.2b) prescribes the so-called 'radiation' boundary condition.

For our numerical solution $T$ will be approximated and we can minimize the error thus introduced by weighting the governing equation by a new function $T^*$. This gives,

$$\int_\Omega (k\nabla^2 T - p)T^* \, d\Omega = \int_{\Gamma_2} \left(k\frac{\partial T}{\partial n} + hT - \bar{q}\right) T^* \, d\Gamma - \int_{\Gamma_1} (T - \bar{T})k\frac{\partial T^*}{\partial n} \, d\Gamma \tag{5.3}$$

After integrating by parts twice the terms in the Laplacian, Equation (5.3) becomes,

$$k\int_\Omega (\nabla^2 T^*)T \, d\Omega - \int_\Omega pT^* \, d\Omega$$

$$= \int_{\Gamma_2} (hT - \bar{q})T^* \, d\Gamma - \int_{\Gamma_1} k\frac{\partial T}{\partial n} T^* \, d\Gamma + \int_{\Gamma_2} Tk\frac{\partial T^*}{\partial n} \, d\Gamma + \int_{\Gamma_1} \bar{T}k\frac{\partial T^*}{\partial n} \, d\Gamma \tag{5.4}$$

The function $T^*$ is now assumed to be the fundamental solution of the equation, representing a concentrated heat source at a point $i$, i.e. the solution of

$$\nabla^2 T^* + \Delta_i = 0 \tag{5.5}$$

where $\Delta_i$ is the Dirac delta function. Hence Equation (5.4) can be written as,

$$kT_i + \int_\Omega pT^* \, d\Omega + \int_{\Gamma_2} Tk\frac{\partial T^*}{\partial n} \, d\Gamma + \int_{\Gamma_1} \bar{T}k\frac{\partial T^*}{\partial n} \, d\Gamma$$

$$= \int_{\Gamma_2} \bar{q}T^* \, d\Gamma + \int_{\Gamma_1} k\frac{\partial T}{\partial n} T^* \, d\Gamma - \int_{\Gamma_2} hTT^* \, d\Gamma \tag{5.6}$$

This equation is valid for any point inside the domain, but in order to formulate the problem as a boundary technique one needs to take the point $i$ to the boundary. On doing so, care must be taken because of the nature of the singularity of the integral in $\partial T^*/\partial n$, that must be evaluated in the Cauchy principal value sense. Replacing the boundary near the point by a circular arc of radius $\varepsilon$ (Figure 5.1), the integral can be divided into two parts, i.e.

$$\lim_{i \to B} \int_\Gamma \frac{\partial T^*}{\partial n} T \, d\Gamma = \lim_{\varepsilon \to 0} \left( \int_{\Gamma - \Gamma_{\varepsilon_1}} \frac{\partial T^*}{\partial n} T \, d\Gamma + \int_{\Gamma_\varepsilon} \frac{\partial T^*}{\partial n} T \, d\Gamma \right) \tag{5.7}$$

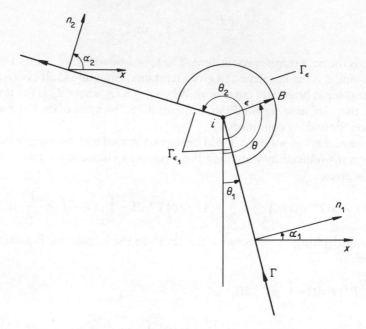

**Figure 5.1**  Definitions for integration purposes

On taking the limit, $\Gamma - \Gamma_{\varepsilon_1} \to \Gamma$ and as $T$ is continuous over the entire region, Equation (5.6) becomes

$$c_i k T_i + \int_\Omega p T^* \, \mathrm{d}\Omega - \int_\Gamma T q^* \, \mathrm{d}\Gamma = - \int_\Gamma q T^* \, \mathrm{d}\Gamma - \int_\Gamma h T T^* \, \mathrm{d}\Gamma \qquad (5.8)$$

where

$$q^* = -k \frac{\partial T^*}{\partial n}, \ q = -k \frac{\partial T}{\partial n} \quad \text{or} -\bar{q}$$

and

$$c_i = 1 + \lim_{\varepsilon \to 0} \int_{\Gamma_\varepsilon} \frac{\partial T^*(\varepsilon)}{\partial n} \, \mathrm{d}\Gamma \qquad (5.9)$$

The same procedure can be applied for three-dimensional problems. Note that we have written the integrals for the whole $\Gamma = \Gamma_1 + \Gamma_2$ boundary but that, depending on which part of the boundary we consider, some values will be prescribed. For instance, on the $\Gamma_1$ part of the boundary we have $T = \bar{T}$ (given), $h = 0$ and the only unknown is $q$; on $\Gamma_2$, both $h$ and $\bar{q}$ are given, the remaining unknown on the equation being $T$.

### 5.2.1 Fundamental solutions

The fundamental solution of Equation (5.5) for three-dimensional problems is[1]

$$T^*_{3D} = \frac{1}{4\pi R} \tag{5.10}$$

which gives

$$q^*_{3D} = \frac{k}{4\pi R^2} R_{,n} \tag{5.11}$$

where $R = [(x - x_i)^2 + (y - y_i)^2 + (z - z_i)^2]^{1/2}$ is the distance from the point of application of the heat source to the point under consideration and $R_{,n} = \partial R / \partial n$. For two-dimensional problems, all quantities are independent of the coordinate $z$. Thus one integration can be performed in advance in Equation (5.8), which means that the point sources are generalized to line heat sources. The fundamental solution is then[1]

$$T^*_{2D} = \frac{1}{2\pi} \ln \left( \frac{1}{R} \right) \tag{5.12}$$

with

$$q^*_{2D} = \frac{k}{2\pi R} R_{,n} \tag{5.13}$$

where $R = [(x - x_i)^2 + (y - y_i)^2]^{1/2}$ now.

For axisymmetric problems, all quantities are independent of the circumferential location and as in the two-dimensional case, one integration can be performed in advance in Equation (5.8). This is equivalent to using ring heat sources as fundamental solutions. Writing the three-dimensional solution in cylindrical coordinates $(r, \theta, z)$, it gives

$$T^*_{AS} = \int_0^{2\pi} T^*_{3D} \, d\theta_i = \frac{1}{4\pi} \int_0^{2\pi} \frac{d\theta_i}{[r^2 + r_i^2 - 2rr_i \cos (\theta - \theta_i) + (z - z_i)^2]^{1/2}} \tag{5.14}$$

and this integral can be evaluated in terms of the complete elliptic integral of the first kind $K(m)$. The axisymmetric fundamental solution can then be written as

$$T^*_{AS} = \frac{K(m)}{\pi [(r + r_i)^2 + (z - z_i)^2]^{1/2}} \tag{5.15}$$

where

$$m = \frac{4rr_i}{(r + r_i)^2 + (z - z_i)^2} \tag{5.16}$$

Differentiating expression (5.15) one finds

$$q^*_{As} = \frac{k}{\pi[(r+r_i)^2+(z-z_i)^2]^{1/2}}\Bigg[\frac{1}{2r}\bigg(K(m)+\frac{r^2-r_i^2-(z-z_i)^2}{(r-r_i)^2+(z-z_i)^2}E(m)\bigg)r_{,n}$$

$$+\bigg(\frac{z-z_i}{(r-r_i)^2+(z-z_i)^2}E(m)\bigg)z_{,n}\Bigg] \tag{5.17}$$

where $E(m)$ is the complete elliptic integral of the second kind. It can be noted that as $r_i \to 0$ the ring source tends to a point source over the axis of revolution.

Fundamental solutions for orthotropic problems are given in Brebbia[1] and for anisotropic ones in Chang, Kang and Chen.[2]

### 5.2.2 Boundary elements

Equation (5.8) can now be applied on the boundary of the domain under consideration. This boundary can be divided into $n$ elements. The points where the unknown values are considered are called 'nodes' and are similar to those of finite elements. The main difference is that now elements and nodes are defined only on the $\Gamma$ boundary. The functions $T$ and $q$ over each boundary element are given by,

$$T = \mathbf{\Phi}^T\mathbf{T}$$
$$q = \mathbf{\Psi}^T\mathbf{q} \tag{5.18}$$

and Equation (5.8) is discretized as follows:

$$c_ikT_i + \sum_{l=1}^{m}\int_{\Omega_l}pT^*\,\mathrm{d}\Omega - \sum_{j=1}^{n}\int_{\Gamma_j}Tq^*\,\mathrm{d}\Gamma = -\sum_{j=1}^{n}\int_{\Gamma_j}qT^*\,\mathrm{d}\Gamma - \sum_{j=1}^{n}\int_{\Gamma_j}hTT^*\,\mathrm{d}\Gamma \tag{5.19}$$

Note that $m$ internal elements or cells need to be defined to compute the integrals in $\Omega$ but these elements do not introduce any further unknown and hence the problem is still a boundary problem.

We can substitute the $T$ and $q$ values given by (5.18) into (5.19) and carry out the integrations. This gives for each node, after assembling, the following equation:

$$\sum_{j=1}^{N}\hat{H}_{ij}T_j - c_ikT_i = \sum_{j=1}^{N}G_{ij}q_j + \sum_{j=1}^{N}h_{ij}G_{ij}T_j + B_i \tag{5.20}$$

where $N$ is the number of nodes and $B_i$ is the result of having integrated the domain term. The whole set can be written in matrix form as

$$(\mathbf{H}-\mathbf{CG})\mathbf{T} = \mathbf{GQ}+\mathbf{B} \tag{5.21}$$

where $\mathbf{C}$ is a diagonal matrix containing the heat transfer coefficients.

Equation (5.21) represents a system of $N$ equations with $2N$ unknowns. As in a well posed problem $N$ of the boundary variables are prescribed, we can reorder the system in such a way that all the unknowns are on the left-hand side, i.e.

$$\mathbf{AX} = \mathbf{F} \tag{5.22}$$

The vector $\mathbf{X}$ contains the unknown $q$ on the $\Gamma_1$ part of the boundary and the unknown $T$ on $\Gamma_2$. If the 'radiation' condition is prescribed, after evaluating $T$ we still have to calculate $q$ applying the boundary condition given by Equation (5.2b).

Each integral in (5.19) can be evaluated using a standard Gaussian quadrature. However, care must be taken when the element $j$ contains the node $i$ since in these cases the integrals become singular and have to be evaluated as their Cauchy principal values.

The singularity of the fundamental solution is such that it is directly integrable. For two-dimensional and axisymmetric problems, these integrals can be evaluated in closed form. Expressions are given in Brebbia[1] for two-dimensional problems. For axisymmetric ones these integrals become more complicated due to the complex nature of the fundamental solution. In order to facilitate the analytical integration, we can write the fundamental solution in terms of Legendre functions as

$$T^*_{AS} = \frac{1}{2\pi (rr_i)^{1/2}} Q_{-1/2}(\gamma) \tag{5.23}$$

where

$$\gamma = 1 + \frac{(r - r_i)^2 + (z - z_i)^2}{2rr_i} \tag{5.24}$$

This Legendre function can be expanded for small values of $\gamma$ as[3]

$$Q_{-1/2}(\gamma) = -\tfrac{1}{2} \ln \left( \frac{\gamma - 1}{32} \right) \tag{5.25}$$

The substitution of expression (5.25) into (5.23) permits the explicit evaluation of the $G_{ii}$ terms. For constant elements, for example, the integration gives,

$$G_{ii} = \int_{-l/2}^{l/2} T^*_{AS} \, d\Gamma = \frac{l}{12\pi b r_i^{1/2}} \left\{ \left[ \frac{2}{3} - \ln \left( \frac{l^2}{256 r_i} \right) \right] [(r_i + b)^{3/2} - (r_i - b)^{3/2}] \right.$$

$$+ (r_i + b)^{3/2} \ln |r_i + b| - (r_i - b)^{3/2} \ln |r_i - b| + 4r_i[(r_i + b)^{1/2} - (r_i - b)^{1/2}]$$

$$\left. - 2r_i^{3/2} \left( \ln \left| \frac{(r_i + b)^{1/2} + r_i^{1/2}}{(r_i + b)^{1/2} - r_i^{1/2}} \right| - \ln \left| \frac{(r_i - b)^{1/2} + r_i^{1/2}}{(r_i - b)^{1/2} - r_i^{1/2}} \right| \right) \right\} \tag{5.26}$$

where $b = -\frac{1}{2}l \sin \alpha$ (Figure 5.2). For $\sin \alpha = 0$, this expression simplifies to,

$$G_{ii} = \frac{l}{2\pi}\left[1 - \ln\left(\frac{l}{16r_i}\right)\right] \tag{5.27}$$

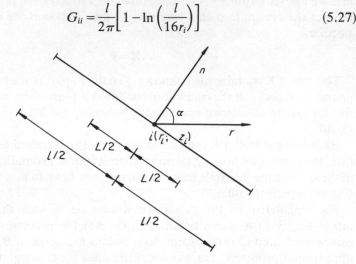

**Figure 5.2**  Portion of axisymmetric element that is integrated analytically

Note that for elements located near the axis of revolution (so with small $r_i$) it is not possible to integrate over the whole element in this way, since the value of $\gamma$ will be large for points far from the singularity and, therefore, approximation (5.25) is no longer valid for these points. To overcome this problem, one can integrate analytically over a short segment near the singularity and integrate numerically over the rest of the element using a standard Gaussian quadrature, as if these parts were separate elements. For computational purposes, the size of that portion of the element near the singularity (Figure 5.2) was assumed to be,

$$\frac{L}{2} \leq \left(\frac{rr_i}{50}\right)^{1/2} \tag{5.28}$$

where $r$ is the distance of the nearest point of that portion to the axis of revolution.

The evaluation of the $c_i$ coefficients is carried out by using Equation (5.9). For two-dimensional problems, we have (Figure 5.1)

$$c_i = 1 - \lim_{\varepsilon \to 0} \int_{\theta_1}^{\theta_2} \frac{1}{2\pi\varepsilon}\varepsilon \, d\theta = \frac{\pi + \alpha_1 - \alpha_2}{2\pi} \tag{5.29}$$

The same result is obtained for axisymmetric problems.

The diagonal elements of the **H** matrix consist of the sum of two terms:

$$H_{ii} = \hat{H}_{ii} - c_i k \tag{5.30}$$

For two-dimensional problems, when straight elements are employed, the first term of the sum will be zero, due to the orthogonality of $R$ and $n$. This is not so for axisymmetric problems, as well as for curved elements, and the integral must be evaluated analytically (at least over a short segment near the singularity) in order to account for its Cauchy principal value. However, we can always calculate the diagonal terms of $\mathbf{H}$ by considering that when a uniform temperature is applied over the whole domain the fluxes must be zero. Hence Equation (5.21) in the absence of heat generation inside the domain becomes,

$$\mathbf{HT} = 0 \tag{5.31}$$

where $\mathbf{T}$ is a uniform temperature. Thus the sum of all the elements of $\mathbf{H}$ in any row must be zero and the value of the coefficient on the diagonal can be easily calculated once the off-diagonal coefficients are all known, i.e.

$$H_{ii} = -\sum_{\substack{j=1 \\ i \neq j}}^{N} H_{ij} \tag{5.32}$$

After solving the system (5.22) and finding the boundary unknowns, the value of $T$ anywhere can be calculated using Equation (5.6). The internal fluxes are obtained by computing the derivatives of (5.6).

## 5.3  TRANSIENT HEAT CONDUCTION

The equation that governs transient heat conduction in a homogeneous, isotropic solid body is,

$$k\nabla^2 T - \rho c \frac{\partial T}{\partial t} = -Q(x, y, z, t) \tag{5.33}$$

where $t$ is time, $\rho$ is the density, $c$ the specific heat and $Q(x, y, z, t)$ is the rate of heat generated per unit volume per unit time at the internal point $(x, y, z)$. Let us consider, for simplicity, that there is no generation of heat inside the domain $\Omega$ and furthermore, that the boundary conditions of the problem will be of the following two types:

$$T = \bar{T} \quad \text{on } \Gamma_1 \tag{5.34a}$$

$$k\frac{\partial T}{\partial n} = \bar{q} \quad \text{on } \Gamma_2 \tag{5.34b}$$

The 'radiation' boundary condition, as well as heat generation, can be introduced into the analysis in the same way as shown in Section 5.2 for steady-state problems.

As the problem is now time dependent, some initial conditions in addition to the boundary conditions must be given and Equation (5.33) will also be integrated with relation to time. Weighting expression (5.33) we have,

$$\int_0^\tau \int_\Omega \left( k\nabla^2 T - \rho c \frac{\partial T}{\partial t} \right) T^* \, d\Omega \, dt$$

$$= \int_0^\tau \int_{\Gamma_2} \left( k \frac{\partial T}{\partial n} - \bar{q} \right) T^* \, d\Gamma \, dt - \int_0^\tau \int_{\Gamma_1} (T - \bar{T}) k \frac{\partial T^*}{\partial n} \, d\Gamma \, dt \qquad (5.35)$$

where $0 \leq t \leq \tau$. Integrating by parts twice we find,

$$\int_0^\tau \int_\Omega \left( k\nabla^2 T^* + \rho c \frac{\partial T^*}{\partial t} \right) T \, d\Omega \, dt - \rho c \left[ \int_\Omega TT^* \, d\Omega \right]_{t=0}^{t=\tau}$$

$$+ \int_0^\tau \int_\Gamma Tq^* \, d\Gamma \, dt = \int_0^\tau \int_\Gamma qT^* \, d\Gamma \, dt \qquad (5.36)$$

where the $\partial T^*/\partial t$ term was obtained integrating by parts with respect to time. The corresponding fundamental solution for this equation is[4]

$$T^* = \frac{1}{[4\pi\alpha(\tau - t)]^{d/2}} \exp\left( -\frac{R^2}{4\alpha(\tau - t)} \right) \qquad (5.37)$$

where $\alpha = k/\rho c$ is the thermal diffusivity and $d$ is the number of spatial dimensionality, i.e. $d = 3$ for three-dimensional problems, etc. Notice that the one- and two-dimensional fundamental solutions are obtained from the three-dimensional one by performing two or one integration in advance in Equation (5.36). The same idea can be applied in order to find the axisymmetric fundamental solution. This gives,

$$T^*_{AS} = \frac{2\pi}{[4\pi\alpha(\tau - t)]^{3/2}} I_0\left( \frac{rr_i}{2\alpha(\tau - t)} \right) \exp\left( -\frac{r^2 + r_i^2 + (z - z_i)^2}{4\alpha(\tau - t)} \right) \qquad (5.38)$$

where $I_0$ is the modified Bessel function of the first kind of order zero.

The fundamental solutions possess the following properties:

$$\nabla^2 T^* + \frac{1}{\alpha} \frac{\partial T^*}{\partial t} = 0 \quad \text{in } \Omega \quad \text{for all } t < \tau \qquad (5.39)$$

and for $t = \tau$:

$$\int_\Omega TT^* \, d\Omega = T_i \qquad (5.40)$$

Substituting this solution into Equation (5.36), one obtains for a point $i$,

$$c_i T_i + \frac{\alpha}{k} \int_0^\tau \int_\Gamma qT^* \, d\Gamma \, dt = \frac{\alpha}{k} \int_0^\tau \int_\Gamma Tq^* \, d\Gamma \, dt + \left[ \int_\Omega TT^* \, d\Omega \right]_{t=0} \qquad (5.41)$$

The last term in the above formula corresponds to the initial conditions at $t = 0$. Since the fundamental solution itself is time dependent, one does not need to propose an iterative scheme to solve time-dependent problems as is usually done in finite elements or finite differences.

One can divide the boundary of the domain under consideration into elements, as shown in Section 5.2. One also needs now to assume a variation with time for the functions $T$ and $q$. As these functions vary much slower than $T^*$ and $q^*$, it is a reasonable approximation to assume that they are constant over small intervals of time and perform the time integrations stepwise. Then, Equation (5.41) becomes,

$$c_i T_i + \frac{\alpha}{k} \int_\Gamma q \int_{t_1}^{t_2} T^* \, dt \, d\Gamma = \frac{\alpha}{k} \int_\Gamma T \int_{t_1}^{t_2} q^* \, dt \, d\Gamma + \left[ \int_\Omega TT^* \, d\Omega \right]_{t=t_1} \tag{5.42}$$

From now on, we will discuss in detail the application of the method to two-dimensional problems, but basically the same ideas can be used for one- or three-dimensional and axisymmetric problems.

Differentiating the fundamental solution (5.27) we find,

$$q^* = \frac{kR}{8\pi\alpha^2(t_2 - t)^2} \exp\left(-\frac{R^2}{4\alpha(t_2 - t)}\right) R_{,n} \tag{5.43}$$

Performing the time integrals in (5.42) gives,

$$\int_{t_1}^{t_2} T^* \, dt = \frac{1}{4\pi\alpha} E_1\left[\frac{R^2}{4\alpha(t_2 - t_1)}\right] \tag{5.44}$$

$$\int_{t_1}^{t_2} q^* \, dt = \frac{k}{2\pi\alpha R} \exp\left(-\frac{R^2}{4\alpha(t_2 - t_1)}\right) R_{,n} \tag{5.45}$$

where $E_1$ is the exponential integral. So Equation (5.42) can be written as,

$$c_i T_i + \frac{1}{4\pi k} \int_\Gamma qE_1[a] \, d\Gamma = \frac{1}{2\pi} \int_\Gamma \frac{T}{R} R_{,n} \, e^{-a} \, d\Gamma + \frac{1}{4\pi\alpha(t_2 - t_1)} \int_\Omega T^{t_1} e^{-a} \, d\Omega \tag{5.46}$$

where

$$a = \frac{R^2}{4\alpha(t_2 - t_1)} \tag{5.47}$$

If we assume that the temperature inside each cell varies according to an interpolation function of the same order as the one that prescribes the variation of temperature on the boundary elements, Equation (5.46) can be written in matrix form as,

$$\mathbf{H}\mathbf{T}_B^{t_2} = \mathbf{G}\mathbf{Q}_B^{t_2} - \mathbf{D}\mathbf{T}_{B+I}^{t_1} \tag{5.48}$$

where $\mathbf{H}$ and $\mathbf{G}$ are $N_B \times N_B$ matrices, $N_B$ is the number of boundary nodes, $\mathbf{T}_B^{t_2}$ and $\mathbf{Q}_B^{t_2}$ are $N_B$ vectors, $\mathbf{D}$ is a $N_B \times (N_B + N_I)$ matrix, $N_I$ is the number of internal nodes, and $\mathbf{T}_{B+I}^{t_1}$ is a $(N_B + N_I)$ vector that accounts for the initial temperature over the domain. The matrices $\mathbf{H}$, $\mathbf{G}$ and $\mathbf{D}$ depend on geometrical data, properties of the medium and the time step. If we assume a constant time step, they can all be computed only once and stored.

Suppose that the boundary conditions of the problem under consideration are prescribed temperatures. As some initial conditions are given, $\mathbf{T}_B^{t_2}$ and $\mathbf{T}_{B+I}^{t_1}$ will be known and we can write the following recurrence relation:

$$\mathbf{Q}_B^{t_2} = \mathbf{G}^{-1}(\mathbf{H}\mathbf{T}_B^{t_2} + \mathbf{D}\mathbf{T}_{B+I}^{t_1}) \qquad (5.49)$$

At the end of each time step, the temperatures at the $N_I$ internal points are given by

$$\mathbf{T}_I^{t_2} = \mathbf{H}_I\mathbf{T}_B^{t_2} - \mathbf{G}_I\mathbf{Q}_B^{t_2} + \mathbf{D}_I\mathbf{T}_{B+I}^{t_1} \qquad (5.50)$$

where $\mathbf{G}_I$ and $\mathbf{H}_I$ are $N_I \times N_B$ matrices and $\mathbf{D}_I$ is a $N_I \times (N_B + N_I)$ matrix, all of them constant in time. Then the solution at any time can be evaluated by the iterative use of Equations (5.49) and (5.50).

The value of the $c_i$ coefficients in Equation (5.46) can be obtained in the same way as for steady-state problems. In fact, it can be shown that on doing so the formula obtained is the same as Equation (5.28), i.e. (Figure 5.1)

$$c_i = 1 - \lim_{\varepsilon \to 0} \int_{\theta_1}^{\theta_2} \frac{1}{2\pi\varepsilon} \exp\left(-\frac{\varepsilon^2}{4\alpha(t_2 - t_1)}\right)\varepsilon \, d\theta = \frac{\pi + \alpha_1 - \alpha_2}{2\pi} \qquad (5.51)$$

Notice that for transient problems one can still calculate the diagonal terms of $\mathbf{H}$ by the application of a uniform temperature over the whole domain, which gives zero fluxes. In this way, we obtain $H_{ii}$ by,

$$H_{ii} = -\left(\sum_{j=1}^{N_B} H_{ij} + \sum_{k=1}^{N_B + N_I} D_{ik}\right) \qquad (5.52)$$

The main disadvantage that appears in calculating the diagonal terms of $\mathbf{H}$ using (5.52) is that one will always need to divide the whole domain into cells, whereas from Equation (5.46) it can be noticed that if part of the domain is kept at zero temperature throughout the analysis there will be no contribution from this part to the temperature elsewhere.

In order to evaluate the $\mathbf{D}$ matrix, we need to compute integrals over triangular cells. These integrals are computed numerically and in previous work the integration scheme employed by the authors was Hammer's quintic quadrature.[5] Using this scheme, the influence of the domain temperature on point $i$ is given by (for linear cells)

$$T_i = \frac{1}{2\pi\alpha\,\Delta t} \sum_{p=1}^{m} A_p \sum_{j=1}^{3} \sum_{l=1}^{7} \exp\left(-\frac{R_{il}^2}{4\alpha\,\Delta t}\right) W_l N_{lj} T_j \qquad (5.53)$$

where $A_p$ is the area of cell $p$, $N_{lj}$ is the interpolation function and $W_l$ is the weight of point $l$. From this expression, one can notice that the influence of point $l$ on the temperature at point $i$ is a function of the square of their distance. Because this scheme distributes the integration points symmetrically inside each cell, its use can lead to numerical problems if the value of $\alpha$ or $\Delta t$ (or both) is very small compared to the geometric dimensions of the problem.

An alternative procedure to evaluate this integral is to transform it to polar coordinates $(R, \theta)$ and perform the integration in $R$ analytically.[6] With relation to Figure 5.3, we can write,

$$
\begin{aligned}
T_i &= \frac{1}{4\pi\alpha\Delta t} \sum_{p=1}^{m} \sum_{j=1}^{3} \int_{\Omega_m} N_j \exp\left(-\frac{R^2}{4\alpha\Delta t}\right) d\Omega \ T_j \\
&= \frac{1}{4\pi\alpha\Delta t} \sum_{p=1}^{m} \sum_{j=1}^{3} \left[\int_{\theta_1}^{\theta_3} \int_{R_2(\theta)}^{R_3(\theta)} N_j(R, \theta) \exp\left(-\frac{R^2}{4\alpha\Delta t}\right) R \ dR \ d\theta \right. \\
&\quad \left. + \int_{\theta_3}^{\theta_2} \int_{R_1(\theta)}^{R_3(\theta)} N_j(R, \theta) \exp\left(-\frac{R^2}{4\alpha\Delta t}\right) R \ dR \ d\theta \right] T_j
\end{aligned}
\tag{5.54}
$$

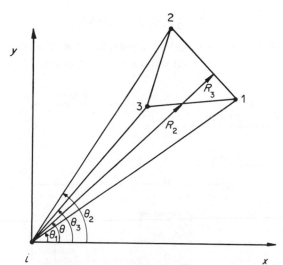

**Figure 5.3** Definitions for polar integration

where

$$
R_j(\theta) = -\frac{2A_p[\zeta_j^p(i)]}{b_j \cos\theta + a_j \sin\theta}
$$

$$
N_j(R, \theta) = \zeta_j^p(i) + \frac{R}{2A_p}(b_j \cos\theta + a_j \sin\theta)
$$

$$
a_j = x_l - x_k
$$

$$
b_j = y_k - y_l
$$

where $j = 1, 2, 3$ for $k = 2, 3, 1$ and $l = 3, 1, 2$ and $\zeta_j^p(i)$ are the triangular coordinates of point $i$ relative to cell $p$. Performing the integration in $R$ gives,

$$
\begin{aligned}
T_i = \sum_{p=1}^{m} \sum_{j=1}^{3} & \left( \frac{\theta_3 - \theta_1}{4\pi} \int_{-1}^{1} \left\{ \zeta_j^p(i) \left[ \exp\left( -\frac{R_2^2(\theta)}{4\alpha\Delta t} \right) - \exp\left( -\frac{R_3^2(\theta)}{4\alpha\Delta t} \right) \right] \right. \right. \\
& \left. + \frac{(\alpha\Delta t)^{1/2}}{A_p} (b_j \cos\theta + a_j \sin\theta) \left( \gamma\left[ \frac{3}{2}, \frac{R_3^2(\theta)}{4\alpha\Delta t} \right] - \gamma\left[ \frac{3}{2}, \frac{R_2^2(\theta)}{4\alpha\Delta t} \right] \right) \right\} d\zeta \\
& + \frac{\theta_2 - \theta_3}{4\pi} \int_{-1}^{1} \left\{ \zeta_j^p(i) \left[ \exp\left( -\frac{R_1^2(\theta)}{4\alpha\Delta t} \right) - \exp\left( -\frac{R_3^2(\theta)}{4\alpha\Delta t} \right) \right] + \frac{(\alpha\Delta t)^{1/2}}{A_p} (b_j \cos\theta \right. \\
& \left. \left. + a_j \sin\theta) \left( \gamma\left[ \frac{3}{2}, \frac{R_3^2(\theta)}{4\alpha\Delta t} \right] - \gamma\left[ \frac{3}{2}, \frac{R_1^2(\theta)}{4\alpha\Delta t} \right] \right) \right\} d\zeta \right) T_j
\end{aligned}
\tag{5.55}
$$

where $\gamma$ is the incomplete Gamma function normalized. The integrals in $\theta$ can be evaluated using Gauss quadrature and experience has shown that four integration points are sufficient for the required accuracy.

## 5.4 APPLICATIONS

Several applications were studied to illustrate how the boundary element method can be used to solve heat conduction problems. They include three

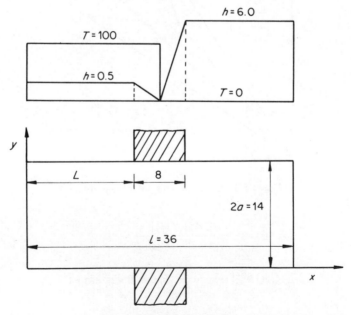

**Figure 5.4** Transverse section of rectangular concrete column with abutting wall separating the inside and outside environments

steady-state problems, the first being a two-dimensional one and the two others axisymmetric, and two transient cases. The results obtained are compared against analytical solutions, when available, or with solutions obtained by using other numerical techniques.

### 5.4.1   Rectangular concrete column

The first application deals with an exposed exterior concrete column of rectangular cross section where part of the boundary surface is subject to an interior ambient conditions, another part is subjected to outside weather

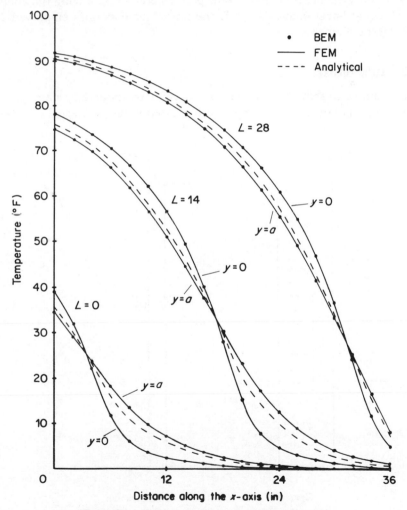

**Figure 5.5**   Steady-state temperature along the *x*-axis

conditions and another part is in contact with an abutting wall which separates both. The temperature and surface heat transfer coefficients on the interior face $(x = 0)$ are $100\,°F$ and $0.5\,Btu\,h^{-1}\,ft^{-2}\,°F^{-1}$, respectively, and at the exterior face $(x = l)$ are $0\,°F$ and $6.0\,Btu\,h^{-1}\,ft^{-2}\,°F^{-1}$. The variation of the temperature and surface heat transfer coefficient on the faces $y = 0$ and $y = 2a$ is indicated in Figure 5.4. The thermal conductivity was assumed to be $1.0\,Btu\,h^{-1}\,ft^{-2}\,°F^{-1}$.

Results corresponding to three different positions for the abutting wall are shown in Figure 5.5, compared against finite elements results[7] and an analytical solution[8] (in terms of a mean temperature over the width of the cross section). The boundary elements analyses were performed by discretizing the column into 50 equal linear elements, while the finite elements ones employed 252 quadrilateral elements.

### 5.4.2 Tube furnace

As an example to show the accuracy of axisymmetric boundary elements, the problem of a hollow cylinder subjected to a discontinuous internal input heat

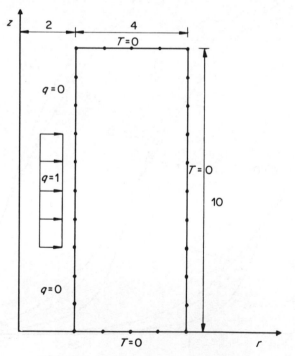

**Figure 5.6** Discretization of hollow cylinder

flow was analysed using both constant and linear elements. This problem is a simple approximation to the one of a tube furnace, for which an analytical solution can be found.[4] The discretization adopted for both analyses is shown in Figure 5.6 (for the constant elements case the nodal points are located in the middle of each element). Results are compared in Tables 5.1 and 5.2 showing a good agreement.

Table 5.1    Temperature at internal points

| Point | BEM (constant) | BEM (linear) | Analytical |
|-------|----------------|--------------|------------|
| 3; 1  | 0.140          | 0.141        | 0.141      |
| 3; 2  | 0.317          | 0.320        | 0.319      |
| 3; 3  | 0.556          | 0.556        | 0.556      |
| 3; 4  | 0.762          | 0.760        | 0.761      |
| 3; 5  | 0.832          | 0.831        | 0.831      |
| 5; 1  | 0.043          | 0.043        | 0.043      |
| 5; 2  | 0.088          | 0.088        | 0.088      |
| 5; 3  | 0.133          | 0.133        | 0.133      |
| 5; 4  | 0.167          | 0.167        | 0.167      |
| 5; 5  | 0.180          | 0.180        | 0.180      |

Table 5.2    Fluxes at $z = 0$ and $z = 10$

| $r$ | BEM (constant) | BEM (linear) | Analytical |
|-----|----------------|--------------|------------|
| 2.0 | —              | 0.148        | 0.155      |
| 2.5 | 0.155          | —            | 0.149      |
| 3.0 | —              | 0.141        | 0.134      |
| 3.5 | 0.112          | —            | 0.113      |
| 4.0 | —              | 0.089        | 0.090      |
| 4.5 | 0.066          | —            | 0.066      |
| 5.0 | —              | 0.043        | 0.042      |
| 5.5 | 0.017          | —            | 0.020      |
| 6.0 | —              | 0.0          | 0.0        |

## 5.4.3   Nuclear reactor pressure vessel

A prototype axisymmetric pressure vessel subjected to an increase of temperature applied on the inside was analysed using 34 constant boundary elements. The resulting temperature distribution is shown in Figure 5.7, together with the discretization adopted in the analysis. The same problem was studied using 96 triangular finite elements[9] and the results, as well as the mesh employed, are shown in Figure 5.8, for comparison.

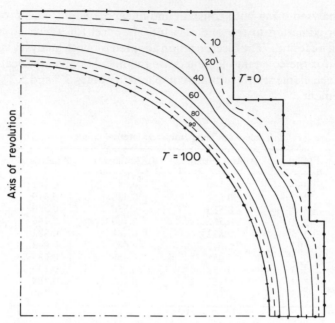

**Figure 5.7**   Axisymmetric pressure vessel; BEM discretization and isothermals

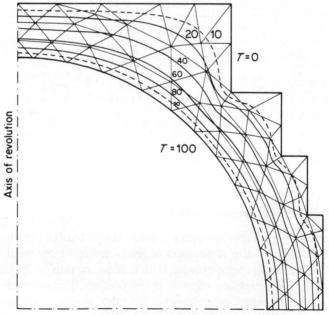

**Figure 5.8**   Axisymmetric pressure vessel; FEM mesh and isothermals

### 5.4.4 Rectangular concrete column (transient analysis)

On this application, a transient analysis of the same rectangular concrete column of the first example was performed for the case when the abutting wall is located at the exterior end of the column. The initial temperature was assumed to be 0 °F and the values of the density and specific heat taken as 125 lb ft$^{-3}$ and 0.2 Btu lb$^{-1}$ °F$^{-1}$, respectively. The same boundary discretization which had been used for the steady-state case was employed, i.e. 50 linear elements, and the domain divided into 196 triangular cells. Results are

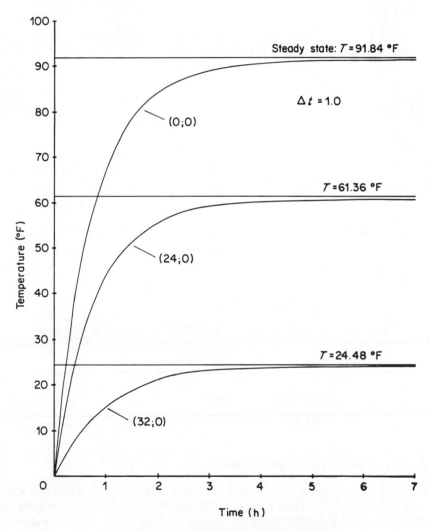

**Figure 5.9** Concrete column; transient temperature

shown in Figures 5.9 and 5.10 and it can be seen that they tend to the steady-state solution as the time tends to infinity.

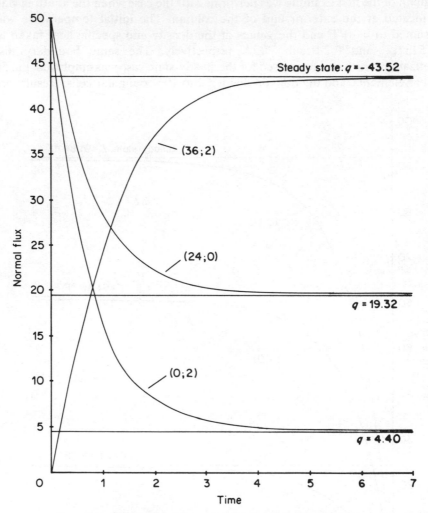

**Figure 5.10**   Concrete column; transient heat flux

### 5.4.5  Fuel element for nuclear reactor

This application considers the problem of transient heat flow in a fuel element for a nuclear reactor. The general configuration of the problem can be seen in Figure 5.11, as well as the symmetric portion of its cross section that was analysed. A constant heat flux of $1\,\mathrm{W\,cm^{-2}}$ is applied at time $t = 0$ along the

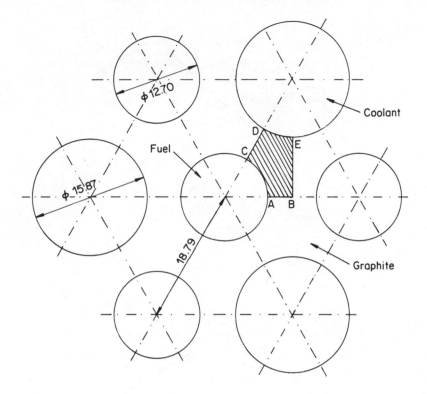

**Figure 5.11** Fuel element for a nuclear reactor

interface A–C between fuel and canning. The numerical values assumed for the material constants are $k = 0.1 \, \text{W cm}^{-1}\,^{\circ}\text{C}^{-1}$, $\rho = 100 \, \text{kg cm}^{-3}$ and $c = 0.01 \, \text{J kg}^{-1}\,^{\circ}\text{C}^{-1}$. The temperature of the coolant material is $0\,^{\circ}\text{C}$ and the surface heat transfer coefficient on the interface D–E equal to $1 \, \text{W cm}^{-2}\,^{\circ}\text{C}^{-1}$.

Isotherms at various times are plotted in Figure 5.12, together with the results obtained with a steady-state analysis. The boundary was divided into 33 linear elements and the domain into 94 cells, with 34 internal nodes. The problem was also studied using finite elements[10] but no results of this kind were presented.

## 5.5 CONCLUSION

This chapter has described the application of the boundary element method to steady-state and transient heat conduction. The main advantage of the technique is that it reduces by one the dimensionality of the problem, resulting in a smaller system of equations to be solved. The method is also well suited for

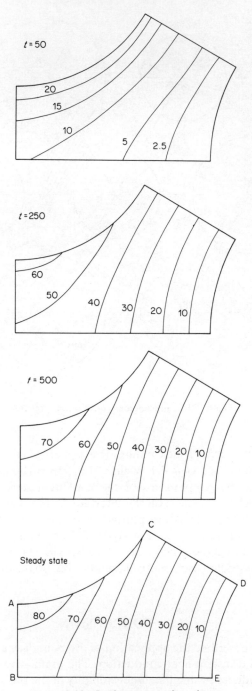

**Figure 5.12** Isotherms at various times

solving problems where high gradients or singularities occur, as well as for problems where the domain extends to infinity.[11]

Transient problems require the use of fundamental solutions which are space and time dependent. This eliminates the need of integrating step-by-step on time using finite differences or similar discretizations. In practice, however, time intervals are recommended in order to obtain accurate results.

Further work still needs to be carried out in order to solve more complex problems, but the examples presented demonstrate some of the potentialities of the method in the solution of thermal problems.

## REFERENCES

1. C. A. Brebbia (1978). *The Boundary Element Method for Engineers*, Pentech Press, London.
2. Y. P. Chang, C. S. Kang and D. J. Chen (1973). 'The use of fundamental Green's functions for the solution of problems of heat conduction in anisotropic media', *Int. J. Heat Mass Transfer*, **16**, 1905–1918.
3. A. Erdelyi *et al.* (1953). *Higher Transcendental Functions*, Vol. 1, Bateman Manuscript Project, McGraw-Hill, New York.
4. H. S. Carslaw and J. C. Jaeger (1959). *Conduction of Heat in Solids*, 2nd edn, Clarendon Press, Oxford.
5. L. C. Wrobel and C. A. Brebbia (1979). 'The boundary element method for steady state and transient heat conduction', *1st Int. Conf. on Numerical Methods in Thermal Problems, University of Wales at Swansea, Swansea, July 1979*.
6. J. C. F. Telles and C. A. Brebbia (1980). 'The boundary element method in plasticity', *2nd Int. Sem. on Recent Advances in Boundary Element Methods, University of Southampton, Southampton, March 1980*.
7. E. L. Wilson and R. E. Nickell (1966). 'Application of the finite element method to heat conduction analysis', *Nucl. Engng Des.*, **4**, 276–286.
8. B. A. Peavy (1965). 'Steady-state heat conduction in an exposed exterior column of rectangular cross section', *J. Res. Natl Bur. Stand.*, **69C**, 145–151.
9. O. C. Zienkiewicz and Y. K. Cheung (1965). 'Finite elements in the solution of field problems', *The Engineer*, **220**, 507–510.
10. J. Donea (1974). 'On the accuracy of finite element solutions to the transient heat conduction equation', *Int. J. Num. Meth. Engng*, **8**, 103–110.
11. L. C. Wrobel and C. A. Brebbia (1981). 'Time-dependent problems', in *Progress in Boundary Elements Vol. 1*, Pentech Press, London, 1981.

Numerical Methods in Heat Transfer
Edited by R. W. Lewis, K. Morgan, and O. C. Zienkiewicz
© 1981 John Wiley & Sons Ltd

Chapter 6

# Boundary Integral Equations Used to Solve Thermoelastic Problems: Application to Standard and Incompressible Materials

Anne Chaudouet and Gilles Loubignac

## SUMMARY

Considering a purely elastic analysis of an homogeneous and isotropic body, if the behaviour of the material follows Hooke's law, then the equilibrium equations can be transformed into an equation depending only upon the displacements and the body forces (Navier's equation). This equation is a linear partial differential equation with constant coefficients, and so, once the fundamental solution is known, can be transformed through a reciprocity theorem into a boundary integral equation in terms of displacement and mechanical tension, thus reducing a three-dimensional problem into a two-dimensional one. The same process is applied to the thermal analysis, leading to Laplace's equation, and then to a boundary integral equation in terms of temperature and flux on the surface. If the temperature is not constant throughout the domain, temperature-dependant terms must be added in the equilibrium equations of the mechanical analysis. Furthermore, if the behaviour of the material still follows Hooke's law, then the boundary integral equation which is obtained, is similar to the one obtained for a purely elastic analysis except that this equation is written in terms of the total displacement (displacement due to the mechanical and the thermal loading) and of the mechanical tension only, the influence of the thermal field being represented by a volume equation. If the thermal loading corresponds to a steady-state problem, then this volume equation can also be transformed into an integral on the surface in terms of temperature and flux. Nowhere in this final equation does the coefficient $1/(1-2\nu)$ appear (where $\nu$ is Poisson's ratio), thus allowing incompressible material to be treated in the same straightforward way as any other material. If the characteristic coefficients of the materials are not constant throughout the domain, but may be considered as piecewise constant, a subregioning process is applied.

The boundary integral equation is solved numerically. The surface is meshed by eight nodes quadrilateral or six nodes triangular elements. The unknowns may vary linearly or quadratically on each element. Under these assumptions, the equation is discretized in each subregion.

Three examples are then described. The first is a hollow cylinder màde of incompressible material, which has a known analytical solution. The second is an industrial example of a body also made of incompressible material. The third is the analysis of a roller in continuous casting of steel slabs when there is a halt in the process. This problem is a transient one and furthermore the characteristics of the materials are varying with respect to temperature and so cannot be considered as constant throughout the whole body.

The main advantages of this method over the classical ones (finite elements or finite differences) are:

(a)  considerable savings on computing time and on input–output operations since only the surface of the domain has to be taken into account;
(b)  good accuracy of the results since only the boundary conditions are approximated, the governing equations being verified exactly in the volume.

## 6.1  INTRODUCTION

The characteristic equation of the elastic problem for material following Hooke's law when thermal behaviour is taken into account, is transformed into a boundary integral equation, thus reducing the dimensionality of the problem by one. In this chapter only three-dimensional problems will be considered. If the unknowns of the analysis are correctly chosen, the final equation, which is to be solved by numerical methods, does not contain the coefficient $1/(1-2\nu)$ (where $\nu$ is Poisson's ratio). Bodies made of incompressible material can then be treated in the same straightforward way as any other material governed by the Navier–Stokes equation.

## 6.2  GOVERNING EQUATIONS OF THE PURELY ELASTIC PROBLEM

Considering an homogeneous and isotropic body, if the temperature is constant throughout the body, the stress tensor $\sigma$ must satisfy the following equilibrium equations:

$$\partial_j \sigma_{ij} + f_i = 0 \quad \text{inside the domain } D \tag{6.1}$$

$$\sigma_{ij} n_j = t_i \quad \text{on the surface } S \text{ of the domain} \tag{6.2}$$

where $f_i$ are body forces, $t_i$ is tension and $n_j$ the outward normal. If the behaviour of the material of the body follows Hooke's law, then the stress tensor $\sigma$ is related to the strain tensor $\varepsilon$ by:

$$\sigma_{ij} = \lambda \varepsilon_{kk} \delta_{ij} + 2\mu \varepsilon_{ij} \tag{6.3}$$

where $\lambda$ and $\mu$ are Lame's coefficients.

The strain tensor can be expressed in terms of displacements as:

$$\varepsilon_{ij} = \tfrac{1}{2}(\partial_i u_j + \partial_j u_i) \tag{6.4}$$

To analyse a given problem, the boundary conditions must be known. Those conditions may be of the following type:

(a)  known displacement on the surface $S_u$;
(b)  known tension on the surface $S_t$;
(c)  relation between displacement and tension such as:

$$t_i = k_{ij} u_j \quad \text{on the surface } S_k$$

with

$$S_u \cup S_t \cup S_k = S$$

(see Figure 6.1).

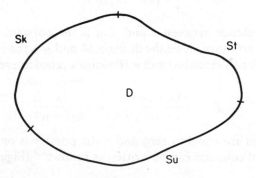

**Figure 6.1**

Using Equations (6.3) and (6.4), Equation (6.1) can be written only in terms of displacements; it is Navier's equation:

$$\Delta^* u = -f \quad \text{inside } D$$

with

$$\Delta^* = (\lambda + \mu) \, \text{grad} \, (\text{div}) + \mu \Delta$$

This can be written as

$$\lambda \, \partial_{ik} u_k + \mu \, \partial_{ji} u_j + \mu \, \partial_{jj} u_i = -f_i \tag{6.5}$$

## 6.3   BOUNDARY INTEGRAL EQUATION

Because Equation (6.5) is a linear partial differential equation with constant coefficients which has a known fundamental solution, Betti's theorem (reciprocity theorem) leads to the following equation:

$$u_i(x) = \int_D U_{ij}(x, y) f_j(y) \, dV_y + \int_S U_{ij}(x, y) t_j(y) \, dS_y$$

$$- \int_S T_{ij}(x, y) u_j(y) \, dS_y \tag{6.6}$$

with

$$U_{ij}(x, y) = \frac{1+\nu}{8\pi E(1-\nu)} \frac{1}{r} \left( (3-4\nu)\delta_{ij} + \frac{(x_i - y_i)(x_j - y_j)}{r^2} \right)$$

and

$$T_{ij}(x, y) = \frac{1}{8\pi E(1-\nu)r^2} \left[ (1-2\nu) \left( n_i(y) \frac{x_j - y_j}{r} - n_j(y) \frac{x_i - y_i}{r} \right) \right.$$

$$\left. + n_s(y) \frac{x_s - y_s}{r} \left( (1-2\nu)\delta_{ij} + 3 \frac{(x_i - y_i)(x_j - y_j)}{r^2} \right) \right]$$

where $r$ is the distance between $x$ and $y$, $n$ is the outward normal, $x$ is the coordinate of an arbitrary inside the domain $D$, and $y$ is the coordinate on the surface $S$. $E$ (Young's modulus) and $\nu$ (Poisson's ratio) are related to $\lambda$ and $\mu$ by:

$$E = \frac{\mu(3\lambda + 2\mu)}{\lambda + \mu} \qquad \nu = \frac{\lambda}{2(\lambda - \mu)}$$

If the body forces are equal to zero and if the point $x$ is on the surface, the boundary integral equation can be written as follows[1,2] (Figure 6.2):

$$C_{ij} u_j(x) + \int_S T_{ij}(x, y) u_j(y) \, dS_y = \int_S U_{ij}(x, y) t_j(y) \, dS_y \tag{6.7}$$

with $C_{ij} = \frac{1}{2}\delta_{ij}$ if the surface is smooth (otherwise it depends upon the solid angle at point $x$).

If the characteristic coefficients of the materials are not constant throughout the domain, but are varying slowly enough to be considered as piecewise constant, then the body is divided into subregions. Equation (6.6) is written for each subregion, and continuity and equilibrium equations are written on the interfaces between the subregions (Figure 6.3):

continuity:   $u_i(x \in a) = u_i(x \in b)$

equilibrium:   $t_i(x \in a) = -t_i(x \in b)$

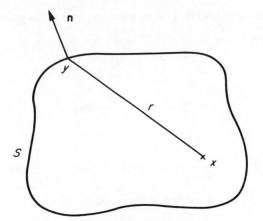

**Figure 6.2**

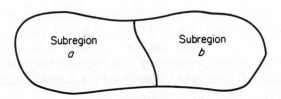

**Figure 6.3**

## 6.4 GOVERNING EQUATIONS OF THE ELASTIC PROBLEM UNDER THERMAL LOADING

If the temperature is not constant throughout the domain, temperature-dependent terms must be added in the equilibrium equations. The equivalent stress tensor in that case is:

$$\Sigma_{ij} = \sigma_{ij} + \frac{\alpha E}{1 - 2\nu} \theta \delta_{ij}$$

where $\alpha$ is the coefficient of linear thermal expansion and $\theta$ is temperature. The relation between the stress and strain tensors is:

$$\Sigma_{ij} = \lambda \varepsilon_{kk} \delta_{ij} + 2\mu \varepsilon_{ij}$$

as in Equation (6.3), but with:

$$\varepsilon_{ij} = \varepsilon_{ij}^{e} + \varepsilon_{ij}^{T}$$

$$\varepsilon_{ij}^{T} = \alpha \theta \delta_{ij}$$

$$\varepsilon_{ij}^{e} = \text{elastic strain tensor of Equation (6.3)}$$

Writing Equations (6.1) and (6.2) in terms of $\Sigma_{ij}$ and $\theta$, the governing equations become:

$$\partial_j \Sigma_{ij} + f_i - \frac{\alpha E}{1-2\nu} \partial_i \theta = 0 \tag{6.8}$$

and

$$\Sigma_{ij} n_j = t_i + \frac{\alpha E}{1-2\nu} \theta n_i \tag{6.9}$$

With

$$F_i = f_i - \frac{\alpha E}{1-2\nu} \partial_i \theta \quad \text{and} \quad T_i = t_i + \frac{\alpha E}{1-2\nu} \theta n_i$$

Equations (6.1) and (6.2), written in terms of the mechanical tensor, become:

$$\partial_j \Sigma_{ij} + F_i = 0 \quad \text{inside the domain } D \tag{6.10}$$

$$\Sigma_{ij} n_j = T_i \quad \text{on the surface } S \tag{6.11}$$

which are written in terms of the equivalent tensor (mechanical + thermal).

Equations (6.1), (6.2) and (6.10), (6.11) have a similar form, then Equation (6.6) can be written in the same way when a thermal loading is taken into account with the following quantities: $T$ (mechanical + thermal) instead of $t$; $u$ (total displacement); $F$ instead of $f$. But when $f$ is equal to zero, it is not true for $F$ and the volume integral does not disappear.

However, for a steady-state problem (from the thermal point of view) with a constant conduction coefficient, this integral in the volume can be transformed into an integral on the surface only.

## 6.5  BOUNDARY INTEGRAL EQUATION FOR THE THERMOELASTIC PROBLEM

In this case Equation (6.7) becomes[3,4]

$$C_{ij} u_j(x) + \int_S T_{ij}(x, y) u_j(y) \, dS_y$$

$$= \int_S U_{ij}(x, y) t_j(y) \, dS_y + \int_S [V_i(x, y)\theta(y) - W_i(x, y)\Psi(y)] \, dS_y \tag{6.12}$$

with

$$V_i(x, y) = \frac{\alpha(1+\nu)}{8\pi(1-\nu)} \frac{1}{r} \left( n_i(y) - \frac{(x_i - y_i)(x_k - y_k)}{r^2} n_k(y) \right)$$

and

$$W_i(x, y) = -\frac{1+\nu}{8\pi(1-\nu)} \frac{1}{r} (x_i - y_i) \frac{1}{\gamma}$$

where $\gamma$ is the conduction coefficient. The kernels $U_{ij}$ and $T_{ij}$ are the same as in Equation (6.7). The unknowns are $u$ (total displacement) and $t$ (mechanical tension) (*not T*).

To analyse a particular problem, the same boundary conditions as those of the purely elastic problem must be known, but furthermore the temperature field ($\theta$) and the flux ($\Psi$) must be known on the whole surface.

Looking at the kernels $U_{ij}$, $T_{ij}$, $V_i$, $W_i$, it is obvious that it is possible to analyse a body which has a Poisson ratio of material of 0.5 (incompressible material) in the same straightforward way as would be done for any other value of $\nu$.

If the temperature and the flux is not known on the whole surface, a thermal analysis must be carried out.

## 6.6 THERMAL ANALYSIS

The characteristic equation of a steady-state problem with constant conduction coefficient is Laplace's equation:

$$\Delta\theta = 0 \quad \text{inside the domain } D$$

On the surface the flux is related to the temperature by:

$$\Psi = -\gamma\frac{d\theta}{dn}$$

where $\gamma$ is the conduction coefficient. Applying Green's theorem to Laplace's equation, the following boundary integral equation (when the equation is written for a point $x$ on the surface $S$) is obtained:[5]

$$C\theta(x) + \int_S H(x, y)\theta(y)\, dS_y = \int_S G(x, y)\Psi(y)\, dS_y$$

with

$$G(x, y) = -\frac{1}{4\pi\gamma}\frac{1}{r}$$

$$H(x, y) = \frac{1}{4\pi r^3}n_i(x_i - y_i)$$

The boundary conditions may be of the same type as those of the elastic problem, i.e. given temperature on the surface, given flux on the surface and heat transfer with the exterior of the domain.

If the conduction coefficient is not constant throughout the domain, the same subregioning process as in the elastic problem is used, and on the interfaces (Figure 6.4):

$$\theta(x \in a) = \theta(x \in b) \qquad \Psi(x \in a) = -\Psi(x \in b)$$

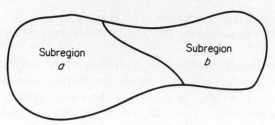

**Figure 6.4**

## 6.7 NUMERICAL TREATMENT

### 6.7.1 Representation of the geometry and functions

The surface is meshed with eight nodes quadrilateral or six nodes triangular elements (see Figure 6.5). The Cartesian coordinates of an arbitrary point of an

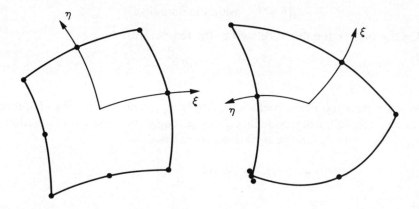

**Figure 6.5**

element are expressed in terms of the nodal coordinates $x^a$ and shape functions of the intrinsic coordinates $\xi$ and $\eta$:

$$x_i(\xi, \eta) = N_a(\xi, \eta)x_i^a \tag{6.13}$$

The displacement, traction, temperature and flux may be considered to vary linearly or quadratically with respect to the intrinsic coordinates:[6]

$$\theta(\xi, \eta) = M_a(\xi, \eta)\theta^a \tag{6.14}$$

Linear:      $M_a = 0.25(1 + \xi\xi_a)(1 + \eta\eta_a)$

Quadratic: $N_a = M_a$

$$M_a = \begin{cases} 0.25(1+\xi\xi_a)(1+\eta\eta_a)(\xi\xi_a+\eta\eta_a-1) & \text{corner node} \\ 0.50(1+\xi\xi_a)(1-\eta^2) & \text{for } \eta_a = 0 \\ 0.50(1+\eta\eta_a)(1-\xi^2) & \text{for } \xi_a = 0 \end{cases} \text{middle node}$$

With the kind of elements mentioned above, any geometrical shape can be described with enough accuracy to have good results on the geometry-dependent terms.

### 6.7.2 Discretization of the boundary integral equation

The boundary integral equation is written for each subregion. For each element: $ds = J \, d\eta \, d\xi$ ($J$: with respect to $N$). Using Equation (6.14) to express the unknowns on each element, then Equation (6.12) becomes:[7,8]

$$C_{ij}u_j(x) + \sum_{b=1}^{n} \sum_{a=1}^{m} u_j^{a(b)} \int_{S_b} T_{ij}(x, \xi, \eta) M^a(\xi, \eta) J(\xi, \eta) \, d\xi \, d\eta$$

$$= \sum_{b=1}^{n} \sum_{a=1}^{m} t_j^{a(b)} \int_{S_b} U_{ij}(x, \xi, \eta) M^a(\xi, \eta) J(\xi, \eta) \, d\xi \, d\eta$$

$$+ \theta^{a(b)} \int_{S_b} V_i(x, \xi, \eta) M^a(\xi, \eta) J(\xi, \eta) \, d\xi \, d\eta$$

$$- \Psi^{a(b)} \int_{S_b} W^i(x, \xi, \eta) M^a(\xi, \eta) J(\xi, \eta) \, d\xi \, d\eta \tag{6.15}$$

where $m = 4$ or 8 (linear or quadratic), the number of unknowns on each element, and $n$ is the number of elements of the surface. Some areas of the surface are either partially or completely fixed. Usually the directions in which each node is fixed or free to move, are not the global directions.

Equation (6.15) must then be written with respect to the local directions of the given displacements. Another problem is the integration of the kernels. Since $T_{ij}$ varies as $1/r^2$, the integration must be carried out with great care in order to have good accuracy for the terms of the matrix, and so have the best possible results on the displacements and on the components of the stress tensors which correspond to the normal of the surface.

Furthermore as the kernel $T_{ij}$ varies as $1/r^2$ on the surface, and as $C_{ij}$ depends upon the geometrical shape (those two terms leading to heavy or sometimes impossible calculation), they are not calculated separately, but this diagonal term of the matrix is calculated as a whole. For that purpose an extra equation considering a rigid body translation (for the elastic analysis) or considering that the temperature is known up to an arbitrary constant (for the thermal analysis) is written between all the terms of each line of the matrix.

## 6.8　ANALYTICAL EXAMPLE OF A BODY MADE OF INCOMPRESSIBLE MATERIAL

Hollow cylinders having the following characteristics have been studied (due to the axisymmetry of the problem, only a 15° sector has been modelled) (see Figure 6.6):

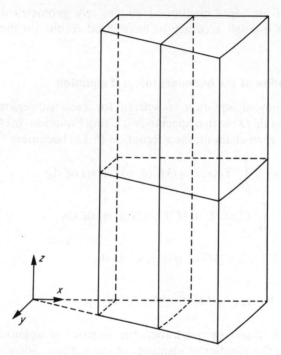

**Figure 6.6**

| Young's modulus | $E = 21000 \text{ daN mm}^{-2}$ |
| Poisson's ratio | $\nu = 0.5$ |
| Coefficient of linear thermal expansion | $\alpha = 10^{-4} \,^{\circ}\text{C}^{-1}$ |
| Conduction coefficient | $\gamma = 5 \text{ kcal mm}^{-1}\,^{\circ}\text{C}^{-1}$ |
| Density | $d = 7.85$ |

The thermal conditions were: constant temperature throughout the cylinder, $\Delta T = -20\,^{\circ}\text{C}$, that leads to no flux on the whole surface ($\Psi = 0$).

The boundary conditions were:

(a) zero normal displacements on the radial surfaces (axisymmetric problem);
(b) no displacement on the outer surface;
(c) zero normal displacements on the surfaces $z = 0$ and $z = h$ (plane strains).

With the above hypothesis, it is possible to calculate the radial displacement on the inner surface:

$$u_r = \tfrac{3}{2}\alpha R_i \left(1 - \frac{R_e^2}{R_i^2}\right)\Delta T$$

if second-order terms are neglected. For $R_e$ and $R_i$ given, three different heights have been studied with quadratic interpolation functions and the results are shown in Table 6.1. The theoretical result is 6.70 mm.

Table 6.1  Radial displacement of the inner surface (mm)

| Height (mm) | 500 | 1000 | 2000 |
|---|---|---|---|
| $z = 0$ | 6.33 | 6.36 | 6.25 |
| $z = h/4$ | 6.33 | 6.33 | 6.35 |
| $z = h/2$ | 6.34 | 6.30 | 6.25 |
| Error | 5.7% | 5.5% | 6.0% |

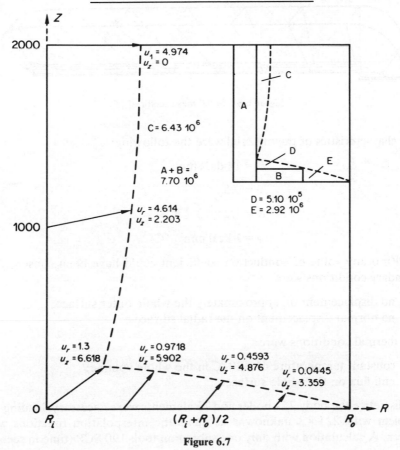

Figure 6.7

Another calculation has been made, for which no analytical solution is known. It is a three-dimensional case corresponding to the examples mentioned first but in which the displacements on the surfaces $z = 0$ and $z = h$ are free. To check the results, the variation of volume was calculated. The theoretical variation of volume is:

$$\Delta V = 3\alpha V \Delta T$$

For $h = 4000$ mm; $\Delta V/2 = 1.78 \times 10^7$ mm$^3$. The BIE calculation gave $1.75 \times 10^7$ mm$^3$ (error = 1.3%) (see Figure 6.7).

## 6.9  INDUSTRIAL EXAMPLE OF A BODY MADE OF INCOMPRESSIBLE MATERIAL

The body studied is shown in Figure 6.8 (only a 15° sector has been modelled).

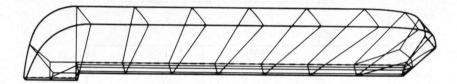

**Figure 6.8**  Initial mesh sector of 15°

The characteristics of the material were the following:

$$E = 14 \text{ daN mm}^{-2}$$

$$\nu = 0.5$$

$$\alpha = 1.2 \times 10^{-4} \,^{\circ}\text{C}^{-1}$$

$$\gamma = 1 \text{ kcal mm}^{-1} \,^{\circ}\text{C}^{-1}$$

(as $\Psi = 0$ any value of conduction coefficient could have been chosen). The boundary conditions were:

(a)  no displacement on approximately the whole outer surface;
(b)  no normal displacement on the radial surfaces.

The thermal conditions were:

(a)  constant temperature of $-20\,^{\circ}$C in the whole volume;
(b)  null flux on the whole surface.

To discretize the body 221 nodes and 75 elements were necessary, leading to a problem with $221 \times 3$ unknowns as quadratic interpolation functions were chosen. A calculation with only one subregion took 190 SCP (time in seconds spent in pure calculation) and 288 STT (total time including time for access to out of core information) on a CDC 7600 computer.

Figure 6.9 shows the deformation of the surface $y = 0$ on which the characteristic cask shape is visible. Comparisons between experimental results,

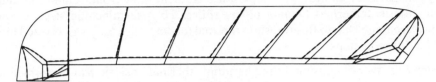

**Figure 6.9** Deformation of the surface $y = 0$. Full lines, initial mesh; dotted lines, mesh after thermal loading

results obtained by an FEM analysis (for which a special treatment was necessary in order to take into account the fact that the Poisson ratio is equal to 0.5), and results obtained by the BIE method are shown in Figure 6.10. The experimental results correspond to a point situated between the points referred as $I$ and $I + 1$.

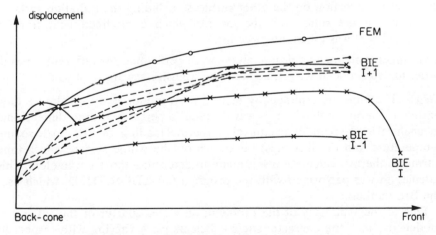

**Figure 6.10** Circumferential displacement. Full lines, results given by programs; dotted lines, experimental results

A second calculation with two subregions was performed. The geometry was discretized with 236 nodes and 85 elements, leading to $261 \times 3$ unknowns. In spite of a greater number of unknowns the computing time was reduced to 153 SCP and 229 STT on a CDC 7600 computer, and the calculation results were nearer the experimental results than with only one subregion.[9] This is due to the fact that the BIE method is most efficient for massive structures.

## 6.10 ANALYSIS OF A ROLLER IN CONTINUOUS CASTING OF STEEL SLABS

The stresses in a roller in continuous casting of steel slabs are known when the problem is a time harmonic one. But when there is a stop in the process, two problems arise: (a) the high temperature gradient in the roller under the slab; (b) transient thermal loading of the roller. To determine the stresses and displacements due to this transient thermal loading, the study was divided into four different steps

*Step 1.* A two-dimensional time harmonic thermal analysis was performed to determine the initial values of step 2. This calculation took into account the radiation conditions.

*Step 2.* A two-dimensional transient thermal analysis using the finite element method was performed with boundary conditions:

(a) given flux on the surfaces corresponding to the contact between the slab and the roller and corresponding to radiation from the slab to the roller (the temperature of the slab being high and constant enough to be considered as giving a constant flux);

(b) given convection on the other surfaces, including the radiation surfaces between the roller and the air for which a modified heat transfer coefficient was taken into account.

Two-dimensional mechanical analyses were performed for different times in order to determine at what time after the stop stresses are the higher.

*Step 3.* The time determined by step 2 corresponding to an almost time harmonic thermal behaviour, it was decided to perform a three-dimensional mechanical analysis under an equivalent thermal loading. In order to do so, the temperature and the flux must be known on the whole surface and a time harmonic thermal calculation was made to determine those parameters. This calculation was performed with the program CA.ST.OR.TH3D, which uses the BIE method.

Due to the symmetry of the problem, only one quarter of the roller was modelled. Since the characteristic coefficients were varying with respect to temperature, and as the gradient of temperature was very high under the slab, the roller was divided into four subregions (see Figures 6.11 and 6.12) each one having different characteristic coefficients.[10] The length of subregion 1 was corresponding to the length of the slab. Unfortunately, due to the temperature field, the shapes and positions of the subregions were the worse possible ones, and the mesh was composed of 206 elements and 509 nodes leading to 689 unknowns.

The boundary conditions were (see Figure 6.13):

(a) zero flux on the surfaces $x = 0$ and $z = 0$ (symmetry);

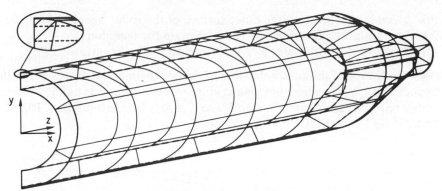

**Figure 6.11** Mesh of the roller

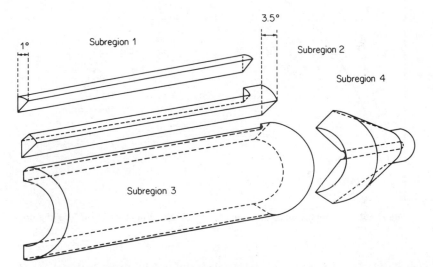

**Figure 6.12** Division into subregions of the roller

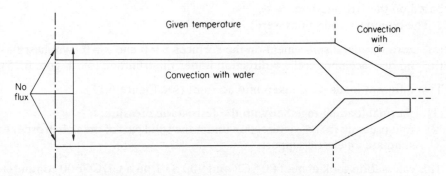

**Figure 6.13** Thermal boundary condition on the roller

(b)  given temperature on the outer surface of the roller (results of step 2);
(c)  heat transfer with the air by convection on the trunnion;
(d)  heat transfer with water by convection on the inner surface.

As a verification of the good behaviour of the approximation, the results on the upper surface $x = 0$, where the temperature was considered as unknown, were compared with results given by step 2 (see Figure 6.14). This run took 163 SCP and 235 STT on a CDC 7600 computer.

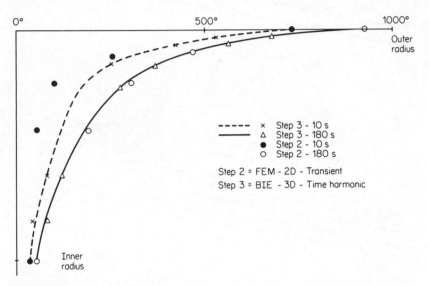

**Figure 6.14**   Comparison of the temperature obtained in steps 2 and 3 on the upper surface $x = 0$

*Step 4.* Under the thermal loading determined by step 3, the three-dimensional mechanical analysis was performed with the program CA.ST.OR.E3D based on the BIE method.

The boundary conditions were:

(a)  zero normal displacement on the surfaces $x = 0$ and $z = 0$ (symmetry);
(b)  no displacement in the $y$-direction under the trunnion (see Figure 6.15).

Two different loads were taken into account (see Figure 6.15):

(1)  thermal loading together with the ferrostatic pressure;
(2)  unit pressure in the $-y$-direction under the solid part of the slab in order to simulate an elastic support.

This calculation took about 600 SCP and 900 STT on a CDC 7600 computer. The results are obtained by combining cases (1) and (2). The coefficient by

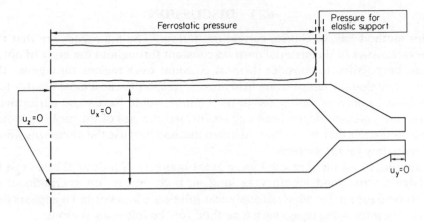

**Figure 6.15** Boundary conditions and mechanical loading of the roller. Reproduced by permission of FCB/Cetim

which to multiply case (2) is obtained by the following relation:

$$\frac{E}{l}(u_1 + \beta u_2) = p_1 + \beta p_2 \text{ (elastic support in the } y\text{-direction)}$$

where $E$ is Young's modulus of the slab at the corresponding temperature, $l$ is the half-thickness of the slab, and $p_1 = 0$ because this surface is not loaded in case (1).

The final deformation of the roller is shown in Figure 6.16, and the displacement that is obtained in the middle upper part of the roller is within 10% of the displacement observed on the plant.

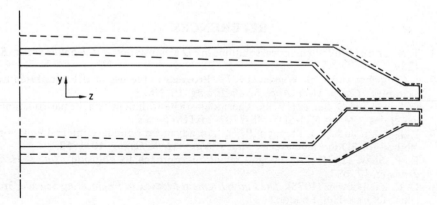

**Figure 6.16** Deformation of the surface $x = 0$. Full lines, initial mesh; dotted lines, mesh after loading. Reproduced by permission of FCB/Cetim

## 6.11 DISCUSSION

This method has, of course, certain limitations. The main ones are that the characteristics of the material must be constant throughout the body (if not, it must be possible to consider them as constant over regions throughout the body) and that the temperature field must correspond to an almost steady-state problem (otherwise an integral in the volume would have to be performed). Even if a volume integral has to be performed the size of the problem would remain smaller than with a finite element method because the unknowns would remain only on the surface.

In spite of all these restrictions, a great majority of industrial cases can be analysed with good accuracy as long as body forces are gravitational or rotational forces (for which an analytical solution is known) and it appears that it is very interesting to use such a method for the following reasons:

(a)    considerable savings in computing time and in input–output operations since only the surface of the domain is taken into account;
(b)    good accuracy of results is obtained since only the boundary conditions are approximate, the governing equations being verified exactly in the volume.

### Acknowledgments

The Société Nationale des Poudres et Explosifs are thanked for having given us the possibility to compare our results obtained with the BIE method, with their experimental and calculated results for the incompressible analysis.

The Fives Cail Babcock Society must be thanked for allowing us to publish the results of their study made at CETIM using our programs CA.ST.OR.TH3D and E3D based on the BIE method.

## REFERENCES

1.  T. A. Cruse (1969). 'Numerical solutions in 3D elastostatics', *Int. J. Solids Struct.*, **5**, 1259.
2.  J. C. Lachat and J. O. Watson (1977). 'Progress in the use of BIE illustrated by examples', *Comp. Meth. Appl. Mech. Engng*, **10**, No. 3.
3.  D. Lange and D. Serres (1978). 'Thermoelasticité tridimensionnelle par équations intégrales', *Report* NTI-SDS-MS-7702-CETIM Senlis.
4.  F. J. Rizzo and D. J. Shippy (1977). 'An advanced boundary integral equation method for 3D thermoelasticity', *Int. J. Num. Meth. Engng*, **10**, 1753.
5.  R. P. Shaw (1974). 'An integral equation approach to diffusion', *Int. J. Heat Transfer*, **17**, 693.
6.  O. C. Zienkiewicz (1978). *The Finite Element Method in Engineering Science*, 3rd edn, McGraw-Hill, London.
7.  J. C. Lachat and J. O. Watson (1976). 'Effective numerical treatment of BIE: a formulation for 3D elastostatics', *Int. J. Num. Meth. Engng*, **10**, 991.

8. A. Chaudouet and D. Lange (1978). 'The use of BIE method in mechanical engineering', *2nd Int. Conf. on Applied Numerical Modelling, Madrid, September 1978.*

9. A. Chaudouet (1978). 'Exemple de calculs tridimensionnels par la méthode des équations intégrales de frontière, de pièces en matériau incompressible soumises à des changements thermiques en régime stationnaire', *Report* NTI-CALCUL-78/002-CETIM Senlis.

10. T. Cruse and R. Wilson (1977). *Boundary Integral Equation Method for Elastic Fracture Mechanics Analysis*, p. 15, Pratt & Whitney Aircraft Group.

*Numerical Methods in Heat Transfer*
Edited by R. W. Lewis, K. Morgan, and O. C. Zienkiewicz
© 1981 John Wiley & Sons Ltd

*Chapter 7*

# On a Boundary Solution Approach for the Dynamic Problem of Thermoviscoelasticity Theory

*M. Predeleanu*

## SUMMARY

The basic integral representations of displacement and temperature fields are deduced for a general class of viscoelastic bodies defined by Riemann–Stieltjes integral convolutions subjected to both mechanical and thermal loading. Formulations are indicated with boundary data only. Two applications of these integral representations are presented for the boundary solution approach. The first is concerned with the boundary integral equation method and the second determines an 'interface' condition for the coupling of the boundary approach with other numerical and analytical methods.

## NOMENCLATURE

| | | | |
|---|---|---|---|
| $\sigma_{ik}$ | stress tensor | $Q$ | heat per unit volume and |
| $\varepsilon_{ik}$ | infinitesimal strain tensor | | unit time |
| $s_{ik}$ | stress deviator tensor | $\nabla^2$ | Laplace operator |
| $e_{ik}$ | strain deviator tensor | $\alpha_T$ | coefficient of linear thermal |
| $\delta_{ik}$ | Kroneker symbol | | expansion |
| $u_i$ | displacement vector | $h$ | Heaviside function |
| $q_i$ | heat flux | $G$ | shear elastic modulus |
| $S$ | entropy | $K$ | elastic bulk modulus |
| $T$ | temperature | $\rho$ | mass density |
| $T_0$ | reference temperature | $F_i$ | body force vector |
| $k$ | thermal conductivity | $n_i$ | unit normal vector |
| | | $\delta$ | Dirac distribution |

## 7.1 INTRODUCTION

Though in the mechanics of solids, more precisely, in linear elasticity theory, the first formulations by integral equations appeared nearly a century ago, this

135

approach became largely workable only when combined with two ideas: the first was to formulate these equations for boundary unknown variables only, and consequently to produce a formulation by boundary integral equations; and the second was to adopt a numerical treatment of these equations by discretization of the boundary into finite elements, geometrically simple, on which the elementary functions are assumed for the unknown variables. These two ideas have been borrowed from potential theory. Beginning with Rizzo,[1] who considered two-dimensional elasticity problems, the boundary integral equation method (BIEM) has been much developed for other problems in linear elasticity theory. Some computer programs were also realized.

The success of this method and the possibility of coupling it with other methods led to intense theoretical work in two directions:

(a)   integral equation formulations of boundary value problems in other fields of mechanics of solids;
(b)   the determination of singular fundamental solutions (Kelvin or Green type) for various classes of materials in quasi-static or dynamic regimes.

The present chapter is concerned with a general class of viscoelastic bodies defined by Riemann–Stieltjes integral convolutions subjected to both mechanical and thermal loading. For this reason, the thermodynamical approach of constitutive equations is used, in which the coupling of thermal and mechanical effects is taken into account. The inertial forces and general loading program, including body forces, heat sources and mixed boundary conditions, are considered to define the governing equations, presented in Section 7.2.

In Section 7.3, the integral representations of displacement field and temperature field are deduced; these are necessary for use in a boundary solution approach.

As is well known, the reciprocal relations of the Green type in potential theory and the Betti type in elasticity theory have been fundamental for the corresponding integral formulation. In the linear theory of viscoelasticity, an analogous reciprocity theorem, including the dynamic and thermal effects, has been given firstly by Predeleanu,[2] who deduced from it an integral representation of the Somigliana–Maisel type for displacement field. It is shown that such integral representations are also valid for coupled theory, and that in the absence of internal inputs (body forces and heat sources), a formulation only with boundary data is possible.

A variant of integral representations of the Green type is presented for the mixed boundary value problems by using fundamental singular solutions with homogeneous boundary conditions. Two applications of these integral representations are presented in Section 7.4. The first formulates the boundary integral equation method for the dynamic problem of thermoviscoelasticity which also contains the case of the quasi-static regime and the case of thermoelasticity theory. The second determines an 'interface' condition for the coupling of the boundary approach with other numerical and analytical methods.

## 7.2 BASIC EQUATIONS

Consider a deformable body, occupying a region $B$ with a regular boundary $\partial B$, in the three-dimensional Euclidian space, referred to a fixed rectangular Cartesian coordinate system $0x_i$, $i = 1, 2, 3$. $x_i$ will denote the coordinates of the point $x \in B$.

Let the time variable be $t$, denoting by $t$ the present time and by $\tau$ the values of $t$ for $\tau \leqslant t$. We assume, for brevity, that every particle of the body is at rest for $\tau < 0$ and that at $\tau = 0$ the body is in a free stress reference configuration. Let $\sigma_{ik}$ be the stress tensor, $\varepsilon_{ik}$ the infinitesimal strain tensor and $s_{ik}$, $e_{ik}$ the deviators, respectively defined by:

$$s_{ik} = \sigma_{ik} - \tfrac{1}{3}\delta_{ik}\sigma_{hh} \qquad e_{ik} = \varepsilon_{ik} - \tfrac{1}{3}\delta_{ik}\varepsilon_{hh} \tag{7.1}$$

where $\delta_{ik}$ is the Kroneker symbol. If $u_i(x, t)$ is the displacement vector, then:

$$\varepsilon_{ik} = \tfrac{1}{2}(u_{i,k} + u_{k,i}) \tag{7.2}$$

where

$$u_{i,k} = \partial u_i / \partial x_k. \tag{7.3}$$

A thermodynamical approach to the linear theory of coupled thermoviscoelasticity can be used to deduce the constitutive relations for stress, $\sigma_{ik}$, heat flux, $q_i$, and entropy difference, $S$, from a constant initial entropy.

Generally, by considering viscoelastic effects in the non-isothermal rheological behaviour of the body, the constitutive relations can be expressed by means of memory functionals defined as a suitable topological space of the histories of independent constitutive variables.

Following Christensen and Naghdi,[3] let the Helmholtz free energy $A$ and heat flux $q_{,i}$ be expressed by:

$$A = f_1[\varepsilon_{ik}(x, \tau), T(x, \tau)]_{\tau=0}^{\tau=t} \tag{7.4}$$

$$q_{,i} = f_2[T_{,i}(x, \tau)]_{\tau=0}^{\tau=t} \tag{7.5}$$

where $T$ is the temperature difference from a constant temperature $T_0$ at a stress-free configuration, $f_1$ a scalar-valued bilinear functional and $f_2$ a vector-valued linear functional.

By using polynomial integral representations of $f_1$ and $f_2$, the following constitutive relations are deduced for the linearized theory:

$$s_{ij} = 2G * de_{ij} \tag{7.6}$$

$$\sigma_{kk} = 3(K * d\varepsilon_{kk} - \varphi * dT) \tag{7.7}$$

$$\rho S = \varphi * d\varepsilon_{kk} + m * dT \tag{7.8}$$

$$q_i = kT_{,i} \tag{7.9}$$

where $G$, $K$, $\varphi$ and $m$ are appropriate memory functions, and $k$ is a constant, in the latter relation, as it results from the satisfaction of the dissipation inequality (Christensen[4]). As a consequence, the coupled heat conduction can be written:

$$k\nabla^2 T - T_0 \frac{\partial}{\partial t}(m * \mathrm{d}T + \varphi * \varepsilon_{kk}) + Q = 0 \qquad (7.10)$$

where $Q$ is the quantity of heat supplied per unit volume and unit time, and $\nabla^2$ is the Laplace operator. In these equations, $\varphi * \mathrm{d}\psi$ is the Stieltjes convolution of two real-valued functions $\varphi$ and $\psi$ defined by:

$$(\varphi * \mathrm{d}\psi)(t) = \int_{-\infty}^{t} \varphi(t-\tau)\,\mathrm{d}\psi(\tau) = \int_{-\infty}^{t} \psi(\tau)\,\mathrm{d}\varphi(t-\tau)$$

under appropriate conditions formulated, for instance, by Gurtin and Sternberg.[5]

*Special cases.* From Equations (7.6)–(7.9), the following special cases of the linear coupled thermoviscoelasticity theory can be obtained:

(a) $$\varphi(t) = 3\alpha_T K(t) \qquad (7.11)$$

where $\alpha_T$ is the time-independent coefficient of linear thermal expansion, and $K$ the bulk memory function. Then Equation (7.7) becomes:

$$\sigma_{kk} = 3K * \mathrm{d}(\varepsilon_{kk} - 3\alpha_T T) \qquad (7.12)$$

(b) $$\varphi(t) = \gamma h(t) \qquad m(t) = mh(t) \qquad (7.13)$$

where $m$, $\gamma$ are constants and $h$ is the Heaviside function. Then Equation (7.7) becomes

$$\sigma_{kk} = -3\gamma T + 3K * \mathrm{d}\varepsilon_{kk} \qquad (7.14)$$

and the coupled heat conduction can be written

$$\nabla^2 T - a\frac{\partial T}{\partial t} - b\frac{\partial \varepsilon_{kk}}{\partial t} + W = 0 \qquad (7.15)$$

where

$$a = \frac{T_0 m}{k} \qquad b = \frac{\gamma T_0}{k} \qquad W = \frac{Q}{k}$$

As was remarked by Eringen[6] who developed a linear thermoviscoelasticity theory by using the hypotheses (7.13), the difference between thermoelasticity and thermoviscoelasticity for isotropic solids occurs only in the stress constitutive equations.

(c)    The thermoelastic case can be obtained from the latter one by setting:

$$G(t) = Gh(t) \qquad K(t) = Kh(t)$$

where $G$ is the shear elastic modulus and $K$ the elastic bulk modulus. Equations (7.1)–(7.10) must be completed by the stress field equations of motion:

$$\sigma_{ij,j} + F_i = \rho \frac{\partial^2 u_i}{\partial t^2} \tag{7.16}$$

where $\rho$ is the mass density, $F_i$ body force vector and

$$\sigma_{ij} = 2G * d\varepsilon_{ij} + \delta_{ij}(\lambda * d\varepsilon_{hh} - \varphi * dT) \tag{7.17}$$

with $\lambda(t) = K(t) - \frac{2}{3}G(t)$.

If it is assumed that the body is originally undisturbed, the initial conditions will be:

$$T(t) = u_i(t) = \frac{\partial u_i(t)}{\partial t} = \sigma_{ij}(t) = 0 \quad \text{on } \bar{B}x(-\infty, 0) \tag{7.18}$$

The mixed boundary conditions are taken as:

$$\sigma_{ij}n_j = \hat{\sigma}_{ni}(x, t) \quad \text{on } \partial B_\sigma x[0, \infty) \qquad u_i = \hat{u}_i(x, t) \quad \text{on } \partial B_u x[0, \infty) \tag{7.19}$$

$$T = \hat{T}(x, t) \quad \text{on } \partial B_T x[0, \infty) \qquad kT_{,i}n_i = -\hat{q}_n(x, t) \quad \text{on } \partial B_q x[0, \infty) \tag{7.20}$$

where

$$\partial B_\sigma \cup \partial B_u = \partial B \qquad \partial B_T \cup \partial B_q = \partial B$$

with

$$\partial B_\sigma \cap \partial B_u = \varnothing \qquad \partial B_T \cap \partial B_q = \varnothing$$

$\hat{\sigma}_{ni}$, $\hat{u}_i$, $\hat{T}$, $\hat{q}_n$ represent prescribed surface functions, and $n_i$ is the unit normal vector.

In the first boundary value problem $\partial B_\sigma$ is empty, and in the second $\partial B_u$ is empty.

## 7.3   INTEGRAL REPRESENTATIONS

A direct method to deduce the integral equations of the dynamic problem of thermoviscoelasticity theory, which generalizes Somigliana's results in elasticity theory, is to use Betti's reciprocal theorem. It is interesting to note that this theorem can be expressed in an analogous form for different classes of linear constitutive relations.

### 7.3.1   Generalized reciprocal theorem

The evolution of the body subjected to two loading programs

$$\{F_i, Q, \hat{\sigma}_{ni}, \hat{u}_i, \hat{T}, \hat{q}_n\} \quad \text{and} \quad \{F'_i, Q', \hat{\sigma}'_{ni}, \hat{u}'_i, \hat{T}', \hat{q}'_n\}$$

satisfies the reciprocity relations between the corresponding field variables $\{u_i, T\}$ and $\{u_i', T'\}$. Such relations can be obtained directly by using the known results on Stieltjes convolutions or by introducing the Laplace transform space.

Let

$$\bar{f}(x, p) = \int_0^\infty e^{-pt} f(x, t) \, dt \qquad (7.21)$$

be the Laplace transform of a numerical function $f(x, t)$. From constitutive and motion equations and Gauss theorem, the following reciprocity relation can be deduced:

$$\int_B (\bar{F}_i \bar{u}_i' - \bar{F}_i' \bar{u}_i) \, dv + p\bar{\varphi} \int_B (\bar{T}\bar{\varepsilon}_{hh}' - \bar{T}'\bar{\varepsilon}_{hh}) \, dv + \int_{\partial B} (\bar{\sigma}_{ni}\bar{u}_i' - \bar{\sigma}_{ni}'\bar{u}_i) \, ds = 0 \qquad (7.22)$$

which is available also in the non-coupled theory. By using the coupled heat conduction equation, one obtains:

$$p^2 T_0 \bar{\varphi} \int_B (\bar{\varepsilon}_{hh}\bar{T}' - \bar{\varepsilon}_{hh}'\bar{T}) \, dv - \int_B (\bar{Q}\bar{T}' - \bar{Q}'\bar{T}) \, dv + \int_{\partial B} (\bar{T}'\bar{q}_n - \bar{T}\bar{q}_n') \, ds = 0 \qquad (7.23)$$

Therefore, in the Laplace transform space, the following Betti type reciprocity relation is given:

$$\int_B (\bar{F}_i \bar{u}_i' - \bar{F}_i' \bar{u}_i) \, dv + \int_{\partial B} (\bar{\sigma}_{ni}\bar{u}_i' - \bar{\sigma}_{ni}'\bar{u}_i) \, ds$$

$$= \frac{1}{pT_0} \int_B (\bar{Q}\bar{T}' - \bar{Q}'\bar{T}) \, dv + \frac{1}{pT_0} \int_{\partial B} (\bar{T}\bar{q}_n' - \bar{T}'\bar{q}_n) \, ds \qquad (7.24)$$

For brevity, the prescribed quantities on the boundary are not specified explicitly in relations (7.20)–(7.24).

It is worth noting that relations (7.22) and (7.24) do not contain the inertia terms and their form is the same in the quasi-static case. These relations are also analogous to those deduced in elasticity theory by Ionescu-Cazimir.[7] A generalized reciprocal relation in anisotropic viscoelasticity has been obtained by Iesan.[8]

### 7.3.2 Integral representations of the Somigliana type

By choosing the displacement field $u_i$ and the temperature field $T$, $\{u_i, T\}$, as unknown fields which define the evolution of the body, the corresponding coupled field equations can be written:

$$G * d(\nabla^2 u_i) + (\lambda + G) * d(u_{i,i}) - \rho \frac{\partial^2 u}{\partial t^2} - \varphi * d(T_{,i}) + F_i = 0 \qquad (7.25)$$

$$k\nabla^2 T - T_0 \frac{\partial}{\partial t} [m * dT + \varphi * d(u_{i,i})] + Q = 0 \qquad (7.26)$$

or, for the Laplace transformed fields $\{\bar{u}_i, \bar{T}\}$, the Laplace transform equations:

$$p\bar{G}\nabla^2\bar{u}_i + p(\bar{\lambda} + \bar{G})\bar{u}_{i,i} - \rho p^2\bar{u}_i - p\bar{\varphi}\bar{T}_{,i} + \bar{F}_i = 0 \qquad (7.27)$$

$$k\nabla^2\bar{T} - T_0 p^2(\bar{m}\bar{T} + \bar{\varphi}\bar{u}_{i,i}) + \bar{Q} = 0. \qquad (7.28)$$

Laplace transforms of initial conditions and boundary conditions (7.18), (7.19) must be added to the partial differential equations (Equations (7.27), (7.28)).

To deduce integral representations analogous to Somigliana's results for the displacement field $u_i$ and the temperature field $T$, the reciprocal relations (7.24) will be used in two different cases.

Firstly, consider $\{U_i^k, \theta^k\}$ a fundamental solution of the Equations (7.25), (7.26), regular at infinity, with homogeneous initial conditions, for the loading:

$$F_i'(x, \xi, t) = \delta(t)\delta(x - \xi)\delta_{ki}Q'(x, t) = 0 \qquad (7.29)$$

where $\delta$ is the Dirac distribution. This problem is the counterpart in thermoviscoelasticity theory of Kelvin's problem in elastostatics, namely the problem of a concentrated load in the $x_k$-direction applied at a point $\xi$ of a viscoelastic medium that occupies the entire space. From (7.24), the following integral representation for the displacement field results:

$$\bar{u}_k(\xi, p) = \int_B \bar{F}_i(x, p)\bar{U}_i^k(x, \xi, p)\, dv(x)$$

$$+ \int_{\partial B} [\bar{\sigma}_{ni}(x, p)\bar{U}_i^k(x, \xi, p) - \bar{A}_{ni}^k(x, \xi, p)\bar{u}_i(x, p)]\, ds(x)$$

$$- \frac{1}{pT_0}\int_B \bar{Q}(x, p)\bar{\theta}^k(x, \xi, p)\, dv(x) - \frac{1}{pT_0}\int_{\partial B} [\bar{T}(x, p)\bar{C}_n^k(x, \xi, p)$$

$$- \bar{\theta}^k(x, \xi, p)\bar{q}_n(x, p)]\, ds(x) \qquad (7.30)$$

where $A_{ij}^k$ is the stress tensor determined by $U_i^k$, and:

$$A_{ni}^k = \bar{A}_{ij}^k n_j \qquad C_n^k = -k\theta_{,i}^k n_i$$

Secondly, consider $\{V_i, \chi\}$ a fundamental solution of the Equations (7.25), (7.26), regular at infinity, with homogeneous initial conditions, for the loading:

$$F'(x, t) = 0 \qquad Q'(x, \xi, t) = \delta(t)\delta(x - \xi) \qquad (7.31)$$

From (7.24), the following integral representation for the temperature field results:

$$\bar{T}(\xi, p) = \int_B \bar{Q}(x, p)\bar{\chi}(x, \xi, p)\, dv(x) - pT_0\int_B \bar{F}_i(x, p)\bar{V}_i(x, \xi, p)\, dv(x)$$

$$- pT_0\int_{\partial B} [\bar{\sigma}_{ni}(x, p)\bar{V}_i(x, \xi, p) - \bar{B}_{ni}(x, \xi, p)\bar{u}_i(x)]\, ds(x)$$

$$+ \int_{\partial B} [\bar{T}(x, p)\bar{d}_n(x, \xi, p) - \bar{\chi}(x, \xi, p)\bar{q}_n(x, p)]\, ds(x) \qquad (7.32)$$

where $B_{ij}$ is the stress tensor determined by $V_i$ and $d_n = -k\chi_{,i}n_i$.

The relations (7.30) and (7.32) may be considered as the generalized Somigliana identity for the dynamic problems of coupled thermoviscoelasticity theory. Their form remains the same for the quasistatic problem and also for every class of viscoelastic constitutive laws, e.g. for the one considered by Boschi[9] and also for elastic media Ionescu-Cazimir,[10] Nowacki.[11] The difference is in the generalized fundamental Kelvin functions $U_i^k$, $\theta^k$, $V_i$, $\chi$ which must be determined as solutions of the corresponding equations to (7.25), (7.26) for the problem at hand.

By inversion of the Laplace transform performed on the relations (7.30) and (7.32), integral representations for $U_i^k(\xi, t)$ and $T(\xi, t)$ can be obtained.

### 7.3.3 Integral representation of the Green type

Important simplifications of the formulae (7.30) and (7.32) can be obtained if singular solutions of the Green type are used for Equations (7.25), (7.26). Various consequences of the integral representations can be deduced, but it depends on the type of the boundary value problem at hand. Thus, it will be pointed out:

(a)  For the first fundamental boundary value problem (Dirichlet type), in which the surface displacements and the temperature are prescribed over the entire boundary $\partial B$ for all time, the fundamental Green's solution $U_i^k$, $\theta^k$ of Equations (7.25), (7.26) for the loading (7.29) will be introduced, with homogeneous initial conditions and the following boundary conditions:

$$U_i^k(x, \xi, t) = \theta^k(x, \xi, t) = 0 \quad \text{on } \partial B x[0, \infty) \qquad (7.33)$$

From (7.30):

$$\bar{u}_k(\xi, p) = \int_B \bar{F}_i(x, p)\bar{U}_i^k(x, \xi, p)\, dv - \frac{1}{pT_0}\int_B \bar{Q}(x, p)\bar{\theta}^k(x, \xi, p)\, dv(x)$$

$$- \int_{\partial B} \bar{A}_{ni}^k(x, \xi, p)\hat{\bar{u}}_i(x, p)\, ds(x) - \frac{1}{pT_0}\int_{\partial B} \hat{\bar{T}}(x, p)C_n^k(x, \xi, p)\, ds(x) \qquad (7.34)$$

Functions prescribed on the boundary are indicated by a carat $(\hat{\ })$ above the character.

Similarly, if the fundamental Green's solutions $V_i$, $\chi$ of Equations (7.25), (7.26) for the loading (7.31) are introduced, with homogeneous initial conditions and the following boundary conditions:

$$V_i(x, \xi, t) = \chi(x, \xi, t) = 0 \quad \text{on } \partial B x[0, \infty) \qquad (7.35)$$

from (7.32):

$$\bar{T}(\xi, p) = \int_B \bar{Q}(x, p)\bar{\chi}(x, \xi, p)\, dv(x) - pT_0 \int_B \bar{F}_i(x, p)\bar{V}_i(x, \xi, p)\, dv(x)$$

$$+ pT_0 \int_{\partial B} \bar{B}_{ni}(x, \xi, p)\hat{\bar{u}}_i(x, p)\, ds(x) + \int_{\partial B} \hat{\bar{T}}(x, p)\bar{d}_n(x, \xi, p)\, ds(x)$$

$$\tag{7.36}$$

(b)   For the second fundamental boundary value problem (Newmann type), in which surface tractions and the temperature are prescribed over the entire boundary $\partial B$ for all time, the fundamental Green's solutions $U_i$, $\theta^k$ of Equations (7.25), (7.26) for the loading (7.29) will be introduced, with homogeneous initial conditions and the following boundary conditions:

$$A_{ni}^k(x, \xi, t) = \theta^k(x, \xi, t) = 0 \quad \text{on } \partial Bx[0, \infty) \tag{7.37}$$

From (7.30):

$$\bar{u}_k(\xi, p) = \int_B \bar{F}_i(x, p)\bar{U}_i^k(x, \xi, p)\, dv(x) - \frac{1}{pT_0}\int_B \bar{Q}(x, p)\bar{\theta}^k(x, \xi, p)\, dv(x)$$

$$+ \int_{\partial B} \hat{\bar{\sigma}}_{ni}(x, p)\bar{U}_i^k(x, \xi, p)\, ds(x) - \frac{1}{pT_0}\int_{\partial B} \hat{\bar{T}}(x, p)C_n^k(x, \xi, p)\, ds(x)$$

$$\tag{7.38}$$

Similarly, if the fundamental Green's solutions $V_i$, $\chi$ of Equations (7.25), (7.26) for the loading program (7.31) are used, with homogeneous initial conditions and homogeneous boundary conditions:

$$B_{ni}(x, \xi, t) = \chi(x, \xi, t) = 0 \quad \text{on } \partial Bx[0, \infty) \tag{7.39}$$

from (7.32):

$$\bar{T}(\xi, p) = \int_B \bar{Q}(x, p)\bar{\chi}(x, \xi, p)\, dv(x) - pT_0 \int_B \bar{F}_i(x, p)\bar{V}_i(x, \xi, p)\, dv(x)$$

$$- pT_0 \int_{\partial B} \hat{\bar{\sigma}}_{ni}(x, p)V_i(x, \xi, p)\, ds(x) + \int_{\partial B} \hat{\bar{T}}(x, p)\bar{d}_n(x, \xi, p)\, ds(x)$$

$$\tag{7.40}$$

It should be noted that analogous integral representations can be easily deduced from (7.30) and (7.32) by using Green's functions, e.g. for the displacement field and heat flux field or for the stress field and heat flux field. The choice of the type of homogeneous boundary conditions is depending on the boundary value problem at hand.

(c)   For the mixed boundary value problem defined by Equations (7.19), some more significant cases only will be considered; the other cases can easily be deduced from the general formulae (7.30) and (7.32).

Thus, if the Green's solutions $\{U_i^k, \theta^k\}$ of Equations (7.25), (7.26) for the loading (7.29) are used, with homogeneous initial conditions and the following homogeneous mixed boundary conditions:

$$U_i^k(x, \xi, t) = 0 \quad \text{on } \partial B_u x[0, \infty)$$
$$A_{ni}^k(x, \xi, t) = 0 \quad \text{on } \partial B_\sigma x[0, \infty)$$
$$\theta^k(x, \xi, t) = 0 \quad \text{on } \partial B_T x[0, \infty) \tag{7.41}$$
$$C_n^k(x, \xi, t) = 0 \quad \text{on } \partial B_q x[0, \infty)$$

then the displacement vector is deduced from Equation (7.30):

$$\bar{u}_k(\xi, p) = \int_B \bar{F}_i(x, p) \bar{U}_i^k(x, \xi, p) \, dv(x) - \frac{1}{pT_0} \int_B \bar{Q}(x, p) \bar{\theta}^k(x, \xi, p) \, dv(x)$$

$$+ \int_{\partial B_\sigma} \hat{\bar{\sigma}}_{ni}(x, p) \bar{U}_i^k(x, \xi, p) \, ds(x) - \int_{\partial B_u} \bar{A}_{ni}(x, \xi, p) \hat{\bar{u}}_i(x, p) \, ds(x)$$

$$- \int_{\partial B_T} \hat{\bar{T}}(x, p) \bar{C}_n^k(x, \xi, p) \, ds(x) + \int_{\partial B_q} \bar{\theta}^k(x, \xi, p) \hat{\bar{q}}_n(x, p) \, ds(x) \tag{7.42}$$

Similarly, if the Green's solutions $\{V_i, \chi\}$ of Equations (7.25), (7.26) for the loading program (7.31) are used, with homogeneous initial conditions and the following homogeneous mixed boundary conditions:

$$V_i(x, \xi, t) = 0 \quad \text{on } \partial B_u x[0, \infty)$$
$$B_{ni}(x, \xi, t) = 0 \quad \text{on } \partial B_\sigma x[0, \infty)$$
$$\chi(x, \xi, t) = 0 \quad \text{on } \partial B_T x[0, \infty) \tag{7.43}$$
$$d_n(x, \xi, t) = 0 \quad \text{on } \partial B_q x[0, \infty)$$

then from (7.32), the temperature field is obtained:

$$\bar{T}(\xi, p) = \int_B \bar{Q}(x, p) \bar{\chi}(x, \xi, p) \, dv(x) - pT_0 \int_B \bar{F}_i(x, p) \bar{V}_i(x, \xi, p) \, dv(x)$$

$$- pT_0 \int_{\partial B_\sigma} \hat{\bar{\sigma}}_{ni}(x, p) \bar{V}_i(x, \xi, p) \, ds(x) + \int_{\partial B_u} \bar{B}_{ni}(x, \xi, p) \hat{\bar{u}}_i(x, p) \, ds(x)$$

$$+ \int_{\partial B_T} \hat{\bar{T}}(x, p) \bar{d}_n(x, \xi, p) \, ds(x) - \int_{\partial B_q} \bar{\chi}(x, \xi, p) \hat{\bar{q}}_n(x, p) \, ds(x) \tag{7.44}$$

The integral representation of Maisel type can be obtained from (7.42) and (7.44), by setting:

$$\hat{\bar{\sigma}}_{ni}(x, p) = 0 \quad \text{on } \partial B_\sigma$$
$$\hat{\bar{u}}_i(x, p) = 0 \quad \text{on } \partial B_u \tag{7.45}$$

If no body forces and no heat sources are acting on the body, then from (7.42) and (7.44):

$$\bar{u}_k(\xi, p) = -\int_{\partial B_T} \hat{\bar{T}}(x, p) C_n^k(x, \xi, p)\, ds(x) + \int_{\partial B_q} \bar{\theta}^k(x, \xi, p)\hat{\bar{q}}_n(x, p)\, ds(x)$$

$$\bar{T}(\xi, p) = \int_{\partial B_T} \hat{\bar{T}}(x, p)\bar{d}_n(x, \xi, p)\, ds(x) - \int_{\partial B_q} \bar{\chi}(x, \xi, p)\hat{\bar{q}}_n(x, p)\, ds(x)$$

(7.46)

If $\partial B_q$ is empty or $\bar{q}_n(x, p) = 0$ on $\partial B_q$:

$$\bar{u}_k(\xi, p) = -\int_{\partial B} \hat{\bar{T}}(x, p)\bar{C}_n^k(x, \xi, p)\, ds(x) \tag{7.47}$$

$$\bar{T}(\xi, p) = \int_{\partial B} \hat{\bar{T}}(x, p)\bar{d}_n(x, \xi, p)\, ds(x) \tag{7.48}$$

It was shown that, for appropriate definitions of the Green's functions, it is possible to express the displacement and temperature fields exclusively in terms of prescribed functions by volume integrals and surface integrals. The resulting formulae can be viewed as generalizations of well known Green's formulae in potential theory.

## 7.4  APPLICATIONS TO BOUNDARY SOLUTION APPROACH

### 7.4.1  Boundary integral equation method

As is well known, an important application of the integral representation theorems of Somigliana or Green type is the formulation of the method for solving continuum media problems by reducing the field equations to the appropriate boundary integral equations.

This approach became very fruitful once combined with the idea of the numerical treatment of integral equations by discretization of the boundary of the body into boundary elements on which the unknown functions are taken in an elementary simple form. This method was first applied and developed for elastostatic problems (Rizzo[1]), and it was then generalized to elastodynamic laws (Cruse and Rizzo[12,13]), to quasi-static problems of viscoelasticity and uncoupled steady-state thermoelasticity (Rizzo and Shippy[14,15]), and to elasticplastic flow (Swedlow and Cruse,[16] Bui[17]).

This chapter contains the necessary elements to apply the BIEM to the coupled dynamic or quasi-static problem of thermoviscoelasticity.

Indeed, the boundary integral equations can be obtained from integral representations of Somigliana or Green type deduced in Section 7.3, if the point $\xi$ inside $B$ approaches an arbitrary point on $\partial B$.

If fundamental Kelvin functions are used, then, from (7.30)–(7.32), the following simultaneous boundary integral equations for $u_k$ and $T$ are obtained for general mixed boundary value problems defined by Equations (7.19):

$$\bar{u}_k(y,p) + {}^*\!\!\int_{\partial B_\sigma} \bar{A}^k_{ni}(x,y,p)\bar{u}_i(x,p)\,\mathrm{d}s(x) - {}^*\!\!\int_{\partial B_u} \bar{U}^k_i(x,y,p)\sigma_{ni}(x,p)\,\mathrm{d}s(x)$$

$$+ {}^*\!\!\int_{\partial B_q} \bar{C}^k_n(x,y,p)\bar{T}(x,p)\,\mathrm{d}s(x) + k {}^*\!\!\int_{\partial B_T} \bar{\theta}^k(x,y,p)\bar{T}_{,n}(x,p)\,\mathrm{d}s(x)$$

$$= \bar{f}^k(y,p) \quad (7.49)$$

$$\bar{T}(y,p) - {}^*\!\!\int_{\partial B_\sigma} \bar{B}_{ni}(x,y,p)\bar{u}_i(x,p)\,\mathrm{d}s(x) + {}^*\!\!\int_{\partial B_u} \bar{V}_i(x,y,p)\sigma_{ni}(x,p)\,\mathrm{d}s(x)$$

$$- {}^*\!\!\int_{\partial B_q} \bar{d}_n(x,y,p)\bar{T}(x,p)\,\mathrm{d}s(x) - k {}^*\!\!\int_{\partial B_T} \bar{\chi}(x,y,p)T_{,n}(x,p)\,\mathrm{d}s(x) = \bar{g}(y,p)$$

$$(7.50)$$

where $\bar{f}^k$ and $\bar{g}$ are known functions given by:

$$\bar{f}^k(y,p) = \int_B \bar{U}^k_i(x,y,p)\bar{F}_i(x,p)\,\mathrm{d}v(x) - \frac{1}{pT_0}\int_B \bar{\theta}^k(x,y,p)\bar{Q}(x,p)\,\mathrm{d}v(x)$$

$$+ {}^*\!\!\int_{\partial B_\sigma} \hat{\bar{\sigma}}_{ni}(x,p)\bar{U}^k_i(x,y,p)\,\mathrm{d}s(x)$$

$$- {}^*\!\!\int_{\partial B_u} A^k_{ni}(x,y,p)\hat{\bar{u}}_i(x,p)\,\mathrm{d}s(x)$$

$$- {}^*\!\!\int_{\partial B_T} \bar{C}^k_n(x,y,p)\hat{\bar{T}}(x,p)\,\mathrm{d}s(x) + {}^*\!\!\int_{\partial B_q} \bar{\theta}^k(x,y,p)\hat{\bar{q}}(x,p)\,\mathrm{d}s(x)$$

$$(7.51)$$

$$\bar{g}(y,p) = -pR_0 \int_B \bar{V}_i(x,y,p)\bar{F}_i(x,p)\,\mathrm{d}v(x) + \int_B \bar{\chi}(x,y,p)\bar{Q}(x,p)\,\mathrm{d}s(x)$$

$$- pT_0 {}^*\!\!\int_{\partial B_\sigma} V_i(x,y,p)\hat{\bar{\sigma}}_{ni}(x,p)\,\mathrm{d}s(x)$$

$$+ {}^*\!\!\int_{\partial B_{ni}} \bar{B}_{ni}(x,y,p)\hat{\bar{u}}_i(x,p)\,\mathrm{d}s(x)$$

$$+ {}^*\!\!\int_{\partial B_T} \bar{d}_n(x,y,p)\hat{\bar{T}}(x,p)\,\mathrm{d}s(x) - {}^*\!\!\int_{\partial B_q} \bar{\chi}(x,y,p)\hat{\bar{q}}_n(x,p)\,\mathrm{d}s(x)$$

$$(7.52)$$

In these relations, the following limit was denoted by the integral symbol with an asterisk:

$${}^*\!\!\int_{\partial B} \bar{\Omega}(x,y,p)\bar{\omega}(x,p)\,\mathrm{d}s(x) = \lim_{\xi\to y}\int \bar{\Omega}(x,\xi,p)\bar{\omega}(x,p)\,\mathrm{d}s(x) \quad (7.53)$$

where $\xi \in B$ and $y \in \partial B$.

For certain singular kernel $\bar{\Omega}$ and regularity conditions for the density $\bar{\omega}$ and the boundary $\partial B$, some additional terms (jumps), of type $\eta\bar{\omega}(y, p)$, can be separated from the singular integral, understood in the sense of the Cauchy principal value (see, for instance, Kudpradze,[18] for Kelvin kernels in elasticity).

Important simplifications of the integral equations can be obtained by using special Green's fundamental functions, instead of Kelvin's functions.

### 7.4.2 Coupling of boundary solution approach with other solution methods

The coupling of BIEM with the finite elements method (FEM) or finite difference methods (FDM) uses the most efficient features from each of these methods. It is known, for instance, that FEM is difficult to adapt for infinite domains, costly in computing time for large domains, and introduces supplementary approximations for problems with singularities, as in fracture mechanics. For all these cases, BIEM is more convenient, and the applications considered will now prove its efficiency. On the other hand, FEM and FDM are very workable for bounded domains or local regions of a body governed by complicated evolution equations, eventually involving material non-homogeneity, physical or geometrical non-linearity, etc. For such problems, BIEM may not be applicable.

Consider the region $B$ occupied by a body, large or unbounded, divided into a bounded (internal) region $B_1$, with the boundary $\Gamma$, and $B_2$, the complement of $B_1$. It can be supposed that in $B_1$ and $B_2$, either *different* governing equations of the body are available, or the *same* field equations are appropriate for the two parts of the body $B$. In this latter case, an arbitrary boundary $\Gamma$ can be introduced to separate the two parts $B_1$ and $B_2$. If in $B_1$, FEM or FDM or other analytical or numerical methods of solution are used, the effects of $B_2$ on $B_1$ must be included in a 'interface condition' on the common boundary $\Gamma$. For instance, if $u_i^{(1)}$ and $u_i^{(2)}$ are respectively the displacement vectors of the points belonging respectively to $B_1$ and $B_2$ it may be required that:

$$u_i^{(1)}(y, t) = u_i^{(2)}(y, t) \quad (y, t) \in \Gamma \times [0, \infty) \tag{7.54}$$

Other interface conditions can be assumed too. It depends on the local behaviour on the boundary $\Gamma$. If $\Gamma$ is a natural boundary between two different media, different interface conditions on stress tensors can be assumed.

Referring to the interface condition (7.54), a *boundary approach* is recommended to take into account the action of $B_2$ on $B_1$, including thus the 'distance effects' for the infinite regions. Let it be assumed that such an approach is possible, via integral representation theorems obtained for linear equations, also considered in this chapter. Two possibilities are to be used:

(a)   One is given by the integral representations of Somigliana type, in which the values of $u_i^{(2)}(\Gamma, t)$ are coupled by a boundary integral equation, as for instance given by (7.29) and (7.31). This boundary integral equation will be used, for instance, in FEM as a natural boundary condition for the variational formulation of the problem to be solved in $B_1$. Obviously, as was remarked by Shaw,[19] *the coupling of the* $u_i^{(2)}(\Gamma, t)$ *by a global condition on* $\Gamma$ *can introduce new numerical problems in solving the resulting algebraic equations system.* This approach has been discussed recently by Zienkiewicz, Kelly and Bettess,[20] Brezzi, Johnson and Nedelec,[21] and Brebbia and Georgiu.[22]

(b)   The second possibility is the use of the Green, or Maisel, type integral representations deduced in Section 7.3. This avoids difficulties connected with the presence of differential terms in $u_i^{(2)}$ (traction kernels), which arise when integral representations of Somigliana type are used. Thus, let us consider the following mixed boundary valued problem defined on $B_2$:

$$u_k^{(2)}(y, t) = u_k^{(1)}(y, t) \quad (y, t) \in \Gamma x[0, \infty)$$

$$T^{(2)}(y, t) = T^{(1)}(y, t) \quad (x, t) \in \Gamma x[0, \infty)$$

$$\sigma_{ni}^{(2)}(y, t) = 0 \qquad\qquad (y, t) \in \partial Bx[0, \infty) \tag{7.55}$$

$$q_n^{(2)}(y, t) = 0 \qquad\qquad (y, t) \in \partial BX[0, t)$$

A suitable boundary solution approach for $B_2$ is to use the integral representation of Green type given by (7.42) and (7.44), making $\xi \to y \in \Gamma$. Then, on $\Gamma$, $u_k^{(1)}$ and $T^{(1)}$ must satisfy the boundary integral conditions:

$$\bar{u}_k^{(1)}(y, p) = \int_{B_2} \bar{F}_i(x, p)\bar{U}_i^k(x, y, p)\, dv(x) - \frac{1}{pT_0}\int_{B_2} \bar{Q}(x, p)\theta^k(x, \xi, p)\, dv(x)$$

$$- \overset{*}{\int}_\Gamma \bar{A}_{ni}(x, y, p)\bar{u}_k^{(1)}(x, p)\, ds(x) + \overset{*}{\int}_\Gamma \bar{T}^{(1)}(x, p)\bar{C}_n^k(x, y, p)\, ds(x) \tag{7.56}$$

$$\bar{T}^{(1)}(y, p) = \int_{B_2} \bar{Q}(x, p)\bar{\chi}(x, \xi, p)\, dv(x) - pT_0\int_{B_2} \bar{F}_i(x, p)\,\bar{V}_i(x, \xi, p)\, dv(x)$$

$$+ \overset{*}{\int}_\Gamma \bar{B}_{ni}(x, y, p)\bar{u}_k^{(1)}(x, p)\, ds(x) - \overset{*}{\int}_\Gamma T^{(1)}(x, p)\bar{d}_n(x, y, p)\, dv(x) \tag{7.57}$$

Obviously, the same relations are valuable if a part of $B$ is allowed to go to infinity. It is interesting to note that if $Q = 0$, $F_i = 0$, then the integral equations contain only boundary quantities.

### 7.5   SOME REMARKS

(a)   The integral relations presented in this paper are concerned with a coupled theory of thermoviscoelasticity, and some of these generalize the results for uncoupled problems of thermoviscoelasticity obtained by Pre-deleanu.[2] Thus, if the constitutive relations (7.11) are used, and the body is subjected to a prescribed temperature field $T^{(2)}(x, t)$, $(x, t) \in B_2 x[0, \infty)$, then the corresponding interface condition (7.54) becomes:

$$\bar{u}_k^{(1)}(y, p) = \int_{B_2} \bar{F}_i(x, p) \bar{U}_i^k(x, y, p) \, dv(x)$$

$$- \int_{\Gamma}^{*} \bar{A}_{ni}(x, y, p) \bar{u}_k^{(1)}(x, p) \, ds(x)$$

$$+ 3\alpha p \bar{K}(p) \int_{B_2} \bar{U}_{i,i}^k(x, y, p) \bar{T}^{(2)}(x, p) \, ds(x) \qquad (7.58)$$

where $\bar{U}_i^k$ are Green's fundamental solutions of Equation (7.25) for the loading:

$$F_i(x, \xi, t) = \delta(t) \, \delta(x - \xi) \delta_{ki} \qquad T = 0 \qquad (7.59)$$

with homogeneous initial conditions and the following mixed boundary conditions:

$$\bar{U}_i^k(y, \xi, p) = 0 \quad \text{on } \Gamma$$
$$\bar{A}_{ni}^k(y, \xi, p) = 0 \quad \text{on } \partial B \qquad (7.60)$$

As in (7.55), it was assumed that $\sigma_{ni}^{(2)}(y, t) = 0$ on $\partial B$. Other similar relations can be deduced too for the non-coupled theory.

(b) As it has been already remarked, the integral representation and consequently integral equations have been expressed, in the absence of internal inputs (body forces and thermal sources), only in terms of the boundary data, and hence discretization for numerical treatment is necessary only for body surfaces. Consequently, the calculations are simplified. As was deduced by Rizzo and Shippy,[15] a similar approach is possible for the non-coupled theory of quasi-static thermoelasticity, but an analogous result can be deduced for viscoelasticity theory.

(c) The Laplace transform with respect to time used in this paper permits the deduction of Kelvin's and Green's fundamental solutions by using the associated elastic problem to the viscoelastic one.

On the other hand, singular fundamental solutions known for the non-coupled problem can be used for coupled problems by using the method proposed by Ionescu-Cazimir.[10]

Nevertheless, further work will again be necessary to complete the list of fundamental solutions for dynamic problems of thermoviscoelasticity, where it will also be necessary to use the boundary solution approach suitably.

## REFERENCES

1. F. J. Rizzo (1967). 'An integral equation approach to boundary value problems of classical elastostatics', *Q. Appl. Math.*, **25**, 83–95.
2. M. Predeleanu (1959). 'On thermal stresses in viscoelastic bodies', *Bull. Math. Soc. Sci. Math. Phys. RPR*, **3** (**51**), No. 2, 223–228.
3. R. M. Christensen and P. M. Naghdi (1967). 'Linear non-isothermal viscoelastic solids', *Acta Mech.*, **3**, No. 1, 1–12.
4. R. M. Christensen (1971). *Theory of Viscoelasticity*, Academic Press, New York.
5. M. E. Gurtin and E. Sternberg (1962). 'On the linear theory of viscoelasticity', *Archs Rational Mech. Anal.*, **11**, No. 4, 291–356.
6. C. Eringen (1967). *Mechanics of Continua*, John Wiley, London.
7. V. Ionescu-Cazimir (1964). 'Problem of linear coupled thermoelasticity. Theorems of reciprocity for the dynamic problem of coupled thermoelasticity (I)', *Bull. Acad. Polon. Sci., Sér. Sci. Tech.*, **9**, No. 12.
8. D. Iesan (1969). 'Sur la théorie de la thermoviscoelasticité couplée', *C.R. Acad. Sci. Paris, Sér. A*, **268**, 6 Jan., 58–61.
9. E. Boschi (1973). 'A thermoviscoelastic model of the earthquake source mechanism', *J. Geophys. Res.*, **78**, No. 32, 7733–7737.
10. V. Ionescu-Cazimir (1964). 'Problems of linear coupled thermoelasticity. Some applications of the theorems of reciprocity for the dynamic problem of coupled thermoelasticity (II)', *Bull. Acad. Polon. Sci., Sér. Sci. Tech.*, **9**, No. 12.
11. W. Nowacki (1966). *Dynamic Problems of Thermo-elasticity*, Noordhoff International Publishing, Leyden; P.W.N. Polish Scientific Publishers, Warszawa.
12. T. A. Cruse and F. I. Rizzo (1968). 'A direct formulation and numerical solution of the general transient elastodynamic problem I', *J. Math. Anal. Appl.*, **22**, 244–259.
13. T. A. Cruse (1968). 'A direct formulation and numerical solution of the general transient elastodynamic problem II', *J. Math. Anal. Appl.*, **22**, 341–355.
14. F. J. Rizzo and D. J. Shippy (1971). 'An application of the correspondence principle of linear viscoelasticity theory', *SIAM J. Appl. Math.*, **21**, No. 2, Sept., 321–330.
15. F. J. Rizzo and D. J. Shippy (1977). 'An advanced boundary integral equation method for three-dimensional thermoelasticity', *Int. J. Num. Meth. Engng*, **11**, 753–768.
16. J. L. Swedlow and T. A. Cruse (1971). 'Formulation of boundary integral equations for three-dimensional elastoplastic flow', *Int. J. Solids Struct.*, **7**, 1673–1684.
17. H. D. Bui (1978). 'Some remarks about the formulation of three-dimensional thermoelastoplastic problems by integral equations', *Int. J. Solids Struct.*, **14**, 935–939.
18. V. D. Kupradze (1963). 'Dynamical problems in elasticity', in *Progress in Solid Mechanics*, Eds I. N. Sneddon and R. Hill, Vol. III. North Holland Publishing Company.
19. R. P. Shaw (1978). 'Coupling boundary integral equation methods to "other" numerical techniques', in *Recent Advances in Boundary Element Methods*, Ed. C. A. Brebbia, Pentech Press, London.
20. O. C. Zienkiewicz, D. W. Kelly and P. Bettess (1977). 'The coupling of the finite element method and boundary solution procedures', *Int. J. Num. Meth. Engng*, **10**, No. 5, 355–375.
21. F. Brezzi, C. Johnson and J. C. Nedelec (1978). 'On the coupling of boundary integral and finite element methods', *Rapport interne No. 39*, *Centre de Mathématiques Appliquées, Ecole Polytechnique, Palaiseau*.
22. C. A. Brebbia and P. Georgiu (1979). 'Combination of boundary and finite elements in elastostatics', *Appl. Math. Modelling*, **3**, 212–222.

*Numerical Methods in Heat Transfer*
Edited by R. W. Lewis, K. Morgan, and O. C. Zienkiewicz
© 1981 John Wiley & Sons Ltd

*Chapter 8*

# Thermal and Stress Analysis of Composite Nuclear Fuel Rods by Numerical Methods

*H. Wolf, L. D. Wills, R. M. Krudener, and M. D. Almond*

## SUMMARY

This chapter describes the application of finite difference methods to the delineation of temperature histories in composite nuclear fuel rods of the type used in fast oxide reactors. The basic equations describing the steady-state and transient cases are given together with the boundary conditions as applicable to the case for a sodium temperature of 1000 °F and a representative power output of 18 kW ft$^{-1}$ for the fuel rod. Physical properties of the composite materials are described and values of the pertinent gap conductances are given. The development of the Crank–Nicolson and Peaceman–Rachford finite difference methods from the basic equations is outlined and application is given to the adiabatic centre void case. The case with radiative transfer in the centre void is outlined and a treatment currently in progress utilizing the finite element method is discussed. Pertinent representative results are given for the adiabatic centre void.

## 8.1  INTRODUCTION

In the nuclear power plants currently in operation and for the foreseeable future, the energy needed to transform the Rankine cycle working fluid into a high-energy content vapour results from the fissioning of heavy atoms. The removal of heat from the fuel is one of the primary design problems of nuclear power generation. The nuclear fuel under consideration for advanced reactors is a mixture of the oxides of uranium and plutonium, which are actually ceramics and have a relatively low thermal conductivity of about 1 Btu h$^{-1}$ ft$^{-1}$ °F$^{-1}$ (1.7 W m$^{-1}$ K$^{-1}$). The carbides of uranium and plutonium have higher thermal conductivities by about a factor of 10, but their swelling characteristics and chemical incompatibility with cladding materials are current obstacles to their use in fast reactors.

The determination of the temperature history, and hence the associated heat flow and stress characteristics for a composite nuclear fuel model requires the solution of non-linear elliptic or parabolic partial differential equations together with the associated boundary conditions. The composite fuel model investigated in this work is too complex to permit closed form solutions; instead, it is necessary to apply numerical techniques which are developed as a finite analogue to the continuous boundary value problems with properties variable with temperature.

## 8.2   PHYSICAL DESCRIPTION

Most current designs for fast oxide reactors have utilized fuel rods from 0.230 to 0.310 inch outside diameter (OD). Accordingly, we have selected 0.250 inch OD fuel rods and 0.224 inch (cold) diameter for the oxide fuel pellet. The auxiliary conduction path is realized by placing a highly conductive washer of molybdenum between the fuel pellets as shown in Figure 8.1. The washer also

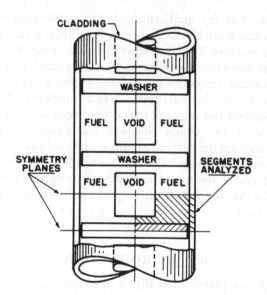

**Figure 8.1.**   Cross section of nuclear fuel rod

serves the very important function of preventing axial redistribution of fuel particles that could spall or break loose from the inner surface of the void during reactor operation. Figure 8.1 also shows the planes of thermal symmetry and the three segments analysed in this work. Parasitic neutron absorption imposes limits on the washer volume; however, neutron economy was not considered.

For the reference design one-sixth of the fuel module length was selected for the washer and five-sixths for the fuel pellet. Further, the washer thickness was selected to be 0.025 inch thick, which for a 5 : 1 ratio in length yielded 0.125 inch for the fuel pellet length. Other physical dimensions of the reference design are summarized as follows:

| | |
|---|---|
| Nominal fuel rod diameter | 0.250 inch |
| Fuel pellet outer diameter | 0.228 inch |
| Rod diameter at surface of clad | 0.254 inch |
| Cladding thickness (stainless steel) | 0.012 inch |
| Module length (pellet plus washer) | 0.150 inch |
| Axial mesh point separation (ZZ) | 0.00625 inch |

Figure 8.2 compares the size of the different washer and fuel pellet arrangements investigated; MOD-2 is the reference design. The reader will note that the figure gives actual pellet length, washer thickness, and centre void height, whereas in the analysis only half needed to be considered due to

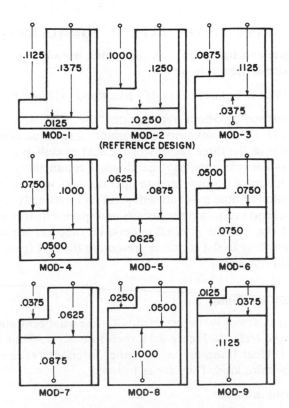

**Figure 8.2**   Fuel pellet geometry variations

thermal symmetry. Two sizes for the centre void were selected for investigation: 0.080 inch and 0.125 inch in diameter by 0.100 inch in axial length. For all MOD's investigated, the axial length of the centre void was chosen so as to have two rows of solid fuel between the conductive washer and the centre void.

### 8.2.1 The conductive washer

In order to more effectively utilize annular fuels (and their reduced temperatures) a means must be proposed to preclude undesirable rearrangement of fuel. A washer of high melting point placed between fuel pellets has been proposed.[1] Fragments of fuel if released from the surface of the centre void would be restrained. The washer would also provide an alternate heat transport path from the pellet to the cladding, thus reducing the fuel temperatures. The dimensions chosen for the centre void and conductive washer represent estimates of dimensions to be expected in practice, and the results indicate significant reductions in the temperature of annular fuels using the conductive washer concept.

## 8.3 BASIC MATHEMATICAL FORMULATION

The most general and basic equation describing the subject problem is the parabolic equation of heat conduction:

$$\nabla \cdot [k(x, u)\nabla u(x, t)] + s(x, t) = \rho(x, u)c(x, u) \, \partial u(x, t)/\partial t \qquad (8.1)$$

with the appropriate equations representing the boundary conditions. In Equation (8.1) $u(x, t)$ is the unknown temperature, a function of position $x$ and time $t$, continuous in the three regions of the fuel module but discontinuous across interfaces between regions. The thermal conductivity $k(x, u)$ of the regions is a function of position and temperature $u$: we note that $k$ is not explicitly a function of time $t$, but does depend on $t$ indirectly through $u$. The quantities $\rho(x, u)$ and $c(x, u)$ are the density and specific heat functions for the substances, and $s(x, t)$ is the internal generation function per unit time and volume; non-zero only in the region representing the fuel (gamma heating is not considered in this work).

### 8.3.1 Boundary conditions

No heat flow exists across symmetry planes, on the axial centreline, or on the surface of the void (refer to Figure 8.1), therefore $\partial u/\partial r = 0$ or $\partial u/\partial z = 0$ on these interfaces. Heat transport at the other interfaces requires boundary conditions of the third kind. They are as follows:

(a)  Clad–coolant interface

$$-k(\partial u/\partial r)_{\text{clad OD}} = h_{\text{rcc}}(u_{\text{clad OD}} - u_{\text{coolant}}) \qquad (8.2)$$

(b)  Fuel–clad interface

$$-k(\partial u/\partial r)_{\text{fuel OD}} = h_{\text{rfc}}(u_{\text{fuel OD}} - u_{\text{clad ID}})$$ (8.3)

(c)  Fuel–washer interface

$$-k(\partial u/\partial z)_{\substack{\text{fuel} \\ \text{face}}} = h_{\text{zfw}}(u_{\substack{\text{fuel} \\ \text{face}}} - u_{\substack{\text{washer} \\ \text{face}}})$$ (8.4)

(d)  Washer–clad interface

$$-k(\partial u/\partial r)_{\text{washer OD}} = h_{\text{rwc}}(u_{\text{washer OD}} - u_{\text{clad ID}})$$ (8.5)

## 8.3.2  Initial conditions

In a strict sense the elliptic heat equation has no initial conditions; however, a finite difference solution approximating the continuous problem requires a starting place to reach steady state. The simplest starting place is probably a uniform temperature with each property value at some initial fixed value. A transient solution (of the parabolic equation) requires that the initial condition be completely specified.[1]

## 8.4  SPATIAL DISCRETIZATION

To obtain a solution we replace the continuous boundary value problem by an analogous finite system of linear difference equations. We integrate both sides of Equation (8.1) over the volume element $V_{mn}$ which symbolizes the mesh region surrounding a mesh point to obtain, after application of the Gaussian divergence theorem,

$$\sum_{s=1}^{6} \int\int_{\sigma_s} k(x, u)(\partial u(x, t)/\partial n)\, d\sigma = \int_v [c(x, u)\rho(x, u)(\partial u(x, t)/\partial t) - s(x, t)]\, dv$$
(8.6)

where $\sigma_s$ represents the six surfaces of the volume $V_{mn}$ and $n$ the outward directed normal. The partial derivatives in Equation (8.6) are approximated as constant at each mesh point with the following values:

$$(\partial u/\partial z)_n \simeq [U(r_i, z_j) - U(r_i, z_{j-1})]/(z_j - z_{j-1})$$
$$(\partial u/\partial r)_m \simeq [U(r_i, z_j) - U(r_{i-1}, z_j)]/(r_i - r_{i-1})$$
(8.7)

In order to compress the lengthy notation, we omit the functional dependencies of the variables and denote discretized values by the capitals ($U$, for example). The entire set of equations for all interior regions and the special equations which represent boundary conditions comprise the finite difference analogy to the continuous boundary value problem. For the steady-state case (elliptic equation) the right-hand side of Equation (8.1) must equal zero. The developments outlined above are discussed in extensive detail by Krudener.[1]

### 8.5   MATRIX REPRESENTATION

The energy balance described by Equation (8.6) was written in terms of axial and radial conductances to simplify the notation. The finite difference equation for an interior point can be written as follows:

$$-[ZK(j-1, i)+ZK(j, i)+RK(j, i-1)+RK(j, i)]U(j, i)$$

$$+ZK(j-1, i)U(j-1, i)+ZK(j, i)U(j+1, i)$$

$$+RK(j, i-1)U(j, i-1)+RK(j, i)U(j, i+1)+S(j, i)$$

$$= C(J, i)\, \mathrm{d}U(j, i)/\mathrm{d}t \tag{8.8}$$

The three regions of interest (as indicated in Figure 8.1) were then split up into mesh regions and nodal points as shown in Figure 8.3. The entire set of linear

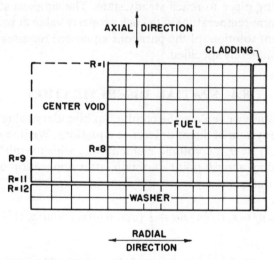

**Figure 8.3**   Mesh regions of fuel, washer, and cladding

equations representing the parabolic boundary value problem can be expressed in matrix form by:

$$-\mathbf{A}\mathbf{U}+\mathbf{S} = \mathrm{d}\mathbf{U}/\mathrm{d}t \tag{8.9}$$

The quantity $\mathbf{U}$ is a column matrix comprising all of the unknown temperatures $U(j, i)$ at every mesh point of the net. Likewise $\mathrm{d}\mathbf{U}/\mathrm{d}t$ is a matrix of all the time derivatives of the unknown temperature distribution. A schematic diagram of the upper left corner of $\mathbf{A}$ is shown in Figure 8.4; we hope that this diagram will enable the reader to visualize the method of construction of normal sections of the complete matrix, more specific details are given in reference 1. It should be pointed out that the notation in the schematic diagram is that employed in the

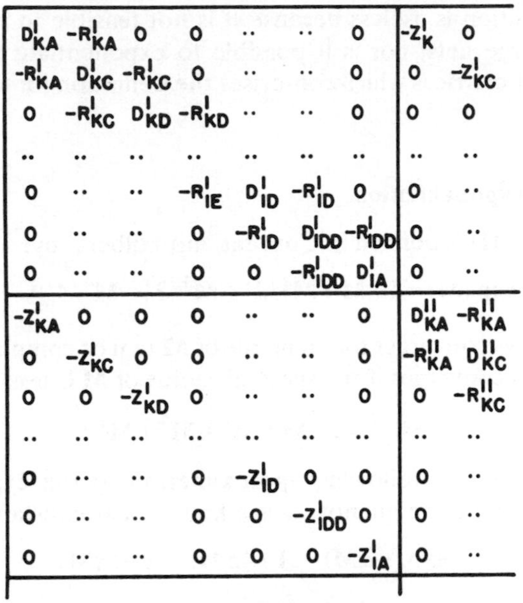

$$
\begin{array}{cccccccc|cc}
D^I_{KA} & -R^I_{KA} & 0 & 0 & \cdots & \cdots & 0 & & -z^I_K & 0 \\
-R^I_{KA} & D^I_{KC} & -R^I_{KC} & 0 & \cdots & \cdots & 0 & & 0 & -z^I_{KC} \\
0 & -R^I_{KC} & D^I_{KD} & -R^I_{KD} & \cdots & \cdots & 0 & & 0 & 0 \\
\cdots & \cdots & \cdots & \cdots & \cdots & \cdots & & & \cdots & \cdots \\
0 & \cdots & \cdots & -R^I_{IE} & D^I_{ID} & -R^I_{ID} & 0 & & 0 & \cdots \\
0 & \cdots & \cdots & 0 & -R^I_{ID} & D^I_{IDD} & -R^I_{IDD} & & 0 & \cdots \\
0 & \cdots & \cdots & 0 & 0 & -R^I_{IDD} & D^I_{IA} & & 0 & \cdots \\
\hline
-z^I_{KA} & 0 & 0 & 0 & \cdots & \cdots & 0 & & D^{II}_{KA} & -R^{II}_{KA} \\
0 & -z^I_{KC} & 0 & 0 & \cdots & \cdots & 0 & & -R^{II}_{KA} & D^{II}_{KC} \\
0 & 0 & -z^I_{KD} & 0 & \cdots & \cdots & 0 & & 0 & -R^{II}_{KC} \\
\cdots & \cdots & \cdots & \cdots & \cdots & \cdots & \cdots & & \cdots & \cdots \\
0 & \cdots & \cdots & 0 & -z^I_{ID} & 0 & 0 & & 0 & \cdots \\
0 & \cdots & \cdots & 0 & 0 & -z^I_{IDD} & 0 & & 0 & \cdots \\
0 & \cdots & \cdots & 0 & 0 & 0 & -z^I_{IA} & & 0 & \cdots \\
\end{array}
$$

**Figure 8.4** Schematic diagram of upper left corner of matrix **A**

computer program described in reference 1 and is not identical but similar to the mathematical notation of Equation (8.8).

The system of equations given above is not in the most convenient form. We introduce the diagonal matrix **D** which is the inverse of the square root of **C**, such that $\mathbf{D} = \mathbf{C}^{-1/2}$ and after assigning the following definitions: $\mathbf{T} = \mathbf{D}^{-1}\mathbf{U}$, $\mathbf{Q} = \mathbf{DS}$, and $\mathbf{B} = \mathbf{DAD}$, we substitute them into Equation (8.9) to obtain:

$$-\mathbf{BT} + \mathbf{Q} = d\mathbf{T}/dt \qquad (8.10)$$

The formal solution of Equation (8.10) can be set forth by standard methods. Let $\mathbf{T}(0)$ be the transformed temperature distribution at time zero. Then for **B**, not a function of time nor of the unknown temperature, we obtain the formal solution as:

$$\mathbf{T}(t) = \mathbf{T}(0) \exp(-\mathbf{B}t) + \exp(-\mathbf{B}t) \int_0^t \exp(\mathbf{B}t)\mathbf{Q}\, dt \qquad (8.11)$$

The result given here is an adaptation of the method due to Varga.[2] For a constant source, the solution can be rearranged to stress the interpretation as a steady-state distribution modified by a transient:

$$\mathbf{T}(t) = \mathbf{B}^{-1}\mathbf{Q} + [\mathbf{T}(0) - \mathbf{B}^{-1}\mathbf{Q}] \exp(-\mathbf{B}t) \qquad (8.12)$$

The formal solution is useless because it is not feasible to invert **B** by direct methods for large nets, nor is it possible to exponentiate $(-\mathbf{B}t)$ due to the infinite series of matrices which comprises the definition for the exponential of a matrix.

### 8.5.1  Matrix exponentiation

The matrix exp $(\mathbf{M})$ is defined in Courant and Hilbert[3] by:

$$\exp{(\mathbf{M})} = \mathbf{I} + \mathbf{M} + \mathbf{M}^2/2! + \mathbf{M}^3/3! + \mathbf{M}^4/4! + \cdots \tag{8.13}$$

This series always converges for elements of **M** real or complex. The following Neumann series converges if the spectral radius of **M** is less than one:

$$(\mathbf{I} - \mathbf{M})^{-1} = \mathbf{I} + \mathbf{M} + \mathbf{M}^2 + \mathbf{M}^3 + \mathbf{M}^4 + \cdots \tag{8.14}$$

Varga[2] introduces the following approximation to the exponential of the matrix $(-a\mathbf{M})$, where the quantity $a$ is a finite scalar constant:

$$\exp{(-a\mathbf{M})} \simeq (\mathbf{I} + \tfrac{1}{2}a\mathbf{M})^{-1}(\mathbf{I} - \tfrac{1}{2}a\mathbf{M}) \tag{8.15}$$

We replace the inverse term by its Neumann series:

$$(\mathbf{I} + \tfrac{1}{2}a\mathbf{M})^{-1} = \mathbf{I} - \tfrac{1}{2}a\mathbf{M} + (\tfrac{1}{2}a\mathbf{M})^2 - (\tfrac{1}{2}a\mathbf{M})^3 + \cdots \tag{8.16}$$

Multiplying termwise by the factor $(\mathbf{I} - \tfrac{1}{2}a\mathbf{M})$ yields:

$$\exp{(-a\mathbf{M})} \simeq \mathbf{I} - a\mathbf{M} + \tfrac{1}{2}a^2\mathbf{M}^2 - \tfrac{1}{4}a^3\mathbf{M}^3 + \cdots \tag{8.17}$$

Comparing this to the definition of the exponential of $(-a\mathbf{M})$

$$\exp{(-a\mathbf{M})} = \mathbf{I} - a\mathbf{M} + \tfrac{1}{2}a^2\mathbf{M}^2 - \tfrac{1}{6}a^3\mathbf{M}^3 + \cdots \tag{8.18}$$

we note that the approximation agrees through quadratic terms, and begins to depart only starting with the cubic term.

## 8.6  THE CRANK–NICOLSON METHOD

The approximation of the exponential of the matrix $(-\mathbf{B}ti)$ by $(\mathbf{I} + \tfrac{1}{2}ti\mathbf{B})^{-1}(\mathbf{I} - \tfrac{1}{2}ti\mathbf{B})$, as expressed by Equation (8.15), when placed in the formal solution of the matrix differential equation leads to the Crank–Nicolson implicit method: accordingly we formulate the constant heat source solution at a time $t_{p+1}$ at a finite time interval $ti$ after an initial time $t_p$, by substituting from Equation (8.15):

$$\mathbf{T}(t_{p+1}) = \mathbf{B}^{-1}\mathbf{Q} + [\mathbf{T}(t_p) - \mathbf{B}^{-1}\mathbf{Q}](\mathbf{I} + \tfrac{1}{2}ti\mathbf{B})^{-1}(\mathbf{I} - \tfrac{1}{2}ti\mathbf{B}) \tag{8.19}$$

Multiplying through by $(2\mathbf{I}/ti + \mathbf{B})$ yields

$$(2\mathbf{I}/ti + \mathbf{B})\mathbf{T}(t_{p+1}) = (2\mathbf{I}/ti - \mathbf{B})\mathbf{T}(t_p) + 2\mathbf{Q} \tag{8.20}$$

This result is a matrix representation of the Crank–Nicolson method. The method requires a simultaneous solution (thus implicit) of the entire system of difference equations at each successive time step. The advantage of the method is that the solutions are unconditionally stable for any magnitude of time step. The drawback of the Crank–Nicolson method is that it is not feasible for use with a two-dimensional net of any appreciable size.

## 8.7 THE PEACEMAN–RACHFORD METHOD

Peaceman and Rachford[4] introduced an alternating direction implicit method for solution of steady-state and transient problems in two dimensions. In the Peaceman–Rachford scheme for a $J \times I$ net, $J$ simultaneous equations are solved $I$ times, then $I$ simultaneous equations are solved $J$ times, for each time step. But for the Crank–Nicolson method, $J \times I$ equations are solved simultaneously (once) at each time step, which quickly becomes time prohibitive compared to the Peaceman–Rachford method for even as few as ten mesh points in each direction.

To obtain the pair of general matrix equations which is known to represent the Peaceman–Rachford method, the matrix of coefficients $\mathbf{B}$ is resolved into two matrices $\mathbf{H}$ and $\mathbf{V}$ such that $\mathbf{B} = \mathbf{H} + \mathbf{V}$, where $\mathbf{H}$ represents the matrix of coefficients (transformed) that arise from heat transfer in the horizontal direction, while $\mathbf{V}$ represents the matrix of coefficients that arise from heat transfer in the vertical direction, taken here as the radial (column) direction. Details of the procedure as applied to the subject investigation are given in Chapter 5 of Krudener.[1]

The steps we now perform are valid if $\mathbf{H}$ and $\mathbf{V}$ commute. For the fuel element being simulated in this investigation, $\mathbf{H}$ and $\mathbf{V}$ cannot precisely commute; however, the results obtained are physically reasonable and an energy balance equating total energy generated to that convected from the surface was in excellent agreement, lending credence to the results.

The formal solution of the matrix differential equation for a constant source is entered with the Crank–Nicolson approximation for the exponential of matrix $(-\mathbf{B}ti)$, but this time in terms of the $\mathbf{H}$ and $\mathbf{V}$ matrices so that

$$\exp(-\mathbf{B}ti) = \exp(-\mathbf{H}ti)\exp(-\mathbf{V}ti)$$

$$\approx (\mathbf{I}+\tfrac{1}{2}ti\mathbf{H})^{-1}(\mathbf{I}-\tfrac{1}{2}ti\mathbf{H})(\mathbf{I}+\tfrac{1}{2}ti\mathbf{V})^{-1}(\mathbf{I}-\tfrac{1}{2}ti\mathbf{V}) \tag{8.21}$$

Upon placing the exponential approximation into the formal solution, the following may be obtained:

$$\mathbf{T}^{(p+1)} = (\mathbf{H}+\mathbf{V})^{-1}\mathbf{Q} + [\mathbf{T}^{(p)} - (\mathbf{H}+\mathbf{V})^{-1}\mathbf{Q}]$$

$$\times [(\mathbf{I}+\tfrac{1}{2}ti\mathbf{V})^{-1}(\mathbf{I}+\tfrac{1}{2}ti\mathbf{H})^{-1}(\mathbf{I}-\tfrac{1}{2}t\mathbf{V})(\mathbf{I}-\tfrac{1}{2}t\mathbf{H})]$$

$$= (\mathbf{I}-\tfrac{1}{2}ti\mathbf{V})(\mathbf{I}-\tfrac{1}{2}ti\mathbf{H})\mathbf{T}^{(p)} + ti(\mathbf{H}+\mathbf{V})(\mathbf{H}+\mathbf{V})^{-1}\mathbf{Q}$$

$$+ (\tfrac{1}{2}ti)^2(\mathbf{VH}-\mathbf{HV})(\mathbf{H}+\mathbf{V})^{-1}\mathbf{Q} \tag{8.22}$$

The last term in the final equation is zero, since $\mathbf{H}$ and $\mathbf{V}$ are assumed to commute. The result may thus be written as:

$$(\mathbf{I}+\tfrac{1}{2}ti\mathbf{V})(\mathbf{I}+\tfrac{1}{2}ti\mathbf{H})\mathbf{T}^{(p+1)} = (\mathbf{I}-\tfrac{1}{2}ti\mathbf{V})(\mathbf{I}-\tfrac{1}{2}ti\mathbf{H})\mathbf{T}^{(p)} + ti\mathbf{Q} \qquad (8.23)$$

The result in Equation (8.23) is of no particular advantage if left in this form. We assume, according to Spanier,[5] that the transient method may be represented by the following pair of matrix equations for the alternating-direction implicit (ADI) method:

$$(\mathbf{I}+\tfrac{1}{2}ti\mathbf{H})\mathbf{T}^{(p+\frac{1}{2})} = (\mathbf{I}-\tfrac{1}{2}ti\mathbf{V})\mathbf{T}^{(p)} + \tfrac{1}{2}ti\mathbf{Q}$$
$$(\mathbf{I}+\tfrac{1}{2}ti\mathbf{V})\mathbf{T}^{(p+1)} = (\mathbf{I}-\tfrac{1}{2}ti\mathbf{H})\mathbf{T}^{(p+\frac{1}{2})} + \tfrac{1}{2}ti\mathbf{Q} \qquad (8.24)$$

In these equations, $\mathbf{T}^{(p+\frac{1}{2})}$ is an interim temperature distribution due to axial heat flow. The quantity $\mathbf{T}^{(p+1)}$ is the result at the end of the $(p+1)$ time interval. It is obtained as a second step due to radial heat flow using axial results as 'knowns'. Finally to obtain the 'Peaceman–Rachford ADI transient method' in a form especially useful for application in this chapter, we multiply each equation through by the scalar $2/t$:

$$(2\mathbf{I}/ti + \mathbf{H})\mathbf{T}^{(p+\frac{1}{2})} = (2\mathbf{I}/ti - \mathbf{V})\mathbf{T}^{(p)} + \mathbf{Q}$$
$$(2\mathbf{I}/ti + \mathbf{V})\mathbf{T}^{(p+1)} = (2\mathbf{I}/ti - \mathbf{H})\mathbf{T}^{(p+\frac{1}{2})} + \mathbf{Q} \qquad (8.25)$$

In these equations, $\mathbf{T}^{(p+\frac{1}{2})}$ is an interim temperature distribution obtained by implicit solution for heat flow in the axial direction. The quantity $\mathbf{T}^{(p+1)}$ is the completed result for the temperature distribution at the end of the $(p+1)$st time interval. It is obtained as a second step by implicit solution for heat flow in the radial direction, using axial heat flow based on the interim temperature distribution as 'knowns'. To verify the ADI pair of equations, one can formally solve for $\mathbf{T}^{(p+\frac{1}{2})}$ in Equation (8.24) by inverting its matrix coefficient, and then inserting the outcome in the second part of Equation (8.24) to obtain Equation (8.22).

### 8.7.1 Steady-state solution

In steady state the right-hand side of Equation (8.9) vanishes and the matrix representation is given by the equation: $-\mathbf{BT} + \mathbf{Q} = 0$. The vector $\mathbf{Q}$ must be constant with respect to time, but its elements are a function of position. The formal solution to the elliptic problem is given by $\mathbf{T} + \mathbf{B}^{-1}\mathbf{Q}$. This solution is not economically feasible for variable properties. There is the possibility of accelerating the transient method by using larger time increments as one approaches steady state, but doing so causes wild gyrations in the net that delay steady state. Varga,[2] p. 219, has shown that iteration with a single optimum time increment is of the same order of efficiency as the method of successive over-relaxation (Gauss–Seidel); however, the ADI scheme allows a remarkable advance in technique by applying sets of time increments which vary in cycles. These sets (acceleration parameters), correspond to the inverses of discrete quantities lying between the maximum and minimum eigenvalues of **H** and **V**. Repetitive application of these sets of acceleration parameters results in fully converged results in relatively few cycles. Determination of acceleration parameters has been the subject of several studies. We have chosen to apply the derivation discovered by Wachspress[6] called the Wachspress–Habetler variant of the ADI steady-state method. The method may be represented by the following pair of matrix equations:

$$(\mathbf{I}w_p + \mathbf{H})\mathbf{T}^{(p+\frac{1}{2})} = (\mathbf{I}w_p - \mathbf{V})\mathbf{T}^{(p)} + \mathbf{Q}$$
$$(\mathbf{I}w_p + \mathbf{V})\mathbf{T}^{(p+1)} = (\mathbf{I}w_p - \mathbf{H})\mathbf{T}^{(p+\frac{1}{2})} + \mathbf{Q}$$

$$(8.26)$$

No physical interpretation is now attached to $w_p$; they are construed merely as a set of numbers applied to achieve a more rapidly converging solution towards steady state. The prolixity involved in a description of the means for calculating the sets of parameters precludes their description in the limited space of this chapter; Krudener,[1] pp. 108–118, gives complete detail of their formulation and application. The results of the subject thermal problems were obtained with the single set of $2^4 = 16$ acceleration parameters.

## 8.8 GEOMETRY VARIATIONS

The investigation of geometrical effects by Almond[7] was done by using a Gauss–Seidel iterative solution (with over-relaxation) applied to the same net of equations as described above but varying the number of mesh regions in each of the segments analysed. Equation (8.9) then reduces to $\mathbf{AU} = \mathbf{S}$ and involves roughly 200 equations in 200 unknowns, depending on the size of the centre void. Although the ADI method is faster in convergence, the Gauss–Seidel method is simpler to program and was selected for this portion of the work. The method of successive over-relaxation, as discussed by Meyers[8] and Stanton,[9] takes a fixed linear combination of $U_j^{i+1}$ and $U_j^i$ to calculate a new $U_j^{i+1}$ (designated $U_j^{i+1*}$) which is then used in calculating $U_{j+1}^{i+1}$ where:

$$U_j^{i+1*} = U_j^i + \lambda(U_j^{i+1} - U_j^i)$$

$$(8.27)$$

and $\lambda$ is called the relaxation factor. For $\lambda = 1$ this method reduces to the Gauss–Seidel method. However, for $\lambda$ greater than 1, the method converges much more rapidly than does the Gauss–Seidel method. By trial and error, an optimum value for the relaxation factor was found to be about 1.8. For example, the reference design set of parameters using $\lambda = 1$ did not converge to $10^{-4}$ degree accuracy in 12 minutes. However, the same program using $\lambda = 1.8$ converged to $10^{-6}$ degree accuracy in four minutes. As an additional check for each computational run the sum of the individual mesh region generation rates was compared to the total surface heat flux from the cladding, agreement to five significant figures was obtained on all calculations.

## 8.9 SELECTION OF PARAMETERS AND THERMAL PROPERTIES

Since the intended application for the subject fuel element model is the sodium cooled breeder reactor, we take the sodium coolant temperature as 1000 °F (538 °C) with a linear power of 18 kW ft$^{-1}$ as representative of the maximum peak power output of the fuel rod. This value of power imposes a somewhat more severe test of the proposed configuration than it would be subjected to under current conditions at this writing, which are about 6 to 9 kW ft$^{-1}$. A value of 5100 °F (2816 °C) has been selected as the nominal melting point of $UO_2$–$PuO_2$ mixture for this study, but is to be recognized as an approximation from within a spread of about 100 °F which encompasses the melting point.[12]

### 8.9.1 Gap conductance

For the fuel–cladding gap initially filled with helium, a nominal value of 1500 Btu h$^{-1}$ ft$^{-2}$ °F$^{-1}$ in accord with current practice[13] has been selected for this study; temperature distributions have also been computed for values of 1000 and 2000 Btu h$^{-1}$ ft$^{-2}$ °F$^{-1}$ and are reported in reference 1.

### 8.9.2 Fuel–washer conductance

The uncertainties in the possible values for the fuel–washer conductance may be considerably greater than for gap conductance. Consequently, parameterization runs have been computed for values of the fuel–washer conductance ranging from 750 to 6000 Btu h$^{-1}$ ft$^{-2}$ °F$^{-1}$. It is estimated[1] that the most likely value is about 3000 Btu h$^{-1}$ ft$^{-2}$ °F$^{-1}$ and this value is taken as nominal when other parameters are varied.

### 8.9.3 Thermal properties

The density, thermal conductivity, and specific heat of the fuel and washer materials are functions of temperature. The variation of specific heat with temperature for molybdenum from reference 10 and mixed urania–plutonia from references 11 and 14 were employed in the numerical calculations. As discussed in detail in reference 1 the data for the mixed oxide fuel were extrapolated above 4580 °F to the assumed melting point of 5100 °F. The value of specific heat for the stainless steel cladding was taken at 0.140 Btu lb$^{-1}$ °F$^{-1}$ and assumed constant, since the cladding temperature varied only over a small range.

The variations of thermal conductivity with temperature for molybdenum from reference 10 and mixed urania–plutonia from reference 11 were utilized in the numerical codes. It must be recognized that a considerable uncertainty exists in the currently accepted values for conductivity of a given mixture of uranium and plutonium oxide fuel material. The conductivity of the stainless steel cladding was taken as invariant with temperature at a value of 12.6 Btu h$^{-1}$ ft$^{-1}$ °F$^{-1}$, corresponding to temperatures in the range of 1100 °F. The temperature variation in the cladding is not wide enough to justify consideration of temperature dependence.

## 8.10   RESULTS FOR ADIABATIC CENTRE VOIDS

Representative results are presented for the subject fuel rod configuration with adiabatic surfaces on the centre void for steady-state and transient conditions for a centre void diameter of 0.080 inches and for washer thicknesses ranging from 0.0125 to 0.1125 inches (see Figure 8.2) for significant values of the fuel–washer gap conductance and the fuel–cladding gap conductance. The most significant series of parameterization runs were those which evaluated the temperature response due to variations of the thermal conductance between the surface of the fuel pellet and the face of the conductive washer. Acting on the information set forth in reference 13, a nominal value of the fuel–washer contact conductance of 3000 Btu h$^{-1}$ ft$^{-2}$ °F$^{-1}$ and a nominal value of the fuel–cladding gap conductance of 1500 Btu h$^{-1}$ ft$^{-2}$ °F$^{-1}$ were selected for the reference design. These nominal values have demonstrated a useful effect of the washer as an auxiliary heat transport path.

### 8.10.1   Reference design

With regard to the considerations above, the reference design for comparison of the temperature distributions calculated in reference 1 for the conceptual fuel element of this work is delineated by the following conditions:

(a)   peak linear power density                18.0 kW ft$^{-1}$ (570 W cm$^{-1}$)
(b)   fuel material: uranium (80%) plutonium
        (20%) dioxide

(c)   fuel pellet dimensions:
        total axial length                        0.125 in (3.18 mm)
        outer diameter (hot)                   0.228 in (5.79 mm)
        centre void length                       0.100 in (2.54 mm)
        centre void diameter                   0.080 in (2.03 mm)
(d)   washer material: molybdenum
(e)   washer dimensions:
        thickness (axial length)              0.025 in (0.635 mm)
        diameter                                    0.228 in (5.79 mm)
(f)   cladding material: type 347 stainless steel

(g)   cladding thickness (cold, nominal)          0.012 in (0.305 mm)
(h)   coolant (liquid sodium) temperature          1000 °F (538 °C)
(i)   fuel–cladding gap conductance          1500 Btu h$^{-1}$ ft$^{-2}$ °F$^{-1}$
                                                                     (2627 W m$^{-2}$ °C$^{-1}$)
(j)   cladding–coolant convective conductance          30,000 Btu h$^{-1}$ ft$^{-2}$ °F$^{-1}$
                                                                     (52,540 W m$^{-2}$ °C$^{-1}$)
(k)   fuel–washer contact conductance          3000 Btu h$^{-1}$ ft$^{-2}$ °F$^{-1}$
                                                                     (5240 W m$^{-2}$ °C$^{-1}$)

## 8.10.2   Steady-state results with 0.080 centre void

Figure 8.5 shows the variation of temperature with radius for radial lines of the finite difference net (i.e. radial blocks of the vertical matrix) denoted by axial row numbers 1, 8, 9, 11, and 12 (see Figure 8.3) for the reference design. The figure shows that the maximum temperature in the module, 4106 °F, occurred

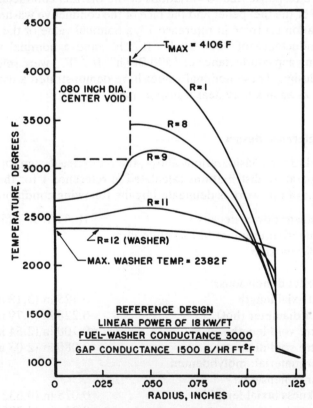

**Figure 8.5**   Steady-state radial temperatures for rows 1, 8, 9, 11 and 12; reference design, small centre void

on the edge of the centre void at the plane of symmetry. This location, as might be expected, was found to be the hot-spot of all steady-state distributions. Row R-9 is on the axial edge of the centre void; it is interesting to note that in the fuel from this row to the face in contact with the washer and outwards to a radius of about 0.50 inch, the heat flows to the washer in a direction inclined away from the cladding and *towards* the fuel centreline! Figure 8.5 also shows that near the fuel–washer–cladding interface, some energy flows (spills over) from the washer back into the fuel before entering the cladding.

### 8.10.3 Effect of gap conductance

Parameterization runs have also been computed for variations in the gap conductance from the surface of the fuel pellet and washer to the inner surface of the cladding, with all other parameters held at their reference design values. The response of the fuel element is roughly linear in that the temperature at every location rises (or falls) approximately the same amount for a given decrease (or increase) in the gap conductance. This fairly predictable behaviour was established from the analysis of steady-state distributions computed for gap conductances of the analysis of steady-state distributions computed for gap conductances of 1000, 1500, and 2000 Btu h$^{-1}$ ft$^{-2}$ °F$^{-1}$, which seems a sufficient range for present purposes.[1]

### 8.10.4 Influence of increasing power

Inasmuch as temperature distributions throughout the subject fuel module for fuel pellets utilizing a centre void of 0.125 inch diameter were far below the melting points of either fuel or washer, it was desirable to test the conductive washer concept at higher peak power densities. Pertinent results from these runs are shown in Figure 8.6. The strict linearity of these plots for crucial maximum temperatures as functions of linear power is rather remarkable. It will be observed that at 27 kW ft$^{-1}$, the maximum fuel temperature at 4880 °F is still more than 200 degrees below the mixed oxide melting point. The maximum washer temperature at this power level is almost 1900 °F below its melting point (4750 °F).

### 8.10.5 Effect of washer thickness

Results were obtained for washer thicknesses ranging from 0.0125 inch (MOD-1) to 0.1125 inch (MOD-9) keeping all other parameters at their reference design values. As the washer thickness was increased, the fuel pellet length was reduced to maintain a constant overall module length, and the heat generation rate in the fuel was increased in appropriate amounts to maintain a constant linear power output from the fuel rod of 18 kW ft$^{-1}$. To examine the

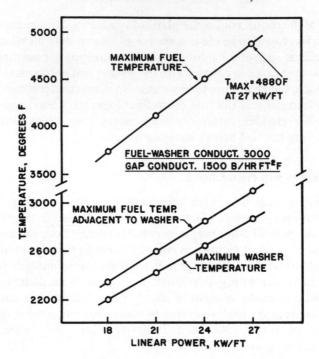

**Figure 8.6** Maximum steady-state temperatures as a function of power; large centre void

effect of fuel–washer conductance on the maximum fuel temperature, computational runs were performed for each value of washer thickness, MOD-1 to MOD-9, to determine the maximum fuel temperature for the same range in values of axial contact conductance from 100 to 6000 Btu $h^{-1}$ $ft^{-2}$ $°F^{-1}$. Those results are shown in Figure 8.7 and they indicate that for values of axial contact conductance greater than about 1000 Btu $h^{-1}$ $ft^{-2}$ $°F^{-1}$, increasing the washer thickness reduces the maximum fuel temperature. If, however, the fuel–washer conductance is below about 1000 Btu $h^{-1}$ $ft^{-2}$ $°F^{-1}$, making the washer thicker produces an increase in maximum fuel temperature.

Figure 8.8 depicts the axial variation of cladding temperature at a radius of 0.114 inch (the fuel–cladding interface) for representative washer thicknesses. As the washer thickness was increased, temperatures in the cladding near the mid-plane of the washer (row 13) decreased significantly while temperatures near the mid-plane of the fuel (row 1) increased slightly. The figure shows that the thinner washer subjects the cladding to a more severe thermal strain than the thicker washers. Since thicker washers result in lower maximum fuel temperature as well as reduced thermal strain, it would be desirable to operate with as thick a washer as possible, consistent with neutron economy constraints.

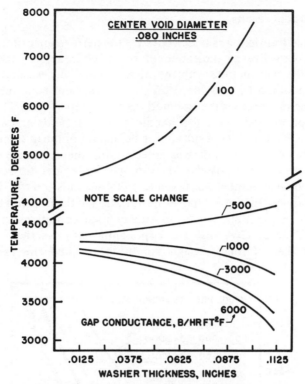

**Figure 8.7** Maximum fuel temperature as a function of washer thickness; small centre void

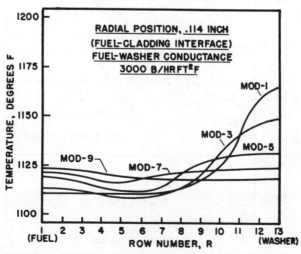

**Figure 8.8** Axial temperatures for MOD's 1, 3, 5, 7, and 9 (see Figure 8.2)

### 8.10.6   Transient results

An exponential transient was selected to test the performance of the numerical code and also to give an indication of the effectiveness of the composite structure of the fuel module with the high-conductivity washer. A relevant question then would be how long can such a transient continue before the hot-spot in the fuel reaches the assumed melting temperature? The numerical code described in reference 1 can handle other transients and also gives an indication of accuracy of the results in the behaviour of the interim solution. A doubling time of two seconds was selected for the reference exponential transient and was superimposed on the steady-state temperature distributions. The iterative time increment was taken as 0.004 seconds and was short enough to ensure accuracy.[1] Figure 8.9 shows the variation of the hot-spot temperature in the fuel with time, starting from steady-state conditions at $18 \text{ kW ft}^{-1}$ $(590 \text{ W cm}^{-1})$ power level with the exponential rise described above. The hot-spot for the reference design condition reached the assumed melt

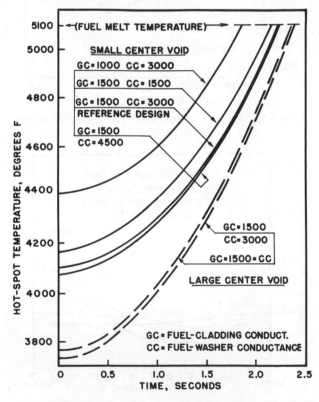

**Figure 8.9**   Transient response of fuel hot-spot for an exponential power excursion (two second doubling time)

temperature in 2.27 seconds at a power level slightly more than twice that of the reference design, fast enough to actuate scram and safety devices.

## 8.11   RESEARCH IN PROGRESS

The results obtained by Krudener[1] and Almond,[2] as shown in Figure 8.5 indicate that there is a significant difference in temperature on the surface of the centre void at location $R = 1$ and $R = 9$. Accordingly, the adiabatic assumption for these surfaces is open to serious question, and it appears that there could be significant radiant heat exchange as well as possibly some minor effects due to free convection by the fission gases. Present work by Wills[15] addresses the radiation problem in the centre void as well as extending the thermal analysis to also include a stress analysis of the fuel module.

Current work pertaining to the coupling of conduction heat transfer with a radiation boundary condition[16] uses a linearized approximation of the quartic temperature dependence. While this method applies with acceptable accuracy when relatively small temperature variations exist on the enclosure surface, the extreme temperature differences being considered in this application preclude the use of such an approach. In order to achieve a reasonable estimation of the thermal behaviour of the fuel element, the problem must be solved with the non-linearity intact.

As in the finite difference approach, the problem can be stated mathematically for the steady state as:

$$\partial(k_r r\, \partial t/\partial r)/\partial r + \partial(k_z r\, \partial t/\partial z)/\partial z + rQ = 0 \tag{8.28}$$

with boundary conditions:

(a)   on the symmetry planes (imposed heat flux)

$$k_r(\partial t/\partial r)n_r + k_z(\partial t/\partial z)n_z = 0 \tag{8.29}$$

(b)   on the cladding surface (convection)

$$k_r(\partial t/\partial r)n_r + k_z(\partial t/\partial z)n_z = -h(t - T_\infty) \tag{8.30}$$

(c)   on the surface of the internal void (radiation)

$$k_r(\partial t/\partial r)n_r + k_z(\partial t/\partial z)n_z = G - J \tag{8.31}$$

and

(d)   at the material interfaces (gap conduction)

$$k_r(\partial t/\partial r)n_r + k_z(\partial t/\partial z)n_z = h_g(t - t_a) \tag{8.32}$$

where $k_r$ and $k_z$ are the radial and axial thermal conductivities, $G$ and $J$ are the irradiation and radiosity functions along the internal void surface, $h_g$ is a linearized gap conductance coefficient (strongly dependent on the internal stress and strain distribution within the element), and $t_a$ is the temperature at the node immediately opposite the interstice or, at the point of three-material intersection, a weighted average of those temperatures.

Following through with the classical finite element treatment,[17,18,19] the axisymmetric domain is divided into a number of eight-node isoparametric serendipity ring elements as shown in Figures 8.10 and 8.11. We assume that

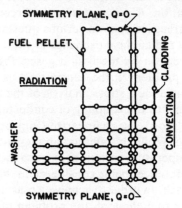

SYMMETRY PLANE, Q=0

FUEL PELLET

RADIATION

CLADDING

CONVECTION

WASHER

SYMMETRY PLANE, Q=0

**Figure 8.10**  Domain division into 24 isoparametric serendipity quadratic elements with 117 nodes

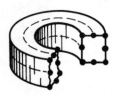

**Figure 8.11**  Axisymmetric configuration of isoparametric element with eight nodes

$r = \mathbf{N} \cdot \mathbf{r}$ and $t = \mathbf{N} \cdot \mathbf{t}$ where $\mathbf{N}$ is a vector consisting of the approximating functions given by Zienkiewicz[19] and $\mathbf{r}$ and $\mathbf{t}$ are, respectively, vectors formed from the element–node radial dimensions and temperatures. Using Galerkin's method with Green's theorem, the problem can be expressed for each element by the integral equation:

$$\iint_{A^e} \{(\mathbf{N} \cdot \mathbf{r})^2[(k_r \, \partial \mathbf{N}/\partial \mathbf{r} \cdot \mathbf{T}) \, \partial N_i/\partial r + (k_z \, \partial \mathbf{N}/\partial \mathbf{z} \cdot \mathbf{T}) \, \partial N_i/\partial z]$$

$$+ (\mathbf{N} \cdot \mathbf{r})[N_i k_r \, \partial \mathbf{N}/\partial \mathbf{r} \cdot \mathbf{T}]\} \, dr \, dz$$

$$+ \oint_{s_a} (\mathbf{N} \cdot \mathbf{r})^2 q N_i \, ds + \oint_{s_b} (\mathbf{N} \cdot \mathbf{r})^2 h \mathbf{N} \cdot \mathbf{T} \, ds$$

$$+\oint_{s_b} (\mathbf{N} \cdot \mathbf{r})^2 h T_\infty N_i \, ds - \oint_{s_c} (\mathbf{N} \cdot \mathbf{r})^2 (G - J) N_i \, ds$$

$$+\oint_{s_d} (\mathbf{N} \cdot \mathbf{r})^2 h_g (\mathbf{N} \cdot \mathbf{T}) \, ds - \oint_{s_d} (\mathbf{N} \cdot \mathbf{r})^2 h_g N_i T_{a_i} \, ds$$

$$-\iint_{A^e} (\mathbf{N} \cdot \mathbf{r})^2 Q N_i \, dr \, dz = 0 \qquad (8.33)$$

where $i = 1, 2, \ldots, 8$.

Since the material interface and radiation boundary conditions will necessitate the inclusion of nodal temperatures not contained in a single element, the classical finite element procedure of forming elemental matrix equations from the integral equation (8.33) and assembling these to form a matrix equation representing the entire domain will not apply directly. However, by proceeding directly to the global matrix equation, as is done in most computer codes in any case, the extra-elemental influence coefficients can be handled with little difficulty. In so doing, a matrix equation of the form:

$$[\mathbf{K}_T + \mathbf{K}_H + \mathbf{K}_G + \mathbf{K}_A]\mathbf{T} - [\mathbf{K}_R]\mathbf{T}^4 = \mathbf{R}_Q - \mathbf{R}_q + \mathbf{R}_{T_\infty} \qquad (8.34)$$

results where the $\mathbf{K}$'s represent square matrices of a rank equal to the number of nodes in the system; the $\mathbf{R}$'s represent column matrices; $\mathbf{T}$ is a column matrix of unknown absolute nodal temperatures; and $\mathbf{T}^4$, those temperatures raised to the fourth power. Individually, $[\mathbf{K}_T]$ is determined by the thermal conductivities; $[\mathbf{K}_H]$ and $\mathbf{R}_{T_\infty}$ by the convection boundary condition; $[\mathbf{K}_G]$ and $[\mathbf{K}_A]$ by the interstices in the domain; $\mathbf{R}_Q$ by element heat generation; $\mathbf{R}_q$ by imposed boundary heat flux (in this case a null matrix); and $[\mathbf{K}_R]$ by the radiation boundary condition. With the exception of the matrix $[\mathbf{K}_R]$ all can be obtained as a consequence of the integral equation (8.33).

The difficulty in obtaining an expression for the matrix $[\mathbf{K}_R]$ lies in the fact that the term expressing the radiation boundary condition in Equation (8.33)

$$\oint (\mathbf{N} \cdot \mathbf{r})^2 (G - J) N_i \, ds \qquad (8.35)$$

is not in terms of an explicit function of temperature and neither the surface radiosity, $J$, nor irradiation, $G$, is, in general, known *a priori*. In order to continue with the solution, these must be expressed as functions of material properties, domain geometry, and nodal temperatures.

### 8.11.1 Radiation formulation

Let us assume that the surface subject to radiation can be divided into $n$ subsurfaces defined by the nodes of the finite element partitioning along that surface as shown in Figure 8.12, and that each of these subsurfaces is opaque,

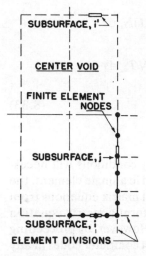

CENTER VOID

FINITE ELEMENT
NODES

SUBSURFACE, j →

SUBSURFACE, i

ELEMENT DIVISIONS

**Figure 8.12**　Orientation of subsurface, $i$, $i'$, and $j$ in the centre void

diffuse, and at a constant temperature equal to the temperature of the defining node. Further assume that the emmisivity, $\varepsilon$, and the reflectivity, $\rho$, are constants for each subsurface. We can then write for the heat flux, $q_j = G - J$ into subsurface $j$:

$$q_j = \sum_{i=1}^{n} \int_{A_i} J_i F_{dA_i \to A_j} \, dA_i - \int_{A_j} J_j \, dA_j \tag{8.36}$$

$$= \sum_{i=1}^{n} J_i F_{ij} A_i - J_j A_j \tag{8.37}$$

$$= \sum_{\substack{i=1 \\ i \neq j}}^{n} J_i F_{ij} A_i + J_j A_j (F_{jj} - 1) \tag{8.38}$$

where $J_i$ is the radiosity of the subsurface $i$, $A_i$ is the area of subsurface $i$, and $F_{ij}$ is the configuration factor from subsurface $i$ to subsurface $j$. In matrix notation this expression can be written $\mathbf{q} = [\mathbf{L}]\mathbf{J}$ where $\mathbf{q}_i = q_i$ and $\mathbf{J}_i = J_i$ and

$$[\mathbf{L}]_{ij} = \begin{cases} F_{ij} A_j & i \neq j \\ A_i (F_{ij} - 1) & i = j \end{cases} \tag{8.39}$$

From the definition of radiosity, we can write

$$J_j = \varepsilon_j \sigma T_j^4 + (\rho_j / A_j) \sum_{i=1}^{n} j_i F_{ij} A_i \tag{8.40}$$

where $\sigma$ is the Stefan–Boltzmann constant, so that

$$\varepsilon_j \sigma T_j^4 = (1 - \rho_j F_{ij}) J_j - (\rho_j / A_j) \sum_{\substack{i=1 \\ i \neq j}}^{n} J_i F_{ij} A_i \tag{8.41}$$

or

$$[\mathbf{D}]\mathbf{T}^4 = [\mathbf{M}]\mathbf{J} \tag{8.42}$$

where

$$[\mathbf{D}]_{ij} = \begin{cases} \varepsilon_i \sigma / \rho_i & i = j \\ 0 & i \neq j \end{cases}$$

$$\mathbf{T}_i^4 = T_i^4$$

$$[\mathbf{M}]_{ij} = \begin{cases} -F_{ji} A_j / A_i & i \neq j \\ (1 - \rho_i F_{ii}) / \rho_i & i = j \end{cases}$$

Substituting (8.42) into $\mathbf{q}$ and eliminating $\mathbf{J}$ we have

$$\mathbf{q} = [\mathbf{L}][\mathbf{M}] - \mathbf{1}[\mathbf{D}]\mathbf{T}^4 \tag{8.43}$$

The radiant exchange boundary condition can then be handled by modifying the global matrix equation with the term $[\mathbf{K}_R]\mathbf{T}^4$ as shown in Equation (8.35) where

$$[\mathbf{K}_R]_{ij} = \begin{cases} \oint (\mathbf{N} \cdot \mathbf{r})^2 [\mathbf{L}\mathbf{M}^{-1}\mathbf{D}]_{ij} N_i \, ds & \text{nodes } i \text{ and } j \text{ on boundary} \\ & \text{subject to radiation} \\ 0 & \text{otherwise} \end{cases} \tag{8.44}$$

To this point we have been dealing with surfaces of arbitrary emmisivity and reflectivity. If one restricts consideration to black surfaces where $\rho = 0$ and $\varepsilon = 1$, a somewhat simplified formulation is possible. In this case it is easily shown that the matrix is given by:

$$[\mathbf{L}\mathbf{M}^{-1}\mathbf{D}]_{ij} = \begin{cases} \sigma F_{ji} A_j / A_i & i \neq j \\ \sigma(F_{ij} - 1) & i = j \end{cases} \tag{8.45}$$

In either case, several modifications to the standard finite element code must be made in order to accommodate the radiation term. Perhaps the most straightforward method is to first calculate the $n \times n$ matrix of configuration factors where $n$ is the number of viewable nodes on the enclosure surface. In this particular case, with axisymmetric geometry, these configuration factors are quite readily obtained using flux algebra and the equation given by Hamilton and Morgan[20] for parallel circular planes. The use of symmetry, while reducing the complexity of the associated conduction problem, makes the radiation condition somewhat more difficult to visualize, since the complete enclosure surface must be considered in the problem solution. This can be accomplished within the bounds of the symmetrical section if a modified configuration factor from some subsurface $i$ to another subsurface, $j$, $F'_{ij}$, is calculated as

$$F'_{ij} = F_{ij} + F_{i'j} \tag{8.46}$$

where, as shown in Figure 8.13, the subsurface $i'$ is the symmetrical counterpart of subsurface $i$. These modified configuration factors can then be used to develop an $m \times m$ matrix where $m$ is the number of nodes on the symmetrical section of the enclosure surface. In our case, since only one-half of the fuel element need be considered, the storage required for this matrix is 25% of the storage necessary were the entire void surface included.

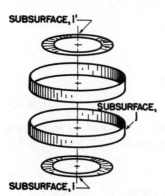

SUBSURFACE, I'

SUBSURFACE, J

SUBSURFACE, I

**Figure 8.13** Subsurface orientation for radiation configuration factors

Having constructed the configuration factor matrix, the matrices $[\mathbf{LM^{-1}D}]$ and $[\mathbf{K}_R]$ follow immediately. Once $[\mathbf{K}_R]$ is constructed, the storage necessary for the configuration factor matrix and the $[\mathbf{LM^{-1}D}]$ matrix are no longer required and may be utilized for the remaining matrices defining the domain. The resulting system of equations, by virtue of the $\mathbf{T}^4$ matrix, is non-linear. The temperature distribution can then be had using the Newton–Raphson technique[21] with the first trial solution that of the linear portion. This not only provides a reasonable starting point for iteration but, since the linear set is obtained by ignoring radiation on the centre void surface, in effect setting up an adiabatic condition as examined by Krudener and Almond, this method automatically shows the results of radiation in the void.

## REFERENCES

1. R. M. Krudener (1973). 'A matrix iterative development of transient and steady temperature distributions in nuclear fuel elements incorporating an auxiliary heat-transport path concept', *Ph.D. Dissertation*, Mechanical Engineering Department, University of Arkansas, Fayetteville, Arkansas.
2. R. S. Varga (1962). *Matrix Iterative Analysis*, Prentice-Hall, Englewood Cliffs, NJ, pp. 262–268.
3. R. Courant and D. Hilbert (1966). *Methods of Mathematical Physics*, Interscience Publishers, New York, Vol. I, p. 10.
4. D. W. Peaceman and H. H. Rachford (1955). 'The numerical solution of parabolic and elliptic differential equations', *J. Soc. Indust. Appl. Math.*, **3**, 28–41.

5. J. Spanier (1967). 'Alternating direction methods applied to heat conduction problems', in *Mathematical Methods for Digital Computers*, Eds A. Ralston and H. S. Wilf, John Wiley, New York, Vol. II, pp. 215–245.
6. E. L. Wachspress (1966). *Iterative Solutions of Elliptic Systems and Applications to the Neutron Diffusion Equation of Reactor Physics*, Prentice-Hall, Englewood Cliffs, NJ, pp. 194–199.
7. M. D. Almond (1976). 'Thermal response of a multi region nuclear fuel pellet with varying thickness auxiliary heat transport path', *M.S.M.E. Thesis*, Mechanical Engineering Department, University of Arkansas, Fayetteville, Arkansas.
8. G. E. Meyers (1971). *Analytical Methods in Conduction Heat Transfer*, McGraw-Hill, New York.
9. R. G. Stanton (1961). *Numerical Methods for Science and Engineering*, Prentice-Hall, Englewood Cliffs, NJ.
10. A. Goldsmith, T. E. Waterman and H. J. Hirschhorn (1961). *Handbook of Thermophysical Properties of Solid Materials*, Macmillan, New York, Vol. I, p. 415.
11. J. G. Conway and A. D. Feith (1969). 'An interim report on a round robin experimental program to measure the thermal conductivity of stoichiometric uranium dioxide', *Gen. Elec. Co. Rep.* GEMP-715.
12. E. A. Aitken and S. K. Evans (1968). 'A thermodynamic data program involving plutonia and urania at high temperatures', *Gen. Elec. Co. Rep.* GEAP-5634.
13. A. C. Rapier, T. M. Jones and J. E. McIntosh (1963). 'The thermal conductance of uranium dioxide/stainless steel interfaces', *Int. J. Heat Mass Transfer*, 6, 397–416.
14. B. F. Rubin (1970), 'Summary of (U, Pu)$O_2$ properties and fabrication methods', *Gen. Elec. Co. Rep.* GEAP-13582.
15. L. D. Wills, Mechanical Engineering Department, University of Arkansas, Fayetteville, Arkansas, personal communication.
16. W. E. Mason (1979). 'Finite element analysis of coupled heat conduction and enclosure radiation', *Int. Conf. on Numerical Methods in Heat Transfer, University College of Swansea, Swansea, UK, July 1979.*
17. K. H. Huebner (1975). *The Finite Element Method for Engineers*, John Wiley, New York.
18. P. Silvester and A. Konard (1973). 'Axisymmetric triangular finite elements for the scalar Helmholtz function', *Int. J. Num. Meth. Engng*, 5, No. 3, 481–497.
19. O. C. Zienkiewicz (1977). *The Finite Element Method*, 3rd edn, McGraw-Hill, London.
20. D. C. Hamilton and W. R. Morgan (1952). 'Radiant interchange configuration factors', *NACA Report No.* TN-2836, December.
21. B. Carnahan, H. A. Luther and J. O. Wilkes (1969). *Applied Numerical Methods*, John Wiley, New York.

*Numerical Methods in Heat Transfer*
Edited by R. W. Lewis, K. Morgan, and O. C. Zienkiewicz
© 1981 John Wiley & Sons Ltd

## Chapter 9

# How to Deal with Moving Boundaries in Thermal Problems

*John Crank*

### 9.1 INTRODUCTION

The study of moving boundary problems in heat flow has become a highly popular subject for research in recent years and also has presented a variety of problems of practical importance. Reports on three recent conferences[1,2,3] give good impressions of the present state of the art and extensive lists of references to previous and current work. Other up to date surveys are given by Furzeland,[4] Hoffmann[5] and Fox,[6] all with useful bibliographies.

Current activities cover a wide range of interests extending from real problems of importance to engineers and other practical people, and the numerical results of interest to computer-oriented numerical analysts, to the other extreme of pure mathematical proofs for establishing existence and uniqueness and for studying error bounds and rates of convergence. Much of this more theoretical work is using the methods and results of functional analysis.

It is the intention in this chapter to concentrate mainly on methods of obtaining numerical solutions which have already been used in two- and three-dimensional problems and approaches to one-dimensional problems which seem promising for extension to higher dimensions. Consideration will also be given to the difficulties posed by more general problems.

Throughout, a moving boundary problem (MBP) or Stefan problem will be taken to mean a time-dependent problem presented by a parabolic partial differential equation together with a prescribed initial condition and boundary conditions, two of which are given on a boundary or boundaries which move in a way that depends on the solution of the partial differential equation. Heat conduction or diffusion problems with phase changes constitute a large class of MPB's.

### 9.2 SIMPLE EXAMPLES OF MPB's

#### 9.2.1 Melting ice

A simple version of the original moving boundary problem studied by Stefan round about 1890 is the melting of a semi-infinite sheet of ice, initially at zero temperature (the melting temperature) throughout, the surface of which is raised at time $t = 0$ to a constant temperature, above zero, at which it is subsequently maintained. A boundary or interface on which melting occurs moves from the surface into the sheet and separates a region of water from a region of ice at zero temperature as in Figure 9.1.

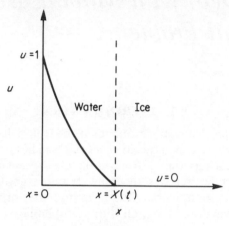

**Figure 9.1**

If $u(x, t)$ denotes the temperature distribution in the water phase and $X(t)$ the position of the interface, both at time $t$, the problem is to find both these unknowns by solving the heat flow equation, in non-dimensional terms,

$$\frac{\partial u}{\partial t} = \frac{\partial^2 u}{\partial x^2} \qquad 0 < x < X(t), \qquad t > 0 \tag{9.1}$$

subject to a fixed boundary condition

$$u = 1 \qquad x = 0 \qquad t > 0 \tag{9.2}$$

and initial conditions

$$u = 0 \qquad x > 0 \qquad t = 0 \tag{9.3}$$

$$X(0) = 0 \tag{9.4}$$

Two conditions are needed on the moving boundary, the second in order to find its position at any time. For melting ice they are

$$\left. \begin{array}{l} u = 0 \\[2mm] \dfrac{\partial u}{\partial x} = -\dfrac{dX(t)}{dt} \end{array} \right\} \qquad x = X(t) \qquad t > 0 \qquad \begin{array}{l} (9.5) \\[4mm] (9.6) \end{array}$$

The first indicates the phase change temperature and the second expresses the heat balance inclusive of the melting of ice on the moving interface. A corresponding two-phase problem in which there is a temperature distribution in the ice as well as in the water is represented by the system

$$\frac{\partial u_1}{\partial t} = k_1 \frac{\partial^2 u_1}{\partial x^2} \qquad 0 < x < X(t) \tag{9.7}$$

$$\frac{\partial u_2}{\partial t} = k_2 \frac{\partial^2 u_2}{\partial x^2} \qquad x > X(t) \tag{9.8}$$

$$-K_1 \frac{\partial u_1}{\partial x} + K_2 \frac{\partial u_2}{\partial x} = L\rho \frac{dX}{dt} \qquad x = X(t), \tag{9.9}$$

$$u = U_1 \qquad x = 0 \qquad t \geq 0 \tag{9.10}$$

$$u = -U_2 \qquad x \to \infty \qquad t \geq 0 \tag{9.11}$$

$$u_1 = u_2 = 0 \quad x = X(t) \quad t \geq 0 \tag{9.12}$$

Condition (9.9) is easily derived by referring to Figure 9.2, which shows the boundary moving a distance $\delta x$ in time $\delta t$. In order to melt the ice contained, per unit area perpendicular to $x$, in the shaded region an amount of heat $L\rho \, \delta x$

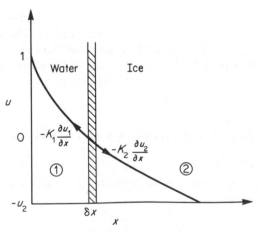

**Figure 9.2**

is required, where $L$ is latent heat and $\rho$ the density of the ice assumed to be the same as the water. Heat enters the shaded element from the water phase equal to $-K_1(\partial u_1/\partial x) \, \delta t$ and heat leaves in the ice phase equal to $-K_2(\partial u_2/\partial x) \, \delta t$. The heat balance of the shaded element implies

$$-K_1 \frac{\partial u_1}{\partial x} + K_2 \frac{\partial u_2}{\partial x} = L\rho \frac{dX}{dt} \tag{9.13}$$

In non-dimensional form and for the single phase problem (9.13) reduces to (9.6).

### 9.2.2  Oxygen diffusion and consumption problem

A slightly different problem which has been used as a testing ground for new methods is the oxygen diffusion problem of Crank and Gupta.[7,8]

The diffusion of oxygen in tissue which simultaneously consumes oxygen occurs in a number of biochemical and medical situations. When oxygen diffuses into or out of a plane sheet of a medium which absorbs oxygen at a constant rate, a moving boundary marks the progress of the innermost limit of penetration. In its simplest, non-dimensional form the problem may be stated:

$$\frac{\partial c}{\partial t} = \frac{\partial^2 c}{\partial x^2} - 1 \qquad 0 \leqslant x \leqslant x_0(t) \tag{9.14}$$

$$\partial c/\partial x = 0 \qquad\qquad x = 0 \qquad t \geqslant 0 \tag{9.15}$$

$$c = \partial c/\partial x = 0 \qquad x = x_0(t) \quad t \geqslant 0 \tag{9.16}$$

$$c = \tfrac{1}{2}(1-x)^2 \qquad 0 \leqslant x \leqslant 1 \quad t = 0 \tag{9.17}$$

Here again we have two conditions (9.16) on the moving boundary but neither of them specifies explicitly the velocity of the boundary. The second of (9.16) is called an implicit condition. It is often possible, however, to transform the explicit conditions of type (9.6) into (9.16) and vice versa[9] by using $v = \partial u/\partial x$ or $v = \partial u/\partial t$, but see Tayler in Ockendon and Hodgkins,[1] p. 125, for a note of caution.

## 9.3  ANALYTIC SOLUTIONS: EXACT AND APPROXIMATE

### 9.3.1  Neumann's solution

Referring to the two-phase system (9.7)–(9.12) above, we observe that (9.7) and (9.10) are satisfied by

$$u_1 = U_1 + A \operatorname{erf} \frac{x}{2(k_1 t)^{1/2}} \tag{9.18}$$

and (9.8) and (9.11) by

$$u_2 = -U_2 + B \operatorname{erfc} \frac{x}{2(k_2 t)^{1/2}} \tag{9.19}$$

where $A$ and $B$ are constants to be determined. Condition (9.12) requires

$$A \operatorname{erf} \frac{X}{2(k_1 t)^{1/2}} = -U_1 \qquad B \operatorname{erfc} \frac{X}{2(k_2 t)^{1/2}} = U_2 \tag{9.20}$$

These two relationships (9.20) can only be satisfied for all $t$ if

$$X = \alpha t^{1/2} \tag{9.21}$$

where $\alpha$ is constant.

By differentiating (9.18) and (9.19) and using (9.9) and (9.20), we find that $\alpha$ is given by the root of

$$\frac{U_1 K_1 \exp(-\alpha^2/4k_1)}{(\pi k)^{1/2} \operatorname{erf}(\alpha/2k_1^{1/2})} - \frac{U^2 K_2 \exp(\alpha^2/4k_2)}{(\pi k_2)^{1/2} \operatorname{erfc}(\alpha/2k_2^{1/2})} = \frac{L\rho\alpha}{2} \tag{9.22}$$

where the physical significance of $L$ and $\rho$ is stated in Section 9.2.1. Graphs which facilitate the numerical solution of (9.22) are given by Crank[10] and some tabulated values of $\alpha$ by Carslaw and Jaeger[11] for typical values of physical constants.

### 9.3.2 Heat balance integral methods

An approximate analytic solution obtained by Goodman[12] satisfies not the partial differential equation itself but its integrated form with respect to the space variable. Assumption of a temperature profile for each phase reduces a system of integral equations to a set of ordinary differential equations in time $t$, which yields the velocity of the moving boundary. We illustrate its use in solving the one-phase, one-dimensional melting ice problem (Equations (9.1) to (9.6) inclusive).

Integrate Equation (9.1) with respect to $x$, from $x = 0$ to $X(t)$

$$\frac{\mathrm{d}}{\mathrm{d}t} \int_0^{X(t)} u \, \mathrm{d}x = -\left(\frac{\mathrm{d}X}{\mathrm{d}t} + \frac{\partial u(0, t)}{\partial x}\right) \tag{9.23}$$

Assume a temperature profile in the water phase

$$u = a(x - X) + b(x - X)^2 \tag{9.24}$$

which automatically satisfies (9.5). Modify condition (9.6) by first writing

$$\frac{\mathrm{d}u}{\mathrm{d}t} = \frac{\partial u}{\partial x}\frac{\mathrm{d}X}{\mathrm{d}t} + \frac{\partial u}{\partial t} = 0 \qquad x = X(t) \tag{9.25}$$

and then eliminating $\mathrm{d}X/\mathrm{d}t$ between (9.6) and (9.25) to give

$$\left(\frac{\partial u}{\partial x}\right)^2 = \frac{\partial u}{\partial t} = \frac{\partial^2 u}{\partial x^2} \qquad x = X(t) \tag{9.26}$$

We now use (9.2) and (9.20) to obtain, after some manipulation,

$$a = (1 - \sqrt{3})/X \qquad b = \frac{aX + 1}{X^2} \tag{9.27}$$

Using these forms of $a$ and $b$ and substituting the resulting profile into the integrated form (9.23), we obtain the ordinary differential equation

$$X\frac{dX}{dt} = \frac{6(3-\sqrt{3})}{7+\sqrt{3}} = \tfrac{1}{2}\alpha^2 \qquad (9.28)$$

say. Since $X(0) = 0$ we have finally

$$X = \alpha t^{1/2} \qquad (9.29)$$

and the corresponding profile can be deduced by working back through the equations.

This is a relatively simple method to use though it is not always clear how accuracy can be improved.[12] Cho and Sunderland[13] have used the method and Noble (in Ockendon and Hodgkins,[1] p. 208) and Bell[14] have suggested introduction of finite elements so that the integral method is applied within each element. Poots[15] and Poots and Rodgers[16] have extended the heat balance method to two space dimensions and Imber and Huang[17] have included temperature-dependent thermal properties.

### 9.3.3 Embedding

Boley[74] (and in Ockendon and Hodgkins,[1] p. 150) treated a domain whose dimensions vary with time in an unknown way by considering it to be embedded or to form a part of a larger domain of constant dimensions. Clearly, only those parts of the larger domain which coincide with the actual domain have any physical meanings and the other parts are fictitious. The conditions on the boundaries of the embedding domain are also fictitious and are constructed so that the actual conditions on and within the embedded body are satisfied. Often, well known methods and solutions for bodies of constant dimensions can be utilized. In practice, any one phase can be considered to be embedded as described.

We shall illustrate the details by referring to Boley's application of his embedding technique to the melting of a solid in one space dimension, when the liquid was instantaneously removed on formation.

The solid originally occupies the space $0 < x < l$; heat enters at a rate $Q(t)$ through the surface $x = 0$, and $x = l$ is insulated. The temperature first reaches $u = u_m$ at time $t = t_m$ on $x = 0$. At a subsequent time $t$, a thickness $s(t)$ is melted, with $s(t_m) = 0$, and the solid remaining occupies the space $s(t) < x < l$. The heat $Q(t)$ always enters through $x = s(t)$.

The embedding solid always occupies the original space $0 < x < l$ and for $t > t_m$ there is a fictitious heat flux on the fictitious surface $x = 0$. The solution for $t < t_m$ is an elementary one for heat conduction in a body of fixed dimensions.

For $t > t_m$ we need to satisfy the conditions:

(a)    the heat flow equation in $0 < x < l$;
(b)    the zero derivative on $x = l$;
(c)    the temperature must be identical with the pre-melting solution at $t = t_m$;
(d)    the arbitrary heat flux $f(t)$ at $x = 0$.

Let $u_1(x, t)$ satisfy conditions (a), (b), (c) and $u_2$ the conditions (a), (b), (d) and the initial condition at $t = t_m$. Then we write the required solution as

$$u(x, t) = u_1(x, t) + u_2(x, t) \tag{9.30}$$

To find $u_1(x, t)$ we can extend the pre-melting solution beyond $t_m$ by continuing to apply the actual heat flux $Q(t)$ or its analytic continuation, $Q_1$, on $x = 0$ and obtaining the series solution

$$\frac{u_1}{u_m} = \sum_{n=0}^{\infty} \sum_{i=0}^{\infty} a_{ni} y^i X^{n-i} \tag{9.31}$$

in terms of the non-dimensional variables

$$y = \frac{t - t_m}{t_m} \quad X = \frac{x}{2(K_1 t_m)^{1/2}} \quad a_{ni} = \frac{1}{i!(n-i)!} \frac{\partial^n (u/u_m)}{\partial y^i \, \partial X^{n-i}} \tag{9.32}$$

where $K_1 = k_1/\rho c_1$ and $c_1$ is the specific heat in the solid phase.
    The temperature $u_2(x, t)$ is given by

$$\frac{u_2}{u_m} = \frac{1}{2} \int_0^y \frac{Q_2}{Q_0} (y - y_1) \frac{\partial u_0(X, y_1)}{\partial y_1} dy_1 \tag{9.33a}$$

where $Q_2 = f - Q_1$ and $Q_0 = \pi^{1/2} k_1 u_m / 2(K_1 l_m)^{1/2}$ and $u_0$ is the temperature satisfying (a) and (b) together with

$$u_0(X, 0) = 0 \quad \text{and} \quad -k_1(\partial u_0/\partial x) = 1 \qquad x = 0 \tag{9.33b}$$

The resulting temperature $u(x, t)$ contains the single unknown quantity $Q_2(y)$ and this, together with the unknown $s(t)$, must be adjusted so that the solution satisfies

$$u = u_m \quad k_1 \frac{\partial u}{\partial x} = -Q + \rho L \frac{ds}{dt} \quad x = s(t) \quad t > t_m \tag{9.34}$$

where $k$ is the thermal conductivity, $\rho$ the density and $L$ the latent heat of the solid. The resulting two integro-differential equations cannot be solved in simple analytic form. For a constant heat input $Q$ and small times, however, the series solution

$$\frac{s(t)}{2(K_1 t_m)^{1/2}} = \frac{2^m y^{3/2}}{3\pi} - \frac{m^2 y^2}{4\pi^{1/2}} - \frac{4m}{15\pi} \left( \tfrac{1}{2} - m^2 - \frac{16m}{3\pi^{3/2}} \right) y^{5/2}$$

$$+ \frac{m^2}{\sqrt{(2\pi)}} \left( \frac{35m}{8\sqrt{\pi}} - m^2 + \tfrac{1}{2} \right) y^3 + \cdots \tag{9.35}$$

applies, where

$$m = \frac{\sqrt{\pi}}{2} \frac{cu_m}{L}$$

For a two-phase problem it is necessary to solve four integro-differential equations. Boley reviews embedding techniques, including extensions to two space dimensions, in Ockendon and Hodgkins,[1] p. 208.

### 9.3.4 Other analytic techniques

Other analytic techniques are outlined by Ockendon and Hodgkins[1] and Furzeland.[4] Hansen and Hougaard[18] used Green's functions to develop an integral equation type solution for the oxygen consumption problem described in Section 9.2.2 which is considered to be the most accurate solution to date. Asymptotic expansions and perturbation solutions have been developed, notably for large or small $x$, $t$ or $L$.[4]

Rathjen and Jiji[19] solved the two-dimensional problem of freezing in a right-angled corner by considering a moving heat source of strength $L\rho \, ds/dt$ on the moving, freezing interface. Budhia and Kreith[20] generalized the method to a wedge with any angle between $0°$ and $360°$.

## 9.4 FINITE DIFFERENCE SCHEMES ON A FIXED GRID

The usual way of solving the heat flow equation over a fixed domain in one dimension, using finite difference replacements for the derivatives, is to evaluate the temperatures $u_{ij}$ at discrete points $(i \, \delta x, j \, \delta t)$ on a fixed grid in the $(x, t)$ plane. The complication associated with a moving boundary is that at any time $j \, \delta t$ it will usually be located between two neighbouring grid points, say $i \, \delta x$ and $(i+1) \, \delta x$. This can be allowed for by using modified finite difference formulae based on unequal space intervals near the boundary. Interpolation formulae of Lagrangian type can be used and if we confine attention to three-point formulae we have for a general function $f(x)$ which takes known values $f(a_0)$, $f(a_1)$, $f(a_2)$ at the three points $x = a_0, a_1, a_2$ respectively

$$f(x) = \sum_{j=0}^{2} l_j(x)f(a_j) \tag{9.36}$$

where

$$l_j(x) = \frac{p_2(x)}{(x - a_j)p_2'(a_j)} \tag{9.37}$$

$$p_2(x) = (x - a_0)(x - a_1)(x - a_2) \tag{9.38}$$

and $p_2'(a_j)$ is its derivative with respect to $x$ at $x = a_j$. It follows that

$$\frac{df}{dx} = l_0'(x)f(a_0) + l_1'(x)f(a_1) + l_2'(x)f(a_2) \tag{9.39}$$

where

$$l_0'(x) = \frac{(x - a_1)(x - a_2)}{(a_0 - a_1)(a_0 - a_2)} \tag{9.40}$$

and similarly for $l_0'(x)$, $l_2'(x)$. Also

$$\frac{1}{2}\frac{d^2 f}{dx^2} = \frac{f(a_0)}{(a_0 - a_1)(a_0 - a_2)} + \frac{f(a_1)}{(a_1 - a_2)(a_1 - a_0)} + \frac{f(a_2)}{(a_2 - a_0)(a_2 - a_1)} \tag{9.41}$$

With reference to Figure 9.3 which shows the moving boundary at time $t = j\,\delta t$ to be a fractional distance $p\,\delta x$ between the grid lines $i\,\delta x$ and $(i + 1)\,\delta x$, we

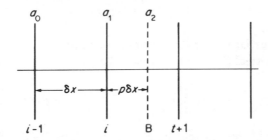

**Figure 9.3**

identify the points $a_0$, $a_1$, $a_2$ with the grid lines $(i - 1)\,\delta x$, $i\,\delta x$ and the moving boundary itself, and correspondingly $f(a_0)$, $f(a_1)$ and $f(a_2)$ with $u_{i-1,j}$, $u_{ij}$ and $u_B$ on the boundary. Then for $x < X(t)$

$$\frac{\partial^2 u}{\partial x^2} = \frac{2}{(\delta x)^2}\left(\frac{u_{i-1}}{p+1} - \frac{u_i}{p} + \frac{u_B}{p(p+1)}\right) \qquad x = i\,\delta x \tag{9.42}$$

and

$$\frac{\partial u}{\partial x} = \frac{1}{\delta x}\left(\frac{pu_{i-1}}{p+1} - \frac{(p+1)u_i}{p} + \frac{(2p+1)u_B}{p(p+1)}\right) \qquad x = X(t) \tag{9.43}$$

These and similar expressions for $x > X(t)$ are substituted in the partial differential equations (9.7) and (9.8) and the boundary condition (9.9), for example, to derive modified finite difference formulae for use near the moving boundary. For other grid points, away from this boundary, the usual difference formulae for equal space intervals are used.

Others, for example Ehrlich[21] and Koh *et al.*,[22] have used Taylor expansions in time and space near the moving boundary. Saitoh[23] used both Crank–Nicolson and fully implicit schemes with unequal intervals near the moving boundary.

Furzeland[4] has reviewed sundry ways of tracking the moving boundary itself, including explicit and implicit replacements of the boundary condition such as (9.6) or (9.9),[24,25] the use of central difference rather than end-on formulae,[26] grid refinement as used by Ciment and Guenther[27] and Schmidt in Hoffmann.[5] For implicit boundary conditions Crank and Gupta[7,8] expressed $ds/dt$ in terms of higher space derivatives by differentiating the implicit condition with respect to $t$ as often as necessary.

The most ambitious extension of irregular grids to two and three space dimensions was made by Lazaridis.[28] He combined quadratic profiles and three-point formulae over unequal intervals near the moving boundary and adopted Patel's[29] form of the moving boundary condition in which a condition

$$[k \; \partial u/\partial n]_1^2 = -\rho L v_n \tag{9.44}$$

where $n$ is the outward normal to the moving boundary and $v_n$ the boundary velocity in the normal direction, is replaced by a relation of the form

$$\left[1 + \left(\frac{\partial s}{\partial x}\right)^2 + \left(\frac{\partial s}{\partial y}\right)^2\right]\left[k \frac{\partial u}{\partial z}\right]_1^2 = -\rho L \frac{\partial z}{\partial t} \quad \text{on } z = s(x, y, t) \tag{9.45}$$

This, together with the corresponding relationships for $\partial x/\partial t$ and $\partial y/\partial t$, is highly convenient in that the velocity component $\partial z/\partial t$ is expressed solely in terms of the space derivative $\partial u/\partial z$ and space derivatives of the front $z = s(x, y, t)$.

## 9.5  VARIABLE SPACE GRID

Murray and Landis[30] kept the number of space intervals constant between say $x = 0$ and $x = X(t)$, i.e. a fixed and a moving boundary, equal to $I$, say. Thus for equal intervals $\delta x = X(t)/I$ is different in each time step. The moving boundary is thus always on the $I$th grid line. They differentiated partially with respect to time, $t$, following a given grid line instead of at constant $x$. Thus for the line $i \, \delta x$

$$\left.\frac{\partial u}{\partial t}\right|_i = \left.\frac{\partial u}{\partial x}\right|_t \left(\frac{dx}{dt}\right)_i + \left(\frac{\partial u}{\partial t}\right)_x \tag{9.46}$$

Murray and Landis assumed a general grid point at $x_i$ to move according to

$$\frac{dx_i}{dt} = \frac{x_i}{X(t)} \frac{dX}{dt} \tag{9.47}$$

The heat equation (9.1) becomes, for example,

$$\left.\frac{\partial u}{\partial t}\right|_i = \frac{x_i}{X}\frac{dX}{dt}\frac{\partial u}{\partial x} + \frac{\partial^2 u}{\partial x^2} \tag{9.48}$$

and finite differences can be applied when $dX/dt$ is determined by a boundary condition.

Gupta[31] used a Taylor expansion in both $x$ and $t$ to obtain function values in successive time steps at points on a grid system which moves bodily with the moving boundary. His equation is effectively (9.48) with $x_i/X = 1$.

Miller *et al.*[32] study the Crank–Gupta oxygen consumption problem (Section 9.2.2) by using finite elements on an adaptive mesh. They maintain elements of standard length for most of the space range, adapting only the two nearest the moving boundary. They have a novel procedure for positioning the adaptive mesh which uses the projective nature of the approximate solution. Thus, they approximate a 'quadratic tail' to the solution by the best straight-line fit in the least-squares sense and conclude that when the last node is correctly positioned at the moving boundary, the zero of the projected approximate solution should be five-sixths along the support of the last element of the mesh.

Bonnerot and Jamet[24,33] used a variable space grid similar to that of Murray and Landis[30] to construct isoparametric finite elements in space and time for the non-rectangular $(x, t)$ grid. Recently[34] thay have extended their method to two space dimensions. They take as an example the following problem:

$$\frac{\partial u}{\partial t} = \frac{\partial^2 u}{\partial x^2} + \frac{\partial^2 u}{\partial y^2} \qquad 0 \leqslant x \leqslant 1, \quad 0 \leqslant y \leqslant s(x,t), \quad t > 0 \tag{9.49}$$

$$\partial u/\partial x = 0 \quad x = 0 \quad x = 1 \quad t > 0 \tag{9.50}$$

$$u = 1 \quad y = 0 \quad 0 \leqslant x \leqslant 1 \quad t > 0 \tag{9.51}$$

$$\left.\begin{array}{l} s(x,0) = 2 + \cos \pi x \\[2mm] u(x,y,0) = 1 - \dfrac{y}{2 + \cos \pi x} \end{array}\right\} \quad \begin{array}{l} 0 \leqslant x \leqslant 1 \\[4mm] 0 \leqslant y \leqslant s(x,0) \end{array} \qquad \begin{array}{l} (9.52) \\[4mm] (9.53) \end{array}$$

Figure 9.4(a) shows a simplified picture of the bases of their triangular elements at time $t = 0$ and Figure 9.4(b) shows how the number of elements remains the same but each is stretched in the $y$-direction as the moving boundary moves in the direction of $y$ increasing. Figure 9.4(c) shows a typical isoparametric finite element in space and time in the time interval $t^n$ to $t^{n+1}$.

More recently, Jamet[35] introduced the idea of finite element approximations which are continuous with respect to the space variables at each time step but allow discontinuities with respect to the time variable between each time step. The elements at each time level can be chosen so that there is no connection with the elements at the previous time level.

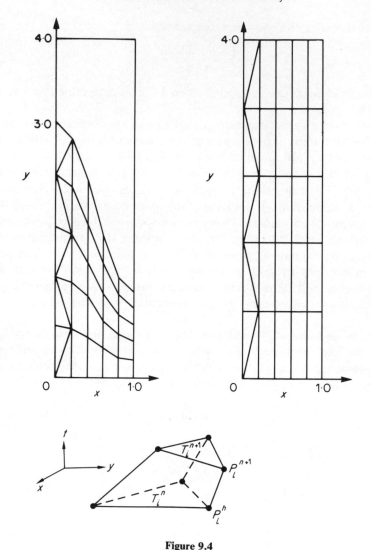

**Figure 9.4**

## 9.6  BODY-FITTED CURVILINEAR COORDINATES

In one space dimensions, various authors[36,37] have fixed the moving boundary for all times by a coordinate transformation. Thus in the simple, one-dimensional melting ice problem (Equations (9.1)–(9.6)) the transformation

$$\xi = x/X(t) \tag{9.54}$$

fixes the boundary at $\xi = 1$ for all $t$. Equations (9.1) and (9.6) become

$$\frac{\partial u}{\partial t} = \frac{1}{X^2}\frac{\partial^2 u}{\partial \xi^2} - \frac{\xi}{X}\frac{dX}{dt}\frac{\partial u}{\partial \xi} \quad 0<\xi<1 \quad t>0 \tag{9.55}$$

$$\frac{1}{X}\frac{\partial u}{\partial \xi} = -\frac{dX}{dt} \qquad \xi=1 \qquad t>0 \tag{9.56}$$

The convenience of a fixed phase-change boundary is obtained at the expense of transforming the linear heat equation (9.1) to the non-linear (9.55) in which the coefficients are functions of $X(t)$ and its time derivative. For an explicit numerical solution this presents no problem because $X$ and $dX/dt$ can be found from (9.56) before the solution of (9.55) is advanced to the next time step.

Such a coordinate transformation is a simple example of the general idea of transforming a curved-shaped region in two or more dimensions into a fixed, rectangular domain, for example. The new curvilinear coordinates are sensibly referred to as body-fitted coordinates.

Following a pioneer paper by Winslow,[38] successive authors have proposed various ways of using a curvilinear grid. Useful lists of references are given by Thompson, Thames and Mastin[39] and by Furzeland.[26] Oberkampf[40] discusses some useful generalized mapping functions.

Figure 9.5 illustrates the transformation of a two-dimensional, four-sided region in the $(x, y)$ plane into a rectangular region in the $(\xi, \eta)$ plane. The transformations may or may not be orthogonal and Furzeland[26] summarizes

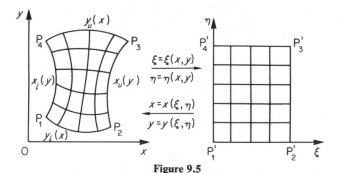

**Figure 9.5**

some of the transformations that have been proposed. We shall illustrate the simplest of these due to Oberkampf[40] which is readily seen to be of the same type as (9.54) above, i.e.

$$\xi = \frac{x - x_l(y)}{x_u(y) - x_l(y)} \qquad \eta = \frac{y - y_l(x)}{y_u(x) - y_l(x)} \tag{9.57}$$

where $x_l, x_u, y_l, y_u$ are the four curved sides of Figure 9.5(a). In general, $x_l$ etc. represent sets of discrete values of boundary points and are chosen so as to give a required mesh spacing, this freedom of choice being one of the advantages of non-orthogonal transformations. Since $\xi$, $\eta$ are then known (discrete) functions of $x$ and $y$, the derivatives $\xi_x$, $\xi_y$, etc. required in solving a problem are available and permit discretized forms of the partial differential equation and the boundary conditions to be formulated and solved on the rectangular $(\xi, \eta)$ grid. We note that in this plane a moving boundary will be fixed.

Furzeland[26] has solved the problem of Bonnerot and Jamet by this and other curvilinear transformations and compared his results with those obtained by other authors. Alternative mappings result from the use of finite elements and bivariate blending functions[41] or isoparametric curvilinear coordinates.[42]

Saitoh[43] uses a version of the transformation (9.57) in cylindrical coordinates and refers to the work of Duda *et al.*[44] More recently, Sparrow *et al.*[45] have applied the same method in each phase of a melting, initially subcooled, region around a circular cylinder. Further applications of the general method are discussed by Hoffmann (reference 5, III).

## 9.7   ISOTHERM MIGRATION METHOD

Traditionally, solutions of the heat flow equation express the time dependence of temperature at a fixed point in space. An alternative, proposed by Chernousko[46] and independently by Dix and Cizek,[47] is to calculate how specified temperatures move through the medium; in other words, how isotherms migrate. We evaluate $x(u, t)$ instead of $u(x, t)$. When a phase change boundary is itself an isotherm, the isotherm migration method (IMM) becomes another way of fixing the boundary and of evaluating $x$ in a fixed domain $0 \leqslant u \leqslant 1$, for example. There are advantages also in dealing with temperature-dependent thermal properties. As an example, Equations (9.1) and (9.6) become

$$\frac{\partial x}{\partial t} = \left(\frac{\partial x}{\partial u}\right)^{-2} \frac{\partial^2 x}{\partial u^2} \quad 0 < u < 1 \quad t > 0 \tag{9.58}$$

$$\frac{dX}{dt} = -\left(\frac{\partial x}{\partial u}\right)^{-1} \quad u = 0 \qquad t > 0 \tag{9.59}$$

Crank and Phahle[48] obtained solutions of (9.58) and (9.59) which agree well with the exact analytical solution of the melting ice problem.

The IMM has been extended into two space dimensions in two ways. Crank and Gupta[49] solved the problem of the solidification of a square prism of fluid, initially at the melting temperature $u = 1$ throughout, and the surface of which is subsequently maintained at $u = 0$, below the melting temperature. They applied the IMM transformation to the $y$ variable only in the heat flow equation

$$\frac{\partial u}{\partial t} = \frac{\partial^2 u}{\partial x^2} + \frac{\partial^2 u}{\partial y^2} \tag{9.60}$$

to obtain

$$\frac{\partial y}{\partial t} = -\left[\frac{\partial^2 u}{\partial x^2} - \frac{\partial^2 y}{\partial u^2}\left(\frac{\partial y}{\partial u}\right)^{-3}\right]\frac{\partial y}{\partial u} \tag{9.61}$$

They then solved discretized equations on a $(u, x)$ grid to obtain values of $y$ at successive times $k\,\delta t$. In order to discretize $(\partial^2 u/\partial x^2)_y$ at the point $(i\,\delta u,\, j\,\delta x)$ they required, at three equally spaced values of $x$, values of $u$ for which $y = y_{ij} = y(i\,\delta u,\, j\,\delta x)$. They interpolated linearly the values of $u$ corresponding to $y_{ij}$ at $x_{j-1}$ and $x_{j+1}$, e.g. at $x_{j-1}$ they used

$$u = \frac{u_{j+1}(y_{i,j-1} - y_{ij}) - u_i(y_{i+1,j-1} - y_{ij})}{y_{i,j-1} - y_{i+1,j-1}} \tag{9.62}$$

The IMM expression for the Patel[29] form of the boundary condition $\partial u/\partial n = -\beta v_n$ on $f(x, y) = 0$ is

$$\frac{\partial y}{\partial t} = \left[1 + \left(\frac{\partial y}{\partial x}\right)^2\right]\left(\frac{\partial y}{\partial u}\right)^{-1} \tag{9.63}$$

A different and novel approach to transient heat flow problems based on the IMM transformation is described by Crank and Crowley.[50] The movements of isotherms along orthogonal flow lines are tracked in successive small intervals of time by solving a locally one-dimensional IMM form of radial heat equation. The determination of the new orientation of the orthogonal system at the end of each time interval is based on geometrical considerations. In a general heat flow problem in two dimensions there will be a family of isotherms and an associated family of flow lines, orthogonal to the isotherms. In an isotropic medium, any point on an isotherm moves along the flow lines normal to the isotherm at that point. Heat flow is everywhere normal to the isotherms and never crosses the flow lines. Crank and Crowley[50] regarded a small element of an isotherm as part of a cylindrical system for a small time and obtained the normal movement of the isotherm from the IMM form of the radial heat equation

$$\frac{\partial r}{\partial t} = \frac{\partial^2 r/\partial u^2}{(\partial r/\partial u)^2} - \frac{1}{r} \tag{9.64}$$

where the coordinate $r$ is identified as the local radius of curvature of the element of an isotherm measured from the local centre of curvature assumed fixed over the time interval $\delta t$. At the end of each interval the new direction of the normal for each element of isotherm is calculated from a geometric construction based on approximating sections of the isotherms by circular arcs. An explicit discretized form of (9.64) is derived to obtain the new position of the isotherm after each time step $\delta t$. More recently, these authors[51] have developed an implicit version of the same method.

## 9.8 ENTHALPY AND OTHER FIXED DOMAIN METHODS

Some problems can be reformulated in a way that avoids the need to solve two partial differential equations, one in each phase of a two-phase problem for example, with their solutions coupled through the conditions on the melting interface. One procedure is to introduce an enthalpy function, $H(u)$, where $H(u)$ is the sum of the specific heat and latent heat contents. The two or more heat equations in different phases can be replaced by the single equation

$$\rho \frac{\partial H}{\partial t} = \frac{\partial}{\partial x}\left(K \frac{\partial u}{\partial x}\right)$$

(9.65)

where $\rho$ and $K$ may be functions of $u$.

The different heat properties in the two phases and the identification of the melting interface are all bound up in the form of the function $H(u)$. There may be a discontinuous jump in $H$ at the value of $u$ for which melting occurs, as in Figure 9.6. Szekely and Themelis[52] replaced the step function by a continuous curve such as the dotted line in Figure 9.6. The essential feature is that, in this approach, the interface effectively disappears from the primary process of numerical solution and its position is given, *a posteriori*, by the values of the space coordinate(s) where $u = u_M$ (Figure 9.6). The enthalpy function was used by Eyres *et al.*[53] and since then by Albasiny,[54] Atthey,[55] Meyer[25] and others.

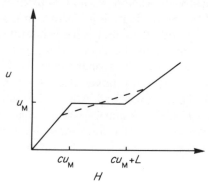

**Figure 9.6**

Thus, instead of an interface condition (9.9), Atthey[55] takes

$$u = \begin{cases} H/c & H \leq cu_M \\ u_M & c_M \leq H \leq cu_M + L \\ (H-L)/c & H \geq cu_M + L \end{cases}$$

(9.66)

where $u_M$ is the melting temperature. He proved that his numerical solution converges, as the intervals tend to zero, to a weak solution of the differential equation. He used the simplest explicit replacement of (9.65) for constant $K$

$$\rho \frac{H_{i,j+1} - H_{ij}}{\delta t} = \frac{K}{(\delta x)^2} (u_{i+1,j} - 2u_{ij} + u_{i-1,j})$$  (9.67)

Atthey[55] applied the form (9.66) to a study of mushy regions where there is no sharp interface between the two phases.

A straightforward extension of the enthalpy method to two dimensions has been made by Crowley.[56] She considers the inward solidification of a square cylinder of fluid with two different surface conditions. First, the surface temperature is lowered at a constant rate and Crowley's results are compared with corresponding experimental data obtained by Saitoh.[57] Second, the surface temperature is taken to fall discontinuously at zero time and these enthalpy results are compared with those of earlier workers. Crowley concludes that the enthalpy formulation provides a simple and accurate scheme for the numerical solution of multidimensional Stefan problems.

Furzeland[26] used explicit and implicit finite differences and in particular suggested a numerical procedure which alleviates the need for a trial and error search in determining which of the three regions defined in (9.66) is applicable at any stage of an iterative implicit scheme.

Ciavaldini[58] used explicit and implicit finite element schemes to solve a discretized weak enthalpy formulation. Comini *et al.*[59] concentrate attention on permitting simultaneous temperature dependence of thermal conductivity, heat capacity, rate of internal heat generation and surface heat transfer coefficients. They find a three-time-level difference scheme to be amenable with an automatically adjusted time step as the solution proceeds. Latent heat effects are approximated by a large heat capacity over a small temperature interval enclosing the melting temperature, expressed by a Dirac type function. For this they find an enthalpy function useful and differentiate to derive the heat capacities. The spacewise discretization uses finite elements. An improved procedure is described by Morgan *et al.*[60]

Elliott[61] has demonstrated links between discretizations of the enthalpy formulation and the corresponding discretized forms expressing moving boundary problems in terms of variational inequalities. A simple one-dimensional example is provided by the oxygen diffusing problem described by Equations (9.14)–(9.17).

If the definition of $u$ is extended to the whole fixed domain by

$$u \equiv 0 \qquad x_0(t) \leqslant x \leqslant 1 \qquad 0 < t < T$$  (9.68)

then it can be shown that $u$ satisfies the variational inequalities

$$\left( \frac{\partial u}{\partial t} - \frac{\partial^2 u}{\partial x^2} - Q \right) u = 0$$

$$\frac{\partial u}{\partial t} - \frac{\partial^2 u}{\partial x^2} - Q \geqslant 0 \qquad u \geqslant 0$$  (9.69)

On suitable discretization, (9.69) leads to a quadratic programming problem of minimization at each time level. Elliott[61] obtained a finite element solution, using linear basis functions and the Crank–Nicolson finite difference scheme in time. The position of the moving boundary was found by quadratic extrapolation on the last two grid points coupled with the moving boundary conditions. Elliott and Janovsky[62] have extended the method to more than one space dimension.

Another, closely related approach to the oxygen consumption problem is the truncation method proposed by Berger *et al.*[63] The problem is solved by either finite difference or finite element methods over a fixed domain $0 \leqslant x \leqslant 1$ with $u = 0$ at $x = 1$. For part of the domain $u$ takes negative values which are set to zero or 'truncated' at each time step. The moving boundary is taken to be at the grid point at which $u$ first becomes zero. Elliott[61] considers his method to be more accurate than truncation because he uses the additional constraint $\partial u/\partial t - \partial^2 u/\partial x^2 + 1 \geqslant 0$ as well as $u \geqslant 0$.

Berger *et al.*[64] have extended the truncation method to two-phase problems by truncating in one plane and then the other in alternate time steps, and Berger[65] has combined truncation with variational inequalities.

Evans and Gourlay[66] have used truncation and hopscotch methods to solve a two-dimensional version of the oxygen consumption problem discussed in Section 9.2.2.

Duvaut[67] used the Baiocchi[68] transformation

$$v(\mathbf{x}, t) = \begin{cases} \displaystyle\int_{l(\mathbf{x})}^{t} u(\mathbf{x}, t)\, \mathrm{d}t & t > l(x) \\ 0 & t \leqslant l(\mathbf{x}) \end{cases} \tag{9.70}$$

to transform the explicit moving boundary condition, e.g. condition (9.6) for a one-phase problem, into the implicit form

$$v = 0 \qquad \partial v/\partial n = 0 \qquad t = l(\mathbf{x}) \tag{9.71}$$

Here the moving boundary position is given by $t = l(\mathbf{x})$, the time at which melting first occurs at $\mathbf{x}$. The corresponding two-phase problem is dealt with similarly by Duvaut (see Ockendon and Hodgkins[1]).

The enthalpy formulation which includes discontinuous functions, e.g. $H(u)$, and the lack of differentiability suggests the use of weak or integral solutions may be profitable. For example, a weak form of Equation (9.65) is obtained by multiplying by an arbitrary test function $\phi$ defined on the domain, $D$, which is twice continuously differentiable with respect to the space variables and once with respect to time and which satisfies

$$\phi(\mathbf{x}, t) = 0 \quad x \in \partial D$$
$$\phi(\mathbf{x}, T) = 0 \quad x \in D \tag{9.72}$$

Integration by parts yields

$$\int_0^T \int_D \left[ u \frac{\partial}{\partial x}\left( K \frac{\partial \phi}{\partial x}\right) + H(u) \frac{\partial \phi}{\partial t}\right] dx \, dt$$

$$= \int_0^T \oint_{\partial D} Kg \frac{\partial \phi}{\partial n} dx \, dt - \int_D H(u_0)\phi(x, 0) \, dx \qquad (9.73)$$

where we have taken

$$u = \begin{cases} g(x, t) & x \in \partial D \\ u_0(x) & t = 0 \end{cases} \qquad (9.74)$$

More progress has been made in establishing the theoretical properties of weak solutions in some cases where corresponding properties are difficult to establish for classical solutions (see Furzeland[4] for references).

## 9.9   METHOD OF LINES

In this method only the time variable is discretized. The partial differential equation is replaced by a sequence of ordinary differential equations at discrete time levels. Meyer[69,25] has developed this approach in a more consistent and generalized way than hitherto, and reference 69 contains references to earlier work. We introduce the ideas here through the one-dimensional, two-phase Stefan problem specified by

$$\frac{\partial}{\partial x}\left( k_1 \frac{\partial u_1}{\partial x}\right) - c_1 \frac{\partial u_1}{\partial t} = F_1 \qquad b_1 < x < s(t) \qquad t > 0 \qquad (9.75)$$

$$\frac{\partial}{\partial x}\left( k_2 \frac{\partial u_2}{\partial x}\right) - c_2 \frac{\partial u_2}{\partial t} = F_2 \qquad s(t) < x < b_2 \qquad t > 0 \qquad (9.76)$$

$$u_1 = \beta_1(t) \frac{\partial u_1}{\partial x} + \alpha_1(t) \qquad x = b_1 \qquad t > 0 \qquad (9.77)$$

$$\frac{\partial u_2}{\partial x} = \beta_2(t)u_2 + \alpha_2(t) \qquad x = b_2 \qquad t > 0 \qquad (9.78)$$

where $k_1, c_1, F_1, k_2, c_2, F_2$ may be functions of $x$ and $t$, e.g. $k_1 = k_1(x, t)$, etc. On the free boundary $s(t)$ we have conditions

$$u = \mu_1(s(t), t) \qquad u_2 = \mu_2(s(t), t) \qquad (9.79)$$

$$k_1 \frac{\partial u_1}{\partial x} - k_2 \frac{\partial u_2}{\partial x} + \lambda(s(t), t) \frac{ds}{dt} = \mu_3(s(t), t) \qquad x = s(t) \qquad t > 0 \quad (9.80)$$

Meyer (in Wilson *et al.*[2]) shows that the introduction of convection terms $\partial u_1/\partial x$, $\partial u_2/\partial x$ causes no further difficulty.

At the $n$th time level we can approximate the system (9.75)–(9.78) by

$$(k_i u_i')' - c_i \frac{u_i - v_i(x)}{\delta t} = F_i \qquad i = 1, 2$$

$$u_1(b_1) = \beta_1 u_1'(b_1) + \alpha_1$$

$$u_2'(b_2) = \beta_2 u_2(b_2) + \alpha_2 \qquad\qquad\qquad (9.81)$$

$$u_1(s) = \mu_1 \quad u_2(s) = \mu_2 \quad k_1 u_1'(s) - k_2 u_1'(s) + \lambda\left(\frac{s - s_{n-1}}{\delta t}\right) = \mu_3$$

where a prime indicates $d/dx$, $\delta t = t_n - t_{n-1}$ and where all parameters are evaluated at $t = t_n$.

The functions $v_i$ denote $u_i$ at the previous time level or their linear extensions beyond their domain of definition in the neighbourhood of the moving interface. Thus if the boundary is moving in the direction of $x$ increasing, so that $s(t_n) > s(t_{n-1})$, we use

$$v_1(x) = u_1(s_{n-1}) + (x - s_{n-1})u_1'(s_{n-1}) \quad x \geqslant s_{n-1} \qquad (9.82)$$

A full description of the method of lines applied to the discretized system (9.81) is given by Meyer.[25] We require to find the positions of the moving boundary at successive times $t = t_n$ and to solve the sequence of second-order, ordinary differential equations. Meyer introduces the Riccati transformations

$$u_1(x) = U(x)\phi_1(x) + w(x)$$

$$\phi_2(x) = R(x)u_2(x) + z(x) \qquad\qquad\qquad (9.83)$$

$$\phi_i = k_i(x, t_n)u_i'$$

We obtain $U$, $w$, $R$ and $z$ by solving the initial value problems

$$\frac{dU}{dx} = \frac{1}{k_1} - \frac{c_1 U^2}{\delta t} \qquad\qquad U(b_1) = \frac{\beta_1}{k_1}$$

$$\frac{dw}{dx} = -\frac{c_1}{\delta t} U(w - v_1) - UF_1 \qquad w(b_1) = \alpha_1$$

$$\qquad\qquad\qquad\qquad\qquad\qquad\qquad\qquad\qquad (9.84)$$

$$\frac{dR}{dx} = \frac{c_2}{\delta t} - \frac{R^2}{k_2} \qquad\qquad R(b_2) = \beta_2 k_2$$

$$\frac{dz}{dx} = \frac{R}{k_2} z - \frac{c_2 v_2}{\delta t} + F_2 \qquad z(b_2) = \alpha_2 k_2$$

By substituting $k_1 u_1'$ and $k_2 u_2'$ from the Riccati relations (9.83) into the discretized boundary condition in the last of (9.81), we see that the position of the interface at $t = t_n$ is the root $x = s_n$ of the equation

$$\frac{\mu_1(x, t_n) - w(x)}{U(x)} \div [R(x)\mu_2(x, t_n) + z(x)] + \lambda(x, t_n)\frac{x - s_{n-1}}{\delta t} - \mu_3(x, t_n) = 0$$

(9.85)

Once $s(t_n)$ is known, the functions $u_i$ are obtained by backward integration of

$$\frac{du_2}{dx} = \frac{1}{k_2}(Ru_2 + z) \qquad u_2(s) = \mu_2(s) \qquad s < x < b_2$$

(9.86)

from the second of (9.83), and

$$\frac{d\phi_1}{dx} = c_1\frac{U\phi_1 + w - v_1}{\delta t} + F_1$$

with

$$\phi_1(s) = \frac{\mu_1(s) - w(s)}{U(s)} \qquad b_1 < x < s$$

from the first equations of (9.81) and (9.83), and $u_1(x)$ follows from the first of (9.83).

Meyer[25] suggests integration of the differential equations by the simple trapezoidal rule coupled with linear interpolation between nodal values to locate the root $x = s_n$ of (9.85). He discusses the relative merits of this and more elaborate methods.

Meyer[70] uses the method of lines along alternate directions in successive time steps in order to deal with problems in two space dimensions. The Riccati transformation is again employed. A method of lines/SOR iterative algorithm is thought to be even more widely applicable than the ADI[71] method. Some work in three dimensions is also reported by Meyer.[72] Furzeland[73] is compiling a general purpose computer package for solving non-linear problems by the method of lines.

## 9.10  SUMMING-UP

It would be premature to attempt a detailed comparison of methods for solving moving boundary problems with the intention of identifying 'the best method'. Maybe such a simple conclusion is a 'will-o-the-wisp' anyway. It is perhaps useful, however, to conclude this paper with one or two remarks of a broad nature.

There seems to be a growing acceptance that the enthalpy approach is the simplest to handle, provided an enthalpy function can be identified. However, the *a priori* determination of the position of the moving boundary may not yield a very accurate result even in one dimension and still less so in higher space dimensions. If the location of the boundary is critical, therefore, the enthalpy method should be avoided or used with caution in its present state of development. Similar remarks apply to other fixed domain methods, including the 'truncation method'.

Where tracking of the moving boundary is important, adaptive meshes, which are modified at each time step to fit the current position of the moving boundary, offer advantages. So also do body-fitted, curvilinear coordinates which transform the curved region of changing shape and size into a fixed domain in one or more dimensions. It looks as though finite elements in space are attractive for implementing the former algorithms, sometimes, though not always, combined with finite differences in time. Other workers prefer the fixed mesh in the transformed domain and, provided the moving boundary is an isothermal surface, the handling of temperature-dependent heat properties is more economically done in the isotherm migration methods, which are special cases of coordinate transformation methods.

A strong attraction of the 'method of lines' is that well established algorithms for solving ordinary differential equations can be used, and the position of the moving boundary is accurately determined.

Formulations in terms of variational inequalities are the subject of intense investigation at the present time. They offer the double attraction of allowing the pure mathematical properties such as existence and uniqueness to be investigated and of leading to computational algorithms. A fairly wide knowledge of functional analysis particularly is needed, however, to understand the literature and to develop appropriate inequalities, where indeed this is possible at all. Such variational approaches are perhaps best left for the mathematicians to explore for the time being until they can systematically produce algorithms of wider practical application.

The soundest advice to engineers and others with practical problems at the moment seems to be, 'If you have an algorithm which works, use it!'

## REFERENCES

1. J. R. Ockendon and W. R. Hodgkins (Eds) (1975). *Moving Boundary Problems in Heat Flow and Diffusion*, Clarendon Press, Oxford.
2. D. G. Wilson, A. D. Solomon and P. T. Boggs (1978). *Moving Boundary Problems*, Academic Press, New York.
3. R. M. Furzeland (1979). *Bull. Inst. Math. Applic.*, **15**, 172–176.
4. R. M. Furzeland (1977a). *Brunel University Math. Report* TR/76.
5. K. H. Hoffmann (1977). *Freie Universität (Berlin), Fachbereich Mathematik*, Vols I (*Preprint No. 22*), II (*No. 28*), III (*No. 34*).
6. L. Fox (1979). in *A survey of Numerical Methods for Partial Differential Equations*, Eds Gladwell, I. and Wait, R., pp. 332–356, Oxford.
7. J. Crank and R. S. Gupta (1972a). *J. Inst. Math. Applic.*, **10**, 19–33.
8. J. Crank and R. S. Gupta (1972b). *J. Inst. Math. Applic.*, **10**, 296–304.
9. A. Schatz (1969). *J. Math. Anal. Applic.*, **28**, 569–580.

10. J. Crank (1975). *The Mathematics of Diffusion*, 2nd edn, Clarendon Press, Oxford.
11. H. S. Carslaw and J. C. Jaeger (1959). *Conduction of Heat in Solids*, Clarendon Press, Oxford.
12. T. R. Goodman (1961). *J. Heat Transfer, Trans. ASME(C)*, **83**, 83–86.
13. S. H. Cho and J. E. Sunderland (1969). *J. Heat Transfer, Trans. ASME(C)*, **91**, 421–426.
14. G. E. Bell (1978). *Int. J. Heat Mass Transfer*, **21**, 1357–1361.
15. G. Poots (1962). *Int. J. Heat Mass Transfer*, **5**, 339–348.
16. G. Poots and G. G. Rodgers (1976). *J. Inst. Math. Applic.*, **18**, 203–217.
17. M. Imber and P. N. S. Huang (1973). *Int. J. Heat Mass Transfer*, **16**, 1951–1954.
18. E. Hansen and P. Hougaard (1974). *J. Inst. Math. Applic.*, **13**, 385–398.
19. K. A. Rathjen and L. M. Jiji (1971). *J. Heat Transfer, Trans. ASME(C)*, **93**, 101–109.
20. H. Budhia and F. Kreith (1973). *Int. J. Heat Mass Transfer*, **16**, 195–211.
21. L. W. Ehrlich (1958). *J. Ass. Comp. Mach.*, **5**, 161–176.
22. J. C. Y. Koh, J. F. Price and R. Colony (1969). *Prog. Heat Mass Transfer*, **2**, 225–247.
23. T. Saitoh (1972). *Mem. Sagami Inst. Technol.*, **6**, 1–14.
24. R. Bonnerot and P. Jamet (1974). *Int. J. Num. Meth. Engng*, **8**, 811–820.
25. G. H. Meyer (1976). *Brunel University Math. Report* TR/62.
26. R. M. Furzeland (1977b). *Ph.D. Thesis*, Brunel University.
27. M. Ciment and R. B. Guenther (1974). *Applic. Anal.*, **4**, 39–62.
28. A. Lazaridis (1970). *Int. J. Heat Mass Transfer*, **13**, 1459–1477.
29. P. D. Patel (1968). *AIAA J.*, **6**, 2454.
30. W. D. Murray and F. Landis (1959). *J. Heat Transfer, Trans. ASME(C)*, **81**, 106–112.
31. R. S. Gupta (1973). *Ph.D. Thesis*, Brunel University; also *Brunel University Math. Report* TR/32.
32. J. V. Miller, K. W. Morton and M. J. Baines (1978). *J. Inst. Math. Applic.*, **22**, 467–477.
33. R. Bonnerot and P. Jamet (1975). *J. Comp. Phys.*, **18**, 21–45.
34. R. Bonnerot and P. Jamet (1977). *J. Comp. Phys.*, **25**, 163–181.
35. P. Jamet (1978). *SIAM J. Num. Anal.*, **15**, 912–928.
36. H. G. Landau (1950). *Q. Appl. Math.*, **8**, 81–94.
37. D. H. Ferriss and S. Hill (1974). *NPL report* NAC 45; also pp. 251–255 in Ockendon and Hodgkins.[1]
38. A. M. Winslow (1967). *J. Comp. Phys.*, **2**, 149–172.
39. J. F. Thompson, F. C. Thames and C. W. Mastin (1974). *J. Comp. Phys.*, **15**, 299–319.
40. W. L. Oberkampf (1976). *Int. J. Num. Meth. Engng*, **10**, 211–223.
41. W. J. Gordon and C. A. Hall (1973). *Int. J. Num. Meth. Engng*, **7**, 461–477.
42. O. C. Zienkiewicz and D. V. Phillips (1971). *Int. J. Num. Meth. Engng*, **3**, 519–528.
43. T. Saitoh (1978). *J. Heat Transfer*, **100**, 294.
44. J. L. Duda, M. F. Malone, R. H. Notter and J. S. Ventras (1975). *Int. J. Heat Mass Transfer*, **18**, 901–910.
45. E. M. Sparrow, S. Ramadhyani and S. V. Patankar (1978). *J. Heat Transfer*, **100**, 395.
46. F. L. Chernousko (1970). *Int. Chem. Engng*, **10**, No. 1, 42.
47. R. C. Dix and J. Cizek (1970). in *Heat Transfer 1970*, Vol. 1, *4th Int. Heat Transfer Conf., Paris, Versailles*, Elsevier, Amsterdam.
48. J. Crank and R. D. Phahle (1973). *Bull. Inst. Math. Applic.*, **9**, 12–14.

49. J. Crank and R. S. Gupta (1975). *Int. J. Heat Mass Transfer*, **18**, 1101–1107.
50. J. Crank and A. B. Crowley (1978). *Int. J. Heat Mass Transfer*, **21**, 393–398.
51. J. Crank and A. B. Crowley (1979). *Int. J. Heat Mass Transfer*, to appear.
52. J. Szekely and N. J. Themelis (1971). *Rate Phenomena in Process Metallurgy*, Wiley-Interscience, New York, Ch. 10.
53. N. R. Eyres, D. R. Hartree, J. Ingham, R. Jackson, R. J. Sarjant and S. M. Wagstaff (1946). *Phil. Trans. R. Soc.*, **A240**, 1.
54. E. L. Albasiny (1956). *Proc. IEE*, **103B**, 158–162.
55. D. R. Atthey (1974). *J. Inst. Math. Applic.*, **13**, 353–366.
56. A. B. Crowley (1978). *Int. J. Heat Mass Transfer*, **21**, 215–218.
57. T. Saitoh (1976). *Technol. Rep., Tohoku Univ.*, **41**, 61–72.
58. J. F. Ciavaldini (1975). *SIAM J. Num. Anal.*, **12**, 464–487.
59. G. Comini, S. del Guidice, R. W. Lewis and O. C. Zienkiewicz (1974). *Int. J. Num. Meth. Engng*, **8**, 613–624.
60. K. Morgan, R. W. Lewis and O. C. Zienkiewicz (1978). *Int. J. Num. Meth. Engng*, **12**, 1191–1195.
61. C. M. Elliott (1976). *D. Phil. Thesis*, Oxford University.
62. C. M. Elliott and V. Janovsky (1977). *Inst. Comp. Math., Brunel University Report* 77-5.
63. A. E. Berger, M. Ciment and J. C. W. Rogers (1975a). *SIAM J. Num. Anal.*, **12**, 646–672.
64. A. E. Berger, M. Ciment and J. C. W. Rogers (1975b). *Séminaire IRIA, Analyse et contrôle des systèmes*.
65. A. E. Berger (1976). *RAIRO Analyse Numérique*, **10**, 29–42.
66. N. T. S. Evans and A. R. Gourlay (1977). *J. Inst. Math. Applic.*, **19**, 239–251.
67. G. Duvaut (1973). *C.R. Acad. Sci. Paris*, **276A**, 1461–1463.
68. C. Baiocchi (1972). *Ann. Mat. Pura Appl.*, **92**, 107–127.
69. G. H. Meyer (1975). *Brunel University Math. Report* TR/52.
70. G. H. Meyer (1977a). *Int. J. Num. Meth. Engng*, **11**, 741–752.
71. G. H. Meyer (1977b). *J. Inst. Math. Applic.*, **20**, 317–329.
72. G. H. Meyer (1978). *Inst. Comp. Math., Brunel University, Report* 78-13.
73. R. M. Furzeland (1979). *Oxford University Computing Lab. Report*, to appear.
74. B. A. Boley (1961). *J. Math. Phys.*, **40**, 300–313.

*Numerical Methods in Heat Transfer*
Edited by R. W. Lewis, K. Morgan, and O. C. Zienkiewicz
© 1981 John Wiley & Sons Ltd

Chapter 10

# Multidimensional Integral Phase Change Approximations for Finite Element Conduction Codes

*E. C. Lemmon*

## SUMMARY

Finite element conduction codes may be used to approximate phase change problems using the techniques described in this chapter. These methods include one-, two-, or three-dimensional transient phase change problems such as the cooling of a liquid below its solidification (or freezing) temperature or heating a solid above its decomposition (or melting) temperature. A method is also developed to account for convective heat transfer at the moving phase-change interface. An important feature of the methods given is that once a mesh has been established it is not changed even though the phase-change interface moves through the mesh. A large variety of phase-change problems have been approximated with multidimensional conduction finite element programs and the results compared favourably with exact solutions or other approximate solutions. The great utility of the methods presented is that they may be applied in one, two, or three dimensions for very complex problems with difficult geometries where, in general, exact solutions do not exist.

## 10.1 INTRODUCTION

At present phase-change problems are approximated with both moving mesh and non-moving mesh techniques. The moving mesh techniques often offer excellent accuracy but are limited to very simple problems and geometries. There are numerous examples of these methods in the literature, for example, see Lewis and Morgan,[2] Section 2. The non-moving mesh techniques have also been very limited. For example, convective heat transfer at the phase-change interface and discontinuous conductivity variations have not been taken into account. The purpose of this chapter is to explain how to use the non-moving finite element mesh technique to overcome these problems. Few details will be given on the basic finite element method (FEM). Such details are given in numerous publications such as Zienkiewicz.[7]

Briefly, in the FEM the temperature field throughout the continuum of interest is approximated by a discretization procedure which divides the solution region into finite regions (elements). The temperature field is then expressed in terms of an assumed approximating function within each element. The approximating or interpolation functions are defined in terms of the value of the temperature at specified points called nodes. These nodes usually lie on the element boundaries where adjacent elements are considered to be connected. Hence, in the FEM, the temperatures at the nodes are the unknowns. Once these unknowns are found, the interpolation functions define the temperature throughout the assemblage of elements.

One of the important features of the FEM is the ability to formulate contributions for individual elements before putting them together (assembly) to represent the entire problem.

Although there is more than one way to formulate the contributions for individual elements, only the method of weighted residuals (MWR) will be here described. The MWR is a technique for obtaining approximate solutions to linear and non-linear partial differential equations. In the MWR the general functional behaviour of the temperature is assumed in some way so as to approximately satisfy the given differential equation and boundary conditions. Substitution of this approximation into the original differential equation and boundary conditions then results in some error called a residual. This residual is required to vanish in some average sense over the entire solution domain.

After the individual element characteristics are determined by the MWR, the element contributions are assembled to obtain the overall system equations. These equations are then solved, giving the temperatures at each node point.

The thrust of this chapter is to outline how the thermal properties (conductivity and heat capacity) can be evaluated so that a typical conduction finite element code can be used to approximate multidimensional phase-change problems such as: (a) liquid solidification under action of surface heat removal such as ice production or solidification of a casting; (b) thermal/chemical decomposition of a solid without removal of degraded material from the remaining virgin material such as charring of wood or reinforced plastics; (c) ablation of solids where the products of decomposition are removed on formation, such as melting glass or subliming teflon. Of prime interest to the method to be outlined is the determination of the amount of material solidified, decomposed, melted or sublimed and the location of the phase-change interface as a function of time as it moves through the one-, two-, or three-dimensional finite element mesh. As the interface moves through each element, the energy involved in the phase-change process and the difference in heat capacity and conductivity of the two phases must be taken into account. A method is also included to account for the convective heat transfer at the moving phase-change interface.

The methods to be outlined depend strongly on two integral averaging techniques. These techniques, which will be explained, are used to determine element thermal properties for an element where there is a large variation of either thermal property throughout the element.

## 10.2  INTEGRAL AVERAGING TECHNIQUE

Let $u = u(T)$ define a function of $T$ on the closed interval $T_1 \leq T \leq T_2$. The area $(w)$ bounded above by the graph of the function, on the sides by the vertical lines through $T = T_1$ and $T = T_2$, and below by the $T$-axis, is

$$w = \int_{T_1}^{T_2} u \, dT = \int_{T_0}^{T_2} u \, dT - \int_{T_0}^{T_1} u \, dT = w_{T_2} - w_{T_1} \tag{10.1}$$

where $T_0$ is some reference value of $T$ which is less than $T_1$. The average value of $u(\bar{u})$ in the interval $T_1 \leq T \leq T_3$ is then

$$\bar{u}_{T_1, T_2} = \frac{w_{T_2} - w_{T_1}}{T_2 - T_1} \tag{10.2}$$

This basic integration method will be used as a tool to determine various property values in each region of interest.

For a property $(u)$ which is dependent on temperature $(T)$ as in Figure 10.1(a), $\bar{u}$, the average value of $u$ in the interval from $T_1$ to $T_3$, is obtained by

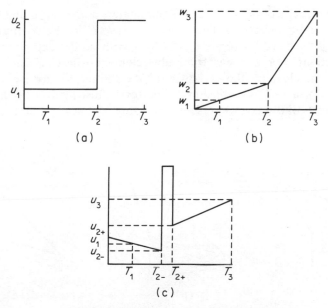

**Figure 10.1**  Integral averaging technique

direct integration. With $w = \int u \, dT$, the $w$–$T$ relationship would be of the form of Figure 10.1(b). The average value of $u$ in the interval can be determined as

$$\bar{u}_{T_1,T_3} = dw/dT|_{T_1,T_3} \equiv (w_{T_3} - w_{T_1})/(T_3 - T_1) \qquad (10.3)$$

The defining symbol ($\equiv$) in Equation (10.3) should be noted. The gradient term is not a usual derivative at every point on the curve of $w$ against $T$. The gradient is defined to be the slope of the line passing through points $w_1$, $T_1$, and $w_3$, $T_3$. An extension of the method that is useful is that if in the interval, $u$ is as illustrated in Figure 10.(c) (where the integral of $u$ from $T_2-$ to $T_2+$ has a known value), the average value of $u$ in the interval is still obtained directly by determining $w$ and then employing Equation (10.3).

For the first illustration given of $u$ against $T$, an inverse or 'series' type of average (analogous to series electrical circuit) of $u$ is defined as

$$\bar{\bar{u}}_{T_1,T_3} = \frac{1}{(T_2 - T_1)/u_1(T_3 - T_1) + (T_3 - T_2)/u_3(T_3 - T_1)} \qquad (10.4)$$

This 'series' average can also be obtained from the relationship

$$1/\bar{\bar{u}}_{T_1,T_3} = dw/dT|_{T_1,T_3} \equiv (w_{T_3} - w_{T_1})/(T_3 - T_1) \qquad (10.5)$$

where

$$w_{T_3} = w_{T_1} + \int_{T_1}^{T_2} (1/u_1) \, dT + \int_{T_2}^{T_3} (1/u_3) \, dT \qquad (10.6)$$

If $T$ is dependent on direction, then the parameters $w$, $dw/dT$, $\bar{u}$ and $\bar{\bar{u}}$ will also be dependent on direction. The value of $dw/dT$ in the $x$-direction, for example, will be $dw/dT = (dw/dx)/(dT/dx)$, with similar expressions for the other directions. For a linear triangular element where $T$ varies linearly throughout the element, the variables $dT/dx$ and $dT/dy$ are not difficult to determine. For a $u$–$T$ relationship of the form illustrated in Figure 10.2(a) where the vertical distance represents the value of $u$ at any point (the dotted

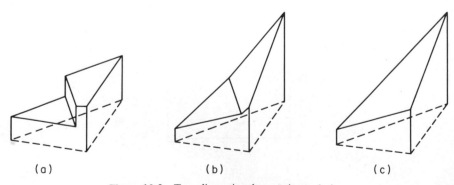

(a)                          (b)                          (c)

**Figure 10.2**   Two-dimensional averaging technique

line is the projection of the element onto the $(x, y)$ plane), the $w$ variation as a function of position throughout the element is illustrated in Figure 10.2(b) where the vertical distance represents the value of $w$ at any point. The terms $dw/dx$ in the $x$-direction and $dw/dy$ in the $y$-direction within the linear element are obtained by passing a plane through the three points defining the values of $w$ at each node as in Figure 10.2(c) and obtaining the gradients of the plane in the directions desired.

A more general way to view the determination of the gradient term $dw/dx$ is to approximate the variation of $w$ within the element in terms of the same assumed approximating function used for $T$. This interpolation function is defined in terms of the values of the variables $w$ or $T$ at the nodal points. The gradients $dw/dx$ and $dT/dx$ are then determined in the same manner. Note that for a linear one-dimensional element, a linear triangular element, or a linear tetrahedral element both $dT/dx$ and $dw/dx$ will be single valued. However, for isoparametric elements $dT/dx$ and $dw/dx$ will not be single valued. Hence, for the simple linear elements, directional values of $\bar{u}$ and $\bar{u}$ will be single valued, but for other elements the values will vary with position.

## 10.3   PHASE-CHANGE APPROXIMATION

A method is outlined below using these integral averaging techniques which allows the conduction finite element code user to approximate phase-change problems without having to change the mesh as the transient progresses. These techniques have been tried on a variety of problems where exact or approximate solutions are available. The results have been quite acceptable.

As a thermal transient progresses through a model composed of one-dimensional linear elements a phase-change interface may move through a particular element. The energy of the phase change and the difference in the conductivity of two phases must be taken into account for this element. With the conductivities of the two phases ($k_a$ and $k_b$), and the interface position ($x^*$), the overall conduction resistance ($R$) to heat flow in the $x$-direction within the element is

$$R = (x^* - x_1)/k_a A + (x_2 - x^*)/k_b A \qquad (10.7)$$

where $x_1$ and $x_2$ are element nodal coordinates and $A$ the cross-sectional area. From the following definition of the effective resistance ($R_e$):

$$R_e = (x_2 - x_1)/k_e A \qquad (10.8)$$

the effective conductivity for the element ($k_e$) is

$$k_e = \frac{x_2 - x_1}{(x^* - x_1)/k_a + (x_2 - x^*)/k_b} \qquad (10.9)$$

or

$$k_e = \frac{T_2 - T_1}{(T^* - T_1)/k_a + (T_2 - T^*)/k_b} \tag{10.10}$$

since $T$ is a linear function of $x$. $k_e$ is observed to be a series average, i.e. $k_e$ replaces $\bar{u}$ where $w = \int (1/k) \, dT$. If $k$ is some known function of temperature such as in Figure 10.3(a) (where $T^*$ is the phase-change temperature), the $w$–$T$ function can be determined.

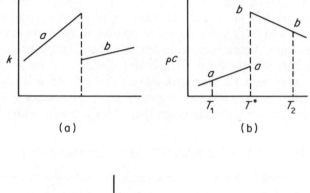

(a)  (b)

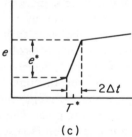

(c)

**Figure 10.3** Property averaging technique

With the value of $w$ and $T$ at each node, the single-valued terms $dw/dx$ and $dT/dx$ for the linear element are determined. The directional effective conductivity within the element is then

$$1/k_e = \frac{dw/dx}{dT/dx} \tag{10.11}$$

This approach will give the same result as previously obtained for $k_e$. However, if $dT/dx = 0$, then $dw/dx = 0$ and the operation is not defined. In this case, $k_e$ is $k$ at the one $T$ value in the element.

The volumetric heat capacity $(\rho c)$ of the material is a function of temperature of the form illustrated in Figure 10.3(b). The volumetric average value for $\rho c$ in the element is

$$\rho c = \frac{(T^* - T_1)(\rho c_a + \rho c_a^*) + (T_2 - T^*)(\rho c_b + \rho c_b^*)}{2(T_2 - T_1)} \quad (10.12)$$

where $\rho c_a$ and $\rho c_b$ are the volumetric heat capacities for the two phases $a$ and $b$. By analogy then, $\rho c$ is observed to be a regular average value, i.e. $\rho c$ replaces $u$ where $w = \int \rho c \, dT$. Thus $\rho c = (dw/dx)/(dT/dx)$. If $dT/dx = 0$, then $dw/dx = 0$, in which case the operation is undefined and $\rho c$ is the $\rho c$ at the single value of $T$ in the element.

In Comini *et al.*[1] it is suggested that the $\rho c$–$T$ variation can be altered to effectively account for the energy of phase change when using a finite element 'conduction only' type of code. This suggestion originates from the idea that in thermodynamics the quantity of enthalpy per unit volume $(e)$ is related to volumetric specific heat as $\rho c = de/dT$ where $e \int \rho c \, dT$. Also, the volumetric heat of fusion or vaporization $(e^*)$ is a volumetric enthalpy 'jump' across the phase-change interface. If $e$ is plotted as a function of $T$ for a material, the relationship would be as shown in Figure 10.3(c) where the small region around $T^*(\pm\Delta T^*)$ represents some small temperature range over which the phase change is assumed to take place. This $\Delta T^*$ is not thought to be a sensitive parameter and it is recommended that it be chosen in the range of $1/100$ to $1/1000$ of the overall temperature difference seen in the problem. An effective volumetric specific heat is defined to include this enthalpy 'jump' as

$$\rho c = (de/dx)/(dT/dx) \quad (10.13)$$

The use of such a definition not only approximates the volumetric average of $\rho c$ in the element but also takes into account the effect of the latent heat term.

These phase-change techniques which have been applied to one-dimensional linear elements can be directly extended to two-dimensional linear triangular elements. For such an element, the interface position is a line. The maximum temperature gradient in the element will be normal to the interface line as the temperature gradient in the direction of the interface line will be zero. Defining the direction $s$ to be normal to the interface line, the temperature gradient in the $s$-direction is

$$dT/ds = [(dT/dx)^2 + (dT/dy)^2]^{1/2} \quad (10.14)$$

and the $w$ $(= \int (1/k) \, dT)$ gradient in the same direction is

$$dw/ds = [(dw/dx)^2 + (dw/dy)^2]^{1/2} \quad (10.15)$$

The resulting effective conductivity in the $s$-direction is then

$$\frac{1}{k_s} = \frac{dw/ds}{dT/ds} = \left(\frac{(dw/dx)^2 + (dw/dy)^2}{(dT/dx)^2 + (dT/dy)^2}\right)^{1/2} \quad (10.16)$$

The principle heat flux in the two-dimensional triangular element will be null in the direction of the interface line and maximum in the direction normal to the interface line. Hence, the component of the principle heat flux in any direction will be the projection of the principle flux in the $s$-direction onto the direction of interest. The directional temperature gradient will vary in the same manner. Hence the appropriate effective conductivity to be used for any direction in the element is $k_s$.

The volumetric enthalpy variation in the $s$-direction is

$$de/ds = [(de/dx)^2 + (de/dy)^2]^{1/2} \tag{10.17}$$

Hence the appropriate $\rho c$ value to be used in the element is

$$(\rho c)_s = \frac{de/ds}{dT/ds} = \left( \frac{(de/dx)^2 + (de/dy)^2}{(dT/dx)^2 + (dT/dy)^2} \right)^{1/2} \tag{10.18}$$

since the principle heat flux is in the $s$-direction.

These phase-change techniques can also be applied to three-dimensional tetrahedral elements. For such an element, the interface position is a plane. The maximum temperature gradient in the element will be normal to the interface plane as the gradient in any direction on the interface plane will be zero. The value of this maximum gradient is

$$dT/ds = [(dT/dx)^2 + (dT/dy)^2 + (dT/dz)^2]^{1/2} \tag{10.19}$$

where $s$ is the direction normal to the interface plane.

Following the same line of thought given for a linear triangular element to define $dw/ds$ and $de/ds$, the appropriate conductivity and $\rho c$ value to be used for a linear tetrahedral element are

$$\frac{1}{k_s} = \frac{dw/ds}{dT/ds} = \left( \frac{(dw/dx)^2 + (dw/dy)^2 + (dw/dz)^2}{(dT/dx)^2 + (dT/dy)^2 + (dT/dz)^2} \right)^{1/2} \tag{10.20}$$

and

$$(\rho c)_s = \left( \frac{(de/dx)^2 + (de/dy)^2 + (de/dz)^2}{(dT/dx)^2 + (dT/dy)^2 + (dT/dz)^2} \right)^{1/2} \tag{10.21}$$

These phase change techniques can also be extended to isoparametric elements. For such elements the gradient terms will not be single valued. In approximating a conduction problem modelled with isoparametric elements numerical integration is required. To approximate phase-change problems with isoparametric elements the expressions previously given are used for $k$ and $\rho c$ where the various gradients are evaluated at the numerical integration points.

Often phase-change problems include convectively controlled heat transfer in the liquid at the solid–liquid interface. Numerically this is a difficult problem to approximate. The element which contains the phase-change interface must take into account the combination of the conduction in the solid part and the convection in the liquid part. For elements completely in the liquid phase which are close to the phase-change element, the directional convection energy transfer to the interface must be accounted for. For liquid elements which are not close to the interface, the nodal temperatures are constrained to be at the liquid temperature $T$. For illustrative purposes, let the conductivity of the solid phase be denoted as $k^s$. In the physical problem, the value of the conductivity in the liquid is usually included in the convection coefficient and the heat transfer is only proportional to the convection coefficient $h$ and the temperature $\Delta T$ between the liquid temperature $(T_\infty)$ and the surface temperature $(T^*)$. The effective conductivity to be used for the liquid portion of the conduction finite element must reflect this convection information. For a one-dimensional linear element this is accomplished by requiring that the effective liquid conductivity $(k_e^1)$ in the fluid portion of the phase-change element be

$$k_e^1 = \frac{h(T_\infty - T^*)}{dT/dx} \tag{10.22}$$

In effect, this requires that the effective conduction flux in the liquid portion of the element $[k_e^1(dT/dx)]$ be approximately the same as the imposed convective flux $[h(T_\infty - T^*)]$. The term approximate is appropriate as the term $dT/dx$ to be used in obtaining $k_e^1$ is the last known value. The problem is non-linear and successive approximations must be employed. The overall effective conductivity in the complete phase-change element is obtained as a series average $k^s$ and $k_e^1$ where

$$w = \int_{T_0}^{T^*} (1/k^s)\, dT + \int_{T^*}^{T} \frac{dT/dx}{h(T_\infty - T^*)}\, dT \tag{10.23}$$

For adjacent elements which are completely in the liquid region, the effective solid conductivity that is to be used must reflect the convection process being approximated. It might be assumed that this could be accomplished by choosing the effective conductivity in the element as per Equation (10.22). However, such is not the case for an all liquid element. Detailed calculations indicate that the process of choosing $k_e$ for the next calculation based only on the last known value of $dT/dx$ is divergent. A process that works well is a relative error reduction technique which approaches the result of Equation (10.22). For example, if the resulting heat flux is 20% low, the value of $k_e$ is increased by 20%. Liquid elements that are not close to the phase-change element may be constrained to be at $T_\infty$. One way to determine if the element is close or not is to initially constrain all liquid nodes to be at $T_\infty$ (except for the nodes on the initial phase change interface), then as the solidification progresses, check within each element and if any one of the nodal temperatures of that element is less than or equal to $T^*$, remove the $T_\infty$ constraint from all nodes of that element. This process assures that at least the fluid nodes between the phase-change element and the adjacent fluid element are not constrained to be $T_\infty$.

In the two- and three-dimensional problems of this type the value of $k_e^1$ must be based on $dT/ds$ instead of $dT/dx$ where $s$ is in the direction of the principal heat flux vector within the element.

## 10.4  INTERELEMENT STEP CHANGE IN CONDUCTIVITY

Solidification of an infinite region of liquid for a case where an exact analytical solution is available is considered to illustrate the effect of the choice of Equation (10.11) to define the effective $k$ of a material undergoing a phase change. An exact analytical solution to this problem is given in Luikov.[4] The problem there considered is the initially liquid infinite region $(x \geq 0)$ at a constant temperature $T_i$, which is greater than the fusion temperature $T^*$. For time $t \geq 0$, the surface at $x = 0$ is maintained at a constant temperature, $T_w$, that is lower than $T^*$. At time $t = 0$, a change of phase occurs at $x = 0$; at time $t > 0$ the phase-change occurs at the phase-change interface, that is expressed by $x = \beta\sqrt{t}$, where $\beta$ is the phase-change interface function. The finite element result of choosing various effective-$k$ relationships to account for the piecewise continuous $k$ function through the element containing the phase-change front on the proportionality factor $\beta$ between the penetration depth and the square root of the time was determined using typical water–ice properties. Choosing $k$ based on the average temperature in the element results in a 10% high $\beta$-value. Choosing a $k$-value based on a regular average value technique results in a $\beta$ that is 5% high. The best results are obtained with the series average method which gives results within 2% of the correct value. For the properties used the ratio of $k$(ice) to $k$(water) is about 4 to 1. As this ratio increases the series average method continues to give good results and the regular average method continues to give relatively poor results.

## 10.5  INTERELEMENT STEP CHANGE IN HEAT CAPACITY

Solidification of a liquid in a quarter-space $(x, y \geq 0)$ for a case where an exact analytical solution is available is considered to illustrate the choice of Equation (10.18) to define the effective $\rho c$ of a material undergoing phase-change. An exact analytical solution to this problem for a special set of conditions is given in Rathjen and Latif.[5] The problem there considered is the initially liquid quarter-space $(x, y \geq 0)$ at a constant temperature, $T_i$, which is greater than the fusion temperature, $T^*$. For time $t \geq 0$, the surfaces $x = 0$ and $y = 0$ are maintained at a constant temperature, $T_w$, that is lower than $T^*$. At time $t = 0$, a change of phase occurs along the surfaces $x = 0$ and $y = 0$; at time $t > 0$, the phase-change occurs at the phase-change interface, that is expressed by $y = y(x, t)$. The conductivity values of the two phases were chosen to be the same. The finite element results of choosing various effective $\rho c$ relationships to account for the phase-change phenomena along the line $y = x$ where determined. Using the average of the two directional ($x$ and $y$) $\rho c$ terms of Equation (10.13) resulted in a 5% error in the intersection of the phase-change interface function and the line $y = x$. However, using the $\rho c$ values determined from Equation (10.18) gave results within 1% of the solution given in Rathjen and Latif.[5]

## 10.6   INTERNAL SOLID TO LIQUID CONVECTIVE HEAT TRANSFER

The uniform convective cooling and accompanying solidification of a plane semi-infinite liquid region ($x \geq 0$) where the heat transfer between the liquid and solid occurs by convection is considered to illustrate the method suggested to include internal convection effects. A solution to this problem for a special set of conditions is given in London and Seban.[3] The problem considered there is an initially liquid infinite region ($x \geq 0$) at a constant temperature, $T_i$, which is greater than the fusion temperature, $T^*$. For time $t \geq 0$ the ambient air temperature at $x < 0$ is maintained at a constant temperature $T_a$ that is lower then $T^*$. Heat transfer between the air and the liquid (or solid) occurs by convection ($h$) at $x = 0$. Also, at the solid–liquid interface, convection ($h_i$) controls the heat transfer. In Figure 10.4 is shown the solidification depth as a function of the other pertinent parameters. The finite element results were obtained relying on the method previously outlined.

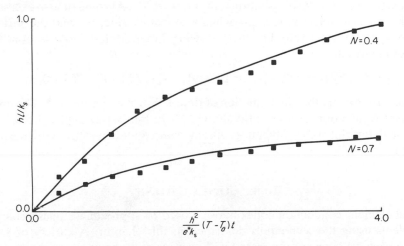

**Figure 10.4**  Penetration depth results as a function of time and convection heat transfer. $N = (h_i/h)[(T_i - T^*)/(T^* - T_a)]$; —, solution of London and Seban (3); ■, finite element results

Note the slight deviation of the finite element results from the curve for $N = 0.4$. The source of this small oscillation appears to be the convectively controlled phase-change interface crossing an element boundary. The oscillation does not turn into an instability but continues to oscillate about the solution of London and Seban.[3] The deviation is within 5% of the expected results.

## 10.7 INTERNAL SOLID TO GAS CONVECTIVE HEAT TRANSFER

Ablation of a semi-infinite solid $(x \geq 0)$ where the thermally degraded or melted material is removed as it forms is considered to further illustrate the method suggested to include internal convection effects. The general problem of predicting material ablation requires consideration of complex material-degradation processes and material–environment interactions.

However, various approximate analyses are available for predicting general behaviour during quasi-steady ablation of materials that degrade in depth. The one to be used for comparison purposes from Sunderland and Grosh[6] is a simplified quasi-steady solution considering in-depth reactions but no internal transpiration or property changes in the decomposition layer (considered to be a surface). In the problem under consideration, it is assumed that a semi-infinite homogeneous solid $(x \geq 0)$ initially at constant temperature, $T_i$, is suddenly (at $t \geq 0$) heated by convection so that eventually a phase-change occurs (at $t^*$ when the surface temperature reaches $T^*$) and the new phase is removed upon formation. The pre-ablation time, $t^*$, is the time required for the surface to attain the decomposition temperature, $T^*$. At some time in excess of $t^*$, the ablation velocity, $v$, approaches a constant value, $v_{ss}$, and the surface recedes at a constant rate. For the following dimensionless groups, set at the values indicated,

$$[h(T_a - T^*)/e^*](t^*/\alpha_s)^{1/2} = 1 \quad \text{and} \quad (T_a - T^*)/(T^* - T_i) = 0.35$$

the penetration depth, $l$, as a function of time is shown in Figure 10.5. The finite element results obtained are also shown. As in the last example, the results obtained for this very difficult problem compare favourably with desired results.

## 10.8 CONCLUSIONS

Finite element conduction codes may be used to approximate phase-change problems using the techniques described in this chapter. A variety of such problems have been approximated with various multidimensional conduction finite element codes and the results compare favourably with exact solutions or other approximate solutions. The great utility of the methods presented is that

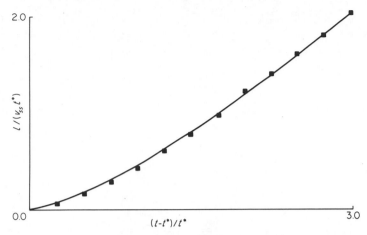

**Figure 10.5**  Ablation depth results against time: —, solution of Sunderland and Grosh (6); ■, finite element results

they may be applied in one, two, or three dimensions for very complex problems (including convective heat transfer at the phase-change interface) with difficult geometries wherein general exact solutions do not exist.

## Acknowledgement

Special thanks are due to Dr H. G. Kraus whose discussion with the author proved helpful throughout this research.

## REFERENCES

1. G. Comini, S. del Guidice, R. W. Lewis and O. C. Zienkiewicz (1974). 'Finite element solution of non-linear heat conduction problems with special reference to phase change', *Int. J. Num. Meth. Engng*, **8**, 613–624.
2. R. W. Lewis and K. Morgan (1979). *Numerical Methods in Thermal Problems*, pp. 133–232, Pineridge Press, Swansea.
3. A. L. London and R. A. Seban (1943). 'Rate of ice formation', *ASME Trans.*, **65**, 771–778.
4. A. V. Luikov (1968). *Analytical Heat Diffusion Theory*, Academic Press, New York, pp. 443–451.
5. K. A. Rathjen and M. J. Latif (1971). 'Heat conduction with melting or freezing in a corner', *J. Heat Transfer*, **93**, 101–109.
6. J. E. Sunderland and R. J. Grosh (1961). 'Transient temperature in a melting solid', *J. Heat Transfer*, **83**, 409–414.
7. O. C. Zienkiewicz (1977). *The Finite Element Method*. McGraw-Hill, New York.

*Numerical Methods in Heat Transfer*
Edited by R. W. Lewis, K. Morgan, and O. C. Zienkiewicz
© 1981 John Wiley & Sons Ltd

Chapter 11

# A Finite Element Solution for Freezing Problems, using a Continuously Deforming Coordinate System

*Kevin O'Neill and Daniel R. Lynch*

## SUMMARY

In a significant class of thermal problems, important quantities vary rapidly or are discontinuous in the vicinity of 'transition zones' whose locations or extents change as the solution evolves. Examples of such zones are areas of phase change or unusually high heat flux. Solution of such problems is facilitated by methods in which the numerical resolution evolves in response to the solution itself. In this chapter, a general method for finite element simulation of these problems is presented, incorporating mesh evolution. Unlike other methods which rely on periodic generation of new finite element meshes during the simulation, the present method is based upon the continuous deformation and/or translation of a single initially specified mesh. The effects of arbitrary mesh motion are included directly in the governing equations. The method is applied to one-dimensional problems involving heat conduction with and without phase change, and is shown to be attractive with respect to stability, accuracy, and economy. Hermite basis functions appear to offer some benefits in the types of problems considered, and a preliminary comparison is made with simple linear elements.

## 11.1  INTRODUCTION

Continuing developments in the numerical solution of transport problems have included progress in such major areas as non-linearity, inhomogeneity, anisotropy, multidimensionality, complex boundary geometry, and arbitrary forcing terms and boundary conditions. A significant additional complication in a number of transport problems is the presence of physically important transition zones wherein the dependent variable and/or the material properties exhibit rapid (and in the limit, step) changes either in value or gradient. In cases where these transition zones are fixed in space and can be identified in advance, numerical methods can be adapted by building in extra detail in these areas. The finite element method has proved to be valuable in these types of problems because: (a) it provides a rationale and a mechanism for generating difference equations on a non-uniform mesh; (b) it allows the use of higher-order elements in regions where they are suited to the physics of the problem. The use of functional coefficients and 'mixed' elements has also facilitated the solution of these types of problems.[1]

When the locations of such transition zones or boundaries are not fixed in space, but depend upon time or upon the values of the dependent variables, the resulting complexity may be severe. In general one may distinguish three problematical domains, which can be treated here under the general term 'transition zone':

(a)  a moving internal–external transition point, i.e. a moving external boundary;
(b)  an internal region which changes in extent, in which disproportionately great numerical attention is transiently warranted to deal with relatively great activity;
(c)  an internal zone changing primarily in location, over which physical or mathematical characteristics change rapidly; in the limit, an internal moving boundary.

In the research reported here, a method is investigated which can be used to attack the difficulties of all three kinds of transition zone. Continuous mesh deformation is incorporated directly in the approximating finite element equations. In particular, an existing set of nodes and basis functions may be convected in a spatially heterogeneous manner. New degrees of freedom are not created through time. Rather, points on the mesh associated with established ones move, and the effects of this are automatically accounted for. This procedure contrasts with those employed in some other methods, in which meshes are repeatedly altered or disassembled, new meshes are constructed based on interpolation or extrapolation around previous mesh points, and new sets of coefficients are enlisted. The method is shown to be accurate, efficient, and convenient in the solution of test problems involving heat conduction in water and soil, with and without phase change. Hermite basis functions allow a direct and precise treatment of these problems, and solutions obtained with these functions are compared with those obtained using simple linear basis functions.

## 11.2 BACKGROUND

### 11.2.1 Deforming and non-deforming grids

Two basic approaches may be followed in attacking changing transition zones:

(a) A straightforward application of existing methods for fixed spatial grids can be adopted, and special precautions taken both to recognize the location of the transition zone at any point in time and to represent the changing physics of the situation. Generally speaking, improved accuracy in the vicinity of a physically important transition zone must be obtained by adding numerical detail in and around it. With nodes fixed in space, this is obtained only at the expense of adding similar numerical detail throughout a much larger portion of the domain. Nevertheless, provided a suitable numerical representation is used for the physics in the transition zone, the approach is attractive insofar as it uses established numerical methods for fixed grid problems.

(b) The alternative approach is to solve the problem using a deforming numerical grid. Grid deformation must be such that appropriate numerical detail is preserved in a changing zone of interest. The zone must be adjusted appropriately in extent or location, and connected properly to its surroundings. In the abstract, this approach has a greater intuitive appeal. The disadvantages are of course the practical and theoretical problems encountered in accounting accurately for the mesh deformation.

### 11.2.2 Stefan-type freezing problems

All three kinds of transition zone difficulties arise in liquid and porous medium freezing problems, where a transition zone occurs between the frozen and unfrozen zones. The location of this zone changes as freezing proceeds, and important quantities change rapidly across it. In addition, at some points in time, for example at the inception of freezing, great numerical detail might be required to describe steep gradients in temperature or descriptive parameters across a frozen layer. However, at other times these gradients across the frozen zone may be quite mild. Perhaps the conceptually simplest case is that of a saturated porous medium with the fluid at rest, and with a pore size distribution such that freezing occurs at a single discrete temperature. The transition zone in this case becomes infinitely thin, i.e. a moving boundary. A simple model of the freezing process, then, is the Stefan problem:

$$c_1 \frac{\partial T}{\partial t} - \nabla \cdot (K_1 \nabla T) \quad \text{in } R_1 \tag{11.1}$$

$$c_2 \frac{\partial T}{\partial t} = \nabla \cdot (K_2 \nabla T) \quad \text{in } R_2 \tag{11.2}$$

where $c_1$, $K_1$ ($c_2$, $K_2$) are the volumetric heat capacity and thermal conductivity of the frozen (unfrozen) portions of the medium, and $T$ denotes the temperature. The domain $R$ is divided into $R_1$ and $R_2$ by a moving boundary $S$, on which the temperature $T$ is a constant (the freezing temperature). The velocity of the moving boundary is given by the boundary condition

$$L\frac{ds}{dt} = K_1 \nabla T(S^-) - K_2 \nabla T(S^+) \qquad (11.3)$$

where $L$ is the volumetric latent heat of fusion of the soil–water mixture, and $s$ is the location of a point on $S$. Temperature is continuous across the boundary, while heat content, material properties, and the temperature gradient are discontinuous. In the limiting case wherein the liquid temperature is constant throughout at the freezing temperature, the problem reduces to the one-phase Stefan problem:

$$c\frac{\partial T}{\partial t} = \nabla \cdot K \nabla T \qquad (11.4)$$

$$L\frac{ds}{dt} = K \nabla T(S) \qquad (11.5)$$

### 11.2.3 Methods of solution

Numerous solutions for Stefan-type problems have been reported in the literature, and several review articles are available.[2,3,4] Crank[2] gives a succinct account of finite difference methods, which need not be repeated here. Generally speaking, both fixed and moving grids have been used. Applications are largely confined to one-dimensional situations and frequently lack generality. The isotherm migration method[5] is in a sense a hybrid, insofar as the roles of the dependent and one of the independent space variables are interchanged, allowing a fixed grid to be created in the temperature domain. Among other things, it appears that straightforward application of the method must be limited to cases in which there can only be a uniquely invertible relationship between space points and temperature values. Among the moving grid approaches, a common (though perhaps not insurmountable) problem would appear to be the distorted finite difference mesh which would result in higher-dimensional situations.

Several finite element solutions have also been reported. Comini *et al.*[6] used a fixed mesh approach and a finite width transition zone. Latent heat effects in the transition zone were accounted for by a special device (the 'apparent heat capacity' approach). Del Giudice *et al.*[7] and Morgan *et al.*[8] have reported improvements of this method. Guymon and Luthin[9] use a fixed grid finite element approach in which freezing occurs simultaneously over a whole element when its hypothetical heat deficit equals the latent heat content.

Bonnerot and Jamet[10] used finite elements in both space and time, creating a spatial mesh which deforms in time. The effects of the mesh deformation are automatically accounted for by the finite element formulation in time.

Lynch and Gray[11,12] developed a general Galerkin finite element approach which, like that of Bonnerot and Jamet, automatically accounts for the effects of mesh deformation. Instead of using finite elements in time, however, they retain the conventional finite difference formulation in time and show one- and two-dimensional applications in the computation of shallow water waves. An extension of this approach appears attractive for numerous parabolic and Stefan-type problems. In particular:

(a) The method appears to be readily applicable to multidimensional situations.

(b) An arbitrary mesh deformation can readily be accommodated. In the case of transition zones of finite thickness, interpolation of rapidly varying functions may be optimized by maintaining appropriate numerical detail within these zones only. In the limit, discontinuous changes across a moving boundary are readily accommodated by requiring element boundaries to coincide with the moving boundary.

(c) In parabolic transport problems, the method is shown below to result in familiar form numerical convective diffusive equations. In the absence of mesh deformation, the method reduces exactly to the conventional Galerkin finite element procedure. The particular method of solution in time may still be specified by the user to suit the details of the specific problem; it is not dictated by the moving mesh feature.

(d) Existing programs which treat only fixed mesh situations can readily be extended to include the effects of a deforming mesh by the addition of a single extra term in the weighted residual equations. For these reasons we have adopted this approach in the research reported herein.

## 11.3 THE NEW METHOD

### 11.3.1 Formulation on a moving grid

The essence of this method is as follows. Since node motion is allowed, the finite element basis functions become implicit functions of time. When the numerical solution to a system such as (11.4), (11.5) is expressed in the form

$$T(\mathbf{x}, t) = T_j(t)\phi_j(\mathbf{x}, t) \tag{11.6}$$

it is apparent that while spatial derivatives retain their usual form, time derivatives acquire an additional term:

$$\frac{\partial T}{\partial t} = \frac{\mathrm{d}T_j}{\mathrm{d}t}\phi_j + T_j\frac{\partial \phi_j}{\partial t} \tag{11.7}$$

(here and elsewhere the subscript summation convention is used). Lynch and Gray[11,12] show that

$$\frac{\partial \phi_j}{\partial t} = -\frac{dx_i}{dt} \, \psi_i \cdot \nabla \phi_j = -\frac{dx}{dt} \cdot \nabla \phi_j \tag{11.8}$$

where $x$ has been expressed using the functions $\psi_i$:

$$x = x_i \psi_i \tag{11.9}$$

O'Neill and Lynch[13] derive an equivalent general result, applicable to finite element, finite difference, and other methods of analysis. The term $dx/dt$ is the mesh deformation velocity and is continuous throughout the spatial domain. Assembling Equations (11.7) and (11.8) yields

$$\frac{\partial T}{\partial t} = \frac{dT_j}{dt} \, \phi_j - \frac{dx}{dt} \cdot \nabla T \tag{11.10}$$

The additional term can be viewed as a correction for the 'convective' effects of mesh motion.

Following the usual Galerkin procedures, one may assume a representation of $T$ as in (11.6), substitute it into (11.4), and take the scalar product of $\phi_i$ with the resulting equation. This yields

$$\langle \phi_i, c\phi_j \rangle \frac{dT_j}{dt} - \left\langle \phi_i, c\frac{dx}{dt} \cdot \nabla \phi_j \right\rangle T_j + \langle \nabla \phi_i, K\nabla \phi_j \rangle T_j - T_j \int_\Gamma \mathbf{n} \cdot \phi_i K \nabla \phi_j \, d\Gamma = 0 \tag{11.11}$$

where $\Gamma(t)$ is the boundary of $R$, and scalar products of the form $\langle a, b \rangle$ denote

$$\langle a, b \rangle \equiv \int_R ab \, dR \tag{11.12}$$

Applying Equation (11.11) at time $t^*$

$$t^* = t_k + \varepsilon(t_{k+1} - t_k) \quad 0 \le \varepsilon \le 1 \tag{11.13}$$

and employing the approximations

$$\frac{dT_j(t^*)}{dt} \simeq \frac{T_j^{k+1} - T_j^k}{t_{k+1} - t_k} \tag{11.14}$$

$$T_j(t^*) \simeq \varepsilon T_j^{k+1} + (1 - \varepsilon) T_j^k \tag{11.15}$$

one may rewrite (11.11) in the form

$$(C^*_{ij} + \varepsilon B^*_{ij}) T_j^{k+1} = [C^*_{ij} + (\varepsilon - 1)B^*_{ij}] T_j^k \tag{11.16}$$

One evaluates the coefficient matrices $B^*{}_{ij}$ and $C^*{}_{ij}$ at $t^*$, taking into account the time dependence of both $dx/dt$ and the region over which the integrations are performed. In the examples described below this was accomplished with trivial additional programming complexity, using a code designed originally for the treatment of general form convective–diffusive problems.

The above equations clarify the nature of the method advanced here. The problem has been recast at the outset in a numerical coordinate system which may grow and deform through time, and this is accounted for explicitly in the governing equations. Coefficients retain their initial attachment to particular basis functions and mesh points as the latter move about. This is to be distinguished from any alternative method in which, at discrete points in time, the mesh is reconstructed and a new set of coefficients is generated by interpolation and extrapolation from previous values on the old mesh. Among other things, any such alternative method will have to construct after each time step, a new mesh which is rather close to the old one, in order to obtain values for the new set of coefficients and to avoid overall loss of accuracy. The time step limitation which this implies may become an important limitation on the overall efficiency of such a method. It will be seen below that this is not generally the case when the mesh is moved as proposed here. The large time steps permitted here are of particular interest because they offset the inconvenience of updating the governing matrices after each mesh movement.

### 11.3.2 Basis functions

Galerkin finite element solutions to parabolic problems in one,[14] two,[15] and three dimensions[16] have been advanced recently, using basis functions derived from Hermite polynomials. In the cases investigated, they appear to offer accuracy and efficiency competitive with or superior to other basis functions, for the same computational effort. When these basis functions are used in one dimension, the collection of coefficients $T_j$ corresponds to the values of $T$ and also of $\partial T/\partial x$ at all node points. This system has the particular attraction that gradient boundary conditions may be specified exactly. Values of $\partial T/\partial x$ are continuous at node points, and all information in terms of those values may be expressed directly and explicitly in terms of the available degrees of freedom of the numerical system, without further approximation, interpolation, etc. These attributes are particularly promising in various diffusion type problems in which fundamental conditions are expressed in terms of gradients or fluxes.

At the same time, use of higher-order basis functions in multidimensional problems may entail formidable programming complexity. A careful study of the strengths and weaknesses of these functions, as compared with simpler basis functions, is thereby warranted. Preliminary work along these lines is reported below.

## 11.4 RESULTS

### 11.4.1 Hermite basis functions

Equation (11.16) was solved using Hermite basis functions for several one-dimensional problems. The first involves pure conduction of heat, without phase change, over a semi-infinite interval $(x \geqslant 0)$. The applicable equation is a one-dimensional form of Equation (11.4) subject to

$$
\begin{array}{ccc}
\tau = 0 & t = 0 & x > 0 \\
\tau = 1 & x = 0 & t > 0
\end{array} \tag{11.17}
$$

where for convenience $T$ has been converted to a non-dimensional temperature $\tau$. The quantities $c$ and $K$ were arbitrarily assigned values of 2 and 8, respectively, and all units are arbitrary. Numerical results for $\varepsilon = 0.5$ are compared in Figure 11.1 to the well known analytical solution

$$
\tau = \text{erfc}\left(\frac{x}{2\sqrt{(\alpha t)}}\right) \qquad \alpha \equiv \frac{K}{c} \tag{11.18}
$$

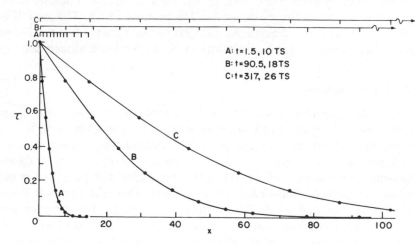

**Figure 11.1** Pure conduction problem with no phase change; semi-infinite domain. Nodal values of $\tau$ and analytic solutions (full lines) are plotted as a function of $x$ for three different points in time ($t$ = time; TS = time steps). Node locations at these points in time are shown at the top of the figure. (Hermite elements)

At first the mesh was held constant as the $\tau$ profile developed over the first ten time steps (mesh A). Subsequent to this the mesh was expanded automatically at each time step such that $\tau$ in the vicinity of the last two elements did not rise above about 0.01. If the location of the last node is taken as $s(t)$, then throughout the domain, for any point $x$ equal to $\beta s$, a continuous velocity field

$\mathrm{d}x/\mathrm{d}t$ is provided by the expression

$$\mathrm{d}x/\mathrm{d}t = \beta \ \mathrm{d}s/\mathrm{d}t \tag{11.19}$$

Thus the velocity is greatest at the point $x = s$, and is always zero at $x = 0$, preserving a stationary left-hand boundary. The excellent results shown in the figure demonstrate some of the potential of this method: one may concentrate numerical attention during one period of time in a region of greatest activity or steepest gradients, without incurring much needless computation in distant, inactive regions. Also, by shifting the mesh so that $\mathrm{d}T_j/\mathrm{d}t$ tends to remain small, one may take enormous time steps without loss of accuracy or stability. The change between curves A and B, entailing an approximately eight-fold increase in the domain, took place over only eight time steps. In all examples described here, the time step size was increased approximately geometrically in time.

Equally good results may be obtained without quite so ideal a choice of mesh deformation rate. Figure 11.2 compares computed and analytical values for the

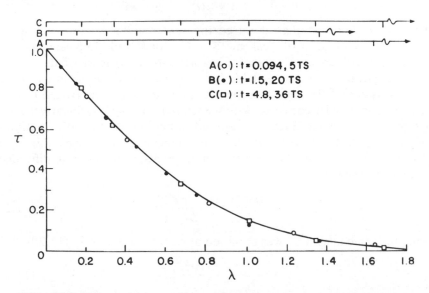

**Figure 11.2**  Pure conduction problem with no phase change; semi-infinite domain. Nodal values of $\tau$ are compared to the analytic solution (full lines) after 5, 10, and 36 time steps, for an accelerating mesh expansion. $\lambda = x/2\sqrt{(\kappa t)}$. (Hermite elements)

same problem, except that here $s$ increased as a quadratic function of time, starting immediately after $t = 0$. At first the mesh expansion rate was less than the rate of progress of the $\tau$ curve, so that temperature change gradually overtook the nodes as they moved away from the origin. Ultimately, as the node movement accelerated, the temperature curve was left behind and $\tau$

values declined at each node after reaching peak values. This and other test runs show that arbitrary mesh deformation may also yield quite good results. Of course, in this case one must avoid unreasonably large time steps, wherein the simple finite difference expressions (11.14) and (11.15) lose accuracy.

Several one- and two-phase Stefan problems were run involving progressive freezing inwards from the origin. In the one-phase cases, the entire mesh was contained within the frozen zone, and the mesh deformation rate depended upon the temperature gradient at the boundary:

$$ L\frac{ds}{dt} = K\frac{\partial T}{\partial x}\bigg|_{x=s} = KT_N \tag{11.20} $$

where $T_N$ is that coefficient corresponding to $\partial T/\partial x_{x=s}$. In this case, Equation (11.4) applies only over the region $0 \leqslant x \leqslant s(t)$, and the boundary conditions are

$$ \begin{aligned} T &= T_b & x &= 0 & T_b &< 0 \\ T &= 0 & x &= s(t), & s(0) &= 0 \end{aligned} \tag{11.21} $$

where it is supposed that the freezing temperature is zero. Values of $0.62$ cal cm$^{-3}$ for $c$, $9.6 \times 10^{-3}$ cal cm$^{-1}$ s$^{-1}$ °C$^{-1}$ for $K$, and $17.68$ cal (cm$^{-3}$ soil) were used, which are appropriate for a water saturated, dense sand.

An analytical solution is available for this case for which the initial frozen thickness is zero.[17] To deal with this situation an initial thin frozen layer of $0.1$ cm was assumed in each case, with the temperature varying linearly from $T_b$ to zero over the layer. Such an assumed initial temperature distribution is conveniently close to the one which would in fact exist over such a thin layer.

To estimate $ds/dt$, and hence $dx/dt$ at all points, a simple finite difference formula was used:

$$ \frac{ds(t^*)}{dt} \simeq \frac{s^{k+1} - s^k}{\Delta t} \tag{11.22} $$

Along the same lines, $s^{k+1}$ was obtained at each time step from (11.20):

$$ s^{k+1} \simeq s^k + \left(\frac{\Delta t K}{L}\right)\bar{T}_N \tag{11.23} $$

where $\bar{T}_N$ may be regarded as an approximation of the value of $T_N$ at some time $\bar{t}$ between $t^k$ and $t^{k+1}$, not necessarily equal to $t^*$. $\bar{T}_N$ was in turn expressed as

$$ \bar{T}_N = \theta T_N^{k+1} + (1-\theta)T_N^k \tag{11.24} $$

In this scheme, $s^{k+1}$ may be estimated from Equations (11.23) and (11.24) at the beginning of each time step with $\theta$ equal to 0, and then either used without modification or improved upon through iterative solution cycles, using a non-zero value of $\theta$.

Some results from this system are shown in Figures 11.3 to 11.6, for which $T_b$ was $-10\,°C$ and one iteration was performed per time step, with a second value of $\theta = 0.5$. When no iterations were performed, the scheme consistently overestimated the frozen zone thickness by about 2 to 3%, even with a four-fold increase in the number of time steps.

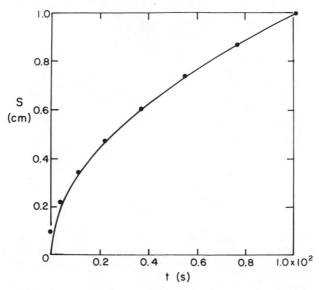

**Figure 11.3** One-phase Stefan problem, Hermite elements. Numerical and analytic (full line) frozen thickness versus time, with an assumed numerical initial thickness $s = 0.1$ cm

One may show analytically that, other things being equal, the effect of the assumed finite initial thickness should disappear rapidly in time, which it clearly does (Figure 11.3). Even though the frozen thickness and hence the mesh expands ten-fold over only 16 time steps, the typical percentage error in thickness stabilizes rapidly and is virtually no greater at $s = 1$ m (Figure 11.5) than it was at $s = 1$ cm. Numerical temperature values are compared to the analytical solution in Figure 11.4 over a frozen thickness range of 1 to 10 cm. The parameter $\varepsilon$ was kept at 0.6 in all runs shown. It was found in some other runs involving long times and large time steps that potentially fatal oscillatory noise tended to accumulate gradually, when the formulation was exactly centred in time ($\varepsilon = 0.5$).

These results depend strongly on the accuracy of the estimation of $ds/dt$, and the great accuracy of values obtained for it using the Hermites is shown in Figure 11.6. This figure pertains to an identical case, except that parameters for pure water were used with only ten time steps per ten-fold mesh expansion.

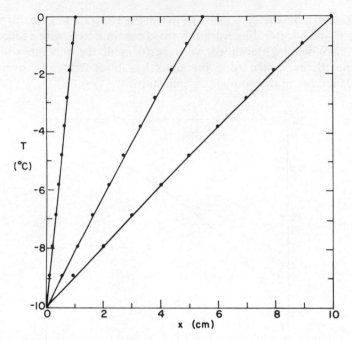

**Figure 11.4** One-phase Stefan problem, Hermite elements. Computed and analytic (full line) temperature profiles across the frozen thickness. Numerical values are plotted at nodes and at mid-element points

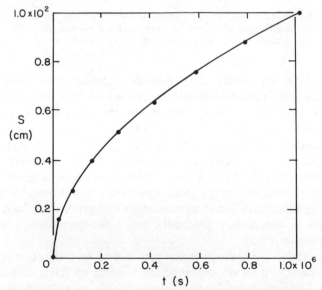

**Figure 11.5** One-phase Stefan problem, Hermite elements. Computed and analytic frozen thickness, up to a thickness of one metre

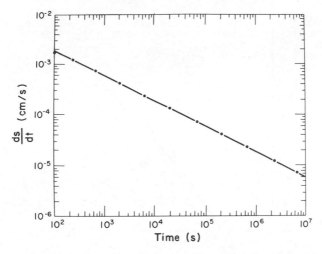

**Figure 11.6** One-phase Stefan problem, Hermite elements. Computed and analytic freezing rate versus time, for a case involving pure water

Although in all these cases the mesh was expanded rapidly, and only five elements were used throughout to span the entire frozen thickness, computed gradients $(T_N)$ were always accurate enough to produce $ds/dt$ values at the level of precision shown. Previously discussed results (Figure 11.1) also show that great accuracy in the temperature solution is not limited to cases in which curvature in the solution is slight.

The method described here may be extended readily to two-phase Stefan problems, in which one must solve for the temperature field in the unfrozen zone as well. Figure 11.7 shows results for such a case. Here also, the analytical and computed solutions agree very well. The reader is referred to a more complete description of the treatment of two-phase cases for more details.[18]

### 11.4.2  Linear basis functions

In order to compare the accuracy of Hermite elements with that obtainable using simpler elements, the one-phase Stefan problem has been solved using linear basis functions. The formulation in this case is identical to that given above, with the exception of the treatment of the moving boundary condition. Since the temperature gradient is not directly available as one of the degrees of freedom in the solution, it is necessary to differentiate the numerical solution, and Equations (11.23) and (11.24) are replaced with

$$s^{k+1} = s^k + \frac{\Delta t K}{L}\left.\frac{\partial \overline{T}}{\partial x}\right|_N \qquad (11.25)$$

$$\left.\frac{\partial \overline{T}}{\partial x}\right|_N = \theta\left(\frac{T_N - T_{N-1}}{X_N - X_{N-1}}\right)^{k+1} + (1-\theta)\left(\frac{T_N - T_{N-1}}{X_N - X_{N-1}}\right)^k \qquad (11.26)$$

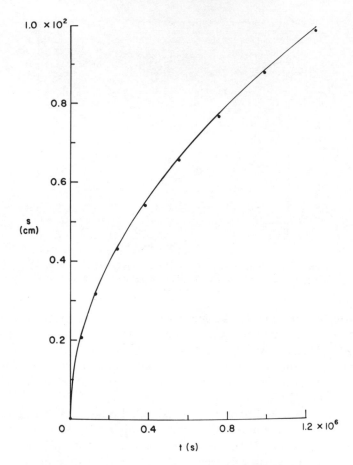

**Figure 11.7** Two-phase Stefan problem, Hermite elements. Numerical and analytic (full line) frozen thickness versus time, up to a thickness of one metre

No other modifications in the formulation are necessary to accommodate the linear elements.

The one-phase Stefan problem described above has been solved using linear elements. Eleven elements were used, giving an identical number of degrees of freedom as in the Hermitian case. All physical parameters remained the same, and identical values of $\Delta t$, $\theta$, and $\varepsilon$ were used. Numerical results along with the analytic solution are reported in Figures 11.8 and 11.9. The same general pattern was observed here as in the Hermitian element runs. When no iteration was performed, the scheme consistently overestimated the value of the frozen thickness, only in this case the error was greater—between 3% and 3.5%. Note however that linear elements with no iteration constitute the simplest use of the

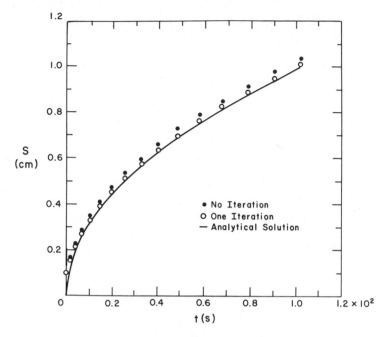

**Figure 11.8** One-phase Stefan problem, linear elements. Numerical and analytic (full line) frozen thickness versus time, with an assumed initial thickness $s = 0.1$ cm

method presented herein, and they produce a level of accuracy which may be quite acceptable for many uses. Iterating once within each time step decreased the error to about 0.5%, which is comparable to the results obtained with the Hermite elements.

## 11.5  SUMMARY AND FUTURE PROSPECTS

A finite element method formulated directly in terms of a deforming numerical coordinate system has been shown to be effective in solving certain simple heat conduction problems with and without phase change. The method also looks very promising for a whole variety of problems involving moving transition zones and deforming regions. Nothing in the formulation limits the method to one dimension or to constant material properties. Preliminary results from applications of the method to problematical convective–diffusive cases[13] and to multidimensional phase change problems[19] are quite encouraging, and work in these areas continues. Additional application to phase change problems without sharp phase boundaries may also prove rewarding. Experimental work is currently underway which will provide data for multidimensional freezing cases for which analytical solutions do not exist.

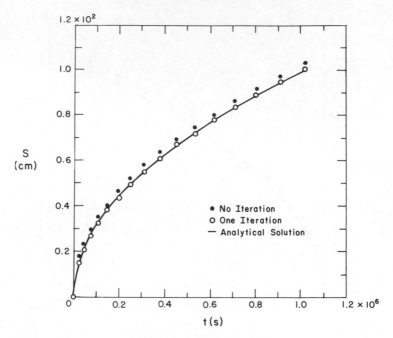

**Figure 11.9** One-phase Stefan problem, linear elements. Computed and analytic frozen thickness, up to a thickness of one metre

In the test cases investigated, both Hermite and linear basis functions produced very good results, the Hermites being slightly more accurate. The Hermites are attractive because the freezing interface condition may be expressed directly and exactly in terms of available degrees of freedom. It may be that in more complex or difficult cases, the advantages of the Hermites would produce a more dramatic enhancement of accuracy, especially if greater curvature of the temperature solution is involved. Future study, especially in higher space dimensions, should bring out better the relative merits of various basis functions for moving mesh applications.

## REFERENCES

1. G. F. Pinder and W. G. Gray (1977). *Finite Element Simulation in Surface and Subsurface Hydrology*, Academic Press, New York.
2. J. Crank (1975). 'Finite difference methods', in *Moving Boundary Problems in Heat Flow and Diffusion*, J. R. Ockendon and W. R. Hodgkins (Eds), Clarendon Press, Oxford.
3. L. Fox (1975). 'What are the best numerical methods?', in *Moving Boundary Problems in Heat Flow and Diffusion*, J. R. Ockendon and W. R..Hodgkins (Eds), Clarendon Press, Oxford.

4. G. H. Meyer (1978). 'The numerical solution of multidimensional Stefan problems—a survey', in *Moving Boundary Problems*, D. G. Wilson, A. D. Solomon and P. T. Boggs (Eds), Academic Press, New York.
5. J. Crank and R. S. Gupta (1975). 'Isotherm migration method in two dimensions', *Int. J. Heat Mass Transfer*, **18**, 1101.
6. G. Comini, S. del Giudice, R. W. Lewis and O. C. Zienkiewicz (1974). 'Finite element solution of nonlinear heat conduction problems with special reference to phase change', *Int. J. Num. Meth. Engng*, **8**, 613.
7. S. del Giudice, G. Comini and R. W. Lewis (1978), 'Finite element simulation of freezing processes in soils', *Int. J. Num. Anal. Meth. Geomech.*, **2**, 223.
8. K. Morgan, R. W. Lewis and O. C. Zienkiewicz (1978). 'An improved algorithm for heat conduction problems with phase change', *Int. J. Num. Meth. Engng*, **12**, 1191.
9. G. L. Guymon and J. L. Luthin (1974). 'A coupled heat and moisture transport model for arctic soils', *Water Res. Rsch.*, **10**, 995.
10. R. Bonnerot and P. Jamet (1977). 'Numerical computation of the free boundary for the two-dimensional Stefan problem by space–time finite elements', *J. Comp. Phys.*, **25**, 163.
11. D. R. Lynch and W. G. Gray (1978). 'Finite element simulation of shallow water problems with moving boundaries', in *Finite Elements in Water Resources, II*, C. A. Brebbia, W. G. Gray and G. F. Pinder (Eds), Pentech Press, London.
12. D. R. Lynch and W. G. Gray (1980). 'Finite element simulation of flow in deforming regions', *J. Comp. Phys.*, May 1980.
13. K. O'Neill and D. R. Lynch (1980). 'Efficient and highly accurate solution of diffusion and convection–diffusion problems, using moving, deforming coordinates', in *Finite Elements in Water Resources, III*, Wang *et al.* (Eds), in the press.
14. M. T. van Genuchten (1977). 'On the accuracy and efficiency of several numerical schemes for solving the convective–dispersive equation', in *Finite Elements in Water Resources, I*, W. G. Gray, G. F. Pinder and C. A. Brebbia (Eds), Pentech Press, London.
15. M. T. van Genuchten, G. F. Pinder and E. O. Frind (1977). 'Simulation of two dimensional contaminant transport with isoparametric Hermitian finite elements', *Water Res. Rsch.*, **13**, 451.
16. K. O'Neill (1978). 'The transient three-dimensional transport of liquid and heat in fractured porous media', *Ph.D. Thesis*, Princeton University.
17. L. R. Ingersoll, O. J. Zobel and A. C. Ingersoll (1954). *Heat Conduction*, University of Wisconsin Press, Madison, Wisconsin, pp. 190–199.
18. D. R. Lynch and K. O'Neill (1980). 'Numerical solution of two-phase Stefan problems using continuously deforming finite elements', *Int. J. Num. Meth. Engng*, in the press.
19. D. R. Lynch and K. O'Neill (1980). 'Elastic grid deformation for moving boundary problems in two space dimensions', in *Finite Elements in Water Resources, III*, Wang *et al.* (Eds), in the press.

*Numerical Methods in Heat Transfer*
Edited by R. W. Lewis, K. Morgan, and O. C. Zienkiewicz
© 1981 John Wiley & Sons Ltd

*Chapter 12*

# Coupled Seepage, Heat Transfer, and Stress Analysis with Application to Geothermal Problems

*M. Borsetto, G. Carradori, and R. Ribacchi*

## SUMMARY

This chapter is devoted to a mathematical model and to a solution procedure for the heat and mass transfer and stress field in transient coupled conditions.

This model has a wide range of applications in the field of civil and reservoir engineering. Specific attention is given to seepage in rock masses, namely in a geothermal context: different approaches to a non-linear equivalent continuum model for jointed or microfractured rocks are investigated.

The derivation of the fundamental differential equations and a short discussion of non-linear parameters and coupling terms are presented. The suggested solution procedure is based on a finite element space discretization and a finite differences time integration. A computer program has been developed on this basis and some numerical results are given.

Applications to geothermal problems are presented at the end of the chapter.

## 12.1  INTRODUCTION

The analysis of the state of stress and of the pore pressure under conditions of transient flow in soils and rock masses is required for a wide variety of engineering problems. A typical problem in geotechnical engineering is, for instance, the prevision of effective stress variations and of rates of settlement following changes in the applied external loads (consolidation). Numerical techniques based on the finite element method have been widely applied for this purpose.[1,2]

Similar equations govern problems connected with reservoir engineering. In these cases, the usual approach is based on the assumption that the total stress remains constant. However, this assumption is often incorrect[3] and could lead to serious errors in some conditions.

In geothermal problems, the flow of heat plays a central role. Heat transfer is strongly coupled with the stress and flow fields, because thermal variations modify the viscosity of the saturant fluids, as well as the permeability of soil (due to the perturbation on the effective state of stress). In turn, these factors influence the flow of the fluid and the convective flow of heat.

A typical problem where the influence of coupling is remarkable is the modeling of the reinjection of cold geothermal fluids. This is necessary in order to refill the system and also to avoid pollution due to the high degree of salinity of the extracted fluid. Water from cooling towers is usually injected at a temperature close to 303 K. Moreover, in case of a large enough flow rate, water heating inside the well is limited, therefore the difference in temperature between injected water and formation may be considerable.

In the following, the need for advanced numerical models in the analysis of well tests and in the simulation of production schemes is emphasized. However a coupled analysis of the state of stress and of heat and mass transfer is needed in other civil engineering problems as well. For instance, temperature variations in low permeability saturated soils can bring about large variations of the pore pressure, altering considerably the stability conditions. In fact, according to recent investigations, these factors have largely influenced the behaviour of some big landslides. Moreover the influence of thermal variations on pore pressure in the foundation rocks of dams was evidenced in some cases through piezometric measurements. Finally similar problems arise in the design of underground disposal of radioactive waste.

The set of non-linear equations governing temperature, pore pressure and stress fields is presented in the following. Numerical results are computed by the finite element computer program TRAITEME. Currently water is the only fluid being considered. This hypothesis is not too limiting, as White[4] estimated that water-dominated reservoirs can provide energy resources twenty times larger than those available from steam-dominated reservoirs. Also the most recently drilled Italian geothermal fields (such as Cesano and Campi Flegrei) are water dominated.

## 12.2  MODELS OF ROCK AND ROCK MASSES

The rock mass is modelled as an equivalent continuum, the global characteristics of which keep account of the presence of joint systems or of diffused fractures. It has been suggested that the behaviour of this equivalent continuum should be similar, at least qualitatively, to that of a brittle, microfissured rock. Non-linear behaviour of the rock can be treated, but treatment of conditions approaching shear failure with the ensuing 'plastic' strains and dilatancy would require further important modifications of the model.

Sometimes, however, it is convenient—or necessary—to introduce in the model individual joint elements too. The choice between the assumption of a fully continuous equivalent medium, or the introduction of individual joints, depends both upon the importance of the joints and upon their spacing with respect to the scale of the problem. For example, for a given set of joints a single fracture crossing a production well should be represented individually, whereas at a certain distance from the well an equivalent continuum would be appropriate enough.

In some applications, the introduction of individual joints makes it possible to tackle the transient flow problem in a more accurate and simpler manner. Widely spaced joints, in fact, influence the equivalent permeability, whereas fluid storage is determined by the porosity of interfracture rock blocks. Otherwise, one should resort to a double porosity continuum characterized by somewhat arbitrary exchange parameters between blocks and fissures.

## 12.3   ISOTROPIC CONTINUUM MODELS

### 12.3.1   Mass flow

The equation of continuity of flow for a completely saturated soil element is

$$-\nabla \cdot (\rho_w \mathbf{u}) = \frac{\partial}{\partial t}(n\rho_w) \tag{12.1}$$

where $\rho_w$ is the fluid density, $\mathbf{u}$ the seepage velocity vector, $t$ the time and $n$ the porosity of the rock. The right-hand side of Equation (12.1), which describes the accumulation of the fluid, can be written

$$\frac{\partial}{\partial t}(n\rho_w) = \rho_w \frac{\partial n}{\partial t} + n \frac{\partial \rho_w}{\partial t} \tag{12.2}$$

Water density is a function of its pressure $p$ and of its temperature $T$:

$$\rho_w = \rho_w(p, T) \tag{12.3}$$

Hence

$$d\rho_w = \rho_w C_w \, dp - \rho_w \beta_w \, dT \tag{12.4}$$

where $C_w$ and $\beta_w$ are the bulk compressibility and thermal volume expansion of water. The numerical values of function (12.3) are summarized by Kennedy and Holsen.[5]

Rock porosity $n$ is function of $p$ and $T$, as well as of the state of stress $\sigma_{ij}$. This state of stress can be decomposed into an isotropic stress, corresponding to the pore pressure $p$, and an effective stress $\sigma'_{ij} = \sigma_{ij} - p\delta_{ij}$, where $\delta_{ij}$ is Kroneker's symbol; therefore:

$$n = n(p, T, \sigma'_{ij}) \tag{12.5}$$

In the case of isotropic rock, only the mean effective stress $\bar{\sigma}'$ influences the porosity. A different decomposition of the state of stress is adopted by other authors. The present definition of effective stresses is convenient because rock's strength depends essentially upon them.

Porosity variations are determined by the difference between volumetric strains of the bulk porous rock and those of the solid matrix. Following the developments carried out, for instance, by Bishop,[6] we have

$$\mathrm{d}n = \frac{\partial n}{\partial \bar{\sigma}'}\,\mathrm{d}\bar{\sigma}' + \frac{\partial n}{\partial p}\,\mathrm{d}p + \frac{\partial n}{\partial T}\,\mathrm{d}T = (C - C^*)\,\mathrm{d}\bar{\sigma}' - nC^*\,\mathrm{d}p + n\beta\,\mathrm{d}T \qquad (12.6)$$

where $C$ is the bulk compressibility, $\beta$ the volume thermal expansion of the rock and an asterisk denotes parameters related to the solid matrix.

Since fluid velocities within the rock are usually very low, the fluid momentum equation is reduced to Darcy's law:

$$\mathbf{u} = -\frac{k_p}{\mu}\nabla(p + \rho_w g h) \qquad (12.7)$$

where the intrinsic permeability $k_p$ is a function of the pore characteristics of the rock, $\mu$ is the viscosity of the fluid and $h$ is the upward coordinate in the vertical direction.

By introducing the approximation

$$\nabla \cdot (\rho_w \mathbf{u}) \cong \rho_w \nabla \cdot \mathbf{u} \qquad (12.8)$$

the mass flow equation (12.1), after the various substitutions and in terms of effective stresses, becomes

$$\nabla \cdot \left(\frac{k_p}{\mu}\nabla(p + \rho_w g h)\right) = -(C - C^*)\frac{\partial \bar{\sigma}'}{\partial t} + n(C_w - C^*)\frac{\partial p}{\partial t} - n(\beta_w - \beta)\frac{\partial T}{\partial t} \qquad (12.9)$$

In terms of total stresses, this can be written

$$\nabla \cdot \left(\frac{k_p}{\mu}\nabla(p + \rho_w g h)\right) = -(C - C^*)\frac{\partial \bar{\sigma}}{\partial t} + [nC_w + C - (1 + n)C^*]$$

$$\times \frac{\partial p}{\partial t} - n(\beta_w - \beta)\frac{\partial T}{\partial t} \qquad (12.10)$$

This form is more convenient for some particular problems or boundary conditions.

In reservoir engineering or groundwater storage problems, Equation (12.10) is usually utilized, even though only the second term of the right-hand side is retained.

### 12.3.2  Heat flow

Heat balance for a saturated soil element is described by the equation

$$[n\rho_w c_w + (1-n)\rho^* c^*]\frac{\partial T}{\partial t} = -\rho_w c_w(\nabla T \cdot \mathbf{u}) + \nabla \cdot (k_T \nabla T) \qquad (12.11)$$

where $c$ is the specific heat and $k_T$ is the thermal conductivity. Equation (12.11) is obtained by assuming that there is a thermal equilibrium between the fluid and the solid matrix, thus assigning the same local temperature $T$ to both. According to Dagan,[7] this assumption can be accepted when the pore fluid moves with Reynolds numbers lower than 1, as generally occurs in practice.

### 12.3.3  Stress analysis

The analysis can be carried out in terms of total or effective stresses: either of these approaches can prove to be the most convenient, according to the type of problem to be solved.

When a linear elastic behaviour of the rock is assumed, the stress–strain relationship in terms of effective stresses is

$$\boldsymbol{\sigma}' = \mathbf{D}(\boldsymbol{\varepsilon} - \boldsymbol{\varepsilon}_T - \boldsymbol{\varepsilon}_p) \qquad (12.12)$$

where $\mathbf{D}$ is the elasticity matrix. The initial isotropic strains $\boldsymbol{\varepsilon}_T$ and $\boldsymbol{\varepsilon}_p$ which take into account the effects of temperature and pressure are given by

$$\boldsymbol{\varepsilon}_T = -\alpha_T \mathbf{m} T \quad \boldsymbol{\varepsilon}_p = S^* \mathbf{m} p \qquad (12.13)$$

where $\mathbf{m}$ is unity for normal components and zero for other components, $S^*$ is the linear compressibility and $\alpha_T$ the linear expansion coefficient ($S^* = C^*/3$ and $\alpha_T = \beta/3$ for isotropic rocks). In addition to the applied forces, the analysis must include the body forces corresponding to the pressure gradient

$$\mathbf{f}_p = -\nabla(p + \rho_w g h) \qquad (12.14)$$

Boundary conditions must obviously be expressed in terms of effective stresses.

If the analysis is carried out in terms of total stresses, the only modification in the stress–strain relation concerns the value of $\boldsymbol{\varepsilon}_p$ which now becomes

$$\boldsymbol{\varepsilon}_p = -(S - S^*)\mathbf{m} p \qquad (12.15)$$

Body forces arising from the pressure gradients should not be included in this case.

When a non-linear elastic behaviour is to be taken into account, the stress–strain relationship must be written in an incremental form. The elastic parameters forming the **D** matrix should now be updated at each time increment.

### 12.3.4  Deformation and transport parameters in the isotropic model

Some of the rock or fluid properties which appear in the preceding equations can be reasonably considered constants; others are strongly dependent on the temperature, pressure or state of stress and, therefore, they introduce couplings between various problems. The situation is summarized in Table 12.1.

Table 12.1   Dependence of the fluid and rock parameters upon pressure, temperature, and effective stress

| | | Pressure $P$ | Temperature $T$ | Stress $\sigma_{ij}$ |
|---|---|:---:|:---:|:---:|
| Bulk compressibility of water | $B_w$ | * | ** | — |
| Thermal volume expansion of water | $\beta_w$ | * | *** | — |
| Water viscosity | $\mu$ | * | *** | — |
| Intrinsic permeability | $k_p$ | — | — | *** |
| Bulk compressibility of rock matrix | $c^*$ | — | * | — |
| Bulk compressibility of rock-mass | $c$ | — | — | ** |
| Thermal conductivity of water and matrix | $k_{Tw}$ $k_T^*$ | — | * | — |
| Specific heat of of rock matrix | $e^*$ | — | * | — |
| Linear expansion coeff. of rock-mass | $d_T$ | — | * | — |

In many geotechnical problems, the compressibility $C_w$ and the viscosity $\mu$ of the water are usually assumed as constants. However, when large temperature variations must be taken into account, this approximation is no longer acceptable; in fact, water compressibility almost doubles passing from a temperature of 300 to 500 K.

For the same conditions viscosity variations approach one order of magnitude. Also the volume expansion coefficient of water is strongly dependent on temperature and varies from about $0.2 \times 10^{-3}$ to $1.2 \times 10^{-3}$ in the range of temperatures mentioned above.

Fluid pressure variations in the 0 to 50 MPa range lead to modifications of the values of the water characteristics generally lower than 10% and therefore negligible in the first approximation.

In the computer program it is assumed that water viscosity depends only on temperature. Numerical values (in mPa s) are calculated by using the empirical relationship

$$\mu = \begin{cases} 42.7/(T-250.0) & 273\ K < T \leqslant 333\ K \\ 27.9/(T-278.8) & T > 333\ K \end{cases} \qquad (12.16)$$

The pressure dependence is neglected because it is practically zero at ambient temperature, and it reaches a maximum of 7% at 500 K.

The values of $C_w$ and $\beta_w$ corresponding to the local values of $p$ and $T$ are calculated by means of numerical differentiation of the function $\rho_w(p, T)$. The permeability of the rocks is related to the volume of voids (porosity) and to their structure, even though a straightforward relationship between these properties cannot be obtained. As a consequence, it must be expected that all factors that modify the porosity of a rock affect its permeability too.

By inspection of Equation (12.6) the influence of effective stresses on permeability be shown to be far greater than the influence of the pore pressure and temperature. Therefore, in practice, the latter variables affect permeability only indirectly through the coupling of the various fields, that is only as much as they modify the effective state of stress in the rock.

The dependence of permeability on the effective state of stress is quite complex. When the effective stresses applied to the rock increase, a decrease in permeability is generally observed. Variations are stronger at low compressive (or at tensile) stresses and become less important at high compressive stresses. However, the behaviour of porous competent rocks (sandstones, for instance) is quite different from that of low porosity microfissured rocks. In the first case, the influence of the effective stresses on permeability is generally small. Experimental data [8,9,10] show that an increase of the effective stress to 30 MPa causes a decrease of permeability to a value of about 75–85% of the initial one. For microfissured rocks, on the other hand, permeability variations of more than one order of magnitude are usually measured.

An empirical relationship, proposed by Morgenstern and Guther,[11] may be used for representing such a behaviour (Figure 12.1):

$$k_p = A(\bar{\sigma}' + B)^{-N} \qquad (12.17)$$

where $A$, $B$, and $N$ are experimental parameters depending on type of rock involved; $B$ has the order of magnitude of the tensile strength of the rock. The influence of deviatoric effective stresses is neglected in this approximate relationship and will be dealt with later. In fractured rock masses treated as an

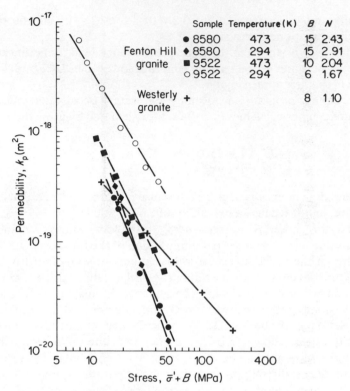

**Figure 12.1**  Stress-dependent permeability of granitic rocks. Data for Westerly granite are taken from Brace *et al.*[12] and for Fenton Hill granite from Fisher[13]

equivalent continuum, the dependence of permeability upon the state of stress is usually strong and quite similar to the one found in brittle microfractured rocks (Figure 12.2).

The solid matrix compressibilities $C^*$ and $S^*$ of the rock (as well as the other elastic constants) can be calculated from the constituent mineral properties by means of averaging procedures.[16] They are almost independent of the state of stress, but they show a more or less marked increase with temperature. The rock compressibility $C$ is often much greater than $C^*$; in brittle rocks a strongly non-linear behaviour is observed which is especially apparent at low compressive stresses, whereas at high stresses the compressibility approaches that of the solid matrix.

This trend may be represented through an empirical relationship similar to the one used for permeability:

$$C/C^* = 1 + A'(\bar{\sigma}' + B')^{-N'} \tag{12.18}$$

where $A'$, $B'$, and $N'$ are empirical constants.

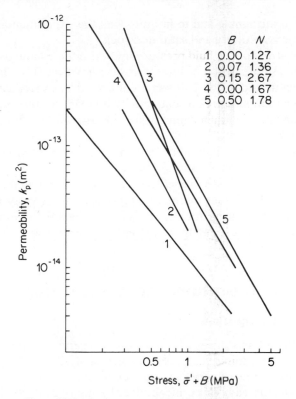

**Figure 12.2** *In situ* permeability of granites and gneisses, from Morgenstern and Guther,[11] Witherspoon *et al.*[14] (curves 1 to 4). The behaviour of a small scale model of a rock mass obtained through artificial fracturing of a marble (Rosengreen and Jaeger[15]) is also indicated (curve 5)

The influence of the state of stress on the 'elastic' constants $E$ and $\nu$, which are required for the stress analysis, can be evaluated by assuming, as a first approximation, that the non-linearity is only due to the closing of initially open microfissures and that no sliding takes place in the closed microfissures. In this case the following relationships hold:

$$\frac{1}{E} = S + 2\frac{\nu^*}{E^*} \qquad \nu = \nu^*\frac{E}{E^*} \qquad (12.19)$$

Among the thermal properties of the rock, global thermal conductivity is the parameter more difficult to evaluate because both the solid matrix and the saturant fluid give a contribution and averaging is not simple. Among the various relationships proposed, the empirical relationship[17]

$$k_T = k_T^{*(1-n)} k_{Tw}^{(n)} \qquad (12.20)$$

appears to be quite simple and to fit quite well the experimental data for the relative conductivity of dry and saturated rock.

The state of stress or the fluid pressure influence the conductivity of porous rocks through a modification of their porosity; however this effect should be very small in saturated rocks. This fact is confirmed by experimental data.[18,19] The influence of temperature on conductivity is however more marked: a 20% decrease of conductivity when temperature rises from 300 to 500 K is standard for granitic rocks.

The specific heat is remarkably uniform for various types of rocks; it increases by about 15% from 300 to 500 K.

A constant thermal dilatation coefficient can be assumed in the temperature range considered. For larger thermal variations, in many types of rocks (and especially in quartz bearing rocks), anelastic irreversible deformations take place because of the dishomogeneity and anisotropy at the scale of the constituent crystals.[20]

## 12.4   ANISOTROPIC CONTINUUM MODELS

### 12.4.1   Governing equations

The introduction of anisotropy in the deformation and transport properties of rock, even though adding complexity to the analysis, makes it possible to represent the behaviour of a rock or of a rock mass in a more accurate manner. For instance, in the case of permeability, the equivalent continuum model corresponding to a rock mass cut by a number of joint systems with arbitrary orientation, requires an anisotropic permeability matrix.[21] For the deformation parameters, the assumption of an orthotropic anisotropy with the same principal directions for the rock and for the solid matrix could be considered accurate enough in practice.

Under these assumptions, an evaluation of the porosity variation carried out with a procedure similar to that employed by Bishop for an isotropic rock, gives the following equation:

$$\mathrm{d}n = -(S_1 - S_1^*)\,\mathrm{d}\sigma_1' - (S_2 - S_2^*)\,\mathrm{d}\sigma_2' - (S_3 - S_3^*)\,\mathrm{d}\sigma_3' - nC^*\,\mathrm{d}p + n\beta\,\mathrm{d}T$$

$$(12.21)$$

where $S_i$ are the linear compressibilities and 1, 2, 3 are the principal directions of elasticity.

Darcy's law for the fluid flow (12.7) and the heat flow equation (12.11) are formally identical in the isotropic and anisotropic cases, but second-order tensors $k_\mathrm{p}$ and $k_T$ replace the respective scalar coefficients.

Finally, in the stress–strain relationships, the elasticity matrix **D** for an anisotropic material must be introduced. Moreover, the 'initial' strains due to temperature variations are anisotropic and are given by

$$\varepsilon_{Ti} = -\alpha_{Ti} \quad i = 1, 2, 3 \tag{12.22}$$

where $\alpha_i$ are the linear expansion coefficients in the principal directions. The initial strains due to pore pressures in the total stress and in the effective stress formulations are respectively

$$\varepsilon_{pi} = \begin{cases} -(S_i - S_i^*)p \\ S_i^* p \end{cases} \quad i = 1, 2, 3 \tag{12.23}$$

## 12.4.2  Deformation and transport anisotropic parameters

Rocks are characterized by three different types of elastic anisotropy:

(a)  anisotropy due to an oriented arrangement of the constituent minerals;
(b)  anisotropy due to the presence of microfissure systems having a preferential orientation;
(c)  stress-induced anisotropy in a rock containing fissures with statistically isotropic orientations.

In rock masses, fracture systems can have an effect quite similar to that played by microfissures in rock materials.

The first type of anisotropy is often unimportant even in strongly banded rocks.[22] It may become significant only when a high content of oriented 'mica' minerals is also present in the rock. The second type of anisotropy is often very important, especially at low compressive stresses, but its effects decrease as the stresses increase (Figure 12.3). Usually, the orientation of the structure in the solid matrix, when present, and in the fissure systems are coincident. Finally,

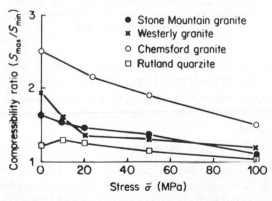

**Figure 12.3**  Anisotropy of linear compressibility as a function of the applied isotropic stress for some rocks (Brace,[16] Simmons *et al.*[23])

the stress-induced anisotropy is due to the progressive closing of microfissures favourably oriented with respect to the principal directions of the state of stress. Therefore, it is actually a consequence of the non-linear behaviour of the rock. Since this cause of anisotropy is present in most kinds of rock, the non-linear behaviour should be modelled at least in an approximate manner (Figure 12.4).

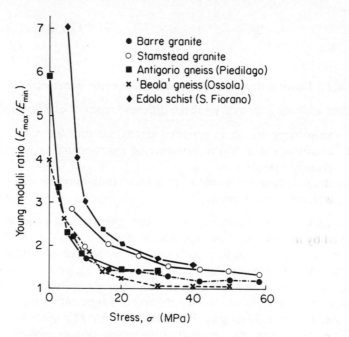

**Figure 12.4** Anisotropy ratios of the Young moduli in uniaxial compression conditions for two granites (Douglass and Voight[24]) and for some metamorphic rocks of the Alpine range

The simplified relationships between the anisotropic rock parameters and its state of stress, which are adopted in the model, are based on the assumption that non-linearity is due to the closing of the microfissures and that sliding along the faces of the closed fissures does not provide a significant contribution to rock deformation.

Three different models can be adopted:

(a) Each principal value of permeability and of linear compressibility is related only to the mean effective stress $\bar{\sigma}'$ through relationships similar to those represented in Equations (12.17) and (12.18). However, different values of the empirical coefficient are assigned to each principal value.

The other elasticity constants are calculated through the equations

$$1/E_i = S_i - \nu_{ij}^*/E_i^* - \nu_{im}^*/E_i^* \quad \nu_{ij} = \nu_{ij}^* E_i/E_i^*$$

$$1/G_{ij} = 1/G_{ij}^* + (1/E_i - 1/E_i^*) + (1/E_j - 1/E_j^*) \tag{12.24}$$

where $i, j, m$ denote the principal directions of anisotropy. The model is simple but cannot take into account the stress-induced anisotropy.

(b) When the stress-induced anisotropy is the most important factor, it is assumed that for each state of stress the principal directions of elasticity and permeability coincide with those of the effective stress and therefore vary during the loading history of the rock.

Taking into account the fact that each principal stress component has the largest effect on the fissures perpendicular to the same principal direction, and no effect at all on the fissures parallel to that direction, it is possible to write

$$S_i/S^* = 1 + A'(B' + \sigma_i')^{-N'}$$

$$i, j, m = 1, 2, 3 \tag{12.25}$$

$$k_{pi} = \tfrac{1}{2}A(B + \sigma_j')^{-N} + \tfrac{1}{2}A(B + \sigma_m')^{-N}$$

where $i, j, m$ denote the principal stress directions which may change with time at each point. The other elastic constants are always calculated according to Equations (12.24).

(c) In a most complete model, the influence of both the initial and the stress-induced anisotropy is accounted for in a simplified manner, by assuming that the initial anisotropy is originated by a fictious set of three fissure systems lying in the principal planes of anisotropy. Under these conditions, equations similar to (12.25) are used. However, different values of the coefficients are taken in the various directions

$$S_i/S_i^* = 1 + A_i'(B_i' + \sigma_i')^{-N_i'}$$

$$k_{pi} = \tfrac{1}{2}A_j(B_j + \sigma_m')^{-N_j} + \tfrac{1}{2}A_m(B_m - \sigma_m')^{-Nm} \tag{12.26}$$

where $i, j, m$ denote the invariant principal directions of anisotropy.

As regards thermal conductivity, the data provided by Johnson and Wenk[25] evidence that rocks with a clearly oriented texture are characterized by anisotropy. Its intensity, however, is much less marked than for permeability or deformability. In fact, typical values of the ratios between maximum and minimum conductivity are about 1.25 for gneisses and about 1.50 for schists. The influence of applied stresses can be neglected in this case.

## 12.5  FINITE ELEMENT DISCRETIZATION

A finite element model of the equations governing seepage, thermal and elastic fields has been implemented for plane and axisymmetric domains. For simplicity only the plane problem will be described in the following.

As usual, the equations are discretized in space through the Galerkin weighted residuals procedure. In the elastic case this is equivalent to applying the virtual work principle. Displacements, pressure and temperature fields are approximated by piecewise quadratic polynomials using triangular and quadrilateral isoparametric elements. However, in order to obtain compatibility with the linear strain approximation in the elastic problem, the pore pressure and temperature fields are linearized in each element before computing the pore pressure and thermal induced strains.

In order to compute the contribution of gravity forces to seepage, the $y$-axis is assumed vertical and directed upward. Hence Equations (12.10) and (12.11) become

$$\left( \int_V (\mathbf{B}_F)^{\mathrm{T}} \mathbf{k}_{\mathrm{p}} \mathbf{B}_F \, dV \right) \mathbf{p} + \int_\Gamma (\mathbf{N})^{\mathrm{T}} u_n \, d\Gamma + \int_V (\mathbf{B}_F)^{\mathrm{T}} \mathbf{k}_{\mathrm{p}} \begin{bmatrix} 0 \\ \rho_{\mathrm{w}} q \end{bmatrix} dV$$

$$= \left( \int_V (\mathbf{N})^{\mathrm{T}} (C - C^*) \mathbf{N} \, dV \right) \dot{\boldsymbol{\sigma}} - \left( \int_V (\mathbf{N})^{\mathrm{T}} d_{\mathrm{p}} \mathbf{N} \, dV \right) \dot{\mathbf{p}}$$

$$+ \left( \int_V (\mathbf{N})^{\mathrm{T}} (n\beta_{\mathrm{w}} - n\beta) \mathbf{N} \, dV \right) \dot{\mathbf{T}} \tag{12.27}$$

$$\left( \int_V (\mathbf{N})^{\mathrm{T}} d_{\mathrm{T}} \mathbf{N} \, dV \right) \dot{\mathbf{T}}$$

$$= -\left( \int_V (\bar{\mathbf{N}})^{\mathrm{T}} \mathbf{U} \mathbf{B}_F \, dV \right) \mathbf{T} - \left( \int_V (\mathbf{B}_F)^{\mathrm{T}} \mathbf{k}_T \mathbf{B}_F \, dV \right) \mathbf{T} - \int_\Gamma (\mathbf{N})^{\mathrm{T}} q_n \, d\Gamma \tag{12.28}$$

where $\mathbf{N}$ is the shape function row vector; $\mathbf{k}_{\mathrm{p}}$ and $\mathbf{k}_T$ are permeability and thermal conductivity matrices in two dimensions; $u_n$ and $q_n$ are the components of seepage velocity and thermal flow normal to the boundary. Moreover

$$\mathbf{B}_F = \begin{bmatrix} \partial \mathbf{N}/\partial x \\ \partial \mathbf{N}/\partial y \end{bmatrix} \qquad \bar{\mathbf{N}} = \begin{bmatrix} \mathbf{N} \\ \mathbf{N} \end{bmatrix} \qquad \mathbf{U} = \rho_{\mathrm{w}} c_{\mathrm{w}} \begin{bmatrix} u_x & 0 \\ 0 & u_y \end{bmatrix}$$

$$d_T = (1-n)\rho^* c^* + n\rho_{\mathrm{w}} c_{\mathrm{w}} \tag{12.29}$$

$$d_p = nC_{\mathrm{w}} + C_- (1+n)C^*$$

With obvious identifications, Equations (12.27) and (12.28) become

$$\mathbf{M}_p \dot{\mathbf{p}} + \mathbf{H}\mathbf{p} + \mathbf{f}_u + \mathbf{f}_\rho + \mathbf{f}_\sigma + \mathbf{f}_T = 0 \tag{12.30}$$

$$\mathbf{M}_T \dot{\mathbf{T}} + \mathbf{E}\mathbf{T} + \mathbf{f}_q = 0 \tag{12.31}$$

It is to be noted that matrix $\mathbf{E}$ is not symmetrical, because of the presence of convective terms.

Finally, the elastic problem leads to the following system of equations:

$$\mathbf{K}\boldsymbol{\delta} + \mathbf{f} = 0 \tag{12.32}$$

where $\boldsymbol{\delta}$ is the nodal displacement vector; taking into account that $\mathbf{b}$ is the vector of body forces, $\mathbf{q}$ is the vector of boundary forces, gives

$$\mathbf{K} = \int_V (\mathbf{B}_s)^{\mathrm{T}} \mathbf{D} \mathbf{B}_s \, \mathrm{d}V \qquad (\mathbf{B}_s)^{\mathrm{T}} = \begin{bmatrix} \partial(\mathbf{N})^{\mathrm{T}}/\partial x & 0 & \partial(\mathbf{N})^{\mathrm{T}}/\partial y \\ 0 & \partial(\mathbf{N})^{\mathrm{T}}/\partial y & \partial(\mathbf{N})^{\mathrm{T}}/\partial x \end{bmatrix}$$
$$\tag{12.33}$$

$$\mathbf{f} = -\int_V (\mathbf{B}_s)^{\mathrm{T}} \mathbf{D}(\boldsymbol{\varepsilon}_T + \boldsymbol{\varepsilon}_p) \, \mathrm{d}V - \int_\Gamma (\bar{\mathbf{N}})^{\mathrm{T}} \mathbf{q} \, \mathrm{d}\Gamma - \int_V (\bar{\mathbf{N}})^{\mathrm{T}} \mathbf{b} \, \mathrm{d}V$$

As usual in rock mechanics, the presence of joints can be modelled using very thin finite elements with appropriate material properties.

## 12.6  TIME DISCRETIZATION

The time variable is discretized using a standard first-order difference scheme. Namely, at time $t_n + \theta \, \Delta t$ (where $0 \leqslant \theta \leqslant 1$) the unknowns and their derivatives are approximated as

$$\dot{\mathbf{p}}(t_n + \theta \, \Delta t) = (\mathbf{p}^{n+1} - \mathbf{p}^n)/\Delta t \qquad \mathbf{p}(t_n + \theta \, \Delta t) = (1-\theta)\mathbf{p}^n + \theta \mathbf{p}^{n+1}$$
$$\dot{\mathbf{T}}(t_n + \theta \, \Delta t) = (\mathbf{T}^{n+1} - \mathbf{T}^n)/\Delta t \qquad \mathbf{T}(t_n + \theta \, \Delta t) = (1-\theta)\mathbf{T}^n + \theta \mathbf{T}^{n+1} \tag{12.34}$$

This leads to a set of coupled non-linear matrix equations

$$(\mathbf{M}_p^{\bar{n}} + \theta \, \Delta t \mathbf{H}^{\bar{n}})\mathbf{p}^{n+1} = [\mathbf{M}_p^{\bar{n}} - (1-\theta) \, \Delta t \mathbf{H}^{\bar{n}}]\mathbf{p}^n - \Delta t(\mathbf{f}_u^{\bar{n}} + \mathbf{f}_\rho^{\bar{n}} + \mathbf{f}_\sigma^{\bar{n}} + \mathbf{f}_T^{\bar{n}}) \tag{12.35}$$

$$(\mathbf{M}_T^{\bar{n}} + \theta \, \Delta t \mathbf{E}^{\bar{n}})\mathbf{T}^{n+1} = [\mathbf{M}_T^{\bar{n}} - (1-\theta) \, \Delta t \mathbf{E}^{\bar{n}}]\mathbf{T}^n - \Delta t \mathbf{f}_q^{\bar{n}} \tag{12.36}$$

$$\mathbf{K}\boldsymbol{\delta}^{n+1} + \mathbf{f}^{n+1} = 0 \tag{12.37}$$

where it is assumed that matrices and 'load' vectors are evaluated using the following estimates of $\mathbf{p}$, $\mathbf{T}$ and $\boldsymbol{\delta}$:

$$\mathbf{p}^{\bar{n}} = (1-\gamma)\mathbf{p}^n + \gamma \mathbf{p}^{n+1}$$
$$\mathbf{T}^{\bar{n}} = (1-\gamma)\mathbf{T}^n + \gamma \mathbf{T}^{n+1} \tag{12.38}$$
$$\boldsymbol{\delta}^{\bar{n}} = (1-\gamma)\boldsymbol{\delta}^n + \gamma \boldsymbol{\delta}^{n+1}$$

The approach is analogous to that used by Morgan *et al.*[26] for solving multiphase flow problems.

The value of the parameters $\theta$ and $\gamma$ may have a strong influence on the accuracy and stability of the method. The problem has been widely discussed in the literature with reference to the particular case of linear and non-linear heat conduction (see, for example, Wood and Lewis,[27] Hughes[28] and Argyris *et al.*[29]). In general, $\theta$ must be not less than 0.5, in order to have unconditional stability. If $\theta = 0.5$, a second-order accuracy is obtained; on the other hand, only the fully implicit scheme ($\theta = 1.0$) prevents any spurious oscillation for arbitrarily large time steps, but its accuracy is only of the first order. Typical values recommended in the literature are $\theta = 2/3$, which corresponds to the so called Galerkin method,[30] or $\theta = 0.878$, which corresponds to achieving an 'exponential fittings', as proposed by Liniger.[31]

In this work, different values of $\theta$ have been experimented. The value $\theta = 0.5$ appears convenient when convection is significant. Treatment of convection by an upwinding scheme,[32] has not been considered essential for the preliminary results presented in this chapter, but will be introduced in an enhanced version of the algorithm. In the case of rapid transients, the value $\theta = 0.5$ leads to spurious oscillations unless a very small time step is used. Therefore, it is convenient to use a value of $\theta$ closer to 1, and a time step of variable length.

As far as $\gamma$ is concerned, clearly during the first iteration $\gamma = 0$ must be used, because no value of the unknowns at time $t_{n+1}$ is available. During the subsequent iterations, it seems more consistent to assume $\gamma = \theta$. In practice, the numerical experience available in using the finite element model described here seems to indicate that iterations are not required. This means that for the range of applications investigated so far and using reasonably large, but not very large, time steps, the equations can be linearized and uncoupled during the time step. This simplification is achieved simply by using the value $\gamma = 0$.

In order to assess the accuracy of the linearized solution, a simple one-dimensional consolidation problem of a layer has been solved numerically.

In this model, because of the application of a constant total vertical load, the time derivative of the mean total stress is proportional to the time derivative of the pressure. The theoretical pressure transient is compared in Figure 12.5 with two numerical solutions computed with the finite element method. The first one is obtained solving, as shown before, Equations (12.35) and (12.37) for this problem, using $\theta = 2/3$ and $\gamma = 0$ (uncoupled solution).

The time derivative of total stress is computed at the beginning of each time step as a backward difference. A second solution is computed solving once again with finite elements the exact equation available in this particular case (fully coupled solution).

In this case, $\dot{\sigma}$ is proportional to $\dot{p}$ at any instant, not only at time $t_n + \gamma \, \Delta t$. Therefore, this solution is more accurate than the one achieved by iterating during the step. Nevertheless, the numerical results shown every other time step are very close. This indicates that error due to uncoupling is comparable to the discretization error.

The same conclusion may hold when non-linearities are taken into account. In fact, the same problem was solved but the permeability was made dependent on the effective stress. The results, plotted every four steps, are shown in Figure

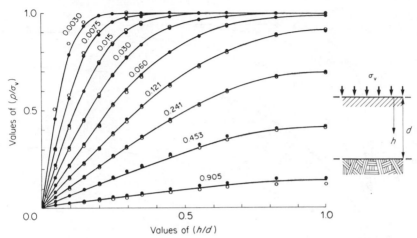

**Figure 12.5** Pressure transient with constant permeability. Equation (12.10) becomes

$$\frac{k_p}{\mu} \nabla^2 p = C\left(-\frac{d\bar{\sigma}}{dt} + \frac{dp}{dt}\right)$$

where

$$\frac{d\bar{\sigma}}{dt} = \frac{1}{3}\frac{dp}{dt}$$

Results are referred to the consolidation of an indefinite layer (plane strain and $\nu = 0.333$)

12.6. No theoretical solution is available, therefore the comparison solution was computed using a smaller time step. Again, a good agreement between the two approaches is found although permeability varies by an order of magnitude during the transient.

## 12.7 SOME APPLICATIONS TO GEOTHERMAL WELL TESTING

The aim of well testing is to produce data on the pressure inside the well and on the corresponding flow rate and to analyse them in order to get information on the structure of the reservoir and on the well itself. Well-testing analysis of geothermal reservoirs is considerably influenced by temperature-dependent effects. Therefore the techniques developed in the oil industry and available in text books[33,34] must be applied with caution. Here this is illustrated by analysing 'injectivity' and 'fall-off' tests for a classical model of a well in a pervious indefinite formation with uniform thickness.

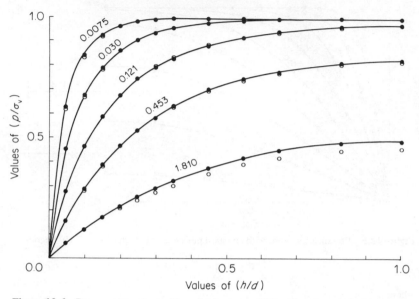

——— Comparison solution at various $\bar{t} = \dfrac{k_p\,t}{\mu C d^2}$

○ Uncoupled solution (no iterations)

● Fully coupled numerical solution

**Figure 12.6**  Pressure transient with variable permeability. Time is scaled with respect to initial permeability, which is equal to that used in Figure 12.5. Permeability varies according to Equation (12.17)

The 'injectivity test' consists in injecting water at a constant flow rate. Numerical results are shown in Figure 12.7. Generally, in isothermal cases, experimental slopes are interpreted on the basis of the following equation:

$$\Delta p = \frac{i\mu}{4\pi k_L h_L}\left[\ln\left(\frac{\bar{t}}{(r/a)^2}\right) + 0.809\right] \qquad \begin{array}{l} \bar{t} = kt/(C_T\mu a^2) \\[4pt] t/r^2 > 25\mu C_T/k_L \end{array} \qquad (12.39)$$

where $\Delta p$ is the pressure rise, $k_L$ is the intrinsic permeability and $h_L$ is the thickness of the pervious layer, $a$ is the well bore radius, $r$ is the radial coordinate, and $C_T$ the total compressibility. Through (12.39), the important well parameter $k_L h_L$ can be obtained. Also the injection of cold water in a hot rock-formation leads to a linear plot (curve C); in order to obtain correct $k_L h_L$ values, it is possible to utilize Equation (12.39), provided that the viscosity $\mu = \mu(303\ \text{K})$ is used. This is suggested by the fact that slopes of curves A and C are the same.

Results can be interpreted by assuming that heat transfer is mainly due to convection and therefore temperature transient is simply the propagation of a step temperature distribution. The most important implication is that

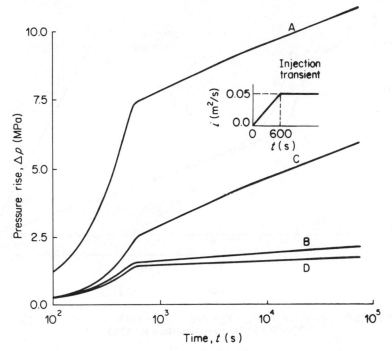

**Figure 12.7** Injectivity test. A, $T_{inj} = T_{pl} = 303$ K; B, $T_{inj} = T_{pl} = 473$ K; C, $T_{inj} = 303$ K, $T_{pl} = 473$ K; D, $T_{inj} = 303$ K, $T_{pl} = 473$ K. (The subscripts inj and pl stand for injection and pervious layer respectively.) Curves A, B, and C are for constant intrinsic permeability $k_L = 0.5 \times 10^{-13}$ m$^2$ while curve D is for stress-dependent permeability. Well bore radius $a = 10$ cm; thickness of pervious layer $h_L = 100$ m. Rock properties: $C = 0.6 \times 10^{-4}$ MPa$^{-1}$, $C^* = 0.18$ M Pa$^{-1}$, $n = 0.03$, $\alpha_T = 0.8 \times 10^{-5}$ K$^{-1}$, $\rho^* c^* = 2.25 \times 10^6$ Jm$^{-3}$ K$^{-1}$, $\lambda^* = 3.35$ Wm$^{-1}$ K$^{-1}$

measured $k_L h_L$ values characterize only the cooled region. If the permeability of the rock is assumed stress dependent, the response of the injectivity test may be quite different. Results using Equation (12.17) and the values $A = 0.95 \times 10^{-7}$, $B = 21.5$, $N = 2.5$, are shown in Figure 12.7 (curve D). Moreover, the similarity between curves B and D indicates that the dependence of permeability on the state of stress and the dependence of viscosity on temperature have counteracting effects. The average value of parameter $k_L h_L$ for the formation may be estimated through a 'fall-off' test, which is performed by shutting the well. After the shut-in time $t_s$, the pressure increment due to the injectivity test decreases according to the equation

$$\Delta p = \frac{i\mu}{4\pi k_L h_L} \ln \left( \frac{t_s + \Delta t}{\Delta t} \right) \quad \frac{\Delta t}{r^2} > 25 \mu C_T / k_L \qquad (12.40)$$

where $\Delta t$ is the time interval from the shut-in. Figure 12.8 shows that, after a short transient, there is no difference between the slopes of the isothermal

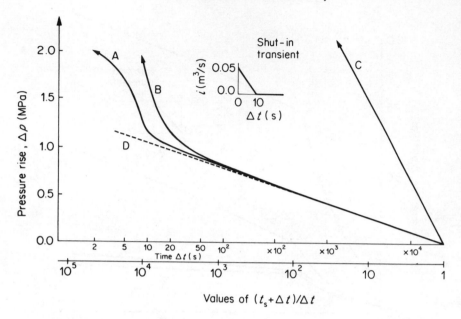

**Figure 12.8** Fall-off test. A, $T_{inj} = T_{pl} = 473$; B, $T_{inj} = 303$ K, $T_{pl} = 473$ K; C, $T_{inj} = T_{pl} = 303$ K; D, theoretical slope for 473 K

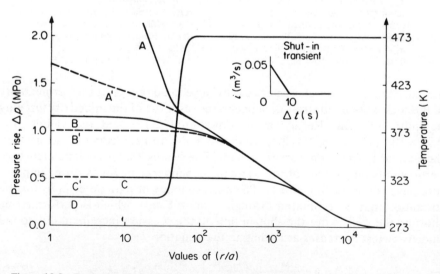

**Figure 12.9** Pressure and temperature near the well during the fall-off test. $T_{inj} = 303$ K, $T_{pl} = 473$ K; A, $\Delta t = 5$ s; B, $\Delta t = 20$ s; C, $\Delta t = 1000$ s; D, temperature at 20 s after the shut-in. $T_{inj} = 473$ K, $T_{pl} = 473$ K; A', $\Delta t = 5$ s; B', $\Delta t = 20$ s; C', $\Delta t = 1000$ s

injection test and the non-isothermal one. This is explained by noting that a few seconds after the shut-in, the pressure gradient and the fluid velocity are very small in the vicinity of the well bore (Figure 12.9). Hence, the well pressure does not depend sensibily on the low value of the ratio $k_L/\mu$ in the cooled region.

When the *in situ* stress is anisotropic, the stress-induced permeability anisotropy may influence well-testing results. For instance, let water be injected into a well at constant inlet pressure $p = 30$ MPa and temperature $T = 348$ K. The temperature in the reservoir is assumed $T = 432$ K and *in situ* stresses are 30 MPa in the vertical direction and 30 MPa and 20 MPa in two orthogonal directions of the horizontal plane. Material properties are isotropic. However, the anisotropic state of stress influences permeability according to the law given by Equation (12.25). A few results in the region near the well are shown at time 1000 s in Figure 12.10. Computations have been repeated assuming that permeability of the rock is not modified by the thermal stress transient during injection. A comparison is shown in Figures 12.10(c) and (d). The total flow rate discrepancy is almost 15%.

## 12.8 A TWO-WELL GEOTHERMAL SYSTEM

The energy recoverable from a deep aquifer depends upon the production scheme and upon the hydraulic and thermal recharge of the system. In some situations, mass recharge is mainly provided by injection of extracted fluid; therefore a cool region grows around the inlet well.

In order to estimate the recoverable energy, it is necessary to determine at what time the temperature of the extracted fluid decreases to below the economical value.

A simplified recovery scheme, featuring all the basic characteristics of a general one, is solved in the following (Figure 12.11). Two wells are drilled in a confined aquifer initially subjected to an hydrostatic pressure of 20 MPa and to a temperature of 473 K; these values are maintained constant at the outer boundary. A mass flow of 30 kg s$^{-1}$ is injected into one well at a temperature of 303 K, while the same amount is extracted from the other. The reservoir permeability is $k_p = 2 \times 10^{-14}$ m$^2$, except for zones near to the wells, where permeability is assumed as much larger, in order to avoid excessive pumping power.

The results presented in Figures 12.12 and 12.13 are referred to a period of 31 years. At very early times, two independent transients develop around the wells. Due to different viscosity, the injection pressure change is much higher than the extraction one. Therefore, the three-year temperature distribution is centred around the inlet. Convection governs heat transfer so that a major

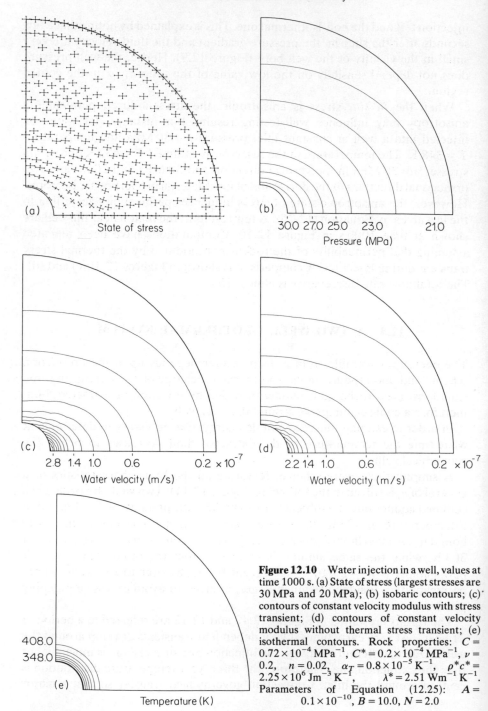

**Figure 12.10**  Water injection in a well, values at time 1000 s. (a) State of stress (largest stresses are 30 MPa and 20 MPa); (b) isobaric contours; (c) contours of constant velocity modulus with stress transient; (d) contours of constant velocity modulus without thermal stress transient; (e) isothermal contours. Rock properties: $C = 0.72 \times 10^{-4}$ MPa$^{-1}$, $C^* = 0.2 \times 10^{-4}$ MPa$^{-1}$, $\nu = 0.2$, $n = 0.02$, $\alpha_T = 0.8 \times 10^{-5}$ K$^{-1}$, $\rho^* c^* = 2.25 \times 10^6$ Jm$^{-3}$ K$^{-1}$, $\lambda^* = 2.51$ Wm$^{-1}$ K$^{-1}$. Parameters of Equation (12.25): $A = 0.1 \times 10^{-10}$, $B = 10.0$, $N = 2.0$

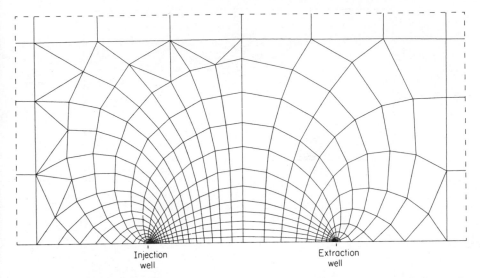

**Figure 12.11** Two-wells geothermal system: 'window' of the finite element mesh. Reservoir dimensions: $4000 \times 4000 \times 100$ m; well bore radius: 15 cm. Two-wells' distance: 500 m. Rock properties: $C = 0.146 \times 10^{-3}$ MPa$^{-1}$, $C^* = 0.2 \times 10^{-4}$ MPa$^{-1}$, $n = 0.1$, $\alpha_T = 0.8 \times 10^{-5}$, $\rho^* c^* = 2.093 \times 10^6$ Jm$^{-3}$ K$^{-1}$, $\lambda^* = 3.47$ Wm$^{-1}$ K$^{-1}$

portion of the cooled region is at 303 K. Far from the well, velocities decrease and conduction smoothens the transition to the hot region. At larger times, the cold front becomes asymmetric and the mass flow along the symmetry axis decreases. At the final computed time, cold isothermal lines reach the extraction well, but a large amount of mass is still provided by the hot region. Near the extraction well radial flow and convection dominate, and isothermal contours get closer.

Finally, it seems worthwhile to notice that an inadequate space and time discretization could have led to a quite inaccurate estimate of the total recoverable energy.

## Acknowledgements

Part of this research has been supported by the Italian National Research Council (C.N.R.) under contract No. 76.01661.92. Part of the numerical results have been supported by the Italian Electricity Board (E.N.E.L./D.S.R.–C.R.I.S.).

The authors would like to thank Mr. L. Giambanini for his contribution to the implementation of the computer program and to the computation of the results.

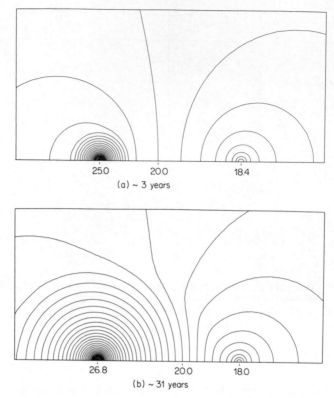

(a) ~ 3 years

(b) ~ 31 years

**Figure 12.12**   Isobaric contours (pressures are expressed in MPa)

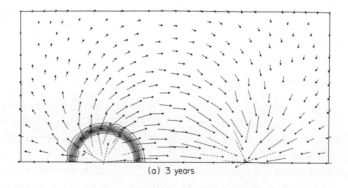

(a) 3 years

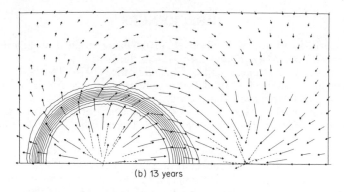

(b) 13 years

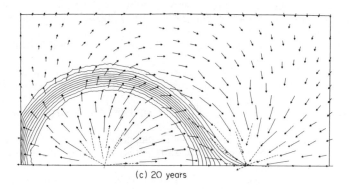

(c) 20 years

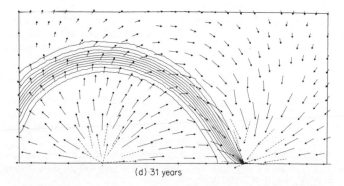

(d) 31 years

**Figure 12.13** Isothermal contours (ranging from 313 K to 463 K) and water velocities (vectors larger than $2 \times 10^{-7}$ m s$^{-1}$ are dashed)

## REFERENCES

1. J. T. Christian and J. W. Boehmer (1970). 'Plane strain consolidation by finite elements', *J. Soil Mech. Found. Div. ASCE*, **96**, 1435–1475.
2. R. W. Lewis, G. K. Roberts and O. C. Zienkiewicz (1976). 'A non-linear flow and deformation analysis of consolidation problems', *Proc. 2nd Int. Conf. on Numerical Methods in Geomechanics, Blacksburg*.
3. J. Geerstma (1974). 'Survey of rock mechanics problems associated with the extraction of mineral fluids from underground formations', *Proc. 3rd Congr. Engng Soc. Rock Mech., Denver*.
4. D. E. White (1970). 'Geochemistry applied to the discovery, evaluation and exploitation of geothermal energy resources', *Geothermics, Spec. Issue 2*, **1**, 58.
5. G. C. Kennedy and W. T. Holsen (1966). 'Pressure–volume–temperature and phase relations of water and carbon dioxide', in *Handbook of Physical Constants*, Vol. 97, pp. 371–383. Geological Society of America.
6. A. W. Bishop (1973). 'The influence of an undrained change in stress on the pore pressure in porous media of low compressibility', *Géotechnique*, 435–442.
7. G. Dagan (1972). in *IAHR, Fundamentals of Transport Phenomena in Porous Media*, Elsevier, Amsterdam.
8. I. Fatt and D. H. Davis (1952). 'Variation in permeability with overburden pressure', *Trans. AIME*, 195.
9. M. Mordecai and L. H. Morris (1970). 'An investigation into the changes of permeability occurring in a sandstone when failed under triaxial stress conditions', *Proc. 12th Rock Mech. Symp., Rolla*.
10. D. H. Gray, I. Fatt and G. Bergamini (1963). 'The effect of stress on permeability of sandstones cores', *Trans. AIME*, 95–100.
11. N. R. Morgestern and H. Guther (1972). 'Seepage into an excavation in a medium possessing stress-dependent permeability', *Proc. Int. Symp. on Percolation through Fissured Rock, Stuttgart*.
12. W. F. Brace, I. B. Walsh and W. T. Francos (1968). 'Permeability of granite under high pressure', *J. Geophys. Res.*, **73**, 2225–2236.
13. H. N. Fisher (1977). 'An interpretation of the pressure and flow data for two fractures of the Los Alamos Hot Dry Rock (HDR) geothermal system', *Proc. 18th US Rock Mech. Symp., Keistone*.
14. P. A. Witherspoon, J. E. Gale and N. G. W. Cook (1977). 'Radioactive waste storage in argillaceous and crystalline rock masses', *Rockstone 77*, **3**, 805–810.
15. K. J. Rosengreen and J. C. Jaeger (1968). 'The mechanical properties of an interlocked low porosity aggregate', *Géotechnique*, **18**, 317–326.
16. W. F. Brace (1965). 'Some new measurements of linear compressibility of rocks', *J. Geophys. Res.*, **70**, 391–398.
17. W. H. Somerton (1958). 'Some thermal characteristics of porous rocks', *Trans. AIME*, **231**, 375–378.
18. S. P. Clark (1966). 'Thermal conductivity', in *Handbook of Physical Constants*, Vol. 97, pp. 459–482. Geological Society of America.
19. W. L. Sibbit (1976). 'Preliminary measurements of the thermal conductivity of rocks from LASL geothermal test holes GT1 and GT2', *Los Alamos Scientific Laboratory Report*, LA 6199 MS.
20. R. Richter and G. Simmons (1974). 'Thermal expansion behaviour of igneous rock', *Int. J. Rock Mech. Min. Sci.*, **11**, 403–411.
21. I. A. Caldwell (1972). 'The theoretical determination of the permeability tensors for jointed rocks', *Proc. Int. Symp. on Percolation through Fissured Rock, Stuttgart*.

22. W. F. Brace (1965). 'Relation of elastic properties of rocks to fabric', *J. Geophys. Res.*, **70**, 565–567.
23. G. Simmons, T. Fodd and W. Scott Baldridge (1975). 'Toward a quantitative relationship between elastic properties and cracks in low porosity rocks', *Am. J. Sci.*, **275**, 318–345.
24. P. M. Douglass and B. Voight (1969). 'Anisotropy of granites: a reflection of microscopic fabric', *Géotechnique*, **19**, 376–398.
25. L. R. Johnson and H. R. Wenk (1974). 'Anisotropy of physical properties in metamorphic rocks', *Tectonophysics*, **23**, 79–98.
26. K. Morgan, R. W. Lewis, K. H. Johnson and I. White (1978). 'The flow of multiphase fluids in porous media', *Proc. Symp. sur les effects d'échelle en milieu poreux, Thessaloniki.*
27. W. L. Wood and R. W. Lewis (1975). 'A comparison of time marching schemes for the transient heat conduction equation', *Int. J. Num. Meth. Engng*, **9**, 679–689.
28. T. J. R. Hughes (1977). 'Unconditionally stable algorithms for non-linear heat conduction', *Comp. Meth. Appl. Mech. Engng*, **10**, 135–139.
29. J. H. Argyris, L. E. Vaz and K. J. William (1977). 'Higher order methods for transient diffusion analysis', *ISD Report 222, Stuttgart.*
30. J. Donea (1974). 'On the accuracy of finite element solution to the transient heat—conduction equation', *Int. J., Num. Meth. Engng*, **8**, 103–110.
31. T. J. D. Lambert (1973). *Computational Methods in Ordinary Differential Equations*, John Wiley, London.
32. J. C. Heinrich and O. C. Zienkiewicz (1977). 'Quadratic finite element schemes for two-dimensional convective transport problems', *Int. J. Num. Meth. Engng*, **11**, 1831–1844.
33. C. S. Matthews and D. G. Russel (1967). *Pressure Buildup on Flow Test in Wells*, Storm Printing Corp., Dallas.
34. R. C. Earlougher (1977). *Advances in Well Test Analysis*, Storm Printing Corp., Dallas.

*Numerical Methods in Heat Transfer*
Edited by R. W. Lewis, K. Morgan, and O. C. Zienkiewicz
© 1981 John Wiley & Sons Ltd

## Chapter 13

# Coupled Convective and Conductive Heat Transfer in the Analysis of Hot, Dry Rock Geothermal Sources

*R. J. Hopkirk, D. Sharma, and P.-J. Pralong*

### SUMMARY

Within the framework of an investigation into the geothermal potential of deep, hot, dry rock funded by the International Energy Agency, an existing series of computer codes has been adapted to predict the performance of such geothermal reservoirs with time.

The subject of this chapter is the geothermal reservoir initiated by a hydraulic fracture. Due to the strongly different time and length scales of the crack and the adjacent rock, this has been modelled with two separate programs coupled together—a flow program and a conduction program.

Both of these programs are similar. They both use the same strongly implicit finite difference technique and use many of the same routines.

During operation of a geothermal reservoir, the rock is expected to shrink, causing the crack to expand and eventually allowing secondary cracking.

In this study, the rock mechanics has been decoupled from the thermo- and hydrodynamics, but in modelling the creation and development of the geothermal reservoir it is of course vitally important. Thus a tandem running of solid mechanics and heat flow programs had to be achieved. A simplified method of treating the rock shrinkage has been used in an example to demonstrate the effects.

### NOMENCLATURE

**Variables and coefficients**

  $A$   surface area of a cell in the crack flow model
  $C_D$  drag coefficient
  $C_p$   specific heat at constant pressure
  $C_v$   specific heat at constant volume

$D_T$   diffusive terms for thermal energy transport equation   $\left.\begin{matrix}\\\\\end{matrix}\right\}$ see Appendix 13.1
$D_U$   diffusive terms for $x$-momentum equation
$D_V$   diffusive terms for $y$-momentum equation
 $h$    convective solid/fluid heat exchange coefficient
 $k$    thermal conductivity
 $\dot{M}$   flow rate of a mass source
 $p$    pressure in fluid
 $\dot{q}''$   heat flux per unit area and time
 $St$   Stanton number for the crack
 $t$    time from start of operation
 $t_0$   $t = 0$
 $T$    temperature
$U, V$   flow velocities in the crack in the directions $x$, $y$ respectively
$x, y$   dimensions in the plane of the crack(s)
 $z$    direction normal to the crack(s)
 $\alpha$   thermal diffusivity $(k/\rho C_p)$
 $\delta$   distance from rock/water interface to centre of first rock cell
 $\varepsilon$   crack thickness
 $\mu$   fluid viscosity
 $\theta$   angle between $x$-axis and projection of the vertical direction on to the crack plane
 $\rho$   density
 $\tau$   shear stress
 $\phi$   angle between the crack plane and the vertical direction

**Suffices**

 I   for cells with injection, extraction or leakage
 O   reference values in the fluid flow domain
 R   in rock
 S   at surface of rock, at water/rock interface
 W   in water

## 13.1   INTRODUCTION

The International Energy Agency has funded recently a study whose purpose is to examine the possibilities of tapping the huge quantities of heat stored in the rock deep in the Earth's crust via artificially created heat exchange surfaces.

The special domain of interest for the authors within the scope of this study has been the thermodynamic behaviour of the underground system. Because the flux of radial heat losses through the Earth's crust in most places is fairly small (of the order of 0.065 W m$^{-2}$ at the surface) it is clear that huge surfaces would need tapping before an energy source of the order of megawatts could be regarded as a renewable resource. In general we must think in terms of 'mining' heat. In this case certain points show up: for example the thermal conductivity of rock is low so that local temperatures and hence the usefulness of the reservoir can be degraded very quickly by too rapid a heat removal. On the other hand the thermal stresses induced by cooling of the rock matrix could cause useful extensions of heat exchange surface.

It must be of great interest therefore, when predicting possible reservoir lifetimes and potentials, to be able accurately to model all possible configurations and physical processes. This chapter describes part of the work undertaken to this end, and in particular a mathematical model developed for the study of the thermodynamic and hydrodynamic processes involved.

Before describing the work itself the authors wish to extend their very special thanks to the computer centre of the FIDES Trust Company in Zurich, without whose generosity in providing free computer time on their CDC 6500 installation many important calculations necessary for the presentation included in this chapter would not have been possible.

## 13.2 A SURVEY OF THE PROBLEM

At the start of the work on this project the most promising method of initiating a heat exchange surface underground appeared to be hydraulic fracturing. Now, more than a year later some sort of artificially created crack still seems to be the most useful basic configuration.

Consequently, attempts at modelling the thermohydrodynamic processes in the underground reservoir have been centred around the concept of the crack. Perhaps a reservoir will consist of a multitude of cracks intersecting and leaking into each other. The basic unit however remains a crack in a rock matrix. Various possible features were regarded as essential in creating a model general enough to be useful:

(1) The crack although roughly planar and hence amenable to two-dimensional treatment must be able to take up any orientation. It seems most likely that the plane of propagation would be normal to the direction of smallest compressive stress. From existing but rather sparse evidence this implies probably a near-vertical crack propagation plane.
(2) Local departures from the plane of the crack are almost certain to occur in practice. It was therefore desirable to allow for such imperfections in the model.
(3) Fluid buoyancy in an operating reservoir is certain to be an important, if not the governing, factor in determining the flow field. The rock/fluid heat transfer in its turn is highly dependent upon the fluid flow speed. A satisfactory treatment of buoyancy therefore was essential.
(4) Another important factor in flow field determination is the frictional drag on the fluid from the sides of the crack. This also then had to be included.

(5)   Leakage of fluid from or to the crack which is likely to be operated at a pressure different from the local pore pressure in the surrounding rock matrix was desired. Even if mass exchanges with the rock itself are not important in the long term, as indeed seems possible, leakages through porous or cracked zones between neighbouring cracks could be interesting.

(6)   For generality there is no alternative to three-dimensional, time-dependent treatment of heat conduction in the rock with heterogeneous material properties.

Examination of the penetration into the rock of temperature disturbances due to crack heat transfer will show that over the operating life of the plant, estimated at 30 years, the thermal influence of the crack extends to distances of the order of 100 m whereas the crack itself is only of the order of millimetres thick. This disparity in distance scale between the fluid and solid media is backed by a difference in the respective time or inertia scales. The flow field adapts itself practically instantly to the prevailing rock temperature field, but this latter requires a great deal of fluid flow to change it.

In situations where the disparities in time and length scales in the two domains are not so extreme, the harmonic averaging technique for the diffusive terms explained in the paper by Sharma, Hopkirk and Pralong[1] permits treatment of heat fluxes across fluid–solid boundaries in one sweep. Because of the peculiarities of the current problem domains, however, it was decided to use two separate models, one for fluid flow in the crack and the other for heat conduction in the solid. The two models are coupled at each time step. Both models are versions of the authors' family of implicit, finite difference codes 'TURF' (turbulent, unsteady, recirculating flow dynamics). For more details see reference 1.

This chapter concentrates on the numerical approaches to the fluid flow and heat conduction problems and on the techniques used in coupling them.

## 13.3   MATHEMATICAL FORMULATION

In what follows consideration is given to the processes of flow and heat transfer within a single crack coupled with heat transfer in the adjoining rock masses. The processes within the crack itself are presumed to vary in two space dimensions and the heat transfer process in the rock to vary in three space dimensions. Furthermore, both sets of processes and the coupling thereof are presumed to vary strongly with time. A listing of the appropriate governing equations now follows. The symbols used in these equations are defined in the nomenclature list. The coordinate system used is illustrated in Figure 13.1. Although one or even several cracks may be included in the solid, this diagram and the examples shown later consider one vertical crack on the front face of

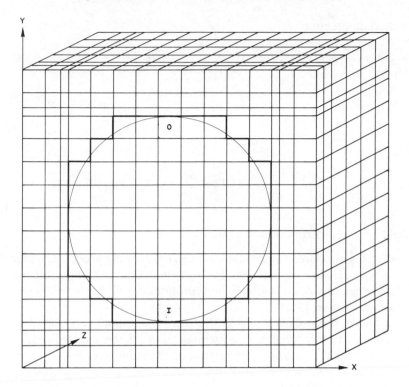

**Figure 13.1** Common coordinate system for the rock mass and a vertical crack

the solid block. This face $(z = 0)$ is therefore a plane of symmetry for the problem.

### 13.3.1 Governing equations

(a) *Crack flow and heat transfer*

Mass continuity

$$\frac{\partial \varepsilon}{\partial t} + \frac{\partial}{\partial x}(\varepsilon U) + \frac{\partial}{\partial y}(\varepsilon V) = \frac{\dot{M}_\mathrm{I}}{\rho A_\mathrm{I}} \tag{13.1}$$

*x*-direction momentum

$$\frac{\partial}{\partial t}(\varepsilon U) + \frac{\partial}{\partial x}(\varepsilon UU) + \frac{\partial}{\partial y}(\varepsilon VU) = -\frac{1}{\rho}\frac{\partial}{\partial x}(\varepsilon p) + \frac{\dot{M}_\mathrm{I}}{\rho A_\mathrm{I}}(U_\mathrm{I} - U) - \frac{1}{\rho}C_\mathrm{D}U$$

$$- \varepsilon g\left(\frac{1}{\rho}\frac{\partial \rho}{\partial T}\right)(T - T_0)\cos\theta\cos\phi + D_U \tag{13.2}$$

$y$-direction momentum

$$\frac{\partial}{\partial t}(\varepsilon V) + \frac{\partial}{\partial x}(\varepsilon UV) + \frac{\partial}{\partial y}(\varepsilon VV) = -\frac{1}{\rho}\frac{\partial}{\partial y}(\varepsilon p) + \frac{\dot{M}_I}{\rho A_I}(V_I - V) - \frac{1}{\rho}C_D V$$

$$- \varepsilon g\left(\frac{1}{\rho}\frac{\partial \rho}{\partial T}\right)(T - T_0)\sin\theta\cos\phi + D_V \tag{13.3}$$

Thermal energy

$$\frac{\partial}{\partial t}(\varepsilon C_p T) + \frac{\partial}{\partial x}(\varepsilon U C_p T) + \frac{\partial}{\partial y}(\varepsilon V C_p T) = \frac{\dot{M}_I}{\rho A_I}C_p(T_I - T) + \frac{h_W}{\rho}(T_{R,S} - T) + D_T \tag{13.4}$$

It should be noted that in these equations:

(i) A Cartesian coordinate system has been used for the processes within the crack merely for convenience. Such a system does in fact permit circular and irregular shaped cracks to be represented sufficiently well. (See grid tests in Section 13.5).

(ii) As required (see Section 13.2) the crack plane may be inclined at any angle ($\phi$) to the vertical. The inclination ($\theta$) of the $x$-coordinate in the plane to the projection of the vertical on the plane may also be chosen at will. The crack is also quasi-planar in that local variations of the angle $\phi$ may be accounted for.

(iii) The integral forms of the Navier–Stokes equations are employed and the crack thickness variations are accounted for by integrating velocities, states, and properties across the crack thickness, $\varepsilon$.

(iv) Variations of properties of the water flowing in the crack are permitted. Here, for the sake of simplicity, variations of density are accounted for only in the buoyancy term for which the Boussinesq approximation has been used.

(b) *Rock heat transfer*

$$\frac{\partial}{\partial t}(\rho_R C_{vR} T_R) = \frac{\partial}{\partial x}\left(k_R \frac{\partial T_R}{\partial x}\right) + \frac{\partial}{\partial y}\left(k_R \frac{\partial T_R}{\partial y}\right) + \frac{\partial}{\partial z}\left(k_R \frac{\partial T_R}{\partial z}\right) \tag{13.5}$$

At this stage it should be noted that for the hot rock problem the coordinate system in the rock uses the same $x$-, $y$-coordinates as in the crack flow formulation. The $z$-direction is that normal to the interface.

The coupling between the crack and rock processes is accommodated in the manner outlined below via time-dependent boundary conditions to the respective enthalpy balance equations. The conditions represent a heat-flux balance together with a continuity of temperature at the interface.

(c)  *Initial and boundary conditions*

Crack

$t < 0$:    $U = 0 = V$ for all $x, y$

$\qquad\quad T_W = T_{R,t_0}$

$t > 0$:    $U = 0 = V$ at edge of crack $\hspace{3cm}$ (13.6)

$\qquad\quad T_R = T_{R,S}(T_W, T_{R,\delta})$

Rock

$t < 0$:    $\qquad\qquad T_R = T_{R,t_0}(x, y, z)$

$t > 0; z = 0$:    $T_R = T_{R,S}(T_W, T_{R,\delta})$ $\hspace{2.5cm}$ (13.7)

(d)  *Auxiliary relations*

The following auxiliary relations are supplied in order to complete the problem specification:

$$\varepsilon = \varepsilon(x, y) \tag{13.8}$$

This relationship implies that temporal variations in crack width, which if known can be taken into account, are in fact neglected here. (See the end of Section 13.4 for the approach used in this project). The further relationships employed are:

$$\dot{q}_S'' = h_W(T_{R,S} - T_W) \tag{13.9a}$$

$$\dot{q}_S'' = h_R(T_{R,\delta} - T_{R,S}) \tag{13.9b}$$

$$\dot{q}_S'' = h_{eff}(T_{R,\delta} - T_W) \tag{13.9c}$$

$$h_W = \rho_W C_{pW} |(U_W^2 + V_W^2)^{1/2} St \tag{13.10}$$

$$h_R = \begin{cases} k_R/\delta & t > 0 \\ 0 & t \leqslant 0 \end{cases} \tag{13.11}$$

In the above, the convective heat transfer coefficient $h_W$ is computed from specified Stanton numbers for parallel-plate flow; the conductive heat transfer coefficient $h_R$, employed on the rock side of the interface, is derived by comparison of Equation (13.9b) with the one-dimensional heat conduction equation:

$$q = -k_R \frac{\partial T}{\partial z} \tag{13.12}$$

expressed through first-order finite differences. The effective heat transfer coefficient $h_{eff}$ incorporating both convective and conductive influences is used basically as a computational device. It provides an efficient means of computing the flux at the interface since it becomes unnecessary to compute the surface temperature $T_{R,S}$ and the instabilities which tend to occur when $T_W$ and $T_{R,S}$ approach each other are avoided.

Eliminating $T_{R,S}$ between relations (13.9a) and (13.9b) and using (13.9c), we obtain

$$h_{\mathrm{eff}} = \frac{h_W h_R}{h_W + h_R} \tag{13.13}$$

Assuming that the flow through the crack thickness cross section is symmetrical, the frictional shear forces on the two side walls are equal. The drag coefficient $C_D$ may then be defined as follows (Figure 13.2). The velocity profile is:

$$U = \frac{\varepsilon^2}{\mu}\left(\frac{\mathrm{d}p}{\mathrm{d}x}\right)\left[\frac{z}{\varepsilon} - \left(\frac{z}{\varepsilon}\right)^2\right]$$

so that integrating across the crack, the average velocity is $\qquad$ (13.14)

$$\bar{U} = -\frac{\varepsilon^2}{6\mu}\frac{\mathrm{d}p}{\mathrm{d}x}$$

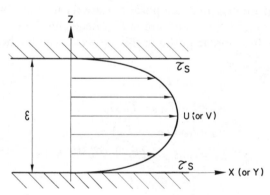

**Figure 13.2** Fully developed (Couette) flow in the crack

and in equilibrium flow, the momentum balance gives

$$2\tau_S = -\varepsilon\frac{\mathrm{d}p}{\mathrm{d}x} = \frac{6\mu}{\varepsilon}\bar{U} = -C_D\bar{U} \tag{13.15}$$

so the value used finally is

$$C_D = \frac{6\mu}{\varepsilon} \tag{13.16}$$

## 13.4 NUMERICAL SOLUTION PROCEDURE

A brief description is provided in this section of the numerical procedure devised to solve the mathematical problem outlined above. The procedure has much in common with procedures for predicting convective momentum, heat, and mass transfers previously reported by the authors[1,2] and others.[3] The novelties of the procedure, as applied to the present problem, are outlined here.

The discretized equations in both fluid and solid domains are derived by the integrated finite difference (IFD) procedure. The manner of treatment of the fluid flow and transport equations is identical to that described in reference 1. Conservation of mass, momenta and enthalpy are strictly observed for each respective micro-control volume.

The manner of problem formulation, the development of coefficients, variable nomenclature, and the use of the principle of scalar variable storage at the cell centres are all common between the two programs.

Strongly implicit forms of the algebraic equations and especially the coupling between them are employed in the solution algorithm. This algorithm is based upon efficient use of the tridiagonal matrix solver in the $x$- and $y$-directions in the fluid and in the $x$-, $y$-, and $z$-directions in the solid.

Novel numerical approaches have been used for increasing the computational efficiency by a significant reduction in computer storage of auxiliary variables necessary to the solution algorithm, for a reduction of unnecessary calculation effort, and for the fluid/solid coupling technique.

### 13.4.1  Numerical handling of the three-dimensional solid domain

For the solution of the three-dimensional heat conduction equation a numerical technique has been employed which is designed to reduce computer storage requirements. The method is incorporated into the 'strip' solution algorithm, so called because solutions are obtained over a series of one-dimensional strips of grid cells. Figure 13.3 illustrates the concept.

Strips of rectangular cells passing right across the three-dimensional region are considered (Figure 13.3(a)). The three-dimensional heat conduction equations for each strip are solved by a tridiagonal matrix algorithm (TDMA), accounting by means of coefficients for the connectivity with the neighbouring cells along each side. The equations having been solved for one strip, the calculation proceeds to the next one, and so on until the entire region has been so visited. In Figure 13.3(b) the illustrated calculation sequence is column-by-column. This diagram indicates how each new strip uses the latest available values of the variable under solution in the neighbouring cells.

The linear heat conduction equation requires but a single application of the TDMA to a strip for a solution. However, several iterations over the whole

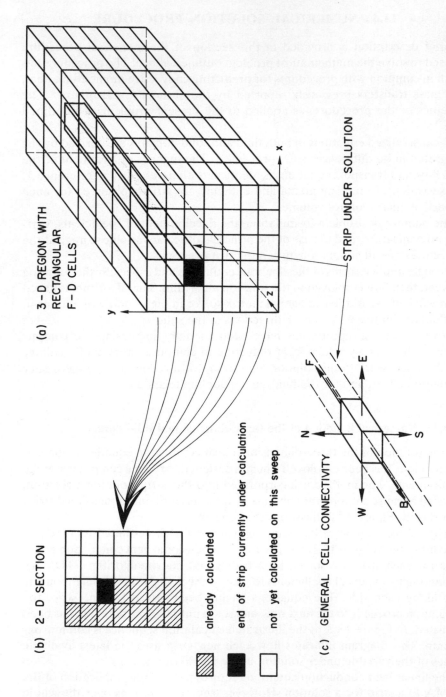

(a) 3-D REGION WITH RECTANGULAR F-D CELLS

STRIP UNDER SOLUTION

(b) 2-D SECTION

(c) GENERAL CELL CONNECTIVITY

already calculated

end of strip currently under calculation

not yet calculated on this sweep

**Figure 13.3** The 'strip' technique. (a) Three-dimensional region with rectangular FD cells; (b) two-dimensional section; (c) general cell connectivity

solid region are needed to account for the non-linearity of boundary conditions. Alternating column-wise and row-wise processing of the strip calculation over the field assists convergence.

The finite difference version of the heat conduction equation is manipulated to fit the standardized form for variable $\phi$ at point P:

$$\sum A_i^\phi \phi_P = \sum A_i^\phi \phi_i + (S_0^\phi + S_N^\phi \phi_P) \qquad i = \text{E, W, N, S, F, B} \qquad (13.17)$$

This same form, for flow equations has been presented by Sharma, Hopkirk, and Pralong.[1] In this case it has proved possible to integrate the side connection coefficients $A_N$, $A_S$, $A_E$, and $A_W$ (see Figure 13.3(c)) into the components $S_0^\phi$ and $S_N^\phi$ of the linearized source term. Thus in an extreme case only the following storage arrays are needed:

three three-dimensional arrays for:
   temperature
   old values of temperature
   thermal diffusivity $(k/\rho C_v)$
six one-dimensional arrays for:
   two coefficients used in the TDMA
   the two source components $S_0^\phi$, $S_N^\phi$,
   the two in-line connection coefficients $A_F$, $A_B$

This technique enables very large regions to be treated using minimum computer storage, and in fact lends itself ideally to the use of out-of-core storage.

The strips in this case are chosen to run in the $z$-direction, since it is in this sense that the strongest temperature gradients are to be expected. This choice results in an improved rate of convergence.

### 13.4.2 Dynamic determination of the solution domain

As soon as three-dimensional problems are tackled limitations in computer storage and long calculational times become an obsession for the numerical modeller. In the previous paragraph we have seen how computer storage can be saved with very little penalty on the time of calculation. The problem under study here is suited to a technique known as 'dynamic determination of the solution domain' (DDSD), and in particular when the 'strip' algorithm is used.

The temperature field in the solid is subject to penetration by a perturbation originating from the crack and travelling in the $z$-direction. It is thus useful and economical in such a case to solve the equations only in that part of the solid domain which is likely to be affected by the perturbation in the current time step. This can be very easily arranged by selecting the length of the strips for that time step, according to some suitable criterion.

Use of this technique brings very considerable savings in coefficient computation and equation solving and has been used here for the coupled
calculations.

### 13.4.3   Coupling of the fluid and solid domains

The concept of an effective surface heat transfer coefficient has been described
in Section 13.3 whereby this value $h_{eff}$ is the harmonic mean of the fluid-side
convective heat transfer coefficient (13.9a) and the conductive heat transfer
coefficient in the first half of the solid control volume adjacent to the interface
(13.9b). This idea is used in the solutions to both the fluid thermal energy
transport equation (13.4) and the solid heat transfer equation (13.5).

Because of the shorter time constants reigning in the fluid domain it is
possible to dispense with the time-dependent terms in Equations (13.2), (13.3),
and (13.4) and regard each calculation as a quasi-steady state 'snapshot'.
Figure 13.4(a) then illustrates the computational sequence.

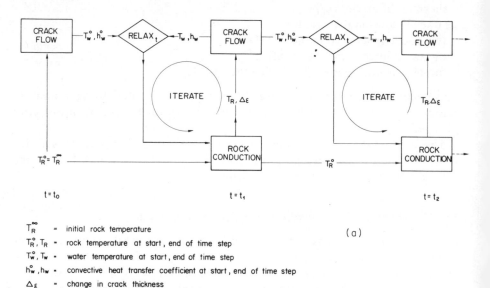

| | | |
|---|---|---|
| $T_R^{\infty}$ | = | initial rock temperature |
| $T_R^{o}, T_R$ | = | rock temperature at start, end of time step |
| $T_w^{o}, T_w$ | = | water temperature at start, end of time step |
| $h_w^{o}, h_w$ | = | convective heat transfer coefficient at start, end of time step |
| $\Delta\varepsilon$ | = | change in crack thickness |

(a)

**Figure 13.4(a)**   Coupling of the two programs

At the beginning of each time step both the solid and fluid temperatures and
the flow field are known from initial conditions or from the situation at the end
of the previous time step. Using this information then the new time step
computation starts with the solid conduction calculation to give an estimate of
the solid temperature field at the end of the step. If the crack thickness (see
Sections 13.4.4 and 13.6) is to be coupled to the rock temperature field, a new

thickness is computed at this stage. The fluid flow program then takes over and a flow field with the associated convective heat transfer coefficients is calculated.

A relaxation is now performed in the flow domain on the water temperatures $T_W$ and convective heat transfer coefficients $h_W$. This permits a new iteration of the solid temperature computation.

This iteration cycle is repeated until satisfactory convergence is attained. The two programs run alternately in separate overlays, the typical numbers of iterations required per time step being:

fluid domain:     initialization requires up to 250 iterations, depending upon the geometrical complexity, to achieve a mass residual of 5%. Subsequent passes through the fluid domain require only about 20 iterations to achieve: a mass residual of 5%; a temperature residual of 0.1% in the first passage of a time step.

solid domain:     between 2 and 5 iterations are needed to produce a temperature residual of 0.001%.

coupling:     10 to 15 iterations between the solid and fluid domains are necessary at each time step to attain a degree of global convergence (i.e. convergence of the solutions for the temperatures at the fluid/solid interface) at least as good as is achieved for the two domains separately.

Figure 13.4(b) demonstrates the effect of the choice of number of global (interdomain) iterations as the energy balance between solid and fluid.

### 13.4.4 Coupling of fluid and solid mechanics

To save programming effort and to enable individual parameters and effects to be studied more economically, automatic coupling of the fluid and solid domains at the level of mechanical interactions has not been attempted. Instead manual interventions have been made at discrete intervals in problem-time. This technique is rendered feasible by the slow changes in rock temperatures. A calculation sequence will begin with a crack profile and thickness distribution being given by solid mechanics considerations. These are used in the thermodynamic/hydrodynamic computation which yields a fluid pressure distribution and rock temperatures, which again serve as input to a stress and fracture mechanics calculation.

### 13.5 TESTING

It has been possible to carry out several tests and validation exercises separately in both physical domains.

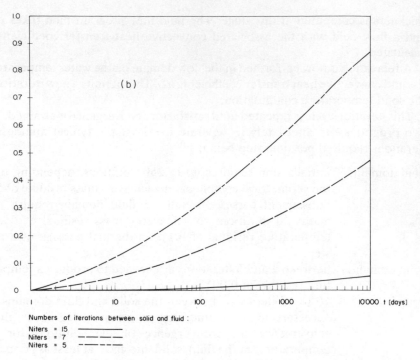

Numbers of iterations between solid and fluid :

Niters = 15　————————
Niters = 7　——— ———
Niters = 5　—— —— ——

**Figure 13.4(b)**　Variation with the number of global iterations per time step of the solution accuracy expressed by the error in overall heat balance between heat removed from the rock and heat transported away by the fluid

### 13.5.1　Grid dependence of the two-dimensional crack flow

For this test a simple round, disc-shaped vertical crack was chosen. The crack is 1000 m in diameter and its thickness profile is such that the cavity forms an ellipsoid of rotation whose minor axis is 3 mm long. The initial ground temperature is 250 °C at mid-height and is subject to a geothermal gradient of $40\,^{\circ}\text{C}\,\text{km}^{-1}$.

Water at 65 °C enters near the bottom and leaves near the top of the crack. The flow rate is $100\,\text{m}^3\,\text{hr}^{-1}$, and for this independent fluid-side test a constant and uniform heat exchange coefficient of $0.79\,\text{J}\,\text{m}^{-2}\,^{\circ}\text{C}^{-1}$ was used. This is a value typical of the early stages (1 month after start-up) of operation.

Four grid systems were used in the test. These contained in each case a uniform, square grid of such a cell size as to use respectively 9, 19, 29, and 39 cells to cover the 1000 m crack diameter. Figure 13.5 shows the four grids together. Figure 13.6 shows the corresponding vector flow patterns and Figure 13.7 the resulting isotherm distributions.

Velocity and temperature profiles respectively are plotted in Figures 13.8 and 13.9. Figure 13.8 gives the *U*-velocity profile across the horizontal diameter (a) and the V-velocity profile along the vertical diameter (b) between inlet and outlet points. Figure 13.9 shows the water temperature profiles over the same cross sections. It can be seen that with increasing grid density a solution is approached, and that this approach is more rapid along the vertical axis (in this case the principle flow path with the strongest pressure gradients) than it is along the horizontal diameter of the disc.

It is clear that the higher resolution of the finer grids results in higher velocity peaks but it is also clear that there is no great benefit to be gained by using a grid finer than the 'grid 3' version as far as temperature distribution is concerned.

The conclusion to be drawn from this test is that for cracks of most forms, as long as no strong fluid recirculations are expected, about 20 to 30 calculational cells are needed across the maximum crack dimension in order to obtain a reasonable spatial resolution.

### 13.5.2 Dependence of heat conduction results on time step treatment

Criteria for stability, convergence, and errors in heat conduction calculations are well documented (see for example Smith[4]), but it was proposed in this case to try to achieve some economies in grid size selection and time step treatment. For this reason tests were undertaken.

The basis of the independent tests on the conduction program is the domain illustrated in Figure 13.10. This rectangular block of dimensions 1030 m × 1030 m × 200 m is assumed to be initially at a uniform temperature of 250 °C. On one of the square vertical faces a central area of 100 m × 100 m is then held at a constant 50 °C. All other surfaces of the block are subject to a constant gradient boundary condition. Temperatures are examined at certain locations and times along the central axis Z'Z shown in Figure 13.10, thus simulating one-dimensional conduction. The analytical solutions used for comparison were taken from Carslaw and Jaeger.[5]

Three aspects of the treatment of time in Equation (13.5) have been investigated:

(a)  the degree of time step relaxation to employ;
(b)  the dependence of the solution on the length of the first time step; and
(c)  its dependence upon the expansion ratio between successive time steps.

Some of the test results are presented in Figures 13.11, 13.12, and 13.13. Figures 13.11 and 13.12 must be compared with each other. They represent the same calculations, executed in the case of Figure 13.11 with an initial time step of 1 day and in the case of Figure 13.12 with an initial time step of 6 hours. Each diagram contains two graphs showing normalized temperatures plotted: (a) against distance at a pair of fixed times; and (b) against time at a pair of fixed distances from the cooled surface. At each time and at each distance the analytical result for one-dimensional conduction in a semi-infinite solid is

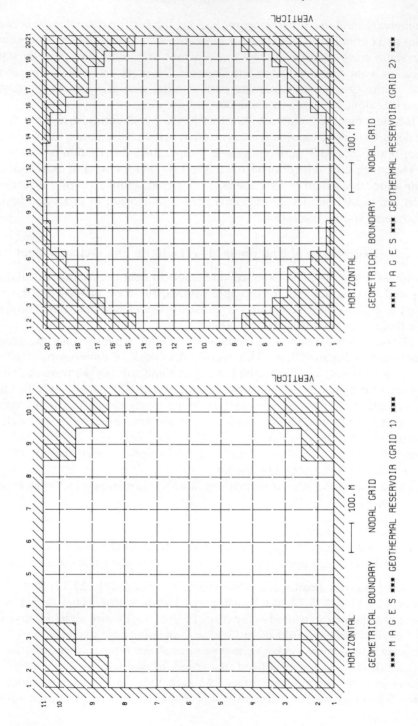

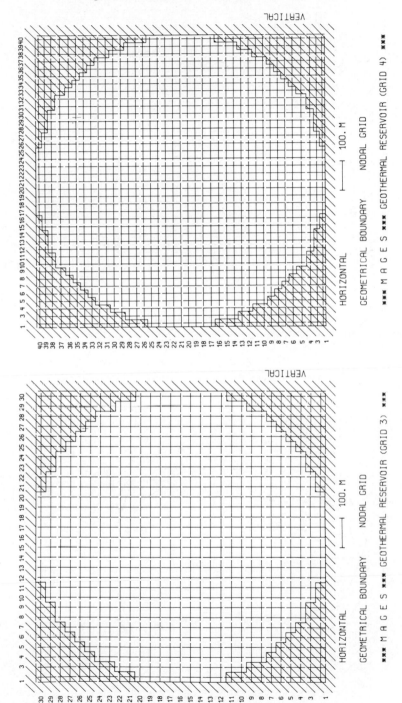

**Figure 13.5** Finite difference grids used for testing the crack flow solution

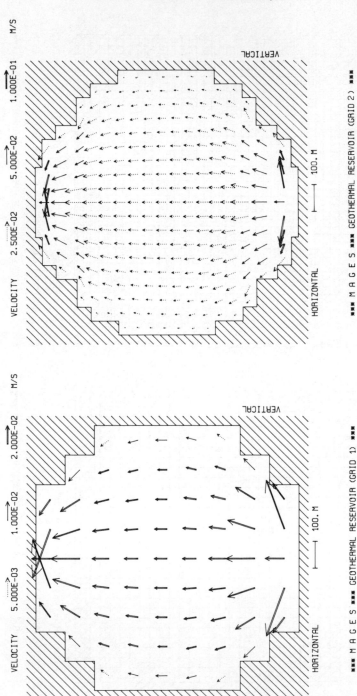

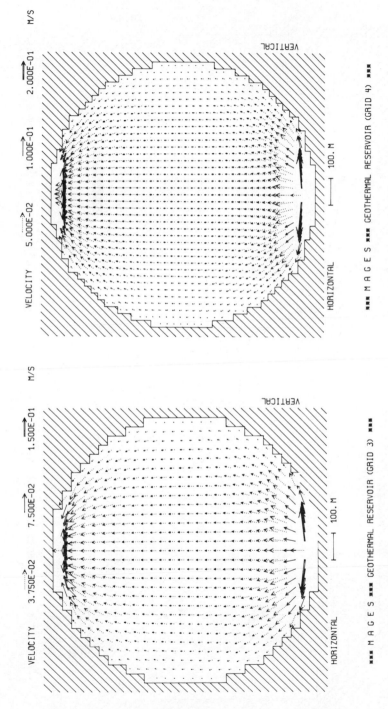

**Figure 13.6** Comparison of vector flow patterns in crack flow using the four grids shown in Figure 13.5

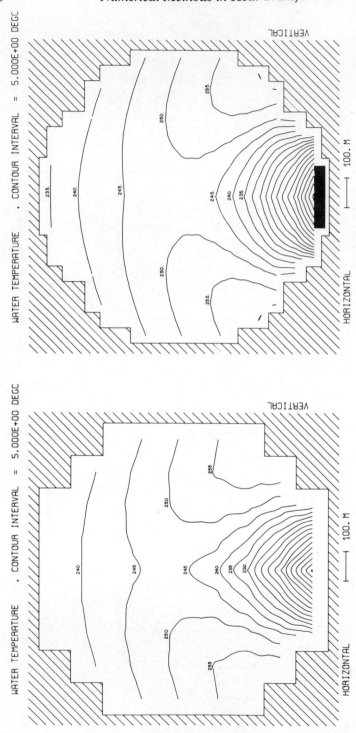

WATER TEMPERATURE , CONTOUR INTERVAL = 5.000E+00 DEGC

*** M A G E S *** GEOTHERMAL RESERVOIR (GRID 2) ***

WATER TEMPERATURE , CONTOUR INTERVAL = 5.000E+00 DEGC

*** M A G E S *** GEOTHERMAL RESERVOIR (GRID 1) ***

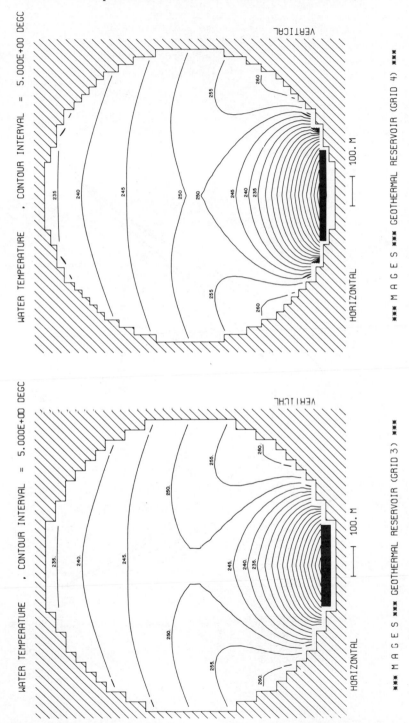

**Figure 13.7** Comparison of water temperature isotherm distributions using the four grids shown in Figure 13.5

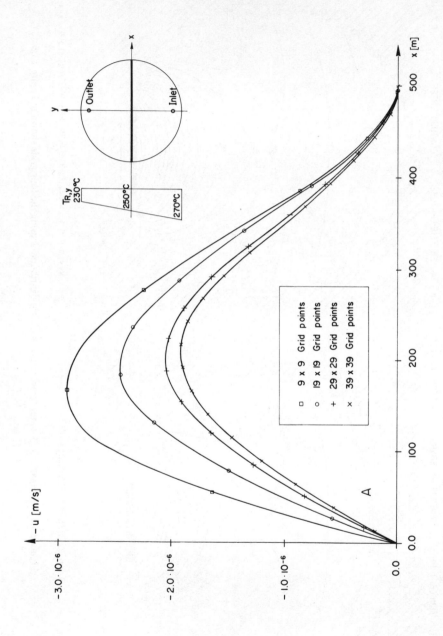

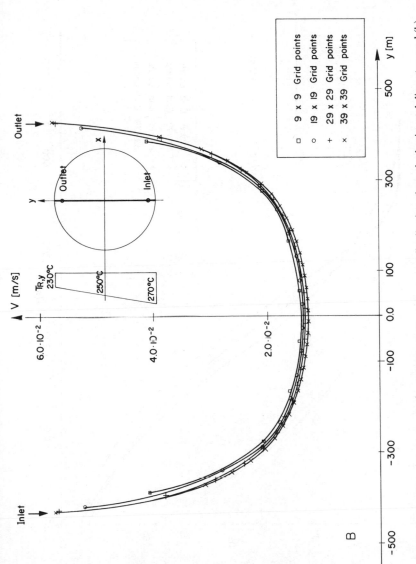

**Figure 13.8** Effect of grid density on: (a) *U*-velocity component distribution along the horizontal diameter; and (b) *V*-velocity component distribution along the vertical diameter between inlet and outlet

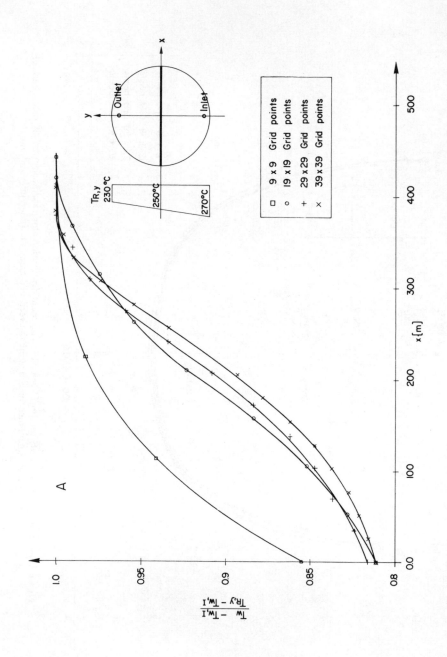

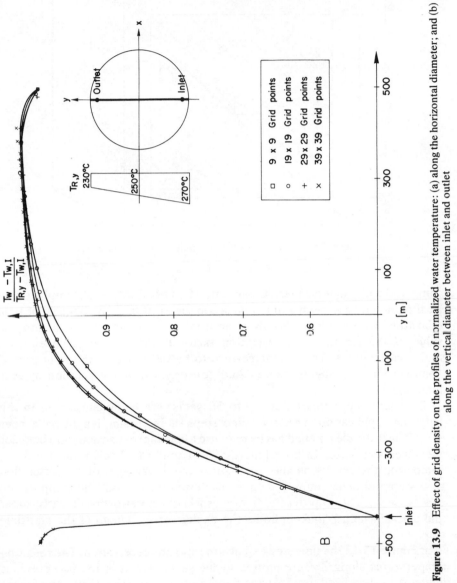

**Figure 13.9** Effect of grid density on the profiles of normalized water temperature: (a) along the horizontal diameter; and (b) along the vertical diameter between inlet and outlet

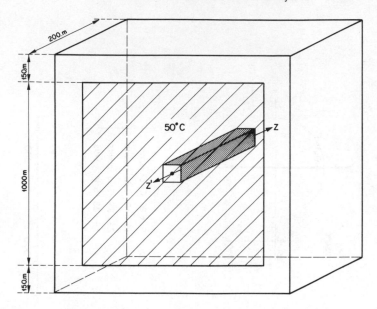

**Figure 13.10**    Test geometry for conduction program trials

compared with numerical results computed first of all with a fully implicit time treatment and again with a time relaxation factor of 0.5. The benefits of time relaxation by the Crank–Nicolson approach[4,6,7] are clearly seen and the value of 0.5 for the time relaxation factor is that which has been used in our calculations. The advantages are to be felt not only in the improved accuracy but in a slightly more rapid convergence of the solution at each time step.

Since large operating times (up to 30 years) are being considered in this particular application successive time steps of increasing length have been used. The principle adopted has been to use a constant expansion ratio between the successive steps. In both Figures 13.11 and 13.12 this ratio is 1.3. By comparing the results on the two figures the advantages of a shorter first time step and hence improved time resolution throughout the computation can be seen. The improvements are especially noteworthy at early times and are sufficient to make 6 hours the choice for the length of the first time step.

In Figure 13.13 the time step expansion ratio has been varied. The resulting temperatures along Z′Z are plotted in the same way as in the two previous figures at certain fixed times (a) and fixed distance (b). Since the system seems to be insensitive to the expansion ratio, the value of 1.5 has been used for coupled calculations.

### 13.5.3 Grid dependence of three-dimensional heat conduction

The purpose of this exercise was to test the dependence on grid spacing in the $z$-direction, that is to say normal to the cooled wall, so three grid arrangements were selected. These are non-uniform, the grid points being concentrated near the cooled wall with the cell size expanding smoothly in the $z$-direction up to the extremity of the calculated domain at a distance of 200 m. Temperatures are calculated at the cell centres, but on the faces of the domain a zero volume cell allows a surface temperature to be handled. The three grids used respectively 11, 15, and 19 points between $z = 0$ and $z = 200$ m.

In Figure 13.14 the computed temperatures, using an initial time step of approximately 6 hours, a time step expansion ratio of 1.3 and time step relaxation are computed: (a) against distance at a pair of fixed times; and (b) against time at a pair of fixed distances from the cooled surface. There is little sensitivity to grid spacing under these conditions.

## 13.6 A SAMPLE CALCULATION

In order to demonstrate the type of results to be expected, plots from a series of calculations are presented in Figures 13.15 to 13.19.

The shape of the crack shown in Figure 13.15(a) has been derived from considerations of crack propagation by hydraulic fracturing.[8] It is a vertical crack ($\phi = 0$), is 1000 m from top to bottom, and has an initial maximum thickness of 1 mm. The particular case chosen here is one in which the flow is forced downwards from a top inlet to a bottom outlet against the buoyancy forces. The flow rate is $100 \text{ m}^3 \text{ h}^{-1}$.

For this test an estimation has been made of the importance of mechanical fluid/solid interactions, by treating crack thickness as a function of rock temperature. Separate tests have shown that the presence of pressurized fluid in a crack produces an approximately elliptic cross section, the length of whose minor axis varies with the pressure. Figure 13.15(b) shows how the hydrostatic pressure dominates in the fluid, resulting in practically horizontal isobars. Thus horizontal cross sections through the crack at $t = 0$ are basically elliptic. Superposed upon this shape at later times is an expansion of the cross section due to cooling of the adjacent rock. It was assumed for the calculation that the rock contraction is one dimensional only, along strips normal to the plane of the crack. This brings with it an underestimation of the true cross section expansion, which has been compensated by using an artificially high linear thermal expansion coefficient for the rock. The result is not quite correct but does enable the order of magnitude of the solid/fluid interaction to be demonstrated.

There is another fluid/solid interaction effect. As the crack expands with time, the pressure drop between inlet and outlet points will become smaller.

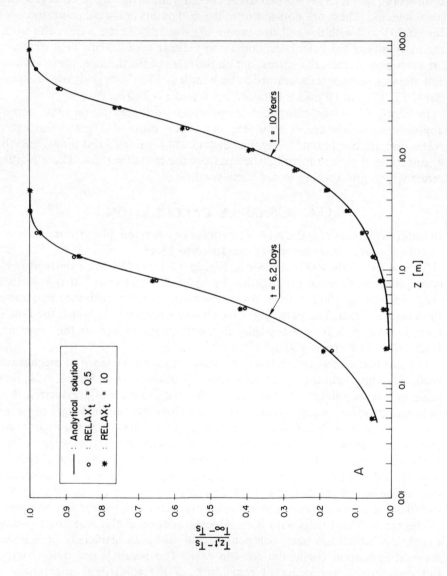

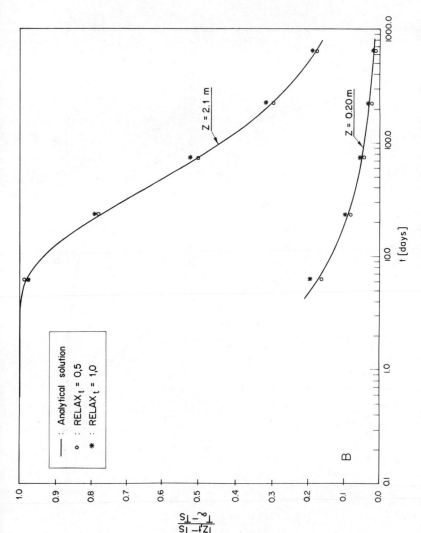

**Figure 13.11** Comparisons of analytically and numerically calculated temperatures in a semi-infinite solid with: (a) temperature a function of distance from the cooled face at two times; (b) temperature a function of time at two distances from the cooled face, initial time step, 1 day; time step expansion ratio, 1.3

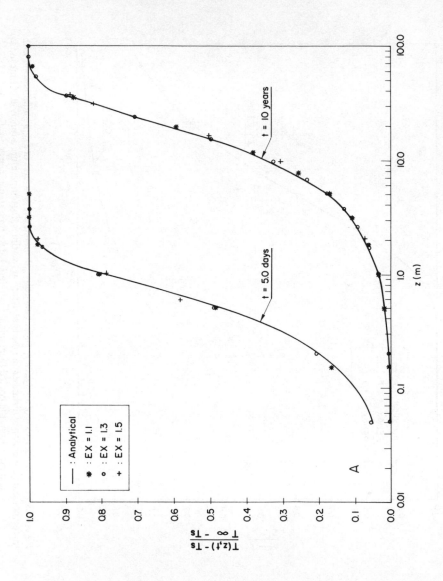

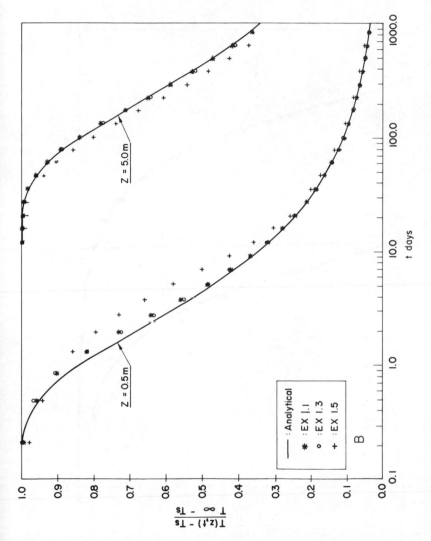

**Figure 13.12** Comparisons of analytically and numerically calculated temperatures in a semi-infinite solid with: (a) temperature a function of distance from the cooled face at two times; (b) temperature a function of time at two distances from the cooled face, initial time step, 6 hours; time step expansion ratio, 1.3

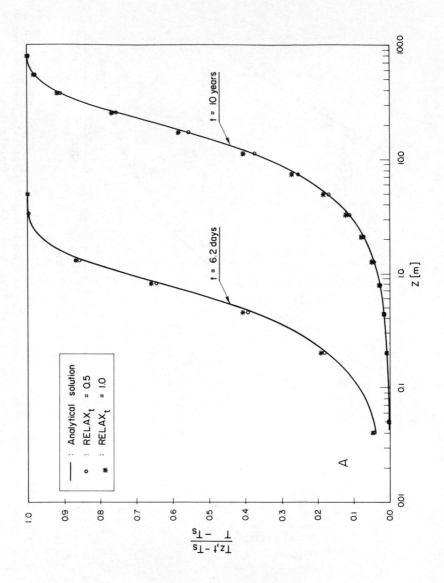

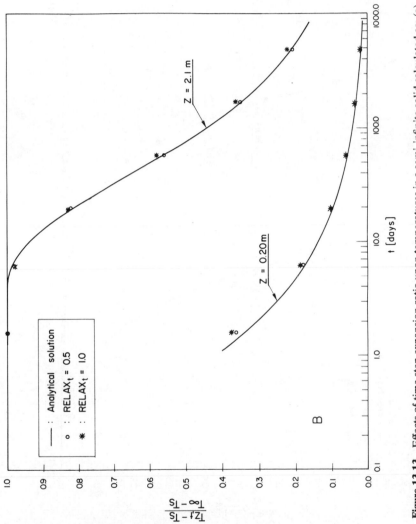

**Figure 13.13**   Effects of time step expansion ratio upon temperatures in a semi-infinite solid calculated as: (a) functions of distance from the cooled face at two times; (b) functions of time at two distances from the cooled face, initial time step, 6 hours

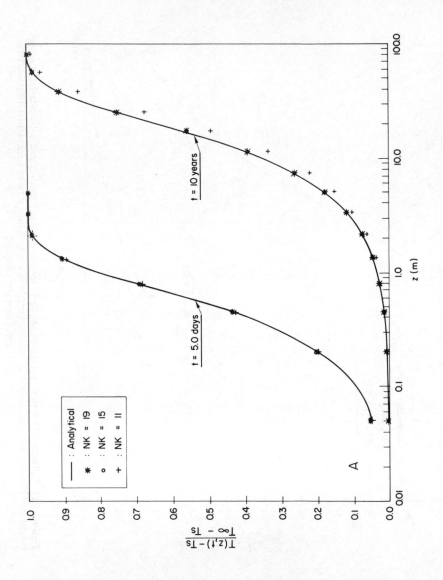

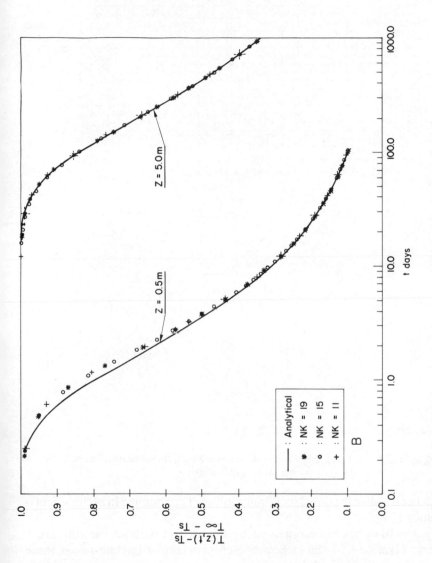

**Figure 13.14** Effect of grid density upon temperatures in a semi-infinite solid calculated as: (a) functions of distance from the cooled face at two times; (b) functions of time at two distances from the cooled face, initial time step, 6 hours; initial step expansion ratio, 1.3

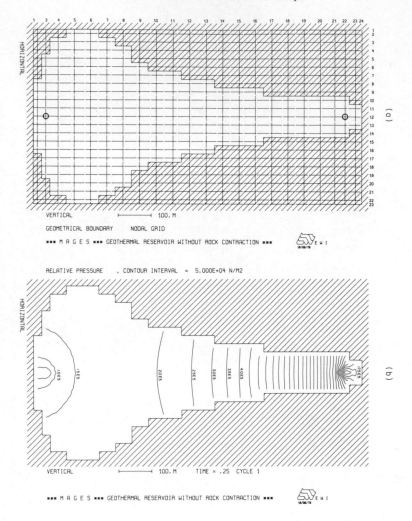

**Figure 13.15** (a) Crack shape and finite difference grid; (b) isobar distribution during the first time step

This will in turn alter the cross-sectional profiles. The coupling in of this effect is not included.

The calculation has been repeated both with and without variable crack geometry. Figure 13.16 shows how the cross-sectional thicknesses at three points on the vertical axis of the crack vary with time. The flow vector and fluid temperature fields are shown at discrete times in Figures 13.17 and 13.18 respectively. Finally Figure 13.19 gives the outlet water temperature and the cumulative heat removal time histories, the results which are finally useful for system design purposes.

## 13.7 CONCLUDING REMARKS

A method of coupling unsteady convective and conductive heat transfer across a fluid/solid interface has been developed and applied to heat transfer in cracks. In the interests of realism the method has been incorporated into a model of considerably greater detail than has previously been attempted.[9] Although the individual program operations have been checked for stability and accuracy, it has only been possible so far to check the coupled calculations for the plausibility of their results.

The crack is only the basic element in a hot, dry rock heat reservoir. At present we do not know exactly what form a reservoir will take—single cracks, multiple intersecting cracks, natural crack system or fractured regions. Experiments which are at present getting under way will reveal what further developments in modelling are still necessary for this field of application.

## APPENDIX 13.8  DIFFUSION IN THE FLUID

The diffusive terms in the fluid transport equations for momentum and thermal energy (13.2), (13.3), and (13.4) have been lumped into the symbols $D_U$, $D_V$, and $D_T$ respectively. In the case of a thin crack these terms are small, transport being dominated by frictional drag and buoyancy forces in the momentum equations and by fluid transport and surface heat transfer in the thermal energy equation. In fact in the 'TURF' programs the diffusion terms are usually included, so are listed here for completeness.

The assumptions used for the fluid flow are: constant density; viscosity $\mu = \mu(T)$; and laminar flow; which result in the following diffusive terms:

$$D_U = \frac{1}{\rho}\left\{2\frac{\partial}{\partial x}\left(\mu\varepsilon\,\frac{\partial U}{\partial x}\right) + \frac{\partial}{\partial y}\left[\mu\varepsilon\left(\frac{\partial U}{\partial y} + \frac{\partial V}{2\partial x}\right)\right]\right\}$$

$$D_V = \frac{1}{\rho}\left\{2\frac{\partial}{\partial y}\left(\mu\varepsilon\,\frac{\partial V}{2\partial y}\right) + \frac{\partial}{\partial x}\left[\mu\varepsilon\left(\frac{\partial V}{\partial x} + \frac{\partial U}{\partial y}\right)\right]\right\}$$

$$D_T = \frac{1}{\rho}\left[\frac{\partial}{\partial x}\left(k\varepsilon\,\frac{\partial T}{\partial x}\right) + \frac{\partial}{2\partial y}\left(k\varepsilon\,\frac{\partial T}{\partial y}\right)\right]$$

## APPENDIX 13.9  VALUES USED FOR THE PROPERTIES OF ROCK AND FLUID IN THE TESTS AND EXAMPLES REPORTED

**Rock**

| | |
|---|---|
| density | $2700 \text{ kg m}^{-1}$ |
| specific heat ($C_v$) | $1000 \text{ J kg}^{-1}\,^{\circ}\text{C}^{-1}$ |

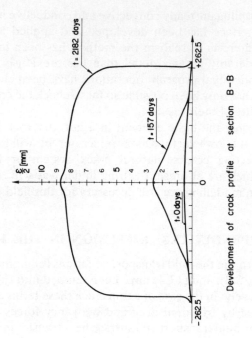

Development of crack profile at section B–B

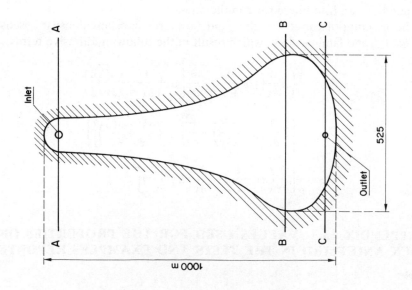

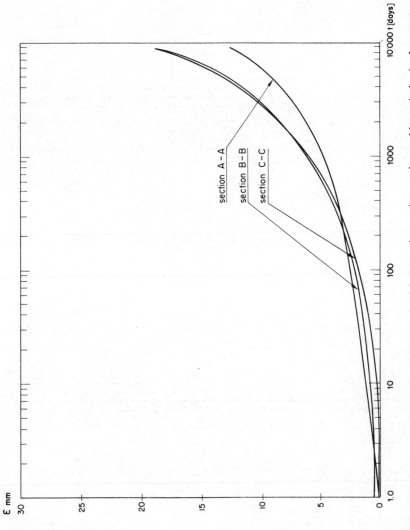

**Figure 13.16**  Variations with time of the crack thickness at three points on the crack's vertical axis of symmetry

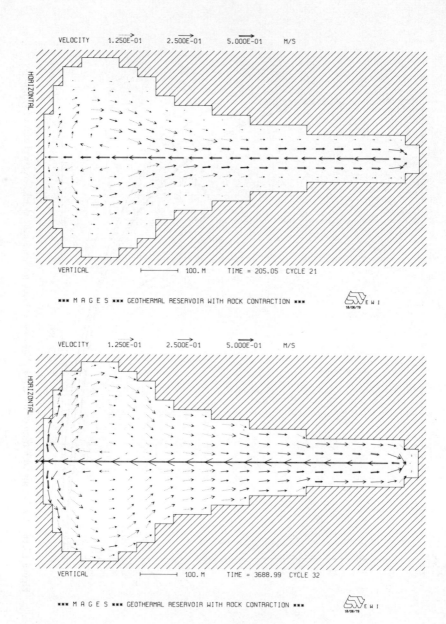

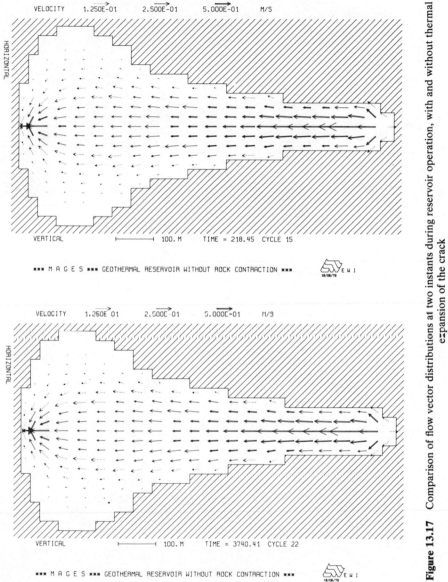

**Figure 13.17** Comparison of flow vector distributions at two instants during reservoir operation, with and without thermal expansion of the crack

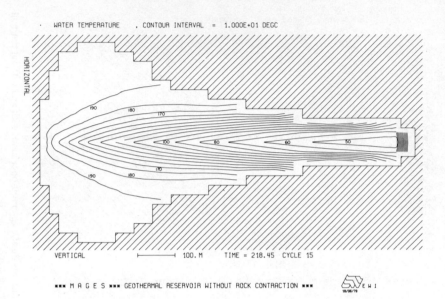

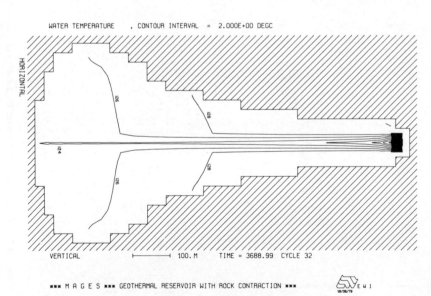

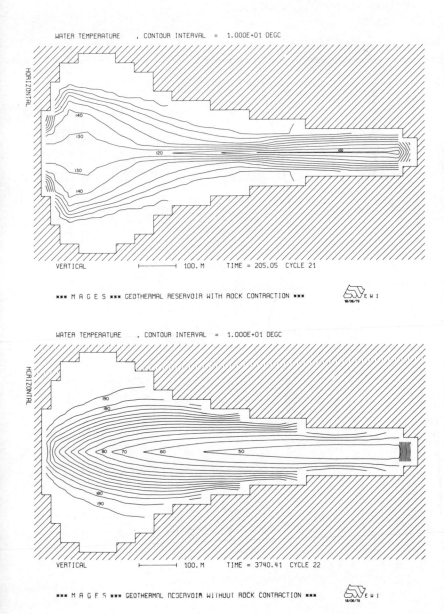
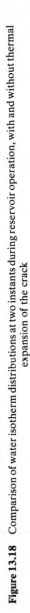

**Figure 13.18** Comparison of water isotherm distributions at two instants during reservoir operation, with and without thermal expansion of the crack

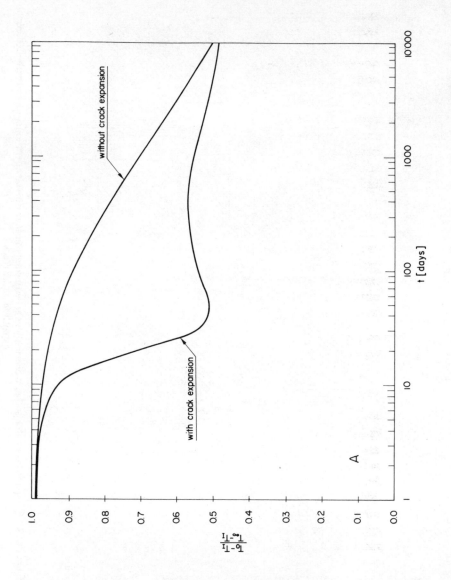

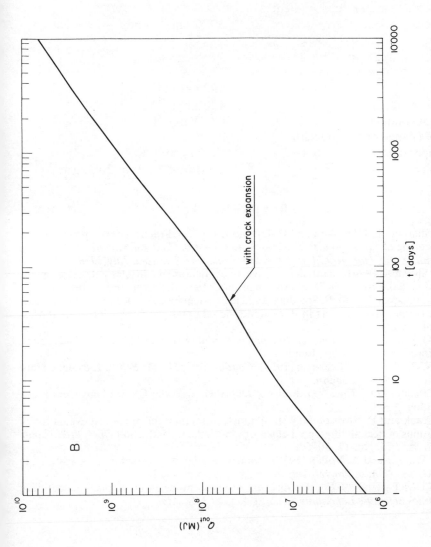

**Figure 13.19** (a) Time histories of water outlet temperature during geothermal reservoir operation. Demonstration of the effect of thermal expansion of the crack. (b) Cumulative heat removal history for the case with consideration of rock thermal contraction

thermal conductivity
   for tests                           $2.0 \text{ W m}^{-1}\,^{\circ}\text{C}^{-1}$
   for production runs        $(3.02-0.003\,T_R) \text{ W m}^{-1}\,^{\circ}\text{C}^{-1}$
geothermal temperature gradient
   in rock                      $30\,^{\circ}\text{C km}^{-1}$ depth

## Fluid

density                          $1000 \text{ kg m}^{-3}$
specific heat $(C_p)$            $4200 \text{ J kg}^{-1}\,^{\circ}\text{C}^{-1}$
thermal conductivity        $0.6 \text{ W m}^{-1}\,^{\circ}\text{C}^{-1}$
rate of change of density with
   temperature               $-0.2 \text{ kg m}^{-3}\,^{\circ}\text{C}^{-1}$
viscosity                    $1/[1000+32(T_W-20)] \text{ kg m}^{-1}\text{ s}^{-1}$

## REFERENCES

1. D. Sharma, R. J. Hopkirk and P.-J. Pralong (1978). 'Practical developments in the modelling of turbulent heat and mass transfer', *Int. Conf. on Numerical Methods in Laminar and Turbulent Flow, University College of Swansea, July 1978.*
2. D. Sharma (1974). 'Turbulent convective phenomena in straight, rectangular-sectioned diffusers', *Ph.D. Thesis*, Imperial College, London University.
3. S. V. Patankar and D. B. Spalding (1972). 'A calculational procedure for heat, mass and momentum transfers in three-dimensional parabolic flows', *Int. J. Heat Mass Transfer*, **15**, 1787–1806.
4. G. D. Smith (1975). *Numerical Solution of Partial Differential Equations*, 2nd reprint, Oxford University Press, London.
5. H. S. Carslaw and J. C. Jaeger (1959). *Conduction of Heat in Solids*, 2nd edn, Oxford University Press, London.
6. J. Crank (1975). *The Mathematics of Diffusion*, 2nd edn, Oxford University Press, London.
7. J. Crank and P. Nicolson (1947). 'A practical method for numerical evaluation of solutions of partial differential equations of the heat conduction type', *Proc. Camb. Phil. Soc.*, **43**, No. 17, 50–67.
8. Th. Wacker and A. Vollan (1979). 'Model calculations of fracture propagation', in *MAGES, Annual Progress Report for 1978* (unpublished).
9. R. D. McFarland (1975). 'Geothermal reservoir models—crack plane model', *Los Alamos Scientific Laboratory Report* LA 5947 MS, April.

*Numerical Methods in Heat Transfer*
Edited by R. W. Lewis, K. Morgan, and O. C. Zienkiewicz
© 1981 John Wiley & Sons Ltd

*Chapter 14*

# A Survey of Some Simulation Models for Industrial Processes

*Helge Moen*

## SUMMARY

This chapter presents a survey of models which have been developed for some processes in the metallurgical industry. The generation, transport, and dissipation of heat is of fundamental importance for these processes. All models are based on a finite difference formulation of the heat conduction equation, which is solved in its dynamic form. The solution techniques are based on the same principles for all models. The formulation and treatment of the boundary conditions, choice of geometry, and practical adaptation of empirical and semi-empirical relations, characterize the different models from a computational view point.

The models have been used extensively in industrial environments, and seem to represent a successful set of applications of mathematical modelling to complex practical problems.

The chapter describes the salient features of the processes and the corresponding modelling assumptions. Some elements from the relevant experience is summarized.

## 14.1 INTRODUCTION

Development of mathematical models for industrial applications has been carried out at the Institute for Energy Technology for more than 10 years. Initially, this work had the character of a spin-off activity, in the sense that knowledge, methods, and techniques which were demonstrated as successful on nuclear problems were applied on a smaller scale for practical problems in more conventional areas. Over the years, this type of work has increased in volume. It has also gradually attained an industrial and commercial character, and is today an activity in its own right. The activity includes a wide range of applications and methods. From the beginning thermal problems have been an important field for this work. This is not surprising, since many important industrial processes are basically thermal in nature. Understanding of thermal energy balance and the detailed temperature picture is therefore instrumental to any basic, quantitative process understanding. This chapter intends to give a survey of some models which have been developed for thermal processes. The survey does not pretend to be complete, but is rather an overview of typical applications of models in this area. Some of the models have, during their lifetime, been subject to significant revisions, extensions, and improvements.

## 14.2   MODEL PHILOSOPHY

A rational and fruitful philosophy for process model design should be derived from the practical functions and purposes of the individual model, as this is related to the operational characteristics of the corresponding process. The purpose of model applications may differ somewhat from case to case, but the following motivations are usually present:

(a)   to find a set of controlled parameters which give a near optimal operational condition;
(b)   to improve the design of the process equipment;
(c)   to improve fundamental understanding of the process.

These objectives will usually imply that a large number of calculations have to be carried out. Moreover, the operating requirements and demands will usually change in the course of time leading to a corresponding need for repeated calculations. We should therefore regard the process model as an operational tool, to be used very frequently, rather than a means for providing, once and for all, the answer to a well specified problem. This model concept can best be realized by a very accurate tailoring of the model to the process at hand. The practical characteristics of this concept include:

(a)   very low computation costs;
(b)   the concepts used in the model are identical to those used by personnel responsible for process operation;
(c)   input/output is well adapted to the practical problem and therefore easily understood;
(d)   adjustment of model parameters may be carried out with a minimal effort.

The inherent value of the process understanding which may be derived from the active use of a model of high quality, is of course very difficult to assess in terms which may be used in a profit analysis of the model development project. Still, there is little doubt that the substitution of qualitative, sketchy process knowledge for quantitative, systematic understanding is the primary objective for model development. The appreciation of this fact is one of the key elements in the planning and implementation of a successful model development project.

## 14.3 THERMAL MODELLING

Transport of heat by conduction is governed by the well known equation

$$\rho c \frac{\partial T}{\partial t} = \nabla \cdot (k \nabla T) + Q \tag{14.1}$$

where $\rho$ is density, $c$ specific heat capacity, $T$ temperature, $k$ thermal conductivity, and $Q$ heat source. The equation is here written in its dynamic form. The source term $Q$ may include heat generation, heat transferred by mass transport, and similar effects. Actually, the behaviour and treatment of the various forms of the $Q$ term is one of the characteristics for some of the models to be described in this chapter.

The boundary conditions may be formulated generally by

$$\alpha T + \beta \frac{\partial T}{\partial n} = \gamma \tag{14.2}$$

where $T$ is the boundary temperature at the point on the surface where Equation (14.2) applies. $\partial T/\partial n$ is the normal derivative out of the domain. $\beta \, \partial T/\partial n$ is thus an expression for the heat flux out of this volume at the appropriate point on the surface. $\alpha$, $\beta$, and $\gamma$ may be functions of space coordinates, time, and temperature. Within the framework of this formalism, heat exchange by convection and radiation between the domain of interest and its environment may be treated with a high degree of flexibility.

Equations (14.1) and (14.2) are the basis for models described here. Although these models cover a wide range of applications, some features are common to all of them:

(a)  The heat equation is solved over a domain which is composed of elements with regular (i.e. rectangular or cylindrical) shape.
(b)  The domain may be composed of several materials, but the boundaries between these must be parallel to the surface.
(c)  The discretization is generally made by imposing a mesh composed by straight lines on the domain. There are exceptions to this rule. In certain cases, a triangular mesh is applied in order to accommodate special process geometries.
(d)  The characteristic properties of each material are functions of temperature. This is of particular interest in the case of phase transitions, since the transition energy is represented by a particular non-linear behaviour of $c$ as function of temperature.

The mathematical methods and techniques applied are similar for the applications described. These methods are outlined by E. E. Madsen,[1] pp. 81–89.

### 14.4  STEELTEMP

#### 14.4.1  Scope

The production of steel slabs is regarded as a sequence of processes, as illustrated in Figure 14.1. Subsequent to the teeming, the mould is left to cool

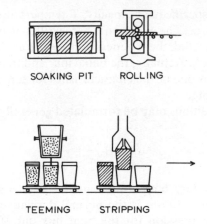

SOAKING PIT     ROLLING

TEEMING     STRIPPING

**Figure  14.1**  Process  sequence  in steel production

until a solid shell is formed, sufficiently strong to support the ingot weight. The mould is then stripped off, and the ingot is left to cool. After a suitable cooling period, the ingots are loaded into a soaking pit for heating to a uniform temperature.

When the ingot has attained a desired temperature, it may be removed from the pit for rolling or forging.

The STEELTEMP model is developed in order to predict the time-dependent temperature distribution in the ingot (and finally, the slab), during the sequence of thermal processes outlined above.

#### 14.4.2  Purpose

The STEELTEMP model[2] may be used to study the time-dependent thermal characteristics of the ingots, as these are influenced by the scheduling of the different events in the production process. The total time spent by the ingot in the different stages from the initial casting to the exit from the rolling mill, is, of course, decisive to the productivity of this operation.

Further, different loading patterns within and the energy input to the soaking pit may be studied to obtain the desired temperature profile in the ingots with a minimal energy consumption.

Finally, the influence from exchange of heat between the slab and the tool during forging or rolling may be studied.

### 14.4.3 Assumptions

The two-dimensional approximation assumes that the ingot length is large relative to the dimensions in the plane where the calculations are made.

With reference to the Equation (14.1) the STEELTEMP model assumes that: the $\partial/\partial z$ term is 0; the source term $Q$ is 0. The ingot may be rectangular or cylindrical. Cylindrical asymmetric ingots may also be treated.

The boundary conditions may specify radiation, convection, or a fixed boundary temperature, or combinations of these. The air gap which is formed between the ingot and the mould during cooling is treated separately. The heat transfer across the air gap is largely radiative, but convective transfer is also included.

Heat transfer between the slab and the tool during forging or rolling is divided into two parts: cooling of the slab while in contact with the tool; heat generated in the slab by deformation and friction. For each part, a corresponding temperature increment is calculated, based on relations which have been experimentally correlated. The mesh is redistributed linearly after each deformation.

When the ingots are left for cooling after stripping, it is assumed that several ingots are standing together in a regular array. The mutual shielding of radiation is then taken into account. The soaking pit may have a rectangular or cylindrical cross section, and may be heated with electricity, oil or gas.

In order to reduce the complexity of the calculation pertaining to the condition in the soaking pit, all ingots in the pit are assumed to have an identical thermal history. A typical ingot is then assumed to radiate heat to its neighbouring ingots, to the pit walls, and to the gas. The flow pattern of gas in the pit is very complex, and precludes estimation of a correct value for the linear gas velocity. The heat transfer coefficient for convective heat transfer is therefore determined by adjustment to experimental data.

The thermal condition of the pit is characterized by three temperature values, i.e. for the gas, the inner pit wall, and the outer pit wall, respectively. The gas and the inner pit wall temperatures are used in the calculation of the boundary condition to surface i of the ingot, with average temperature $T_i$:

$$\lambda \frac{\partial T_i}{\partial t} = h(T_g - T_i) + \sigma \varepsilon_g (T_g^4 - T_i^4) + \sigma \varepsilon_w (T_w - T_i^4) \qquad (14.3)$$

where $h$ is the convective heat transfer coefficient, $\sigma$ is the radiation constant, $\varepsilon_g$, $\varepsilon_w$ are emissivity coefficients for the gas and the wall, respectively, and $T_g$, $T_w$ are gas and wall temperatures, respectively.

Equation (14.3) is valid for an ingot surface facing the pit wall. As the thermal histories of the ingots are assumed identical, there is no radiative transfer between them.

The heat balance for the pit is expressed by a set of differential equations for the three temperatures mentioned. These equations are integrated separately, using a Runge–Kutta scheme. Due to the low heat capacity of the gas, the time step in this integration is kept considerably smaller than that used in the integration of the heat equation for the ingot.

### 14.4.4  Results

The resulting temperatures are given in tabular or graphical form. Isotherm plots are used in order to present the geometric distribution, and curves for selected points display the local temperature transients. For a large part of the practical applications, graphical output in the form of line-printer isothermal plots are adequate. Figure 14.2 shows an example of isothermal line-printer plots of the results from the simulation of rolling of a steel slab.

## 14.5  DYCAL

### 14.5.1  Scope

Casting of aluminium ingots is a semi-continuous process. The principles of the process are illustrated by Figure 14.3. The DYCAL model calculates the temperature distribution in the ingot during the casting process.

The casting starts with the casting shoe in a fixed, uppermost position. The shoe is then filled by molten aluminium, and the shoe and the table are subsequently pulled downwards. The speed of this movement is gradually increased to a predetermined, fixed casting velocity. The shoe is then pulled downwards until it reaches the bottom of the water pool.

The model is fully dynamic and includes the start-up phase. Both rectangular and cylindrical ingots may be studied within the framework of a two-dimensional approximation.

### 14.5.2  Purpose

In a practical situation the material parameters, as well as the ingot dimensions, are given. The casting conditions, i.e. the vertical velocity of the ingot, the cooling rate, and the casting temperature, may to a certain extent be subject to choice. The process is used for a wide range of aluminium alloys. The thermal parameters of these alloys vary over a quite wide range. The casting conditions should therefore be carefully adjusted for each individual alloy. The analysis of casting conditions for an unfamiliar alloy is therefore one of the primary purposes of the model.

As indicated in Figure 14.2, the locations of the liquidus and solidus isotherms are important characteristics of the stationary stage of the process. The profile of the sump, i.e. the lower part of the liquidus–solidus interface, determines the risk for break-through, and must therefore be determined with a high degree of reliability.

The dynamic properties of the model also allow for detailed simulations of metal solidification in the start-up phase. Different designs of the casting shoe may be studied in this context.

### 14.5.3 Assumptions

The temperature is calculated in a central two-dimensional, vertical plane through the ingot. This implies that

(a)  for $x$–$y$ geometry, the ingot should have a strongly overquadratic cross section, as the effects from the heat flux in the direction of the largest dimension are neglected;
(b)  for cylindrical geometry, the boundary conditions must be symmetrical.

With reference to the general heat conduction equation, this implies that the $\partial/\partial z$ term is omitted. The mould which surrounds the ingot is included in the geometry. The model is making the distinction internally between the start-up and the stationary phases, as the two are treated differently.

In the initial phase, the geometry of interest includes the metal down to and including the casting shoe, as well as the mould. Thus, the initial phase is characterized by a variable, and in effect, expanding geometry. During this phase there is no source term ($v = 0$), but the boundary conditions are dynamically changed as the geometry is expanding.

In the stationary part of the process, the casting shoe is moving outside the region of interest. This situation is readily modelled by a fixed geometry with a constant velocity mass flow in the downward direction. The source term is then given by

$$Q = v\frac{\partial}{\partial x}(\rho H) \tag{14.4}$$

where $v$ is the mass flow velocity and $H$ is enthalpy. The heat conduction equation may be integrated up to the stationary state, where the $\partial/\partial t$ term is zero.

As indicated on Figure 14.2, the ingot is cooled by water spray from the mould.

The mechanisms which govern the heat transfer from the ingot are assumed to be either natural convection, or nucleate boiling, or film boiling, dependent upon the surface temperature of the wall. The heat transfer coefficient $h$ is then

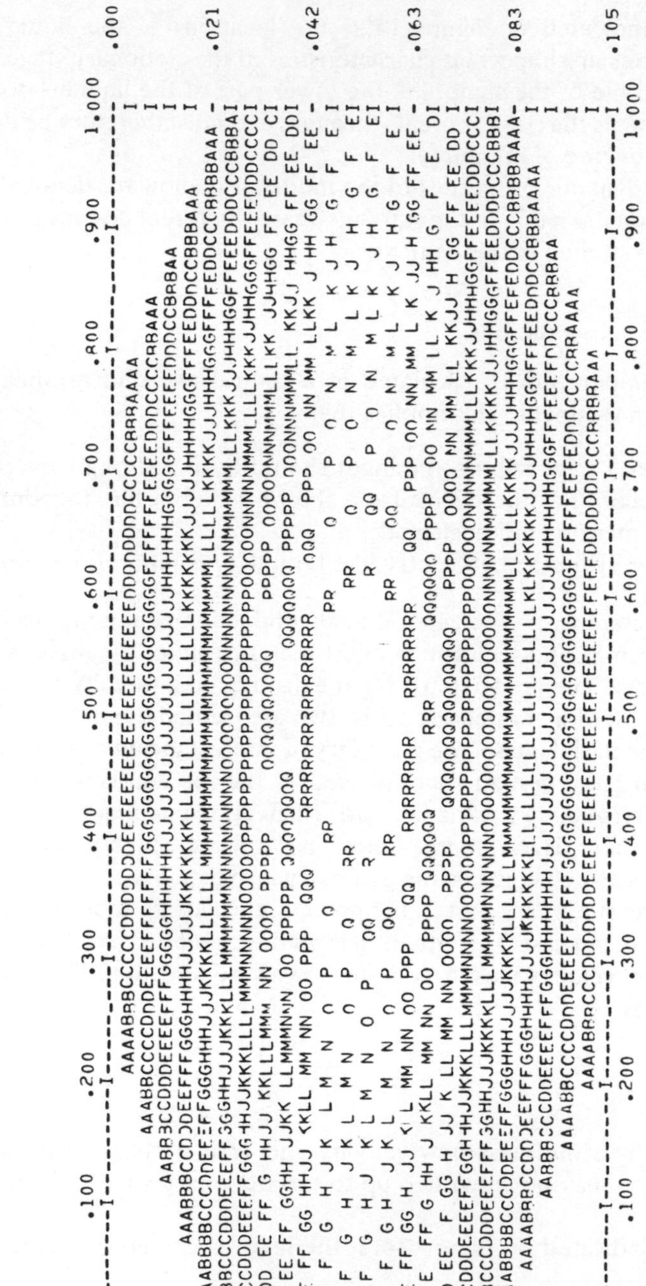

PLOTT AV ISOTERMER       TG =       0 H 1 M 50.0 S

| | | | | |
|---|---|---|---|---|
| AAA = 900. 0C | BBB = 920. 0C | CCC = 940. 0C | DDD = 960. 0C | EEE = 980. 0C |
| FFF = 1000. 0C | GGG = 1020. 0C | HHH = 1040. 0C | JJJ = 1060. 0C | KKK = 1080. 0C |
| LLL = 1100. 0C | MMM = 1120. 0C | NNN = 1140. 0C | OOO = 1160. 0C | PPP = 1180. 0C |
| QQQ = 1200. 0C | RRR = 1220. 0C | SSS = 1240. 0C | TTT = 1260. 0C | UUU = 1280. 0C |

**Figure 14.2**   Results from simulation of rolling of a steel slab

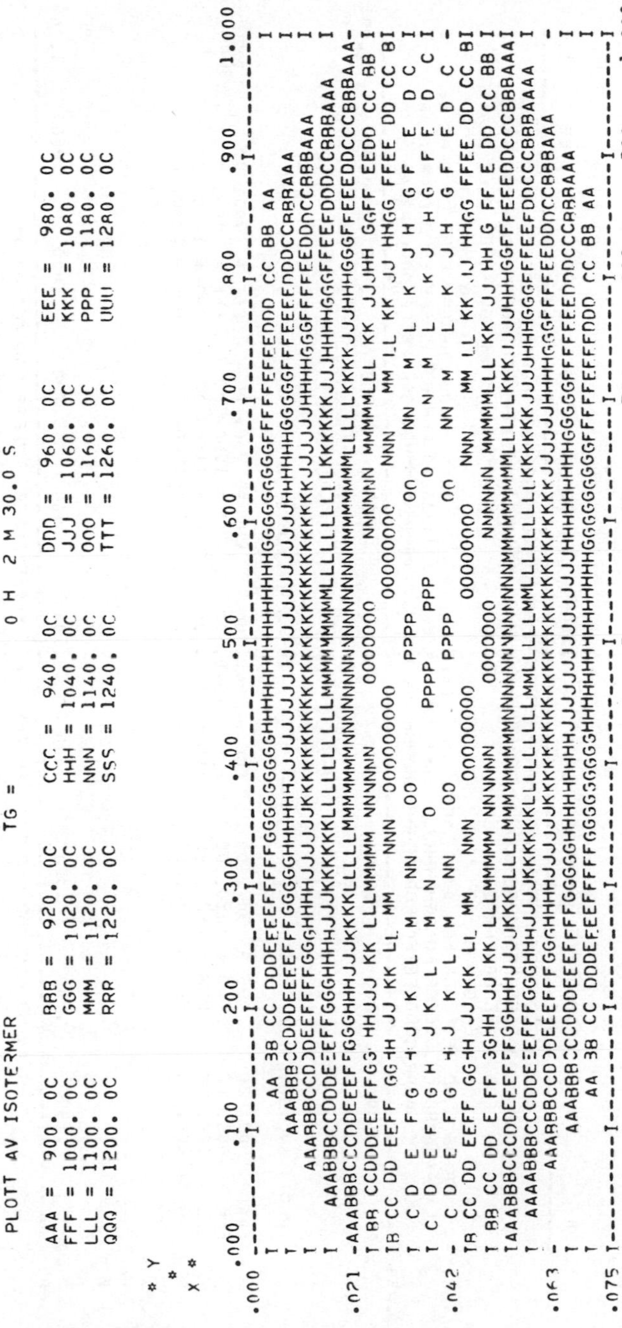
PLOTT AV ISOTERMER
TG = 0 H 2 M 30.0 S
AAA = 900. 0C BBB = 920. 0C CCC = 940. 0C DDD = 960. 0C EEE = 980. 0C
FFF = 1000. 0C GGG = 1020. 0C HHH = 1040. 0C JJJ = 1060. 0C KKK = 1080. 0C
LLL = 1100. 0C MMM = 1120. 0C NNN = 1140. 0C OOO = 1160. 0C PPP = 1180. 0C
QQQ = 1200. 0C RRR = 1220. 0C SSS = 1240. 0C TTT = 1260. 0C UUU = 1280. 0C
Y
X
.000 .100 .200 .300 .400 .500 .600 .700 .800 .900 1.000
.000 .021 .042 .063 .075

**Figure 14.2** Continued

```
PLOTT AV ISOTERMER                TG =               0 H  3 M 10.0 S

AAA =  900. 0C     BBB =  920. 0C     CCC =  940. 0C     DDD =  960. 0C     EEE =  980. 0C
FFF = 1000. 0C     GGG = 1020. 0C     HHH = 1040. 0C     JJJ = 1060. 0C     KKK = 1080. 0C
LLL = 1100. 0C     MMM = 1120. 0C     NNN = 1140. 0C     OOO = 1160. 0C     PPP = 1180. 0C
QQQ = 1200. 0C     RRR = 1220. 0C     SSS = 1240. 0C     TTT = 1260. 0C     UUU = 1280. 0C

* Y
 *
X *

     .000      .100      .200      .300      .400      .500      .600      .700      .800      .900      1.000
.000 ----------I---------I---------I---------I---------I---------I---------I---------I---------I----------   .000
     I          AAABBCCCDDDEEEFFFFGGGGGGGHHHHHHHHHHHHHHHHHHHHHHHHGGGGGGGFFFFEEEEDDDCCCBBBAA                 I
     I     AAABBBCCDDDEEEFFFFGGGHHHHHJJJJJJKKKKKKKKKKKKKKKKKKKJJJJJJHHHHHGGGFFFFFEEDDDCCBBBAAA             I
     I   AA BB CCDD EE FF GGG HH JJJJ   KKKK       LLLLLLL       KKKKK  JJJJ HH GGG FF EE D CC BB AA       I
     I  A  B  C  D  EIF  G    H   J    KK        LYLL          LLLL        K   J   H   G  F E  D  C  B  A  I
.021 -   AA BB C DD EE F  GG HHH JJJ   KKKK        LLLLLLLLL        KKKK   JJJ  HH GG  F EE DD C BB AA    -   .021
     I     AAAAABBBCCDDEEEFFFFGGGHHHHHJJJJJJJJJJKKKKKKKKKKKJJJJJJJJJJHHHHHGGGGFFFFFEEDDCCCCBAAAAAA          I
     I          AAABBCCDDDEEEEFFFFGGGGGHHHHHHHHHHJJJJJJJJJJJHHHHHHHHHHHGGGGFFFFFEEEFDDDCBBBBAA             I
.033 I---------I---------I---------I---------I---------I---------I---------I---------I---------I---------I   .033
     .000      .100      .200      .300      .400      .500      .600      .700      .800      .900      1.000


PLOTT AV ISOTERMER                TG =               0 H  3 M 40.0 S

AAA =  900. 0C     BBB =  920. 0C     CCC =  940. 0C     DDD =  960. 0C     EEE =  980. 0C
FFF = 1000. 0C     GGG = 1020. 0C     HHH = 1040. 0C     JJJ = 1060. 0C     KKK = 1080. 0C
LLL = 1100. 0C     MMM = 1120. 0C     NNN = 1140. 0C     OOO = 1160. 0C     PPP = 1180. 0C
QQQ = 1200. 0C     RRR = 1220. 0C     SSS = 1240. 0C     TTT = 1260. 0C     UUU = 1280. 0C

* Y
 *
Y *

     .000      .100      .200      .300      .400      .500      .600      .700      .800      .900      1.000
.000 ----------I---------I---------I---------I---------I---------I---------I---------I---------I----------   .000
     I                 AA BB CCC DDD EEEEE   FFFFFFFFFFF   FFFFFFFFFFF   EEEEE DDD CCC BB AA               I
     I                AA B CC  DD  EE    FFFF       GGGGGGGGG       FFFF    EE  DD   CC B AA              I
     I                A B  C  D   E    FF     GGGGGG          GGGGGG     FF    E   D   C  B A             I
     I                A B C   D   E   F      GG                    GG       F   E   D    C B A            I
.021 -                AABB C  DD  EE   FFF     GGGGGGGGGGGGGGGGGGGG      FFF   EE  DD  C BBAA             -   .021
.025 I---------I---------I---------I---------I---------I---------I---------I---------I---------I---------I   .025
     .000      .100      .200      .300      .400      .500      .600      .700      .800      .900      1.000
```

**Figure 14.2** Continued

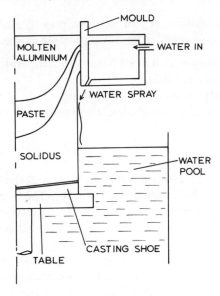

**Figure 14.3** Schematic illustration of the
semicontinuous casting process

given by,

for natural boiling: $\quad h_c = \dfrac{k}{L/Nu}$ (14.5)

where

$$L = 0.302\left(\frac{3\nu}{g}\right)^{1/3} Re^{8/15}$$ (14.6)

for nucleate boiling: $\quad H_b = a(T_w - T_b)^3$ (14.7)

for film boiling: $\quad h_f = \dfrac{b + c(T_b - T_f)}{T_w - T_f}$ (14.8)

Here, $T_{w,b,f}$ are the temperatures of the wall, the boiling point, and the cooling fluid, respectively; $\nu$ is the viscosity; $g$ is the gravitational constant; $Nu$, $Re$ are the Nusselt and Reynolds numbers, respectively; and $a$, $b$, and $c$ are constants determined from experience.

It is apparent that the film boiling mechanism will represent a limit to the heat flux. The formulae for natural convection and nucleate boiling are only valid in certain temperature intervals. In practice, the heat flux is usually calculated as the largest of the two, subject to the limitation given by film boiling. Some of the numerical coefficients given in the above formulae are derived from experimental results, and are used as default values in the computation.

The boundary conditions are in general very flexible. Besides convection/boiling in the cooling water film, it is possible to specify combinations of

(a)  radiation, convection, and fixed boundary temperatures;
(b)  fixed heat transfer coefficient;
(c)  calculated air gap between mould and ingot;
(d)  heat transfer through a lubricating layer between the ingot and the mould.

### 14.5.4  Results

It is often desirable to investigate a certain section of the ingot in more detail than is practical from a full calculation. For this purpose, a 'focusing option' is available. This implies that after a full calculation to a stationary state, the mesh in the interesting section is respecified, and the calculation is restarted for this section alone.

The temperature distributions are given in the form of isocurves or in tabular form, both as a function of time. The energy balance for selected volumes may also be calculated. Several detailed elements in the heat balance may be selected, such as:

(a)  the fraction of heat transferred by mass transport;
(b)  the corresponding fraction transferred by conduction;
(c)  the heat flux in the horizontal and vertical directions, respectively.

## 14.6  BAKING FURNACE MODEL

### 14.6.1  Scope

Anodes to be used in aluminium cells may be made in the form of carbon blocks. A baking procedure is part of the process of manufacturing these blocks. The baking takes place in a large furnace, which is divided in a number of identical cells. The blocks are piled up in a cell, imbedded in petroleum coke, and heated by gas which is led through cavities and perforations in the cell walls. The cells are interconnected and the gas is thus circulated through several cells. After baking, the blocks are cooled by passing cold gas through the cells in the same manner. The layouts of the furnace cells are given in principle by Figure 14.4. The model of this furnace treats the dynamic temperature distribution in the block pile, together with the heat balance in the gas.

### 14.6.2  Purpose

The baking effect is attained by keeping the carbon blocks above a certain temperature for a given period in time. The model is used to study dynamic

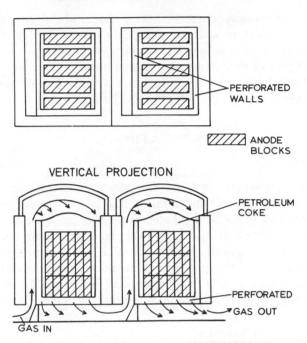

HORIZONTAL PROJECTION

PERFORATED WALLS

ANODE BLOCKS

VERTICAL PROJECTION

PETROLEUM COKE

PERFORATED

GAS OUT

GAS IN

**Figure 14.4**  Principle layout of a vertical baking furnace

temperature profiles in the block piles, as a function of energy supply and gas distribution pattern.[4]

### 14.6.3  Assumptions

The cells, the petroleum coke layer, and the anode blocks are assumed to have a regular three-dimensional geometry. The thermal conditions in the gas and in the blocks are treated separately, and the characteristic temperature of the gas flowing in the $x$-direction in a certain cross section is governed by

$$cw\frac{\partial T}{\partial x} + Q = 0 \tag{14.9}$$

where $w$ is mass flow in the gas channel cross section and $T$ is gas temperature in the cross section.

With reference to the general heat conduction equation, $Q$ is the source term in this equation and hence is the coupling term between the equations for the gas and the solid parts of the system. The boundary conditions may specify radiation, given surface temperature or given heat flux, all as functions in time.

### 14.6.4   Results

For the solid parts of the system the temperatures may be given in tabular form, or as isothermal plots in selected planes. Dynamic plots of the transient temperature behaviour at selected points may also be given.

For the gas part of the system, the temperatures and the heat balance for specified gas volumes may be given, together with the fuel consumption. The gas temperature at specified points may also be presented.

## 14.7   CATHODE

### 14.7.1   Scope

The aluminium electrolysis cell is a complicated construction composed of different construction materials: cathode and anode carbon material, aluminium, aluminium oxide, and other components in liquid as well as solid phase. The process of electrolysis is also very complex, where several interacting chemical and physical phenomena take place simultaneously. The CATHODE model[5] is developed to calculate the temperature distribution in a two-dimensional cross section of an electrolysis cell, when the electric energy supply is known. A simplified picture of the cell is given in Figure 14.5.

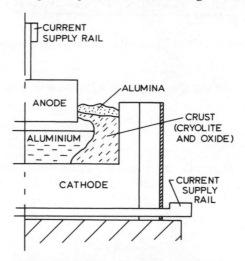

**Figure 14.5** Simplified cross section of an aluminium electrolysis cell

### 14.7.2   Purpose

The model may be used for investigation of the influence on the thermal conditions from changes in the construction of the cell. Furthermore, the relations between the details in the electric current distribution may be evaluated. Heat flow through the wall of the cell leads to cooling and subsequent solidification of the cryolite, which is present as a fluid in the cell. It is of particular interest to determine the liquid–solid interface of this material, and to study where this interface is localized as a function of details in the construction of the cell.

Apart from the studies of constructional changes and improvements, transient behaviour of the cell may be investigated by the dynamic version of the model. After a steady state is obtained, a typical disturbance may be introduced in the model to study the corresponding transients until a steady state is re-established.

### 14.7.3  Assumptions

The temperature is generated by ohmic heating. The local heat source is therefore given by

$$Q = \sigma \left( \frac{\partial \phi}{\partial y} \right)^2 + \sigma \left( \frac{\partial \phi}{\partial z} \right)^2 \tag{14.10}$$

in a two-dimensional representation. Here $\phi$ is electric potential and $\sigma$ is electric conductivity. To determine the heat generating source, it is therefore necessary to solve the equation governing the electrical potential:

$$-\frac{\partial}{\partial y} \left( \sigma \frac{\partial \phi}{\partial y} \right) - \frac{\partial}{\partial z} \left( \sigma \frac{\partial \phi}{\partial z} \right) = 0 \tag{14.11}$$

with the general boundary conditions

$$\alpha \phi + \beta \frac{\partial \phi}{\partial n} + \gamma = 0 \tag{14.12}$$

The model may thus be regarded as a coupled electric–thermal model for the cell. This coupling is emphasized by the fact that $\sigma$ is strongly temperature dependent.

The coupled equations are solved in a domain which must be made up by adjacent rectangular areas. It is possible to specify areas inside the domain which may be excluded from the calculations, and internal boundary conditions may also be prescribed.

With reference to the general heat conduction equation, the model assumed that the $\partial/\partial x$ term is omitted and that the source term $Q$ is as given by Equation (14.9). The model may also be used optionally for cylindrical $(r, z)$ or circular $(r, \phi)$ geometries.

The boundary conditions allow for a high degree of flexibility. The electrical potential or the electric current density may be specified, as well as the temperature or the heat flow on or out of the boundary.

The model was originally developed for stationary calculations only. A revised version was later developed which performed the dynamic calculations, from a given set of initial conditions, to a steady-state situation. In this connection, the dynamic treatment of the non-linear material constants was given special attention. As mentioned previously, the parameters $c$, $k$, and $\sigma$ are dependent on temperature in a strongly non-linear fashion. To take this into account, iteration at each time step is called for in order to ensure sufficient accuracy. The over-all stability of the dynamic calculation may be taken care of by an automatic step-control algorithm. Stability and accuracy may be influenced by practical choices in the iterative procedure. To arrive at a practical scheme, the discrete form

$$D\frac{\partial T}{\partial t} = AT^* + S \qquad (14.13)$$

of the heat conduction equation may be considered. $D$, $A$, and $S$ are the matrices which correspond to the $\rho\sigma$ term, the $\nabla k \nabla$ operator, and the $Q$ term of Equation (14.1). The equation is solved by an implicit Euler scheme. The equation is considered at non-linear iteration number $s + 1$ at time $t^{n+1}$. Thus, the temperature vector $T$ at the left-hand side may be indexed $T^{n+1}(s + 1)$. The schemes shown in Table 14.1 for selection of temperature-dependent arguments for the calculation of the different matrices have been studied.[9] Numerical experiments indicate that scheme 1 shows the greatest stability (in the sense that the largest time step is allowed), but it also leads to the greatest inaccuracy.

Table 14.1

| Scheme | D | A | T* |
|---|---|---|---|
| 1 | $D(t^{n+1}(s))$ | $A(t^{n+1}(s))$ | $T^{n+1}(s+1)$ |
| 2 | $D\left(\dfrac{t^n + t^{n+1}(s)}{2}\right)$ | $A\left(\dfrac{t^n + t^{n+1}(s)}{2}\right)$ | $T^{n+1}(s+1)$ |
| 3 | $D\left(\dfrac{t^n + t^{n+1}(s)}{2}\right)$ | $A\left(\dfrac{t^n + t^{n+1}(s)}{2}\right)$ | $\dfrac{T^n + T^{n+1}(s+1)}{2}$ |

Scheme 3 seems to require smaller time steps with a high degree of accuracy, while scheme 2 is deemed the best compromise for practical purposes. In the model, all three schemes are implemented and optionally available to the user.

### 14.7.4 Results

All desired values for temperature, electric potential, heat currents, and electric currents may be given in tabular form. These values may also be given in terms of isocurves.

## 14.8 ANODE

### 14.8.1 Scope

This model[6] was originally designed to study coupled electrical–thermal phenomena in the anode of an aluminium electrolysis cell. The model is, however, given a very general form, and may therefore be used for a wide range of coupled electrical–thermal problems. The model works in three-dimensional, Cartesian geometry, and provides a dynamic solution.

### 14.8.2 Purpose

The wide scope of this model implies that it may be used for several different purposes, whenever the study of three-dimensional, dynamic diffusion processes are of interest.

### 14.8.3 Assumptions

The equations used in this model are best described when seen in relation to the CATHODE model. The source term and the potential equations are the same as in this model, but solved in full three-dimensional geometry. The heat conduction equation is also here solved in its complete dynamic form. The electrical and thermal boundary conditions may in principle be specified corresponding to those in CATHODE. Additionally, a time dependence may be prescribed for the case of thermal radiation.

### 14.8.4 Results

The values for temperature, electric potential, heat flux or electric current may be presented at desired points. These values may also be given graphically by isocurves in selected planes.

## 14.9 EXPERIENCE

The use of these models over several years has, of course, led to a number of detailed improvements. The results of the models have also been compared with industrial experiments. The possibilities for making measurements of high reliability are very limited under the environmental conditions which are characteristic for the processes described here. The measurements are therefore very limited in number. The practical limitations also imply that measurements rarely can be made of the parameters of prime interest for model verification. There are, however, exceptions.

The model of the rolling process (STEELTEMP) which is included in Section 14.4, has been verified by measurement of surface temperatures by rolling an ingot of AISI 316 stainless steel from $680 \times 680$ mm down to $275 \times 275$ mm.[7] The campaign consists of 33 passes. The transient temperature reduction during a single pass is in the range 200–300 °C, and the total temperature reduction at the ingot surface during the campaign is around 150 °C. The comparison with model calculations showed very good agreement, with a maximum deviation between measurement and calculation of 12 °C.

Another case of experimental verification is reported in Fossheim and Madsen.[3] The DYCAL model has been compared to actual casting measurements, where two quite different sets of casting conditions have been used. The solidus–liquidus transition profile has been measured directly, and the results have not been evaluated in terms of temperature. Some discrepancies exist between the measured and calculated profiles, but these can be explained from known simplifications in the selected boundary conditions as well as the size of the grid mesh in regions with high temperature gradients.

Verification of the baking furnace model as well as the models for the electrolysis cell have also been made, but the detailed results are not available for discussion.

Another analysis of the STEELTEMP model may warrant special attention.[8] In this case, the ingots were left to cool, after stripping, in a rectangular array made up of eight ingots, and surface temperatures were measured. In general, the agreement between measurements and model results was very good. The rectangular array can be thought of as a $3 \times 3$ regular pattern, with the central ingot missing. The temperature of the ingot surface facing this empty central position was considerably lower than calculated by the model, corresponding to a significantly higher cooling rate at the surface. The reason for this is believed to be that the natural convection mechanism included in STEELTEMP is insufficient for prediction of the actual convective heat transfer which occurs when several ingots are located in a manner which implies that the cooling air velocity along one ingot surface is significantly increased by radiant heat from the others. In this particular case, the eight ingots are put in a configuration which superficially resembles a funnel, where the air is heated by radiation from sources covering a very wide angle.

This particular case is believed to illustrate a general and important point, which in principle is valid for all models described here, although to a varying degree of importance. The models include, in several instances, boundary conditions which are formulated to model convective heat transfer mechanisms. The available classical model formulae have a rather limited range of validity, which is quickly surpassed when practical industrial situations are to be described. In some cases, where the geometrical configurations are reasonably simple and not subject to parametric variation, special model relations may be included, if necessary in connection with experimental verification. For situations more remote from the idealized conditions, where convective heat transfer due to natural or forced cooling by air or liquids in complex geometries is significant, it is apparent that more applied research is necessary to improve the modelling possibilities for practical purposes.

## 14.10  CONCLUDING REMARKS

The main reason for describing these models has been to document some successful applications of mathematical and numerical methods to industrial processes. A review of the practical usefulness of these models, as experienced by industrial users, is clearly outside the scope of this paper. But frequent contact and communication with the user groups gives a clear indication of a high degree of user satisfaction.

The models described have been designed for frequent use in industrial environments. In this context, certain characteristic features are believed to be instrumental to the practical value of the models. Without attempting a complete and detailed analysis of these features, in conclusion it may be appropriate to review a few of the most important ones.

Low operating costs, achieved by a fast calculational procedure, will always be decisive when repeated use of the model for parameter studies is the objective. The numerical methods selected here are apparently very suitable for solution of the dynamic heat conduction equation, as very short execution times are obtained even for complex problems.

Simplicity is a keyword in industrial software. The formalism and methods used in the input/output parts of the model must therefore be carefully tailored to the concepts of the process at hand. Thus, the formalism may be more easily understood, learned and practised by the user. The development of this strongly user-oriented part of the model must be carried out in close cooperation with the user groups involved.

Another, somewhat more vague, but nevertheless important feature, has to do with the functional quality of the model. More specifically, the model should represent the fundamental, governing phenomena in the process in an adequate manner, without overdoing the detailing of effects which are of negligible importance. When necessary, approximations and assumptions should therefore be made with this consideration in mind.

## REFERENCES

1. R. W. Lewis and K. Morgan (Eds) (1979). *Numerical Methods in Thermal Problems*, *Proc. 1st Int. Conf.*, Pineridge Press, Swansea.
2. E. C. Madsen, J.-E. Jacobsen and E. Nitteberg (1979). 'STEELTEMP documentation', *Institutt for atomenergi (IFA)*, *Internal Memos* PM 289, 261–264.
3. H. Fossheim and E. E. Madsen (1979). *Application of a Mathematical Model in Level-Pour DC Casting of Sheet Ingots*, *Part 2*, *Light Metals*, pp. 695–720, The Metallurgical Society of AIME, Warrendale PA, USA.

4. E. E. Madsen and J.-E. Jacobsen (1976). 'Baking furnace model documentation', *IFA Internal Memos* PM 246–249.
5. A. Ek and I. Martinussen (1975). 'CATHODE documentation', *IFA Internal Memo* SD-171.
6. I. Martinussen, J.-E. Jacobsen and E. E. Madsen (1977). 'ANODE documentation', *IFA Internal Memos* PM 253–255.
7. B. Leden and S. Rutqvist (1978). 'Temperaturförlopp vid plattvalsning och friformssmidning', Jernkontorets forsking D234, *MEFOS, Luleå, Sweden* (in Swedish).
8. B. Lindstrand, SKF Stål, Hofors, Sweden, private communication.
9. I. Martinussen, IFA, Kjeller, Norway, private communication.

*Numerical Methods in Heat Transfer*
Edited by R. W. Lewis, K. Morgan, and O. C. Zienkiewicz
© 1981 John Wiley & Sons Ltd

## Chapter 15

# Heat Flow Modelling in Underground Coal Liquefaction

*Duane R. Skidmore and Fred K. Fong*

## SUMMARY

*In situ* coal liquefaction involves introducing a solvent or comminution reagent into a coal seam and converting monolithic coal into coal particles in a slurry or into coal-derived oils. The solvents considered were anthracene oil, super-critical toluene, caustic solution, and CO/steam. Ammonia was used as a comminution reagent.

Temperatures and heat flux in the *in situ* coal liquefaction system were modelled by considering solvents flowing in a channel embedded in a coal seam. Three conjugated equations with a moving boundary are involved.

A new method of numerical analysis, the method of alternating variables, was used. The method combines the ADI iteration method for the parabolic equation and the implicit central difference method for the hyperbolic equation by solving for alternate variables at successive time intervals. The new method gave very fast convergence.

## NOMENCLATURE

$C_f$ heat capacity of the fluid
$D$ channel diameter
$H_1, H_2$ parameters
$h_0$ heat transfer coefficient of the model system
$i$ index in $Z$-direction
$j$ index in $r$-direction
$k_p$ thermal conductivity of the medium
$k_0$ thermal conductivity of the medium of the model system
$m$ index for the number of iterations

$n$ index in $t$-direction
$R$ radial distance from the channel surface
$r$ radial distance from the centre of the channel
$r_0$ initial channel radius
$\dot{r}$ radial growth rate of the channel
$\dot{r}_0$ boundary moving rate of the model system
$T_\infty$ temperature of the medium at infinite distance
$T_b$ temperature at the boundary

$T_f$    temperature of the fluid
$T_0$    temperature of the fluid at inlet
$T_p$    temperature of the surroundings
$T'$ $T''$    temperature of the countercurrent fluids
$t$    time
$V$    linear velocity of the fluids
$V_1, V_2$    linear velocities of the countercurrent fluids
$Z$    longitudinal distance, i.e. channel length
$\alpha_p$    thermal diffusivity of the medium
$\rho_p$    density of the medium
$\rho_f$    density of the fluid

## 15.1   INTRODUCTION

*In situ* coal liquefaction (ISCL) involved introducing solvents or chemical reagents into a coal seam and converting monolithic coal into coal particles in a slurry or into coal-derived oils. Heat transfer involved in ISCL processes was analysed by modelling the unsteady-state conductive heat losses to a surrounding coal seam from solvent in a cylindrical channel with a moving boundary. Three conjugated hyperbolic–parabolic equations describe the system.

This chapter uses a new numerical method which we call the method of alternating variables, to solve a set of conjugated hyperbolic–parabolic equations with a moving boundary. The numerical methods used in this chapter consist of a combination of two stable numerical methods, namely the ADI iteration method and the implicit central difference method.

The method of alternating variables can also be used for other combinations of conventional numerical methods where each method is suitable for solving its constituent differential equation.

## 15.2   ADI ITERATION METHOD

Frankel[1] studied many iteration methods to solve the set of implicit finite difference equations derived from an unsteady-state parabolic equation and found the successive over-relaxation method to be one of the best. Peaceman and Rachford[2] proposed the ADI method and concluded that the ADI method was approximately seven times faster than the Crank–Nicolson method and twice as fast as successive over-relaxation. The procedure to be used in this chapter for solving a particular heat conduction problem is very similar to the method proposed by Douglas *et al.*[3]

The following heat conduction equation in its cylindrical form is selected for illustration:

$$\frac{\partial T}{\partial t} = \alpha_p \left( \frac{\partial T}{\partial Z^2} + \frac{\partial^2 T}{\partial r^2} + \frac{1}{r} \frac{\partial T}{\partial r} \right) \tag{15.1}$$

with boundary conditions

$$r = r_0 \qquad T = 1 \qquad T = 0 \quad \text{at } r = \infty$$
$$Z = 0, Z_0 \qquad r > r_0 \qquad T = 0$$

(15.2)

Two sets of equations which approximate Equation (15.1) according to the ADI iteration method are given as follows:

$$\left(\frac{T_{i-1,j,n+1} - 2T_{i,j,n+1} + T_{i+1,j,n+1}}{(\Delta Z)^2}\right)^{(m+\frac{1}{2})}$$

$$+ \left(\frac{T_{i,j+1,n+1} - 2T_{i,j,n+1} + T_{i,j-1,n+1}}{(\Delta r)^2} + \frac{1}{r_j}\frac{T_{i,j+1,n+1} - T_{i,j-1,n+1}}{2\Delta r}\right)^{(m)}$$

$$= \frac{1}{\alpha_p \Delta t}\left(T_{i,j,n+1}^{(m+\frac{1}{2})} - T_{i,j,n}^{(0)}\right)$$

(15.3)

and

$$\left(\frac{T_{i-1,j,n+1} - T_{i,j,n+1} + T_{i+1,j,n+1}}{(\Delta Z)^2}\right)^{(m+\frac{1}{2})}$$

$$+ \left(\frac{T_{i,j+1,n+1} - 2T_{i,j,n+1} + T_{i,j-1,n+1}}{(\Delta r)^2}\right) + \left(\frac{1}{r_j}\frac{T_{i,j+1,n+1} - T_{i,j-1,n+1}}{2(\Delta r)}\right)^{(m+1)}$$

$$= \frac{1}{\alpha_p \Delta t}\left(T_{i,j,n+1}^{(m+1)} - T_{i,j,n}^{(0)}\right)$$

(15.4)

The superscript $(m)$ denotes the number of iterations, the value at the old time level is used as the initial estimation, $T_{i,j,n+1}^{(0)} = T_{i,j,n}$.

Equations (15.3) and (15.4) comprise one full time step. Upon using the ADI iteration technique, only one iteration is required to agree with the analytical solution. Results are shown in Figure 15.1 and Table 15.1.

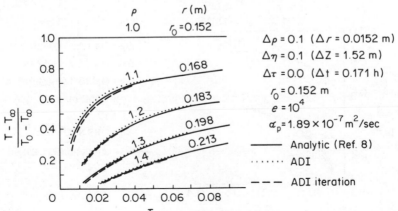

**Figure 15.1** Temperature profiles of the surroundings for a unit step change at the surface of the pipe

Table 15.1 Numerical results as presented in Figure 15.1

| Distance, $\rho$ | Time, $\tau$ | Analytic solution | ADI method | ADI iteration method |
|---|---|---|---|---|
| 1.1 | 0.02 | 0.5891 | 0.6024 | 0.5568 |
|  | 0.04 | 0.6912 | 0.7060 | 0.6781 |
|  | 0.06 | 0.7385 | 0.7543 | 0.7313 |
|  | 0.08 | 0.7672 | 0.7835 | 0.7625 |
|  | 0.10 | 0.7870 | 0.8037 | 0.7837 |
| 1.2 | 0.02 | 0.2901 | 0.3047 | 0.2695 |
|  | 0.04 | 0.4389 | 0.4573 | 0.4229 |
|  | 0.06 | 0.5164 | 0.5376 | 0.5059 |
|  | 0.08 | 0.5657 | 0.5887 | 0.5583 |
|  | 0.10 | 0.6005 | 0.6248 | 0.5951 |
| 1.3 | 0.02 | 0.1172 | 0.1293 | 0.1180 |
|  | 0.04 | 0.2541 | 0.2704 | 0.2441 |
|  | 0.06 | 0.3404 | 0.3609 | 0.3311 |
|  | 0.08 | 0.3996 | 0.4231 | 0.3921 |
|  | 0.10 | 0.4432 | 0.4691 | 0.4372 |
| 1.4 | 0.02 | 0.0382 | 0.0467 | 0.0480 |
|  | 0.04 | 0.1332 | 0.1458 | 0.1315 |
|  | 0.06 | 0.2106 | 0.2277 | 0.2054 |
|  | 0.08 | 0.2696 | 0.2904 | 0.2641 |
|  | 0.10 | 0.3157 | 0.3395 | 0.3106 |

The ADI method has the optimum value at $\Delta t$ for fixed mesh sizes to minimize the error growth and attain fast convergence rates. The truncation error of the ADI method becomes unacceptable as shown in Figure 15.2 when $\Delta \tau = 0.025$, while the ADI iteration method still maintains satisfactory results.

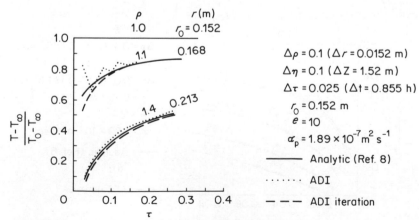

**Figure 15.2** Temperature profiles of the surroundings for a unit step change at the surface of the pipe

## 15.3  ALTERNATING VARIABLE METHOD

Herron and von Rosenberg[4] applied an implicit central differencing scheme which had been specially designed for solving hyperbolic equations with inherent discontinuity. Their results were in excellent agreement with the analytical solution except for some distortion at the discontinuity.

In this section, a new numerical method, the method of alternating variables (AV),[5,6] is to be used to solve the hyperbolic fluid flow equation. For the case being studied, this procedure provides solutions as accurate as that of the implicit central difference equations,[4] with high calculation speeds, as shown below.

This new numerical method is especially designed in conjunction with the ADI iteration method,[5] for the parabolic heat conduction equation in the solid medium to solve the conjugated non-linear heat transfer problems encountered in the *in situ* coal liquefaction process.

An illustrative example will be used to demonstrate the process. The example is the heat transfer problem of a heat exchanger. The equations which govern the temperature of the countercurrent fluids can be rewritten as:

$$\frac{\partial T'}{\partial t} = H_1(T'' - T') - V_1 \frac{\partial T'}{\partial z} \tag{15.5}$$

$$\frac{\partial T''}{\partial t} = -H_2(T'' - T') + V_2 \frac{\partial T''}{\partial Z} \tag{15.6}$$

The boundary conditions are

$$T' = 1 \quad \text{at } Z = 0 \qquad T'' = 0 \quad \text{at } Z = Z$$

and initial conditions are

$$T', T'' = 0 \quad \text{at } t = 0$$

The numerical approximations are written as follows. At the time level $n - \frac{1}{2}$, central difference approximations are written for $T'$ and $\partial T'/\partial Z$ in Equation (15.5) while $T''$ in this equation is approximated by $(T''_{i,n-\frac{1}{2}} + T''_{i-1, n-\frac{1}{2}})/2$:

$$\frac{T'_{i,n} - T'_{i,n-1} + T'_{i-1,n} - T'_{i-1,n-1}}{2\Delta t}$$

$$= H_1\left(\frac{T''_{i,n-\frac{1}{2}} + T''_{i-1,n-\frac{1}{2}}}{2} - \frac{T'_{i,n} + T'_{i-1,n} + T'_{i,n-1} + T'_{i-1,n-1}}{4}\right)$$

$$- V_1 \frac{T'_{i,n} - T'_{i-1,n} - T'_{i-1,n-1} + T'_{i,n-1}}{2\Delta Z}$$

After rearranging,

$$T'_{i,n}\left(\frac{1}{2\Delta t}+\frac{H_1}{4}+\frac{V_1}{2\Delta Z}\right) = T'_{i-1,n}\left(-\frac{1}{2\Delta t}-\frac{H_1}{4}+\frac{V_1}{2\Delta Z}\right) + T'_{i,n-1}\left(\frac{1}{2\Delta t}-\frac{H_1}{4}-\frac{V_1}{2\Delta Z}\right)$$

$$+ T'_{i-1,n-1}\left(\frac{1}{2\Delta t}-\frac{H_1}{4}+\frac{V_1}{2\Delta Z}\right)$$

$$+ (T''_{i,n-\frac{1}{2}} + T''_{i-1,n-\frac{1}{2}})H_1/2 \qquad i = 2, 3, \ldots, N$$

$$(15.7)$$

where $T'_{1,n}$ is known as the boundary condition. Note in Equation (15.7), $T'_{i,n}$ for $i = 2, \ldots, N$ can be solved explicitly provided that $T'$ at time level $n-1$ and $T''$ at time $n-\frac{1}{2}$ are known.

As time advances to the new level $n$, similar approximations are written for Equation (15.6):

$$T''_{i-1,n+\frac{1}{2}}\left(\frac{1}{2\Delta t}+\frac{H_2}{4}+\frac{V_2}{2\Delta Z}\right) = T''_{i,n+\frac{1}{2}}\left(-\frac{1}{2\Delta t}-\frac{H_2}{4}+\frac{V_2}{2\Delta Z}\right)$$

$$+ T''_{i,n-\frac{1}{2}}\left(\frac{1}{2\Delta t}-\frac{H_2}{4}+\frac{V_2}{2\Delta Z}\right)$$

$$+ T''_{i-1,n-\frac{1}{2}}\left(\frac{1}{2\Delta t}-\frac{H_2}{4}-\frac{V_2}{2\Delta Z}\right)$$

$$+ (T'_{i,n} + T'_{i-1,n})H_2/2 \qquad i = N, N-1, \ldots, 3, 2$$

$$(15.8)$$

### 15.3.1   Results and discussion

The numerical values of $T'$ and $T''$ by using Equations (15.7) and (15.8) have been compared with analytical solutions shown in Table 15.2. Their differences are not discernible.

Table 15.2   Comparison of central difference, alternating variables, and analytical solutions for a test case

| Time (s) | Distance | Analytic (Ref. 4) | Central difference (Ref. 4)* | Alternating variables |
|---|---|---|---|---|
| 8 | 0 | 100.00 | 100.00 | 100.00 |
|   | 2 | 81.92 | 81.91 | 81.91 |
|   | 4 | 67.08 | 67.08 | 67.08 |
|   | 6 | 54.91 | 54.91 | 54.94 |
|   | 8 | 44.93 | 56.97 | 56.97 |
|   | 10 | 0.00 | 0.01 | 0.01 |

* Reproduced by permission of Pergamon Press Limited.

The formulations of Equations (15.7) and (15.8) are of implicit nature. However, the presented numerical schemes solve $T'$ and $T''$ alternately at successive new time levels ($n - \frac{1}{2}$, $n = 1, 2, 3, \ldots$), and thus avoid the need for solving them simultaneously, as is done by the implicit central difference method.[4] The resulting effect of the alternating variable method is that high calculating speed can be attained without sacrificing accuracy.

## 15.4 MODELLING OF *IN SITU* COAL LIQUEFACTION AND NUMERICAL PROCEDURE

At time $t = 0$, a solvent flows into a circular channel embedded in a medium. If the medium is dissolved by the solvent and the heat of reaction is negligible, the governing equations in its dimensionless form for the system can be written as:

Medium:
$$\frac{\partial^2 \Omega}{e\, \partial \eta^2} + \frac{\partial^2 \Omega}{\partial \rho^2} + \left(\frac{1}{1 + \rho + \bar{r}\tau} + \bar{r}\right)\frac{\partial \Omega}{\partial \rho} = \frac{\partial \Omega}{\partial \tau} \quad \text{for } \rho > 0 \qquad (15.9)$$

Boundary:
$$-\frac{\phi}{\bar{h}}\frac{\partial \Omega}{\partial \rho} = \theta - \Omega \quad \text{at } \rho = 0 \qquad (15.10)$$

Fluid:
$$-\bar{F}\frac{\partial \Theta}{\partial \eta} - \bar{D}\bar{h}(\Omega - \Theta) = \bar{D}^2\frac{\partial \Theta}{\partial \tau} + 4\bar{r}\bar{D}\theta \qquad (15.11)$$

Interfacial moving rate $\quad \dfrac{\mathrm{d}\bar{D}}{\mathrm{d}\tau} = 2\bar{r}, \qquad \bar{D} = \bar{D}_0 + 2\bar{r}\tau$

Initial and boundary conditions:
$$\bar{D} = 2, \qquad \Omega = 0, \qquad \theta = 0 \quad \text{at } \tau = 0$$
$$\theta = 1 \qquad\qquad\qquad\qquad\qquad \text{at } \eta = 0, \quad \tau > 0$$

where

$$\eta = Z/Z_0 \qquad\qquad \rho = R/r_0 \qquad\qquad \tau = \alpha_\mathrm{p}t/r_0^2$$
$$\bar{D} = D/r_0 \qquad\qquad \bar{F} = \bar{D}^2\bar{V} \qquad\qquad \bar{V} = r_0^2 V/Z_0\alpha_\mathrm{p}$$
$$\bar{r} = \dot{r}r_0/\alpha_\mathrm{p} \qquad\qquad \bar{h} = hr_0/\rho_\mathrm{f}C_\mathrm{f}\alpha_\mathrm{p} \qquad\qquad \phi = \rho_\mathrm{p}C_\mathrm{p}/\rho_\mathrm{f}C_\mathrm{f}$$
$$e = Z_0^2/r_0^2 \qquad\qquad \theta = (T_\mathrm{f} - T_\infty)/(T_0 - T_\infty)$$
$$\Omega = (T_\mathrm{p} - T_\infty)/(T_0 - T_\infty)$$

Note that, in Equation (15.9), a moving coordinate, defined as $R = r - r_0 - \dot{r}t$ has been introduced in the $r$-direction.

The model system of the coal–anthracene oil system was chosen as a basis for the numerical studies (Table 15.3). The constant interfacial moving rate, $\dot{r}$, represents the coal comminution rate in the process. A constant interfacial

**Table 15.3** Numerical values of the parameters in the model anthracene/coal system

| | |
|---|---|
| $Z_0 = 15$ m | $r_0 = 0.15$ m |
| $V = 3$ m s$^{-1}$ | $\rho_f = 1100$ kg m$^{-3}$ |
| $\mu = 0.001$ kg m$^{-1}$ s$^{-1}$ | $h = 500$ cal m$^{-2}$ s$^{-1}$ °C$^{-1}$ |
| $\alpha_p = 1.89 \times 10^{-7}$ m$^2$ s$^{-1}$ | $\rho_p = 1300$ kg m$^{-3}$ |
| $e = 10{,}000(-)$ | $k_p = 0.056$ cal m$^{-1}$ s$^{-1}$ °C$^{-1}$ |
| $T_0 = 500$ °C | $D = 0.3$ m |
| $T_\infty = 0$ °C | $C_p = 230$ cal kg$^{-1}$ °C$^{-1}$ |
| $C_f = 800$ cal kg$^{-1}$ °C$^{-1}$ | |

moving rate is a reasonable assumption in the model system, since the boundary temperature of the coal/fluid interface only varies slightly (from 0.949 to 0.944) along the channel in 25 s as will be shown later. A boundary moving rate of 0.02 m h$^{-1}$ which is the fracturing rate on a lump of coal by anthracene oil[7] is adopted for the model system.

Equations (15.9) through (15.11) are solved numerically by the AV method, whereas Equation (15.11) is approximated by the central difference method and Equations (15.9) and (15.10) are approximated in accordance with the ADI iteration method.[5]

The grid points are shown in Figure 15.3.

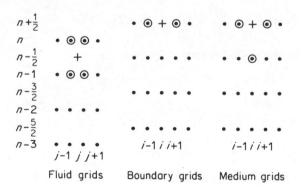

Fluid grids          Boundary grids          Medium grids

**Figure 15.3** Grid points of the heat transfer problem

Solvent ($\tau = n\Delta\tau$, $n = 2, 3, \ldots$):

$$\theta_{i,n}(\bar{D}_{n-\frac{1}{2}}^2/2\Delta\tau + \bar{D}_{n-\frac{1}{2}}^2\bar{r} + \bar{F}/2\Delta\eta + \bar{D}_{n-\frac{1}{2}}\bar{h}/4)$$

$$= \theta_{i-1,n}(-\bar{D}_{n-\frac{1}{2}}^2/2\Delta\tau - \bar{D}_{n-\frac{1}{2}}\bar{r} + \bar{F}/2\Delta\eta - \bar{D}_{n-\frac{1}{2}}\bar{h}/4)$$

$$+ \theta_{i,n-1}(\bar{D}_{n-\frac{1}{2}}^2/2\Delta\tau - \bar{D}_{n-\frac{1}{2}}\bar{r} - \bar{F}/2\Delta\eta - \bar{D}_{n-\frac{1}{2}}\bar{h}/4)$$

$$+ (\Omega_{i,1,n-\frac{1}{2}} + \Omega_{i-1,n-\frac{1}{2}})\bar{D}_{n-\frac{1}{2}}\bar{h}/2 \tag{15.12}$$

Boundary ($\tau = (n = \frac{1}{2})\Delta\tau$, $n = 1, 2, \ldots$):

$$\Omega^{(m)}_{i,1,n+\frac{1}{2}}\left(\frac{\phi}{\bar{h}\Delta\rho}+1\right)-\frac{\phi}{\bar{h}\Delta\rho}\,\Omega^{(m)}_{i,2,n+\frac{1}{2}}=\theta_{i,n+\frac{1}{2}} \tag{15.13}$$

and

$$\Omega^{(m+1)}_{i,1,n+\frac{1}{2}}\left(\frac{\phi}{\bar{h}\Delta\rho}+1\right)-\frac{\phi}{\bar{h}\Delta\rho}\,\Omega^{(m+1)}_{i,2,n+\frac{1}{2}}=\theta_{i,n+\frac{1}{2}} \tag{15.14}$$

Surrounding ($\tau = (n + \frac{1}{2})\Delta\tau$, $n = 1, 2, \ldots$)

$$(\Omega_{i+1,j,n+\frac{1}{2}}-2\Omega_{i,j,n+\frac{1}{2}}+\Omega_{i-1,j,n+\frac{1}{2}})^{(m+\frac{1}{2})}/e(\Delta\eta)^2$$
$$+\,(\Omega_{i,j+1,n+\frac{1}{2}}-2\Omega_{i,j,n+\frac{1}{2}}+\Omega_{i,j-1,n+\frac{1}{2}})^{(m)}/(\Delta\rho)^2$$
$$+\left(\bar{\bar{r}}+\frac{1}{1+\rho_j+\bar{\bar{r}}\tau}\right)(\Omega_{i,j,n+\frac{1}{2}}-\Omega_{i,j-1,n+\frac{1}{2}})^{(m)}/2\Delta\rho$$
$$=(\Omega^{(m+\frac{1}{2})}_{i,j,n+\frac{1}{2}}-\Omega^{(0)}_{i,j,n-\frac{1}{2}})/\Delta\tau \tag{15.15}$$

and

$$(\Omega_{i+1,j,n+\frac{1}{2}}-2\Omega_{i,j,n+\frac{1}{2}}+\Omega_{i-1,j,n+\frac{1}{2}})^{(m+\frac{1}{2})}/e(\Delta\eta)^2$$
$$+\,(\Omega_{i,j+1,n+\frac{1}{2}}-2\Omega_{i,j,n+\frac{1}{2}}+\Omega_{i,j-1,n+\frac{1}{2}})^{(m+1)}/(\Delta\rho)^2$$
$$+\left(\bar{\bar{r}}+\frac{1}{1+\rho_j+\bar{\bar{r}}\tau}\right)(\Omega_{i,j+1,n+\frac{1}{2}}-\Omega_{i,j-1,n+\frac{1}{2}})^{(m+1)}/2\Delta\rho$$
$$=(\Omega^{(m+1)}_{i,j,n+\frac{1}{2}}-\Omega^{(0)}_{i,j,n-\frac{1}{2}})/\Delta\tau \tag{15.16}$$

$\theta_{i,n+\frac{1}{2}}$ in Equation (15.14) may be approximated by the Taylor expression at $\theta_{i,n}$: $1.5\theta_{i,n}-0.5\theta_{i,n-1}$. Since $\theta$ varies slightly with time, evaluation of $\theta$ at the time level $n$ is satisfactory.

The first set of the values of $\Omega_{i,j,\frac{1}{2}}$ may be computed by use of the explicit method with sufficiently small $\Delta t$. However, the approximation of taking $\Omega_{i,j,\frac{1}{2}}$ as the initial values has been shown satisfactory. The errors introduced by this approximation damps out after three or four time steps.

Equations (15.12) to (15.16) were solved by the use of the AV method. Figures 15.4 to 15.6 show the temperature distributions of the model system in terms of various interfacial moving rates, heat transfer coefficients, and heat

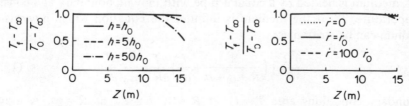

**Figure 15.4**  Solvent temperature distribution

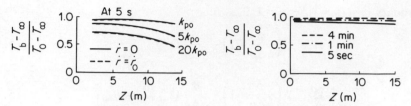

**Figure 15.5**  Boundary temperature distributions

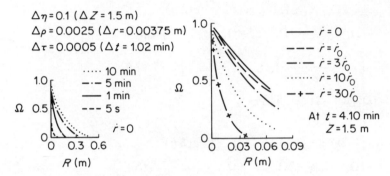

**Figure 15.6**  Medium (coal) temperature distributions

conductivity of coal. One or two iterations are sufficient for accurate solution (±1 °C).

### 15.5  QUASI-STATIONARY STATE

When the boundary moving rate gradually increases, the temperature in the medium will eventually reach a quasi-stationary state. In other words,

$$\frac{\partial T}{\partial t}\bigg|_R = 0$$

An analytical solution follows based on the quasi-stationary assumption. The solution will be derived and be compared with the solution of the strict formulations in the preceding section.

A medium is heated by a circular pipe with moving boundary at a constant temperature $T_b$. If the medium is infinite, the equation which governs the medium can be written as:

$$\frac{\partial^2 T}{\partial R^2} + \left(\frac{1}{R + r_0 + \dot{r}t} + \frac{\dot{r}}{\alpha_p}\right)\frac{\partial T}{\partial R} = \frac{1}{\alpha_p}\frac{\partial T}{\partial t} \tag{15.17}$$

Boundary conditions are: $T = T_b$ at $R = 0$; $T = T_\infty$ at $R = \infty$. A moving coordinate defined as $R = r - r_0 - \dot{r}t$, has been introduced in the $r$-direction.

If a quasi-stationary state is assumed

$$\frac{\partial^2 T}{\partial R^2} + \left(\frac{1}{R + r_0 + \dot{r}t} + \frac{\dot{r}}{\alpha_p}\right)\frac{\partial T}{\partial R} = 0 \tag{15.18}$$

Boundary conditions are: $T = T_b$ at $R = 0$; $T = T_\infty$ at $R = \infty$.
Equation (15.18) is solved for $T$:

$$\frac{T - T_\infty}{T_b - T_\infty} = 1 - \frac{\int_0^R [e^{-aR}/(1 + bR)]\,dR}{\int_0^\infty [e^{-aR}/(1 + bR)]\,dR} \tag{15.19}$$

Figure 15.7 shows the results of Equation (15.19). Results of Equations (15.12) to (15.16) are also presented for comparison. It is seen that when $\dot{r}$ reaches $0.41\ \text{m h}^{-1}$, the results from the two systems are in very good agreement.

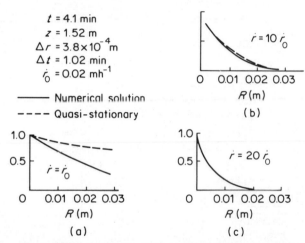

Figure 15.7 Application of the quasi-stationary assumption

The meaning of the agreements of the two systems at high $\dot{r}$ is significant. An analytical solution of Equations (15.9) to (15.11) is very difficult to attain. The results of this section prove that the numerical methods devised in this paper give accurate solutions to Equations (15.9) through (15.11), at least at high values of $\dot{r}$.

## 15.6 REMAINING SOLVENT–COAL INTERACTION

In this section, the heat transfers of various additional solvents to the coal seam are shown. The new solvents to be considered are supercritical toluene, caustic solution, $CO/H_2O$, and liquid ammonia. Figures 15.8 through 15.12 depict the heat transfer of these four solvents. The assumption of a stationary boundary is made for these calculations.

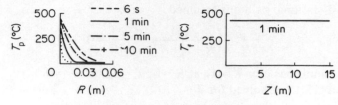

Coal seam temperature = 12 °C
Channel diameter = 0.15 m
Solvent = 50/50, CO/steam, 450 °C, 10340 kPa

**Figure 15.8**   Temperature distributions in the coal/CO and steam system

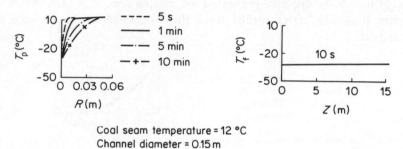

Coal seam temperature = 12 °C
Channel diameter = 0.15 m
Solvent = liquid ammonia, -33 °C

**Figure 15.9**   Temperature distributions in the coal/ammonia system

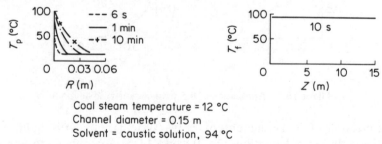

Coal steam temperature = 12 °C
Channel diameter = 0.15 m
Solvent = caustic solution, 94 °C

**Figure 15.10**   Temperature distributions in the coal/caustic solution system

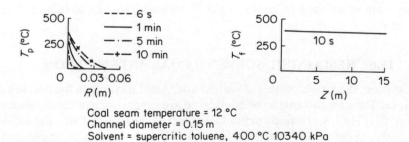

Coal seam temperature = 12 °C
Channel diameter = 0.15 m
Solvent = supercritic toluene, 400 °C 10340 kPa

**Figure 15.11**   Temperature distributions in the coal/supercritical toluene system

```
C          ------------------------------------------------------
C
C          CONJUGATED HEAT TRANSFER PROBLEM BY
C
C          ITERATION AND ALTERNATING VARIABLE METHODS.
C
C             SM—S MODEL
C
C          ------------------------------------------------------
C          COMPUTER USAGE:
C          I         COLUMN INDIX.
C          J         ROW INDIX.
C          DIMENSION REQUIRED:
C                    U(M),AA(M),BB(M),CC(M),DD(M),BETA(M),GAMA(M), AND
C                    T(I,M,3) WHERE M = LARGEST IN I OR J.
C                    TF(I,3) WHERE I = DIMENSION IN I DIRECTION.
C          INPUT DATA CARDS:
C                    FIRST CARD— ROW,COLUMN,TCU,DRAD,DZ,TMAX,RADIUS,RATE1
C                    SECOND CARD— VELO,DEN,VIS,CP,K,DENS,CPS,KS.
C                    THIRD CARD— IDM,ITR1,IPRINT,TSD,TIN
C          ------------------------------------------------------
1          DIMENSION U( 80),TF(15,3)
2          DIMENSION BETA( 80),GAMA( 80)
3          DIMENSION AA( 80),BB( 80),CC( 80),DD( 80)
4          COMMON/TTT/T(15, 80,3)
5          REAL NN,K,K3
6          INTEGER ROW,COLUMN
7          COMMON/AAA/TOU,DTIME1,H,H1,D,DX,ET1,ETT2,ETT3,ZO,FLOW,ARFA,TOU3
8          COMMON/BBB/A11,A1,A2,C22,C32,C,AA3,AB3,AC3,BB3
9          COMMON/CCC/DR,RADIUS,IDO
10         COMMON/RRR/DRAD,RR,IDM
11         COMMON/TIM/TTOU2,TIME5,TIME1
C          ------------------------------------------------------
C          INPUT DATA SECTION
12         READ(5,36)ROW,COLUMN,TOU,DRAD,DZ,TMAX,RADIUS,RATE1
13         READ(5,37)VELO,DEN,VIS,CP,K,DENS,CPS,KS
14         READ(5,38)IDM,ITR1,IPRINT,TSD,TIN
15    36   FORMAT(2I5,6F10.5)
16    37   FORMAT(8F10.4)
17    38   FORMAT(3I5,2F10.5)
C          ------------------------------------------------------
18         WRITE(6,12)
19         WRITE(6,12)
20         WRITE(6,25)
C          ------------------------------------------------------
C          DEFINE PARAMETERS
21         RR = 1.
22         JS = 1
23         TOU1 = 0.
24         TLIMT = 0.0005
25         ITOP = ROW − 1
26         IDD = COLUMN − 1
27         DMED = DRAD*ITOP
28         JST = JS + ITOP
29         JST1 = JST − 1
```

**Figure 15.12**   Computer listings and example output

```
30          JSS = JS + 1
31          JS1 = JS - 1
32          IED = IDD + 1
33          ZO = DZ*IDD
34          DX = DZ/ZO
35          DR = DRAD/RADIUS
36          IH = 1
37          IH2 = IH + 1
38          IH3 = IH + 2
39          IH4 = IH + 3
40          IED1 = IED - 1
41          DIA = 2.*RADIUS
42          DO = DIA
43          FLOW = VELO*3.1416*RADIUS*RADIUS
44          D = DIA/RADIUS
45          C = ZO*ZO/RADIUS/RADIUS
46          ARFA = KS/DENS/CPS
47          TIME5 = RADIUS*RADIUS/ARFA
48          RATE = RATE1*ARFA/RADIUS
49          TOU2 = DX*ZO*ARFA/VELO/RADIUS/RADIUS
50          TTOU2 = TIME5*TOU2
51          TTOU = TIME5*TOU
52          ET1 = 4.*FLOW/3.1416/ZO/ARFA
53          ETT2 = 4.*RADIUS/ARFA/DEN/CP
54          ETT3 = KS/RADIUS
55          AC3 = 0.5*ET1/DX
C  -----------------------------------------------------------------
C           COMPUTE H
56          IF(TSD - TIN)101,101,102
57    101   NN = 0.3
58          GO TO 103
59    102   NN = 0.4
60    103   H1 = 0.023*(DEN**0.8)*(VIS**(NN - 0.8))*(CP**NN)*(K**(1 - NN))
61          H = H1*(DIA**(-0.2))*VELO**0.8
C  -----------------------------------------------------------------
62          A2 = ETT2*H*D/4.
C  -----------------------------------------------------------------
63          WRITE(6,18)RADIUS
64          WRITE(6,17)ZO
65          WRITE(6,602)RATE,RATE1
66          WRITE(6,600)KS,CPS,DENS,ARFA
67          WRITE(6,601)K,CP,DEN,VIS,H,VELO
68    1     FORMAT(/,' --------------------------------------------- ',/)
69    2     FORMAT(' TRANSIENT SOLUTIONS',//)
70    3     FORMAT(1X, 6F12.5)
71    4     FORMAT(/,' TEMP ARE (AT RAD FROM CEN = ',F9.5,'FT, RAD FROM SURFACE
            C = ', F9.5,'FT):')
72    8     FORMAT(1/X,10E12.3)
73    9     FORMAT(/,' COEFF OF AA, BB, CC, DD, AT ROW ',I2,' ARE:')
74    10    FORMAT(/,' INITIAL TEMP DISTRIBUTION IS:')
75    11    FORMAT(' FINAL SOLUTIONS OF THE MEDIUM:', 2I5)
76    12    FORMAT(1H1)
77    17    FORMAT(/,' CHANNEL LENGTH = ',F10.3)
78    18    FORMAT(/,' CHANNEL RADIUS = ',2F10.3)
79    24    FORMAT(///)
```

**Figure 15.12**   Continued

```
80    25    FORMAT(/,' ITERATIVE A D I METHOD AND ALTERNATING VARIABLE
            CMETHOD',/,
            C',///)
81    26    FORMAT(/,' TIME = ',F12.8,' HR',' DIMENSIONLESS TIME = ',F12.8,/)
82    29    FORMAT(//,' SOLVENT TEMPERATURES ARE:')
83    39    FORMAT(/,' FLUID INLET TEMPERATURE = 1')
84    600   FORMAT(//,' MEDIUM',/,' CONDUCTIVITY = ', F10.4./,' CAPACITY  = ',
            C F10.4,/,' DENSITY  = ',F10.4, /,' DIFFUSITY  = ', F10.4)
85    601   FORMAT(//,' SOLVENT',/,' CONDUCTIVITY = ', F10.4,/,' CAPACITY  = '
            C,F10.4 ,/,' DENSITY  = ',F10.4,/,' VISCOSITY  = ', F10.4,/,' HEAT
            CT TRANS COEFF = ', F10.4,/,' FLOW VELO  = ',F10.4)
86    602   FORMAT(//,' BOUNDARY MOVING VELO = ',F10.4,'(FT/HR)',F10.4,'(−)')
87          C21 = 1./DX/DX/C
88          C23 = 1./DX/DX/C
 C          INITIAL TEMPERATURE
89          TF(IED + 1,3) = 0.
90          TF(IED + 1,2) = 0.
91          TF(IED + 1,1) = 0.
92          DO 111 I = 1,IED
93          DO 111 J = 1,JST
94          T(I,J,3) = 0.
95          T(I,J,2) = 0.
96    111   T(I,J,1) = 0.
97          T(IH,JS,1) = 1.
98          T(IH,JS,2) = 1.
99          T(IH,JS,3) = 1.
100         DO 115 K1 = 1,3
101         DO 115 I = 1,IED
102   115   TF(I,K1) = 0.
103         DO 114 I = 1,3
104   114   TF(IH,I) = 1.
105         WRITE(6,10)
106         WRITE(6,39)
107         WRITE(6,3)(TF(I, 1),I = 1,IED)
108         WRITE(6,24)
 C          PRINT TIME
109         IDO = 0
110         ICR = 0
111         TIME1 = 0.
112         DTIME1 = TOU2
113         TTIME = ZO/VELO
114         C22 = −(2./DX/DX/C + 1./DTIME1)
115         A1 = 0.5*D*D/DTIME1
116         A11 =  D*RATE1
117   200   ICR = ICR + 1
118         TTM = TTOU2*(ICR − 1)
119         IF(TTM .GT. TTIME)CALL GROW(RATE1)
120         TIME1 = TIME1 + DTIME1
121         TIME = TIME1*TIME5
122         IF(IPRINT.EQ.0)GO TO 403
123         WRITE(6,12)
124         WRITE(6,1)
125         WRITE(6,26)TIME,TIME1
126         WRITE(6,1)
127   403   CONTINUE
```

**Figure 15.12**   Continued

```
128    500    CONTINUE
 C            COMPUTE FLUID TEMPERATURE
129           DD 540 I = 2,IED
130           CC1 = A1 + A11 + AC3 + A2
131           CC2 = −A1 − A11 + AC3 − A2
132           CC3 =  A1 − A11 − AC3 − A2
133           CC4 =  A1 − A11 + AC3 − A2
134           CC5 = 4.*A2*T(I,JS,1)
135    540    TF(I,3) = ( TF(I−1,3)*CC2 + TF(I,2)*CC3 +  TF(I−1,2)*CC4 +  CC5)/CC1
136           DO 525 I = 2,IED
137           TF(I, 1) = TF(I, 2)
138    525    TF(I, 2) = TF(I, 3)
139    504    WRITE(6,29)
140           WRITE(6,3)(TF(I, 2), I = IH,IED)
 C            COMPUTE MEDIUM AND BOUNDARY TEMPERATURES
141    503    CONTINUE
142    527    ITR = 0
143           ITR2 = 0
144    202    ITR = ITR + 1
145           ITR2 = ITR2 + 1
146           FON = ETT3/H/DR
147           C32 = −2. /DR/DR − 1./DTIME1
148           N = IED − 2
149           DO 123 I = 1,N
150           AA(I) = C21
151           BB(I) = C22
152    123    CC(I) = C23
153           AA(1) = 0.
154           CC(N) = 0.
155           DO 121 J = JSS,JST1
156           R = (J − JS)*DR + RR
157           TI = TIME + 0.5*DTIME1*TIME5
158           RJ = R + RATE*TI /RADIUS
159           RI = RADIUS*RATE/ARFA + 1./RJ
160           DO 124 II = 1,N
161           I = II + 1
162           C24 = −(1./DR + RI/2. )*T(I,J+1,2)/DR  + (2. /DR/DR)
              C *T(I,J,2)  − (1./DR − RI/2. )*T(I,J−1,2)/DR  − T(I,J,1) /DTIME1
163           IF(II.NE.1)GO TO 125
164           DD(1) = C24 − C21*T(1,J,2)
165           GO TO 124
166    125    IF(II.NE.N)GO TO 126
167           DD(N) = C24 − C23*T(IED,J,2)
168           GO TO 124
169    126    DD(II) = C24
170    124    CONTINUE
171           IF(IPRINT.NE.1)GO TO 401
172           WRITE(6,9)J
173           WRITE(6,3)(AA(M),M = 1,N)
174           WRITE(6,8)(BB(M),M = 1,N)
175           WRITE(6,3)(CC(M),M = 1,N)
176           WRITE(6,8)(DD(M),M = 1,N)
177    401    CALL TRI(AA,BB,CC,DD,N,U,BETA,GAMA)
178           DO 122 I = 1,N
179           II = I + 1
```

**Figure 15.12**   Continued

```
180   122    T(II,J,3) = U(I)
181   121    CONTINUE
182          DO 131 I = 1,IED
183          DO 131 J = JSS,JST
184   131    T(I,J,2) = T(I,J,3)
185          IF(IPRINT.EQ.3)GO TO 406
186          IF(IPRINT.NE.1)GO TO 402
187   406    IF(TIME1.LT.TOU1)GO TO 402
188          WRITE(6,24)
189          WRITE(6,2)
190          DO 127 J = JS,JST
191          IF(T(3,J,2).LT.TLIMT)GO TO 405
192          DEP1 = (J – JS)∗DRAD
193          RAD = DEP1 + RADIUS + RATE∗TIME
194          DEP = RAD – RADIUS
195          WRITE(6,4)RAD,DEP1
196   127    WRITE (6,3)(T(I,J,2),I = 1,IED)
197   402    CONTINUE
198   405    CONTINUE
199          N = JST – JS1 – 1
200          DO 132 I = 2,IDD
201          DO 133 J = 1,N
202          JJ = J + JS1
203          R = (JJ – JS)∗DR + RR
204          RJ = R + RATE∗TI /RADIUS
205          RI = RADIUS∗RATE/ARFA + 1./RJ
206          C31 = (1./DR – RI/2. )/DR
207          C33 = (1./DR + RI/2. )/DR
208          C34 = –T(I + 1,JJ,2)/DX/DX/C + (2./DX/DX/C)∗T(I,JJ,2) –
             C T(I – 1,JJ,2)/DX/DX/C – T(I,JJ,1)/DTIME1
209          AA(J) = C31
210          BB(J) = C32
211          CC(J) = C33
212   133    DD(J) = C34
213          AA(1) = 0.
214          CC(N) = 0.
215          BB(1) = FON + 1.
216          CC(1) = –FON
217          DD(1) = TF(I,3)
218          R = (JST1 – JS)∗DR + RR
219          RJ = R + RATE∗TI /RADIUS
220          RI = RADIUS∗RATE/ARFA + 1./RJ
221          C33 = (1./DR + RI/2. )/DR
222          DD(N) = DD(N) – C33∗T(I,JST,2)
223          IF(IPRINT.NE.1)GO TO 400
224          WRITE(6,7)I
225   7      FORMAT(/,' COEFF OF AA, BB, CC, DD, AT COLUMN ',12,' ARE:')
226          WRITE(6,3)(AA(M),M = 1,N)
227          WRITE(6,8)(BB(M),M = 1,N)
228          WRITE(6,3)(CC(M),M = 1,N)
229          WRITE(6,8)(DD(M),M = 1,N)
230   400    CALL TRI(AA,BB,CC,DD,N,U,BETA,GAMA)
231          DO 134 J = 1,N
232          JJ = J + JS1
233   134    T(I,JJ,3) = U(J)
```

**Figure 15.12**   Continued

```
234   132   CONTINUE
235         IF(T(3,JST − 3,3).LT.0.0005)GO TO 136
236         CALL SHIFT(JS,JST1,JST,IH,IED)
237         ITR = 1
238         GO TO 137
239   136   IF(ABS(T(IH2 ,2,3) − T(IH2 ,2,2)).LE.0.0001 )ITR = ITR1
240   137   CONTINUE
241         DO 201 I = 1,IED
242         DO 201 J = JS ,JST
243   201   T(I,J,2) = T(I,J,3)
244         IF(IPRINT.NE.2)GO TO 407
245         IF(ITR.LT.ITR1)GO TO 404
246   407   IF(TIME1.LT.TOU1)GO TO 404
247         WRITE(6,24)
248         WRITE(6,11)ITRS,JST
249         DO 135 J = JS,JST
250         IF(T(2,J,2).LT.TLIMT)GO TO 404
251         DEP1 = (J − JS)*DRAD
252         RAD = DEP1 + RADIUS + RATE*TIME
253         DEP = RAD − RADIUS
254         WRITE(6,4)RAD,DEP1
255   135   WRITE(6,3)(T(I,J,2),I = 1,IED)
256   404   CONTINUE
257         IF(ITR.LT.ITR1)GO TO 202
258         DO 203 I = 1,IED
259         DO 203 J = 1,JST
260   203   T(I,J,1) = T(I,J,2)
261         IF(TIME.LE.TMAX)GO TO 200
262         STOP
263         END


264         SUBROUTINE GROW(RATE1)
265         COMMON/AAA/TOU,DTIME1,H,H1,D,DX,ET1,ETT2,ETT3,ZO,FLOW,ARFA,TOU3
266         COMMON/BBB/A11,A1,A2,C22,C32,C,AA3,AB3,AC3,BB3
267         COMMON/CCC/DR,RADIUS,IDO
268         COMMON/TIM/TTOU2,TIME5,TIME1
269         IDO = IDO + 1
270         IF(IDO.GT.1)GO TO 3
271         IF(TOU.GT.TIME1)GO TO 3
272         STOP
273   3     IF(IDO.EQ.1)DTIME1 = TOU − TIME1
274         IF(IDO.GT.1)DTIME1 = TOU
275         IF(RATE1.GT.1.E − 7)GO TO 1
276         IF(IDO.GT.1)RETURN
277         C22 = −(2./DX/DX/C + 1./DTIME1)
278         C32 = −2. /DR/DR − 1./DTIME1
279         A1 = D*D/DTIME1
280         RETURN
281   1     D = (IDO − 1)*DX + D
282         DIA = RADIUS*D
283         VELO = 1.2732*FLOW/DIA/DIA
284         H = H1*(DIA**(−0.2))*VELO**0.8
285         A1 = D*D/DTIME1
286         A2 = ETT2*H*D/4.
```

**Figure 15.12**    Continued

```
287        A11 = D*RATE1
288        IF(IDO.GT.1)RETURN
289        C22 = -(2./DX/DX/C+1./DTIME1)
290        C32 = -2. /DR/DR-1./DTIME1
291        RETURN
292        END


293        SUBROUTINE SHIFT(JS,JST1,JST,IH,IED)
294        WRITE(6,1)
295     1  FORMAT(//,' SPECIFIED ROWS ARE NOT ADEQUATE.',//)
296        STOP
297        END


298        SUBROUTINE TRI(A,B,C,D,N,U,BETA,GAMA)
299        DIMENSION A(1),B(1),C(1),D(1),U(1),BETA(1),GAMA(1)
300        IF(N-2)4,3,5
301     5  BETA(1) = B(1)
302        GAMA(1) = D(1)/B(1)
303        DO 1 I=2,N
304        BETA(I) = B(I) - A(I)*C(I-1)/BETA(I-1)
305        IF(GAMA(I-1).LE.1.E-60)GAMA(I-1) = 0.
306     1  GAMA(I) = (D(I) - A(I)*GAMA(I-1))/BETA(I)
307        U(N) = GAMA(N)
308        M = N - 1
309        DO 2 I=1,M
310        J = N - I
311        IF(U(J+1).LE.1.E-60)U(J+1) = 0.
312     2  U(J) - GAMA(J) - C(J)*U(J+1)/BETA(J)
313        RETURN
314     3  DTM = B(1)*B(2) - A(2)*C(1)
315        BETA(1) = D(1)*B(2) - D(2)*C(1)
316        BETA(2) = D(2)*B(1) - D(1)*A(2)
317        U(1) = BETA(1)/DTM
318        U(2) = BETA(2)/DTM
319        RETURN
320     4  WRITE(6,6)
321     6  FORMAT(//,' N = 1 N OR JAT SPECIFIED IS NOT BIG ENOUGH.')
322        STOP
323        END
```

ITERATIVE A D I METHOD AND ALTERNATING VARIABLE METHOD

CHANNEL RADIUS = 0.500

CHANNEL LENGTH= 50.000

BOUNDARY MOVING VELO = 0.0000(FT/HR) 0.0000(−)

MEDIUM
CONDUCTIVITY = 0.1354
CAPACITY = 0.2300
DENSITY = 80.5000
DIFFUSITY = 0.0073

SOLVENT
CONDUCTIVITY = 0.1000

**Figure 15.12** Continued

```
CAPACITY      =  0.8000
DENSITY       = 68.6000
VISICOSITY    =  2.4200
HEAT TRANS COEFF = 482.5454
FLOW VELO = 36000.0000
```

INITIAL TEMP DISTRIBUTION IS:

    FLUID INLET TEMPERATURE = 1

| 1.00000 | 0.00000 | 0.00000 | 0.00000 | 0.00000 | 0.00000 |
|---------|---------|---------|---------|---------|---------|
| 0.00000 | 0.00000 | 0.00000 | 0.00000 | 0.00000 | |

--------------------------------------------------------------------

        TIME = 0.00138889 HR DIMENSIONLESS TIME = 0.00004063

--------------------------------------------------------------------

SOLVENT TEMPERATURES ARE:

| 1.00000 | 0.99976 | 0.99949 | 0.99919 | 0.99885 | 0.99847 |
|---------|---------|---------|---------|---------|---------|
| 0.99802 | 0.99748 | 0.99684 | 0.99560 | 0.96389 | |

FINAL SOLUTIONS OF THE MEDIUM: 2 30

    TEMP ARE (AT RAD FROM CEN = 0.50000FT, RAD FROM SURFACE = 0.00000FT):

| 1.00000 | 0.95228 | 0.94980 | 0.94697 | 0.94371 | 0.93991 |
|---------|---------|---------|---------|---------|---------|
| 0.93543 | 0.93008 | 0.92358 | 0.91521 | 0.00000 | |

    TEMP AREA (AT RAD FROM DEN = 0.50250FT, RAD FROM SURFACE = 0.00250FT):

| 0.00000 | 0.52931 | 0.50709 | 0.48170 | 0.45239 | 0.41818 |
|---------|---------|---------|---------|---------|---------|
| 0.37780 | 0.32949 | 0.27094 | 0.19801 | 0.00000 | |

    TEMP ARE (AT RAD FROM CEN = 0.50500FT, RAD FROM SURFACE = 0.00500FT):

| 0.00000 | 0.24335 | 0.22091 | 0.19703 | 0.17173 | 0.14511 |
|---------|---------|---------|---------|---------|---------|
| 0.11744 | 0.08922 | 0.06134 | 0.03530 | 0.00000 | |

    TEMP ARE (AT RAD FROM CEN = 0.50750FT, RAD FROM SURFACE = 0.00750FT):

| 0.00000 | 0.09563 | 0.08180 | 0.06817 | 0.05494 | 0.04239 |
|---------|---------|---------|---------|---------|---------|
| 0.03081 | 0.02058 | 0.01209 | 0.00570 | 0.00000 | |

    TEMP ARE (AT RAD FROM CEN = 0.51000FT, RAD FROM SURFACE = 0.01000FT):

| 0.00000 | 0.03310 | 0.02664 | 0.02074 | 0.01548 | 0.01094 |
|---------|---------|---------|---------|---------|---------|

**Figure 15.12**   Continued

## 15.7   CONCLUSION

(1)   A stable numerical method was derived and used to solve three conjugated equations. The method gave extremely fast calculation speed. One or two iterations were generally sufficient. This method, the alternating variable method, is well suited for multidependent variable problems. It greatly simplifies the calculation procedures of implicit approximations. When applied to hyperbolic equations with inherent discontinuity, the alternating variable method even solves the implicit approximations explicitly with high accuracy.

(2)   When the simulation technique was applied to five coal solvents, namely anthracene oil, ammonia, CO/steam, supercritical toluene, and caustic solution, it was concluded that: (a) within five minutes the outlet solvent temperature and the boundary temperature are 16 °C less than the inlet solvent

temperature; (b) the boundary velocity and thermal conductivity of coal strongly influence the boundary temperature.

## REFERENCES

1. S. P. Frankel (1950). *Mathematical Tables and Other Aids to Computation*, **4**, 65–75.
2. D. W. Peaceman and H. H. Rachford Jr (1955). *J. Soc. Ind. Appl. Math.*, **3**, 28.
3. J. Douglas Jr, D. W. Peaceman and H. H. Rachford Jr (1959). *Petrol. Trans.*, **216**, 297. (Cited by D. U. von Rosenberg (1969). *Methods for the Numerical Solution of Partial Differential Equations*, p. 93, Elsevier, Amsterdam.)
4. E. G. Herron and D. U. von Rosenberg (1966). *Chem. Engng Sci.*, **21**, 337.
5. F. K. Fong (1978). *Ph.D. Thesis*, West Virginia University.
6. F. K. Fong and D. R. Skidmore (1979). in *Advances in Computer Methods for Partial Differential Equations: III, Proc. 3rd IMACS Int. Symp. on Computer Methods for Partial Differential Equations, Lehigh University, Penn., June 1979*, R. Vichnevetsky and R. S. Stepleman (Eds), IMACS.
7. D. R. Skidmore and D. J. Konya (1974). *National AIME Meeting, Dallas, Texas, February 1974*, pp. 25–28.
8. H. S. Carslaw and J. C. Jaeger (1938). *Phil. Mag.*, **26**, 473.

*Numerical Methods in Heat Transfer*
Edited by R. W. Lewis, K. Morgan, and O. C. Zienkiewicz
© 1981 John Wiley & Sons Ltd

## Chapter 16

# The Effect of Free Convection on Entry Flow between Horizontal Parallel Plates

*T. V. Nguyen, I. L. Maclaine-cross, and G. de Vahl Davis*

### SUMMARY

A numerical study has been made of the simultaneous development of the velocity and temperature distributions in a fluid flowing through a cascade of parallel horizontal plates of specified temperature, the thickness of which is not negligible in comparison with their spacing. The two-dimensional equations for stream function, vorticity, and energy, subject to the Boussinesq approximation, have been solved by Peaceman–Rachford ADI on a mesh which provides extra resolution at the entrance to the channels formed by the plates. High values of vorticity and velocity gradient at the upstream corners of the plates have been accommodated by the use of a special mesh system in these regions.

Results have been obtained for the following ranges of parameter values: $10 \leqslant Re \leqslant 1000$; $0.7 \leqslant Pr \leqslant 7$; $0 \leqslant Ra \leqslant 5 \times 10^6$ and $0 \leqslant$ (plate thickness/plate spacing) $\leqslant 0.3$. Contour plots of streamlines and isotherms show the effects of free convection and of the channel entrance on the flow pattern and temperature distribution. Local and average Nusselt numbers and friction factors have been obtained.

### 16.1 INTRODUCTION

The flow in the passages of a parallel plate rotary heat exchanger or regenerator is laminar and, far downstream, fully developed. In the entrance region, however, the velocity and temperature distributions are developing, leading to modified values of friction factor and heat transfer coefficient in that region. These parameters are further affected by natural convection.

The simultaneous development of velocity and temperature between parallel plates is well known (see, for example, Sparrow,[1] Naito,[2] etc.). The effects of natural convection were studied by Gill and del Casal.[3] In all of these studies, it was assumed that the velocity distribution at the entrance to the passage is uniform. In fact, however, uniform flow only exists some distance

349

upstream of the entrance; at the entrance, the flow is distorted by the effect on the streamlines of the plates themselves, the magnitude of the effect depending upon the thickness of the plates in comparison to the plate spacing.

In this chapter we describe the results of a numerical (finite difference) study of the development of the velocity and temperature distributions in the entrance region of a cascade of parallel horizontal plates, the temperature of which is different from that of the incoming fluid. The flow has been assumed to be uniform some distance upstream of the cascade. The thickness of the plates has not been neglected.

The flow of a fluid between parallel horizontal plates subject to heating from below may be unstable: the temperature gradient in the fluid near the lower plate can cause the generation of longitudinal vortices. On the other hand, heating from above has a stabilizing effect. Although Akiyama, Hwang, and Cheng,[4] observed instability under these latter conditions, this first appeared at, and was apparently due to, the side walls of the channel, the width/height ratio of which was 11. In this project, we have been particularly interested in air-conditioning regenerators; for such an application, this ratio is typically at least 100. Moreover, when as here the wall temperature is constant, the vertical temperature gradient in the fluid decreases monotonically in the downstream direction, thereby diminishing the destabilizing effect. It is therefore relevant and of interest to study the flow and heat transfer using a two-dimensional model. The problem has been studied for a range of parameter values of relevance to a typical two-stream, counter-flow air-conditioning regenerator. Results have been obtained here for the following conditions: $10 \leqslant Re \leqslant 1000$, $0.7 \leqslant Pr \leqslant 7$, $0 \leqslant Ra \leqslant 5 \times 10^6$, $0 \leqslant$ (plate thickness/plate pitch) $\leqslant 0.3$.

## 16.2  THEORY

### 16.2.1  Description of the model

In Figure 16.1 fluid is pictured as flowing from left to right onto an array of

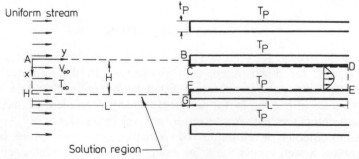

**Figure 16.1**  The geometry of the problem and some of the boundary conditions

horizontal parallel flat plates which are a centreline distance $H$ apart. Both the temperature $T_\infty$ and velocity $V_\infty$ are uniform far upstream from the channel entrance. The plates are of finite thickness $t_p$ and each is kept at the same uniform temperature $T_p$, different from that of the oncoming fluid. Far downstream, the flow becomes fully developed.

Cartesian coordinates are chosen such that the $x$-axis is parallel to the gravity vector and the $y$-axis is horizontal. This system of axes is convenient in setting up the solution field and marching the iterative procedure.

### 16.2.2 Formulation of the governing equations

The equations of motion are written in terms of the stream function $\psi$ and vorticity $\zeta$, defined by

$$u = \frac{\partial \psi}{\partial y} \qquad v = -\frac{\partial \psi}{\partial x} \quad \text{and} \quad \zeta = \frac{\partial v}{\partial x} - \frac{\partial u}{\partial y}$$

where $u$, $v$ are velocity components in the $x$-, $y$-directions.

The non-dimensional governing equations for vorticity, stream function and temperature, subject to the usual Boussinesq approximations, are

$$\frac{\partial \zeta}{\partial t} = -u\frac{\partial \zeta}{\partial x} - v\frac{\partial \zeta}{\partial y} + \frac{1}{Re}\left(\frac{\partial^2 \zeta}{\partial x^2} + \frac{\partial^2 \zeta}{\partial y^2}\right) + \frac{Ra}{Re^2 Pr}\frac{\partial \theta}{\partial y} \tag{16.1}$$

$$0 = \frac{\partial^2 \psi}{\partial x^2} + \frac{\partial^2 \psi}{\partial y^2} + \zeta \tag{16.2}$$

$$\frac{\partial \theta}{\partial t} = -u\frac{\partial \theta}{\partial x} - v\frac{\partial \theta}{\partial y} + \frac{1}{Re\, Pr}\left(\frac{\partial^2 \theta}{\partial x^2} + \frac{\partial^2 \theta}{\partial y^2}\right) \tag{16.3}$$

in which $\theta = (T - T_\infty)/(T_p - T_\infty)$, $Re = V_\infty H/\nu$, $Ra = \beta g(T_p - T_\infty)H^3/\nu\alpha$ and $Pr = \nu/\alpha$. Here $H$, $\nu$, $\alpha$, $\beta$, and $g$ denote respectively the pitch of the plates, the kinematic viscosity, the thermal diffusivity, the coefficient of thermal expansion, and the acceleration of gravity.

### 16.2.3 Boundary conditoins

Figure 16.1 shows the region ABCDEFGHA within which these equations were solved. The solution region extended the same distance $L$ upstream and downstream of the entry to the channels.

On section AH of the boundary of the solution region, uniform flow was imposed: $u = 0$; $v = 1$; $\theta = 0$; $\zeta = 0$.

We assumed that the solution region was associated with two of an infinite number of parallel plates. Hence, on sections AB and HG, periodic boundary conditions were used; the evaluation of the variables at these boundaries was included in the solution procedure for interior points.

On sections BCD and GFE, the velocities are zero; the temperature is specified ($\theta = 1$); the stream function is constant: 0 on BCD and $-1$ on GFE; and the vorticity was determined by the second-order formula

$$\zeta_p = 3(\psi_p - \psi_1)/h^2 - \zeta_1/2 \tag{16.4}$$

where 1 denotes the mesh point next to the plates and $h$ is the mesh size.

The solution region was made sufficiently long (typically, $L$ was set at $100H$) for the flow on section DE to be fully developed: $u = 0$; $\partial v/\partial y = \partial \psi/\partial y = \partial \theta/\partial y = \partial \zeta/\partial y = 0$, and for the solution to be independent of further change in $L$ in either the upstream or the downstream directions. Note that uniform $\theta$ and parabolic $v$ on DE were not assumed, but were obtained in the calculations for suitable parameter values.

## 16.3 METHOD OF SOLUTION

Because of limitations in storage and computing time, and a need for a more detailed solution in the entrance region (where strong variations of flow occur), coupled with the fact that the solution must be obtained for large distances upstream and downstream from the entrance, stretched coordinates were used in the flow direction. This allowed a more suitable distribution of computational grid points. The transformation function used is

$$y_T = (h/\ln a) \ln \left[(a^N - 1)y/A + 1\right] \tag{16.5}$$

where $h$ is the transformed uniform mesh size; $a$ is the transformation coefficient; $N$ is the number of mesh points; and $A$ is the real aspect ratio of the solution region ($L/H$). Typically, the solution was obtained using a uniform $31 \times 61$ mesh in the transformed coordinates, and the streamwise mesh size in the original coordinate ranged from 0.0333 near the entrance to 21.2894 at the ends of the solution region.

A number of changes were introduced into the governing equations, as suggested by Mallinson and de Vahl Davis[5] and also used by de Socio, Sparrow, and Eckert.[6] Firstly, the elliptic equation (16.2) is converted into a parabolic one which can be solved by a marching technique, as are Equations (16.1) and (16.3), by adding a false transient term. Thus (16.2) becomes

$$\frac{\partial \psi}{\partial t} = \nabla^2 \psi + \zeta \tag{16.6}$$

Secondly, different coefficients are inserted into the time derivatives of the equations so that the time rates of change of $\zeta$, $\psi$, and $\theta$ can be controlled separately. The rate of convergence of the numerical solution can be considerably improved by using different time steps for each equation. The true transient solution is lost as a result of these changes but the final steady solution, which is all we are interested in, is still obtained.

The final set of transformed false transient equations is

$$\frac{1}{\alpha_\zeta}\frac{\partial\zeta}{\partial t} = -u\frac{\partial\zeta}{\partial x} - v\left(\frac{dy_T}{dy}\right)\frac{\partial\zeta}{\partial y_T}$$

$$+ \frac{1}{Re}\left[\frac{\partial^2\zeta}{\partial x^2} + \left(\frac{dy_T}{dy}\right)^2\frac{\partial^2\zeta}{\partial y_T^2} + \left(\frac{d^2y_T}{dy^2}\right)\frac{\partial\zeta}{\partial y_T}\right] + \frac{Ra}{Re^2 Pr}\left(\frac{dy_T}{dy}\right)\frac{\partial\theta}{\partial y_T} \qquad (16.7)$$

$$\frac{1}{\alpha_\psi}\frac{\partial^2\psi}{\partial t} = \frac{\partial^2\psi}{\partial x^2} + \left(\frac{dy_T}{dy}\right)^2\frac{\partial^2\psi}{\partial y_T^2} + \left(\frac{d^2y_T}{dy^2}\right)\frac{\partial\psi}{\partial y_T} + \zeta \qquad (16.8)$$

$$\frac{1}{\alpha_\theta}\frac{\partial\theta}{\partial t} = -u\frac{\partial\theta}{\partial x} - v\left(\frac{dy_T}{dy}\right)\frac{\partial\theta}{\partial y_T}$$

$$+ \frac{1}{Re\,Pr}\left[\frac{\partial^2\theta}{\partial x^2} + \left(\frac{dy_T}{dy}\right)^2\frac{\partial^2\theta}{\partial y_T^2} + \left(\frac{d^2y_T}{dy^2}\right)\frac{\partial\theta}{\partial y_T}\right] \qquad (16.9)$$

These equations were approximated by finite difference equations. Forward differences were used for the time derivatives, second-order central or upwind differences for the advection terms in the momentum, and energy equations and central differences for all other space derivatives. The velocities $u$ and $v$ were obtained by differentiation of the stream function. The three finite difference equations were solved by Peaceman–Rachford ADI. For a $31\times61$ mesh, the solution for one set of parameters takes about 20 minutes starting from rest on a Cyber 171 computer.

Second-order upwind differences were used for the $y$-direction advection terms $\partial(\zeta v)/\partial y$ and $\partial(\theta v)/\partial y$ to overcome the flow instability in the upstream portion of the solution region which was present if central differences were used. The approximations used were

$$\frac{\zeta_R v_R - \zeta_L v_L}{h} \qquad \frac{\theta_R v_R - \theta_L v_L}{h}$$

where

$$v_R = \tfrac{1}{2}(v_j + v_{j+1}) \qquad v_L = \tfrac{1}{2}(v_{j-1} + v_j)$$

and

$$\left.\begin{array}{l}\zeta_R = \zeta_j \\ \theta_R = \theta_j\end{array}\right\} \text{ for } v_R > 0 \qquad \left.\begin{array}{l}\zeta_R = \zeta_{j+1} \\ \theta_R = \theta_{j+1}\end{array}\right\} \text{ for } v_R < 0$$

$$\left.\begin{array}{l}\zeta_L = \zeta_{j-1} \\ \theta_L = \theta_{j-1}\end{array}\right\} \text{ for } v_L > 0 \qquad \left.\begin{array}{l}\zeta_L = \zeta_j \\ \theta_L = \theta_j\end{array}\right\} \text{ for } v_L < 0$$

$j$ is the mesh point subscript for the $y$-direction. The '$i$' subscript has been omitted here for clarity.

A special mesh system was applied in the regions near the corners of the plates to allow a more accurate representation of the high values of vorticity and velocity gradients there. The system used was that of Thom and Apelt,[7] p. 126.

The characteristics of interest in the problem under consideration are the Nusselt numbers and the local coefficient of skin friction. These quantities, in dimensionless form, are

Local Nusselt number:   $Nu_1 = \dfrac{2(\partial\theta/\partial x)_{x=0}}{\theta_p - \theta_m}$

where $\theta_m$ is the mixed mean temperature.

Mean Nusselt number:   $Nu_m = \dfrac{2}{y\,\Delta\theta_{lm}} \displaystyle\int_0^y \left(\dfrac{\partial\theta}{\partial x}\right)_{x=0} dy$

where $\Delta\theta_{lm}$ is the dimensionless log mean temperature difference.

Local coefficient of skin friction:   $c_f = \dfrac{2}{Re}\left(\dfrac{\partial^2\psi}{\partial x^2}\right)_{x=0}$

The solution procedure was straightforward, an iteration consisting of the computation, in turn, of $\theta$, $\zeta$, and $\psi$ at internal points and along HG and DE; $\zeta$ on BCD and GFE; $\theta$, $\zeta$, and $\psi$ on AB (assuming periodicity); and finally $u$ and $v$ throughout the region. The program and mesh sizes used were validated by obtaining solutions for thin plates which agreed with previous work.[8]

## 16.4   RESULTS AND DISCUSSION

The effects of natural convection on the flow and heat transfer have been examined for a wide range of values of $Re$, $Ra$, $Pr$, and $t_p$. In general, it was found that these effects are small whenever $Re$ exceeds about 300.

### 16.4.1   Streamlines

The general nature of the flow, and the influence on it of buoyancy, are seen in Figure 16.2 which shows the streamlines for $Re = 10$, $Pr = 0.7$, $t_p = 0.2$, and various $Ra$. In this and later figures, the far upstream and downstream regions have been deleted. In Figure 16.2(a), where the plates are not heated, the streamlines are distorted only by the presence of the plates and the magnitude of the effect depends on their thickness. As $Ra$ increases, the heating has more and more influence on the streamlines. The heat transferred from the hot plates causes the initially parallel streamlines to rise before reaching the channel entrance. Inside the channel, the flow is further affected by the streamwise pressure gradient induced by the buoyancy forces. As expected from boundary

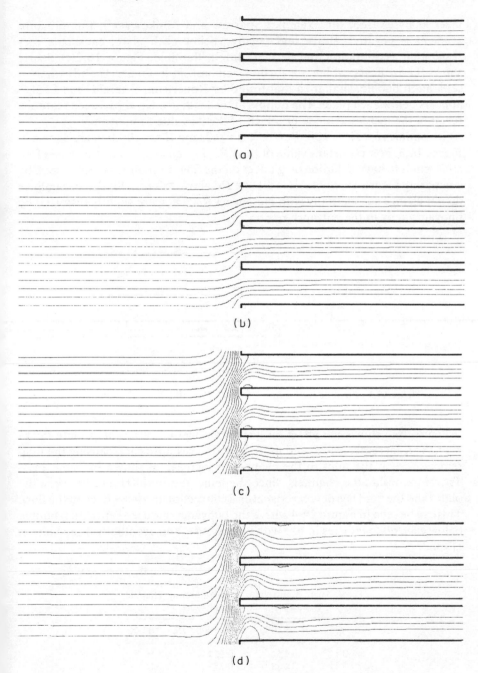

**Figure 16.2** Streamlines for $Re = 10$, $Pr = 0.7$, and $t_p = 0.2$: (a) $Ra = 0$; (b) $Ra = 1000$; (c) $Ra = 5000$; (d) $Ra = 10,000$

layer analyses (see, for example; Sparrow and Minkowycz[9]), this pressure gradient is negative above and positive below each plate, causing a corresponding acceleration and deceleration of the fluid. Figure 16.2(c) clearly shows these effects, which are further emphasized in Figure 16.2(d) where a separation bubble, similar to that found by Robertson, Seinfeld, and Leal,[10] appears below each plate. The bubble at the leading edge above each plate is a result of the increased angle of attack of the flow due to buoyancy.

Streamlines for $Re = 100$, $Pr = 0.7$, $t_p = 0.2$, and $Ra = 100,000$ are shown in Figure 16.3. For a constant value of $(Ra/Re^2 Pr)$, an increase in $Re$ causes free convection to have a diminishing effect on the flow structure, as can be seen by comparing Figure 16.3 with 16.2(b). At $Re$ above about 100, the streamlines do not change substantially with increasing $Ra$. They are slightly distorted by buoyancy only in the entrance region.

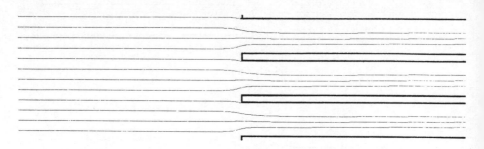

**Figure 16.3**   Streamlines for $Ra = 100,000$, $Re = 100$, $Pr = 0.7$, and $t_p = 0.2$

Figure 16.4 shows the effect on the streamlines of variation of plate thickness. As $t_p$ increases, more and more heat diffuses upstream, causing the fluid temperature at the inlet to the channel to increase as shown in Table 16.1. Therefore, inside the channels, since the temperature difference between the plates and the fluid has decreased, natural convection has less effect on the flow. This can be seen in Figure 16.4 where the buoyancy-induced separation bubble below each plate reduces in size with increasing $t_p$ and eventually disappears at $t_p = 0.3$.

Table 16.1   Centreline temperature at channel entrance; $Re = 10$, $Pr = 0.7$

| $t_p$ | $Ra = 1000$ | $Ra = 5000$ | $Ra = 10,000$ |
|---|---|---|---|
| 0 | 0.154 | 0.227 | 0.316 |
| 0.15 | 0.168 | 0.283 | 0.346 |
| 0.2 | 0.199 | 0.303 | 0.404 |
| 0.3 | 0.262 | 0.337 | 0.470 |

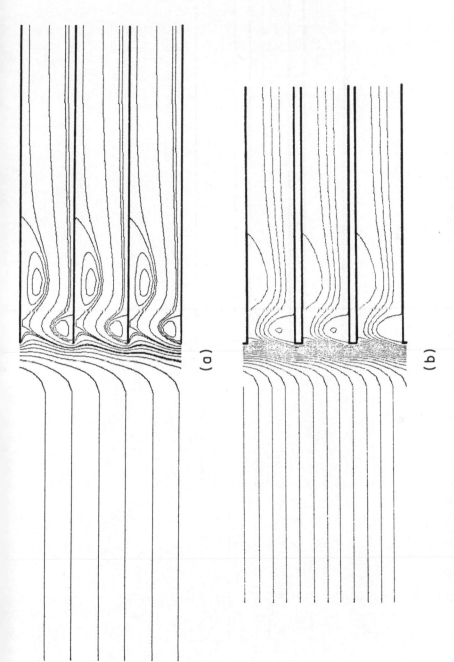

**Figure 16.4**  Streamlines for $Ra = 10,000$, $Re = 10$, and $Pr = 0.7$: (a) $t_p = 0$; (b) $t_p = 0.15$; (c) $t_p = 0.3$

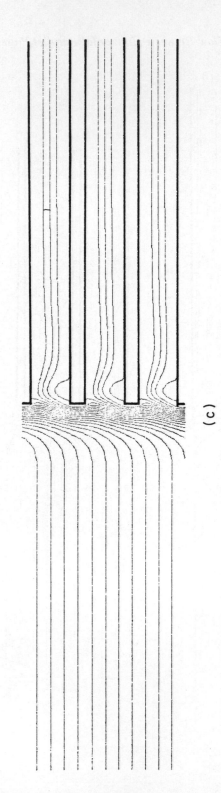

**Figure 16.4** Continued

The effect of buoyancy on the flow is suppressed or enhanced depending upon whether $Pr > 1$ or $Pr < 1$ respectively, i.e. depending upon whether the thermal boundary layer is very thin or very thick relative to the momentum boundary layer. When $Pr > 1$, the buoyancy force is confined essentially to the viscous layer near the plates, and its effect is thereby diminished compared with the case $Pr = 1$. On the other hand, when $Pr < 1$, the buoyancy force is primarily effective outside the momentum boundary layer and its effect is thereby enhanced. This is illustrated in Figures 16.2(d) and 16.5; there is a strong effect of natural convection at $Pr = 0.7$ and almost none at $Pr = 7$.

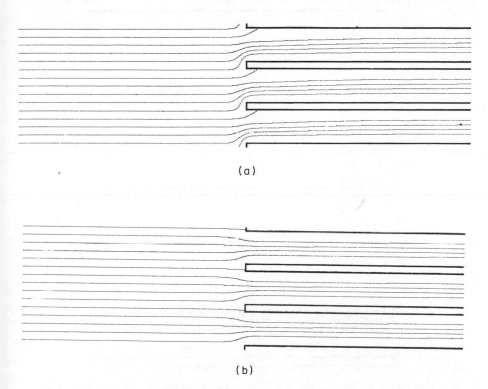

(a)

(b)

**Figure 16.5**   Streamlines for $Ra = 10,000$, $Re = 10$, and $t_p = 0.2$: (a) $Pr = 2$; (b) $Pr = 7$

### 16.4.2   Isotherms

Isotherms for $Pr = 0.7$ and various $Re$ and $Ra$ are presented in Figure 16.6. As can be seen, free convection has little effect on the isotherms, even when, as in Figure 16.6(b) (the streamlines for which are shown in Figure 16.2(d)), there is some reverse flow. The temperature increases on the underside and decreases on the upper side of each plate with increasing $Ra$.

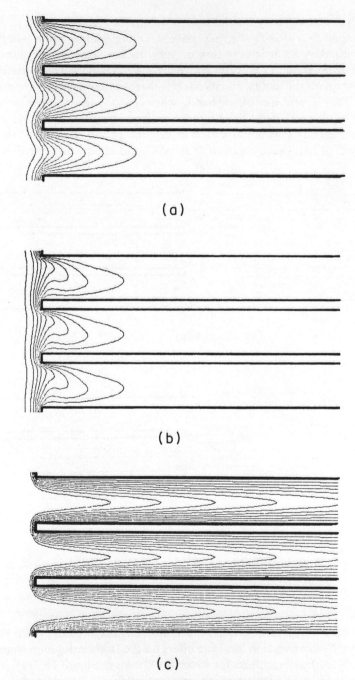

(a)

(b)

(c)

**Figure 16.6** Isotherms for $Pr = 0.7$ and $t_p = 0.2$: (a) $Ra = 0$, $Re = 10$; (b) $Ra = 20,000$, $Re = 10$; (c) $Ra = 100,000$, $Re = 100$

The thermal entrance length $Y_{TEL}$ (defined here as the distance from the entrance at which the centreline temperature reaches 98% of the plate temperature) increases with increasing $Re$ and $Pr$, and decreases with increasing plate thickness. $Y_{TEL}/RePr$ was found to depend linearly with plate thickness as shown in Figure 16.7. There is a very weak variation of this dependence with $Re$ and $Ra$ (but not with $Pr$); the equation $Y_{TEL}/RePr = 0.614 - 0.667 t_p$ is descriptive of our results for all parameter values examined.

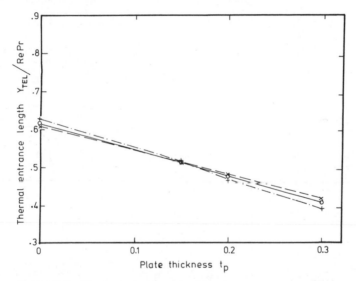

**Figure 16.7** Thermal entrance length as a function of $Re$, $Pr$, and $t_p$. $\bigcirc$: $Re = 10$, $Ra = 0$; $\times$: $Re = 100$, $Ra = 0$; $+$: $Re = 10$, $Ra = 10,000$. The effect of $Pr$ is not distinguishable in this figure

### 16.4.3 Velocity profiles

The streamwise velocity profiles for $Pr = 0.7$, $t_p = 0.2$, and various $Ra$ and $Re$ are shown in Figure 16.8. The figures on the curves denote the downstream location of the profiles in terms of $J$, the column number. The channel entrance is at $J = 31$, and $J = 61$ is the downstream end of the solution region. As seen in Figure 16.8(a) for $Re = 500$ and $Ra = 0$ the velocity profiles are concave in the central portion. The concavity extends for a distance of about 1 downstream and is most severe at the entrance, where the velocity in the peaks near the plates is 11.3% higher than at the centre. At $Ra = 10,000$, $Re = 10$ (Figure 16.8(b), for which case the streamlines are given in Figure 16.2(d)) reverse flow occurs, the streamwise velocity becoming negative around the upstream lower corner of each plate and in the separation bubble above each plate. This bubble causes the location of the velocity maximum to rise; when the bubble is large

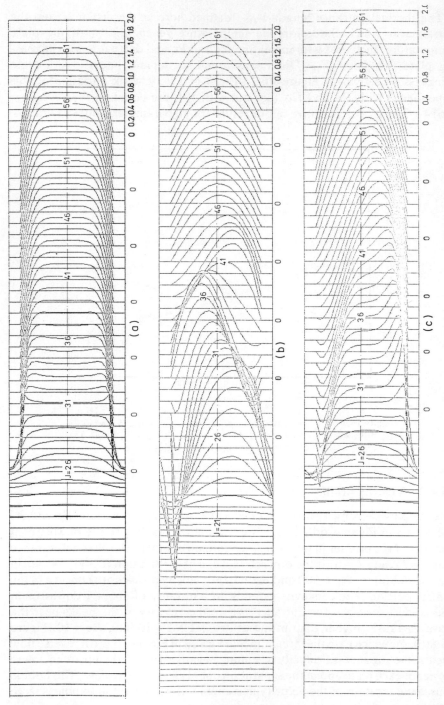

**Figure 16.8**  Streamwise velocity profiles for $Pr = 0.7$ and $t_p = 0.2$: (a) $Ra = 0$, $Re = 500$; (b) $Ra = 10,000$, $Re = 10$; (c) $Ra = 200,000$, $Re = 100$

enough, as in this case, the maximum velocity is pushed upwards above the centre line. Downstream, the free convection effect dominates again and the point of maximum velocity is seen to move downwards before the fully developed profile is established. In Figure 16.8(c) the relatively high $Re$ means that buoyancy has little effect on the flow in the central region of the channel; only the flow near the plates is affected. Notice that the outlet velocity distributions in Figures 16.8(b) and (c) are parabolic, but that in 16.8(a), at a higher $Re$, the flow is still developing.

The hydrodynamic entrance length is defined as the distance from the entrance to the point at which the centreline streamwise velocity reaches a value within $\pm 2$ of its fully developed value. (In Figure 16.8(b), the centreline velocity at first rises considerably above the fully developed value.) Table 16.2 shows the entrance length for various $t_p$, $Ra$, $Pr$, and $Re$. Natural convection and $Pr$ have only a small effect on the hydrodynamic entrance length except at small values of $Re$; i.e. the effect depends, as would be expected from Equation (16.1), on $Ra/Re^2Pr$.

Table 16.2   Hydrodynamic entrance length. $Pr = 0.7$ except (1): $Pr = 2$, (2): $Pr = 7$

| | | $Re$ | 10 | | 100 | |
|---|---|---|---|---|---|---|
| | | $Ra$ | 0 | 10000 | 0 | 20000 |
| | | 0 | 0.455 | 1.100 | 2.670 | 2.720 |
| | | 0.15 | 0.271 | 1.018 | — | — |
| $t_p$ | | 0.2 | 0.254 | 0.881 | 2.425 | 2.708 |
| | | | | 0.601 (1) | | |
| | | | | 0.264 (2) | | |
| | | 0.3 | 0.232 | 0.660 | 2.171 | — |

### 16.4.4   Temperature profiles

Figure 16.9 shows the temperature profiles for various $Re$ and $Ra$. For low $Re$, heat is diffused some distance upstream from the entrance. For example, at $Re = 10$, $\theta$ starts to rise (specifically, $\theta = 0.01$) at a distance of 0.323. However at $Re = 100$ this happens only at $Y = 0.095$ upstream.

### 16.4.5   Heat transfer

Local Nusselt numbers are plotted as a function of distance from the channel entrance in Figures 16.10 and 16.11. All the curves tend to a limiting value far downstream of 7.55 as the entrance and free convection effects die out. This compares very well with the theoretical value of 7.54.[11] As expected from boundary layer analyses, the flow at $Re = 100$ in the boundary layer above each

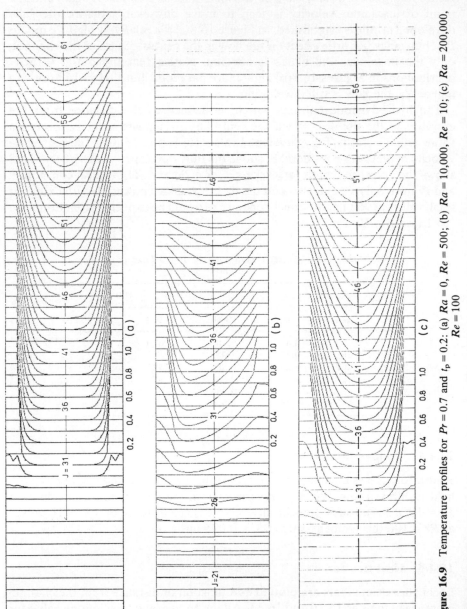

**Figure 16.9**  Temperature profiles for $Pr = 0.7$ and $t_p = 0.2$: (a) $Ra = 0$, $Re = 500$; (b) $Ra = 10,000$, $Re = 10$; (c) $Ra = 200,000$, $Re = 100$

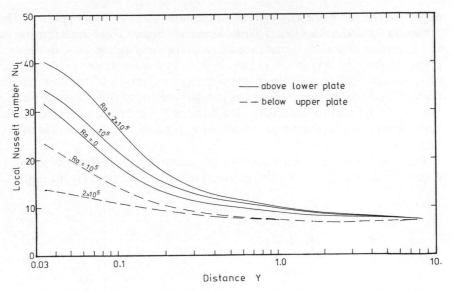

**Figure 16.10**   Local Nusselt number: $Re = 100$, $Pr = 0.7$, and $t_p = 0.2$

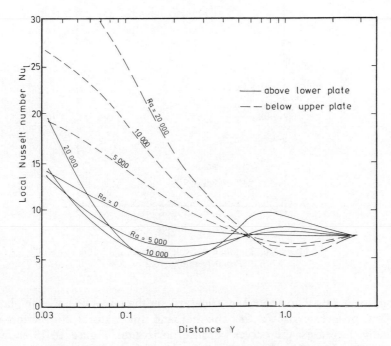

**Figure 16.11**   Local Nusselt number: $Re = 10$, $Pr = 0.7$, and $t_p = 0.2$

plate is accelerated, thereby increasing the heat transfer. The opposite effect occurs on the underside of each plate. As seen in Figure 16.10, heat transfer on the underside of a plate is more sensitive to free convection than above each plate. At $Re = 10$ (Figure 16.11) the bubble at the leading edge above each plate which is a result of the increased angle of attack of the flow due to buoyancy causes the Nusselt numbers to fall there and to rise below the plate in contrast to the effect at higher $Re$. After about $Y = 0.5$ from the entrance, the free convection effects dominate and the effect becomes the same as that at higher $Re$.

The effect of natural convection on the mean Nusselt number is shown in Figures 16.12 and 16.13. Natural convection is quite dominant in the entrance

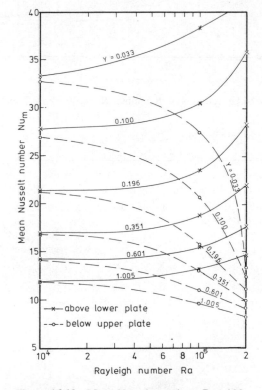

**Figure 16.12**  Mean Nusselt number: $Re = 100$, $Pr = 0.7$, and $t_p = 0.2$

region, for $Y$ less than about 1, particularly on the underside of each plate. As the flow progresses down the channel and the natural convection effect gradually decreases, the curves become horizontal. Figure 16.13 shows the effect on the mean Nusselt number of the separation bubbles which form at low

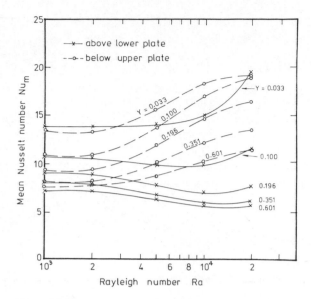

**Figure 16.13**   Mean Nusselt number: $Re = 10$, $Pr = 0.7$, and $t_p = 0.2$

*Re*. In their neighbourhood, the Nusselt numbers generally decrease above and increase below each plate with increasing *Ra*. An exception occurs at $Y = 0.033$ above the plates, which is upstream of the upper separation bubble but near the lower bubble. Here, the Nusselt number on the plate above the bubble rises dramatically with *Ra*; this affects the value of $Nu_m$ for larger values of *Y*.

Nusselt numbers are plotted in Figures 16.14 and 16.15 as a function of Graetz number for Prandtl numbers of 0.7 to 7. The curves for $Pr = 0.7$ (full lines) show the variation of Nusselt number with *Re*. The variations of the local Nusselt number are smaller than those of the mean Nusselt number and the curves can be approximated by a single curve. Mercer, Pearce, and Hitchcock[12] obtained the mean Nusselt number for developing flow between parallel plates with uniform conditions at inlet. Their result is included in Figure 16.15 to show the effects of the plate thickness and of the correct inlet conditions.

Figure 16.16 shows the effect of $t_p$ on local Nusselt number. For pure forced convection ($Ra = 0$), the local Nusselt number decreases with increasing plate thickness. In the case of combined forced and free convection, this trend is only true after some distance from the entrance.

### 16.4.6   Skin friction

The local coefficient of friction is shown in Figure 16.17 for $Re = 10$. When the temperature of the plates is higher than that of the oncoming fluid, the flow in

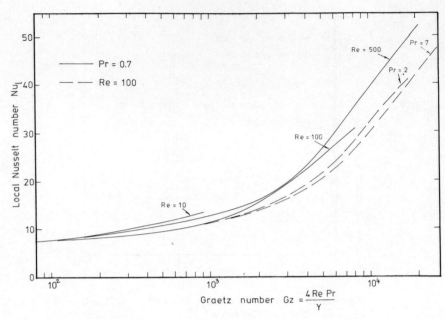

**Figure 16.14**  Local Nusselt number: $Ra = 0$ and $t_p = 0.2$

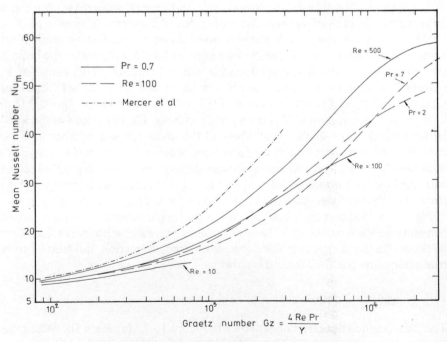

**Figure 16.15**  Mean Nusselt number: $Ra = 0$ and $t_p = 0.2$

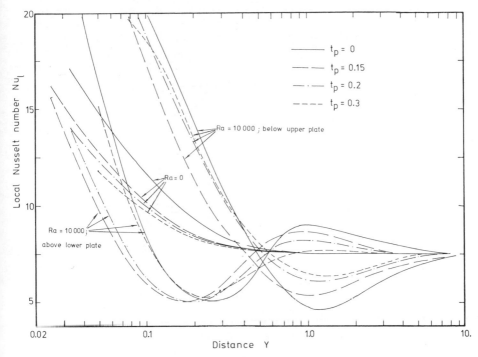

**Figure 16.16**   Local Nusselt number: $Re = 10$ and $Pr = 0.7$

the boundary layer above each plate is accelerated relative to the corresponding forced convection flow ($Ra = 0$), with a resulting increase in the shear stress at the plates and therefore an increase in the local friction. The opposite effect applies to flow below the plates. At $Ra = 1000$, the local friction factor can be seen to increase above and decrease below the plates relative to the $Ra = 0$ value. For sufficiently high $Ra$, reverse flow occurs in the region near the leading edge below each plate as seen in Figure 16.2(c) for $Ra = 5000$; the friction factor therefore becomes negative in this region. Further downstream, the flow below each plate accelerates and then decelerates again as it approaches the growing separation bubble, resulting in an increase and then a decrease of the friction. At $Ra > 10,000$, when (see Figure 16.2(d)) a separation bubble has formed below each plate, the friction factor again becomes negative for a short distance. Above each plate, the separated region near the leading edge grows with $Ra$, causing the friction factor to become negative. Due to the effect of free convection, the flow is then accelerated and the friction factor increases above the $Ra = 0$ value. All curves converge to a constant fully developed friction factor, the value of which is $12/Re(1 - t_p)^2$.

Figure 16.18 shows the friction factor against $Y$ for $Re = 100$. Again, as expected, the effect of free convection is to increase the friction above and

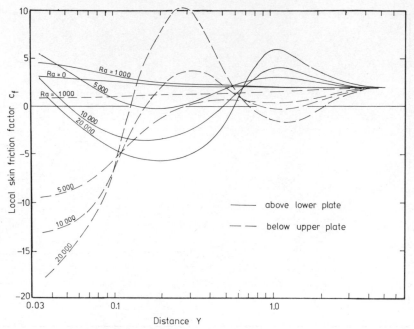

**Figure 16.17**    Friction factor: $Re = 10$, $Pr = 0.7$, and $t_p = 0.2$

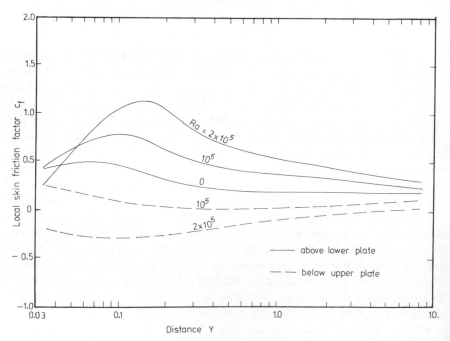

**Figure 16.18**    Friction factor: $Re = 100$, $Pr = 0.7$, and $t_p = 0.2$

decrease it below the plates. Inspection of these figures and heat transfer results reveals that the friction is more strongly affected by free convection than is the heat transfer. This is consistent with the fact that the direct effect of the free convection is an additional force in the momentum balance.

The streamwise velocity for flow between thick plates is higher, for a given upstream $Re$, than for flow between thin plates, resulting in a thinner boundary layer. Consequently, the shear stress and friction factor on thick plates are greater as shown in Figure 16.19. It is seen from Figure 16.19 that for the same $Ra$, the effect of plate thickness on the friction factor in the entrance region is weak above the plates but rather stronger below the plates. This is due to the fact that the bubbles above each plate for different plate thicknesses are of similar size (see Figure 16.4).

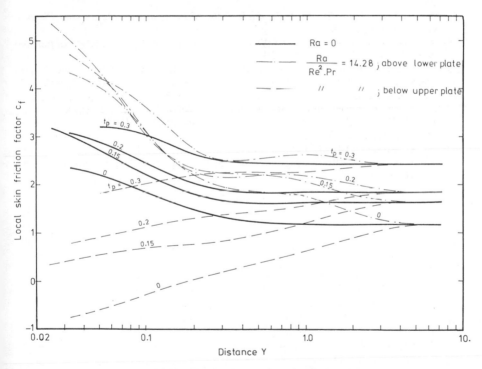

**Figure 16.19**  Friction factor: $Re = 10$ and $Pr = 0.7$

## Acknowledgement

T. V. Nguyen acknowledges the financial support of the Australian Commonwealth Department of Education's special scheme of assistance for Vietnamese and Kampuchean students.

## REFERENCES

1. E. M. Sparrow (1955). 'Analysis of laminar forced convection heat transfer in entrance region of flat rectangular ducts', NACA TN3331.
2. E. Naito (1976). 'Laminar heat transfer in the entrance region of parallel plates. The case of uniform heat flux', *J. Int. Chem. Engng*, **16**, 162–168.
3. W. N. Gill and E. del Casal (1962). 'A theoretical investigation of natural convection effects in forced horizontal flows', *A.I.Ch.E. J.*, **8**, 513–518.
4. M. Akiyama, G. J. Hwang and K. C. Cheng (1971). 'Experiments on the onset of longitudinal vortices in laminar forced convection between horizontal plates', *J. Heat Transfer, Trans. ASME, Series C*, **93**, 335–341.
5. G. D. Mallinson and G. de Vahl Davis (1973). 'The method of the false transient for the solution of coupled elliptic equations', *J. Comp. Phys.*, **12**, 435–461.
6. L. M. de Socio, E. M. Sparrow and E. R. G. Eckert (1973). 'The contrived transient-explicit method for solving steady-state flows', *Comput. Fluids*, **1**, 273–287.
7. A. Thom and C. J. Apelt (1961). *Field Computations in Engineering and Physics*, van Nostrand, London.
8. Y. L. Wang and P. A. Longwell (1964). 'Laminar flow in the inlet section of parallel plates', *A.I.Ch.E. J.*, **10**, 323–329.
9. E. M. Sparrow and W. J. Minkowycz (1962). 'Buoyancy effects on horizontal boundary-layer flow and heat transfer', *Int. J. Heat Mass Transfer*, **5**, 505–511.
10. G. E. Robertson, J. J. Seinfeld and L. G. Leal (1973). 'Combined forced and free convection flow past a horizontal flat plate', *A.I.Ch.E. J.*, **19**, 998–1008.
11. W. M. Kays and A. L. London (1964). *Compact Heat Exchangers*, 2nd edn, McGraw-Hill, New York.
12. W. E. Mercer, W. M. Pearce and J. E. Hitchcock (1967). 'Laminar forced convection in the entrance region between parallel plates', *J. Heat Transfer, Trans. ASME, Series C*, **89**, 251–257.

Numerical Methods in Heat Transfer
Edited by R. W. Lewis, K. Morgan, and O. C. Zienkiewicz
© 1981 John Wiley & Sons Ltd

Chapter 17

# A Characteristics Based Finite Element Method for Heat Transport Problems

W. D. Liam Finn and Erol Varoğlu

## SUMMARY

Heat transport problems involving convection are governed by the diffusion–convection equation. Existing numerical techniques encounter difficulties in the solution of this equation when applied over the entire range from pure diffusion to pure convection. These difficulties include overshooting, undershooting, oscillations, clipping, and the smearing of sharp fronts. Although it is possible to adjust existing solutions so that they give reasonable results for limited ranges of convection, it is desirable to have a method that is stable over the entire range. Such a method has been developed by Varoğlu and Finn.[11,12]

The method is based on a finite element approach which incorporates the method of characteristics. It is capable of solving the diffusion–convection equation over the entire range of convection from pure diffusion to pure convection without any of the numerical difficulties cited above.

The basis of the method suitable for heat transport problems in rivers is presented. The utility and accuracy of the method have been demonstrated by numerical examples.

## NOMENCLATURE

| | | | |
|---|---|---|---|
| $T$ | temperature | $S$ | the net flux of heat into the water surface from atmosphere |
| $A$ | cross-sectional area | | |
| $B$ | top width of the cross section | $\rho$ | density of water |
| $x$ | longitudinal distance | $c$ | specific heat of water |
| $t$ | time | $T_u$ | temperature at the left-hand boundary |
| $L$ | length of the interval along the $x$-axis | | |
| $Q$ | total flow rate | $T_d$ | temperature at the right-hand boundary |
| $E$ | overall longitudinal dispersion coefficient | $T_0$ | initial temperature distribution |

| $T_E$ | equilibrium temperature | P | node in the $(x, t)$ plane |
|---|---|---|---|
| $\varepsilon, \bar{\eta}$ | surface heat exchange coefficients | $k$ | time increment |
| | | $r$ | space increment |
| $\hat{T}$ | approximate temperature | $\eta, \xi$ | local coordinates |
| $w$ | weighting function | $\Delta x$ | uniform space increment |
| $K$ | space–time finite element | $\Delta t$ | uniform time increment |
| $I$ | total number of space–time elements | $C$ | Courant number |

## 17.1  INTRODUCTION

Since the biological and biochemical processes in water are temperature dependent, the prediction of the rise in water temperature due to waste heat discharge from power plants is an important aspect of the environmental impact studies of power plant operations. In the mathematical modelling of heat transfer from condenser–water discharge, a distinction should be made between the near field in the vicinity of discharge, and the far field outside this region. In the far field, the discharged heat is transported by convection and diffusion within the water and is lost to the atmosphere through the water surface.

In most instances, the prediction of the temperatures in the far field requires numerical solution of the diffusion–convection equation. The numerical solution of this equation is based on either finite differences in space and time[1,2] or a combination of Galerkin type finite elements in space and finite differences in time.[3,4]

A review and comparison of numerical techniques for the solution of the unsteady diffusion–convection equation can be found in papers by Lam[5] and Ehlig.[6] These studies show that most of the numerical methods give satisfactory numerical results for diffusion dominated transport problems. However, when convection is the strong component of transport, existing numerical techniques encounter numerical difficulties even in one-dimensional transport problems with constant coefficients for practically acceptable time steps and space meshes. Numerical examples given by Lam[5] show that the central differencing scheme and box scheme are oscillatory; upstream differencing introduces large artificial diffusion and when flux correction is used to control oscillations in upstream differencing schemes, sharp concentration fronts are smeared and clipping errors are created. The classical finite elements are oscillatory even in unsteady one-dimensional convection dominated transport problems.

Similar numerical difficulties are also encountered in the numerical solution of the steady-state, one-dimensional heat transport equation for large values of Peclet number.[7] In the steady-state, one-dimensional case, a remedy for the numerical difficulties is sought by employing upwinding high-order Galerkin methods[8] or employing improved finite differences.[9,10]

A new method for the numerical solution of the diffusion–convection equation has been developed by Varoğlu and Finn.[11,12] The method employs space–time finite elements and incorporates the characteristics of the associated hyperbolic partial differential equation which is obtained by taking the dispersion coefficient as zero in the diffusion–convection equation. These characteristics form the sides of the elements joining nodes at consecutive time levels. This method reduces to conventional finite element analysis in the case of pure diffusion and to the method of characteristics in the case of pure convection. It is capable of solving the diffusion–convection over the entire range of convection from pure diffusion to pure convection without any of the numerical difficulties cited above, even for very large time steps.

A formulation of the method suitable for heat transport problems in rivers or estuaries with variable cross section is presented and the accuracy and utility of the method are demonstrated by numerical examples.

## 17.2   STATEMENT OF THE PROBLEM

### 17.2.1   Governing partial differential equation

The cross-sectional average temperature distribution $T(x, t)$ in a river or estuary in the far field is governed by the heat conservation equation

$$A(x)\frac{\partial T}{\partial t} + Q\frac{\partial T}{\partial x} = E\frac{\partial}{\partial x}\left(A(x)\frac{\partial T}{\partial x}\right) + \frac{B(x)}{\rho c}S(T) \qquad (17.1)$$

in which

$x$ = longitudinal distance along the stream
$t$ = time
$A$ = cross-sectional area
$Q$ = total flow rate
$E$ = overall longitudinal dispersion coefficient
$B$ = top width of the cross section
$S$ = the net flux of heat into the water surface from atmosphere
$\rho$ = density of water
$c$ = specific heat of water

In obtaining Equation (17.1), it is assumed that the vertical stratification in the river or estuary is negligible.[4]

### 17.2.2   Boundary and initial conditions

The upstream and downstream boundaries are taken at $x = 0$, and $x = L$, respectively. The condition at these boundaries can be specified either by temperatures or by temperature gradients. The method will be developed for the case when the temperatures are specified as

$$T(0, t) = T_u(t) \quad t > 0$$
$$T(L, t) = T_d(t) \quad t > 0 \tag{17.2}$$

The initial condition can be expressed as

$$T(x, 0) = T_0(x) \quad 0 \leqslant x \leqslant L \tag{17.3}$$

If the thermal discharge is introduced at the upstream boundary at $t = 0$, then $T_0(x)$ is the natural temperature distribution along the river.

### 17.2.3  Surface heat transfer

The net heat flux into the water surface from the atmosphere is made up of the contributions of several heat transfer processes such as solar and atmospheric radiation, back radiation from the water body, and evaporation and conduction. These heat transfer processes are functions of various meteorological factors. Empirical formulae have been developed[2] for computing each heat transfer contribution to surface flux, $S(T)$, as a function of meteorological conditions and the surface temperature of water. The dependence of the surface heat flux on surface temperature $T$ is non-linear but the linearization of $S(T)$ is common in analysis.[2,4] The surface heat flux is linearized as

$$S(T) = -(\varepsilon T + \bar{\eta}) \tag{17.4}$$

in which $\varepsilon$ and $\bar{\eta}$ are the surface heat exchange coefficients. The ratio $T_E = -\bar{\eta}/\varepsilon$ denotes the equilibrium temperature of the water surface for which, given a set of meteorological conditions, the net surface heat flux is zero. The surface heat exchange coefficients $\varepsilon$ and $\bar{\eta}$ must be chosen such that Equation (16.4) is a good approximation of the surface heat flux for surface temperatures in the range of interest.

### 17.3  SPACE–TIME FINITE ELEMENTS INCORPORATING CHARACTERISTICS

The method of weighted residuals will be employed to solve the initial–boundary value problem defined by Equations (17.1)–(17.3) in which surface heat flux is linearized by Equation (17.4). Let $\hat{T}(x, t)$ be an approximation to the solution $T(x, t)$ of Equations (17.1)–(17.3). The vanishing of the weighted residual of Equation (17.1) with respect to a continuous weighting function $w(x, t)$ defined in $0 \leqslant x \leqslant L$, $t \leqslant 0$ can be expressed as

$$\int_{t^n}^{t^{n+1}} \int_0^L \left[ A \frac{\partial \hat{T}}{\partial t} + Q \frac{\partial \hat{T}}{\partial x} + E \frac{\partial}{\partial x} \left( A \frac{\partial \hat{T}}{\partial x} \right) + \frac{B\varepsilon}{\rho c} (\hat{T} - T_E) \right] w \, dx \, dt = 0 \tag{17.5}$$

for all $0 \leqslant t^n < t^{n+1}$. Here, $Q$, $E$, $\varepsilon$, and $T_E$ are constants. Employing integration by parts, Equation (17.5) can be written as

$$\int_{t^n}^{t^{n+1}} \int_0^L \left(-A\hat{T}\frac{\partial w}{\partial t} + AE\frac{\partial w}{\partial x}\frac{\partial \hat{T}}{\partial x} - Q\hat{T}\frac{\partial w}{\partial x} + \frac{B\varepsilon}{\rho c}(\hat{T} - T_E)\right) dx \, dt$$

$$+ \int_0^L [A\hat{T}w]_{t=t^n}^{t^{n+1}} dx - \int_{t^n}^{t^{n+1}} \left[EAw\frac{\partial \hat{T}}{\partial x} - Q\hat{T}w\right]_{x=0}^L dt = 0 \qquad (17.6)$$

In the last two integrals brackets are used for brevity to denote

$$[\psi(x_1, x_2)]_{x_i=a}^b = \begin{cases} \psi(b, x_2) - \psi(a, x_2) & i = 1 \\ \psi(x_1, b) - \psi(x_1 \, a) & i = 2 \end{cases} \qquad (17.7)$$

The domain of integration $0 < x < L$, $t^n < t < t^{n+1}$ will be discretized by space–time elements $K_0^n, K_1^n, \ldots, K_{I-1}^n$ as illustrated in Figure 17.1. The total

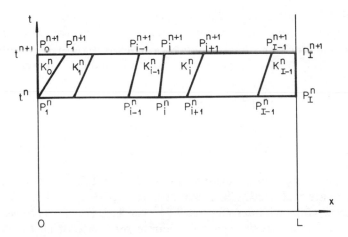

**Figure 17.1** Finite elements in space and time at a typical time step

number of space–time elements at a typical time step is denoted by $I$. The sides of these elements $(P_i^n P_i^{n+1}, i = 1, 2, \ldots, I-1)$ joining the nodes at subsequent time levels $t^n$ and $t^{n+1}$ are oriented along the linearized characteristic lines of the associated hyperbolic equation

$$A(x)\frac{\partial T}{\partial t} + Q\frac{\partial T}{\partial x} = -\frac{B(x)}{\rho c}\varepsilon(T - T_E) \qquad (17.8)$$

and, therefore

$$A_i^n(x_i^{n+1} - x_i^n) = kQ \qquad i = 1, 2, \ldots, I-1 \qquad (17.9)$$

Here, $k$ denotes the time step

$$k = t^{n+1} - t^n \tag{17.10}$$

and $x_i^m$ is the $x$-coordinate of the point $P_i^m$ ($m = n, n+1$) in the $(x, t)$ plane. For brevity, $A(x_i^n)$ is denoted by $A_i^n$.

All the elements are trapezoidal except the left-hand boundary element which is triangular. The triangular element results from using the characteristic through the boundary point $P_1^n$. However, the use of this characteristic is essential in order to transmit the data accurately from the boundary at $x = 0$ expecially in convection dominated boundary value problems.

The problem will be solved step-by-step in time employing isoparametric trapezoidal and linear triangular space–time elements. In local coordinates $(\eta, \xi)$, the approximate solution $\hat{T}$ over a typical trapezoidal element $K_i^n$ will be taken as a polynomial of the form[13]

$$\hat{T}(\eta, \xi) = (1-\eta)(1-\xi)T_i^n + (1-\eta)\xi T_i^{n+1} + \eta(1-\xi)T_{i+1}^{n+1} + \eta\xi T_{i+1}^{n+1} \tag{17.11}$$

in which $T_j^m$ denotes the nodal value, $\hat{T}(x_j^m, t^m)$, $m = n, n+1$, $j = i, i+1$. It should be noted that $\hat{T}(x, t)$ defined on $K_0^n, \ldots, K_I^n$ is linear along the sides of the triangular and trapezoidal elements and, therefore, $\hat{T}(x, t)$ is continuous on $0 \leqslant x \leqslant L$ and $t^n \leqslant t \leqslant t^{n+1}$. The initial condition and the boundary conditions require that

$$T_i^0 = T_0(x_i^0) \qquad i = 1, 2, \ldots, I \tag{17.12}$$

$$T_0^{n+1} = T_u(t^{n+1}) \qquad n \geqslant 0$$
$$T_I^{n+1} = T_d(t^{n+1}) \qquad n \geqslant 0 \tag{17.13}$$

At a typical time step $n$, the nodal values $T_i^n$ ($i = 1, 2, \ldots, I$) are prescribed for $n = 0$ or are evaluated at the previous time step for $n > 0$. Therefore, there is one unknown for each internal node $T_i^{n+1}$ ($i = 1, 2, \ldots, I-1$) to be evaluated at each time step. Let the space $V$ denote the space of all the continuous functions defined on the finite element $K_0^n$ linearly and on $K_1^n, \ldots, K_{I-1}^n$ by Equation (17.11). A function $w$ in $V$ is uniquely determined by its values at all the nodes $P_i^n$ ($i = 1, \ldots, I$) and $P_i^{n+1}$ ($i = 0, \ldots, I$). The weighting function $w(x, t)$ for each $i$ ($i = 1, 2, \ldots, I-1$) is defined as a function in space $V$ such that

$$w^{(i)}(P_j^{n+1}) = \begin{cases} 1 & i = j \text{ or } i = 1, j = 0 \quad j = 0, 1, \ldots, I \\ & \text{or } i = I-1, j = I \\ 0 & \text{otherwise} \qquad i = 1, 2, \ldots, I-1 \end{cases} \tag{17.14}$$

$$w^{(i)}(P_j^n) = \begin{cases} 1 & i = j \text{ or } i = I-1, j = 1 \quad j = 1, 2, \ldots, I \\ 0 & \text{otherwise} \qquad\qquad i = 1, 2, \ldots, I-1 \end{cases} \tag{17.15}$$

Replacing $w(x, t)$ by $w^{(i)}$ $(i = 1, 2, \ldots, I-1)$ in Equation (17.6) and evaluating integrals over the finite elements and the boundaries of the elements by numerical quadrature,[12] $I-1$ equations in the unknowns $T_i^{n+1}$ $(i = 1, 2, \ldots, I-1)$ are obtained as

$$\tfrac{1}{4}[kQ(1+A_1^{n+1}/A_1^n)+2r_1^{n+1}A_1^{n+1}]T_1^{n+1} - \tfrac{1}{2}(kQ+r_1^nA_1^n)T_1^n$$

$$+\tfrac{1}{4}kQ(1-A_2^{n+1}/A_2^n)T_2^{n+1} - \tfrac{1}{2}kQ(1-A_0^{n+1}/A_1^n)T_0^{n+1}$$

$$-\tfrac{1}{4}kE\Big((A_1^n+A_2^n)\frac{T_2^n-T_1^n}{r_1^n}+(A_1^{n+1}+A_2^{n+1})\frac{T_2^{n+1}-T_1^{n+1}}{r_1^{n+1}}$$

$$-2(A_0^{n+1}+A_1^n)\frac{T_1^{n+1}-T_0^{n+1}}{r_0^{r+1}}\Big)$$

$$+\frac{k\varepsilon}{4\rho c}[B_1^n(r_1^n+\tfrac{2}{3}r_0^{n+1})(T_1^n-T_E)+\tfrac{2}{3}B_0^{n+1}r_0^{n+1}(T_0^{n+1}-T_E)$$

$$+B_1^{n+1}(r_1^{n+1}+\tfrac{2}{3}r_0^{n+1})(T_1^{n+1}-T_E)]=0 \tag{17.16}$$

$$\tfrac{1}{4}kQ[(1-A_{i+1}^{n+1}/A_{i+1}^n)T_{i+1}^{n+1}-(1-A_{i-1}^{n+1}/A_{i-1}^n)T_{i-1}^{n+1}]$$

$$+\tfrac{1}{2}[(r_i^{n+1}+r_{i-1}^{n+1})A_i^{n+1}T_i^{n+1}-(r_i^n+r_{i-1}^n)A_i^nT_i^n]$$

$$-\tfrac{1}{4}kE\Big((A_i^n+A_{i+1}^n)\frac{T_{i+1}^n-T_i^n}{r_i^n}-(A_i^n+A_{i-1}^n)\frac{T_i^n-T_{i-1}^n}{r_{i-1}^n}$$

$$+(A_{i+1}^{n+1}+A_i^n)\frac{T_{i+1}^{n+1}-T_i^{n+1}}{r_i^{n+1}}$$

$$-(A_i^{n+1}+A_{i-1}^{n+1})\frac{T_i^{n+1}-T_{i-1}^{n+1}}{r_{i-1}^{n+1}}\Big)+\frac{k\varepsilon}{4\rho c}[B_i^n(r_i^n+t_{i-1}^n)(T_i^n-T_E)$$

$$+B_i^{n+1}(r_i^{n+1}+r_{i-1}^{n+1})(T_i^{n+1}-T_E)]=0 \quad i=2,3,\ldots,I-2 \tag{17.17}$$

$$-\tfrac{1}{4}kQ(1-A_{I-2}^{n+1}/A_{I-2}^n)T_{I-2}^{n+1}+\tfrac{1}{2}(kQ+r_{I-1}^{n+1}A_I^{n+1})T_I^{n+1}+\tfrac{1}{2}(kQ-r_{I-1}^nA_I^n)T_I^n$$

$$-\tfrac{1}{4}[kQ(1-A_{I-1}^{n+1}/A_{I-1}^n)-2(r_{I-2}^{n+1}+r_{I-1}^{n+1})A_{I-1}^{n+1}]T_{I-1}^{n+1}$$

$$-\tfrac{1}{2}(r_{I-2}^n+r_{I-1}^n)A_{I-1}^nT_{I-1}^n+\tfrac{1}{4}kE\Big((A_{I-1}^n+A_{I-2}^n)\frac{T_{I-1}^n-T_{I-2}^n}{r_{I-2}^n}$$

$$+(A_{I-1}^{n+1}+A_{I-2}^{n+1})\frac{T_{I-1}^{n+1}+T_{I-2}^{n+1}}{r_{I-2}^{n+1}}-2A_I^{n+1}\frac{T_I^{n+1}-T_{I-1}^{n+1}}{r_{I-1}^{n+1}}-2A_I^n\frac{T_I^n-T_{I-1}^n}{r_{I-1}^n}\Big)$$

$$+\frac{k\varepsilon}{4\rho c}[B_{I-1}^n(r_{I-1}^n+r_{I-2}^n)T_{I-1}^n+B_{I-1}^{n+1}(r_{I-1}^{n+1}+r_{I-2}^{n+1})T_{I-1}^{n+1}$$

$$+B_I^nr_{I-1}^n(T_I^n-T_E)+B_I^{n+1}r_{I-1}^{n+1}(T_I^{n+1}-T_E)]=0 \tag{17.18}$$

in which $B_i^m = B(x_i^m)$, $m = n, n+1$ and

$$r_i^{n+1} = x_{i+1}^{n+1} - x_i^{n+1} \quad i = 0, 1, \ldots, I-1$$
$$r_i^n = x_{i+1}^n - x_i^n \qquad i = 1, 2, \ldots, I-1 \tag{17.19}$$

In the case of pure diffusion $Q = 0$, and Equation (17.9) yields

$$x_i^{n+1} = x_i^n \quad i = 1, 2, \ldots, I-1 \tag{17.20}$$

Therefore, space–time elements are rectangular and the triangular element $K_0^n$ vanishes. The unknowns of the problem, $T_i^{n+1}$ $(i = 2, 3, \ldots, I-1)$ can be evaluated from Equations (17.17)–(17.18). These equations are equivalent to the conventional finite element equations for pure diffusion equation.

In the case of pure convection with constant coefficients ($A = $ constant, $\varepsilon = 0$), Equations (17.16)–(17.17) yield

$$T_i^{n+1} = T_i^n \quad i = 1, 2, \ldots, I-1 \tag{17.21}$$

Also, it should be noted that characteristics are straight lines and, therefore, the sides of the finite elements are oriented along the exact characteristics by taking (Equation (17.19))

$$x_i^{n+1} = x_i^n + kQ/A \quad i = 1, 2, \ldots, I-1 \tag{17.22}$$

Thus, the solution given by Equation (17.21) at discrete points $P_i^{n+1}$ $(i = 1, 2, \ldots, I-1)$ is exact and the proposed method reduces to the method of characteristics in this case. In the case of pure convection with variable coefficients,[12] it can be shown that the proposed finite element solution is a first-order approximation in $|x_i^{n+1} - x_i^n|$ and $k$ to the solution to be obtained from the method of characteristics.

## 17.4 AUTOMATIC MESH GENERATION

At each time step, a new space–time mesh is automatically generated by the algorithm MESH. The interval $0 \leqslant x \leqslant L$ is initially discretized by $I$ nodes, $x_1^0 = 0$, $x_2^1, \ldots, x_{I-1}^0$, $x_I^0 = L$. These nodes can be equally or variably spaced. The selection of initial discretization is guided by the accuracy desired in the representation of the initial condition and in the evaluation of the solution for $t > 0$.

The time step size is an important parameter which may be prescribed as constant or variable. The accuracy of the representation of the time-dependent boundary conditions and the fineness of the mesh for $t > 0$ is affected by the time step size. At a typical time step $n$, the algorithm MESH computes the $x$-coordinates of the prospective nodes at $t = t^{n+1}$ from Equation (17.9) as

$$x_i^{n+1} = x_i^n + kQ/A_i^n \quad i = 1, 2, \ldots, I-1 \tag{17.23}$$

From these nodes, only the ones which satisfy the conditions

$$0 < x_i^{n+1} < L \quad i = 1, 2, \ldots, I-1 \tag{17.24}$$

and

$$x_{i+1}^{n+1} - x_i^{n+1} > 0 \quad i = 1, 2, \ldots, I-1 \tag{17.25}$$

are retained in order to automatically generate the time–space mesh for $t^n \leqslant t \leqslant t^{n+1}$ as shown in Figure 17.1. The computer program has an option to terminate computations if the total number of nodes retained after a number of time steps are either greater than a prescribed upper bound or less than a prescribed lower bound. The computations can be restarted with a suitable number of nodes within the prescribed limits and computed solution at the last time level of the previous calculation is used as the initial condition for the subsequent calculation.

It should be noted that some nodes may leave the region $0 < x < L$ at each or some time steps. The use of triangular elements next to the boundary at $x = 0$ facilitates entrance of one node into the region at each time step. In general, as the time step size decreases, the mesh generated becomes gradually finer with time. Also, by increasing the time step size, the mesh can be made gradually coarser with time.

In the literature, the time step size for convection dominated flow problems with constant $Q$ and $A$ is chosen satisfying the condition

$$\Delta t \leqslant C(A/Q)\Delta x$$

where $C$ (Courant number) is a constant between 0.1 and $1^{5,6}$ and $\Delta x$ and $\Delta t$ are the uniform space and time increment, respectively. In the following section, large time steps are employed for the numerical solution of the example problems to demonstrate that the method is not restricted by the condition $C \leqslant 1$. In applications to pollution problems, the authors have obtained stable solutions with $C = 60.^{12}$

The processing time necessary for the automatic mesh generation is about 5% of the total processing time required to obtain the numerical solution at each time step.

## 17.5  APPLICATIONS

Three test problems are solved numerically employing the finite element method incorporating characteristics (FEMIC) developed in the previous section.

Problem 1 is related to the winter-regime thermal response of the Mississippi River near Cordova, Illinois. Paily and Macagno[2] gave the numerical results obtained from the closed-form solution and the finite differences for the data set:

$$A = 3050 \text{ m}^2 \quad Q = 0.37 \times 10^7 \text{ m}^3 \text{ h}^{-1} \quad E = 4.85 \times 10^5 \text{ m}^2 \text{ h}^{-1}$$

$$\varepsilon = 28.89 \times 10^3 \text{ cal m}^{-2} \text{ h}^{-1} \,{}^\circ\text{C}^{-1} \quad \bar{\eta} = 98.67 \times 10^3 \text{ cal m}^{-2} \text{ h}^{-1}$$

$$B\varepsilon/(\rho c) = 17.62 \text{ m}^2 \text{ h}^{-1}$$

$$T_u = 0.80 \,{}^\circ\text{C} \qquad T_0 = 0 \,{}^\circ\text{C} \qquad T_d = -3.415 \,[1 - \exp(-0.00578t)] \,{}^\circ\text{C}$$

The results from FEMIC for the same data set are illustrated in Figure 17.2 for $t = 5$, 10, 15, 20, 30, 40, and 54 hours. These results are in such good agreement with the closed-form solution given by Paily and Macagno[2] that it is not possible to show the difference to the scale of Figure 17.2. The steady-state temperature distribution is attained at $t = 54$ hours in both of the solutions.

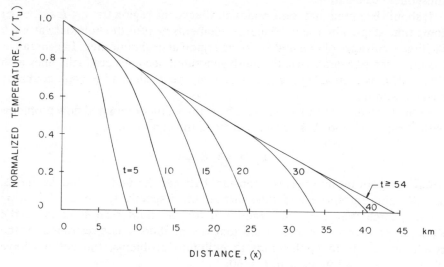

**Figure 17.2**   Problem 1, numerical solution from FEMIC

Initially, the interval $L = 50$ km is discretized by 100 equally spaced nodes. The uniform time step $k = 0.25$ hour is employed in FEMIC.

Problem 2 is chosen as a convection dominated transport problem. The initial condition is prescribed as

$$T_0 = \begin{cases} 1 & 0.10 \leq x/L \leq 0.20 \\ 0 & \text{otherwise} \end{cases}$$

The boundary conditions are

$$T_u = T_d = 0$$

The coefficients in Equation (17.1) are taken as

$$A = 1 \text{ m}^2 \quad Q/L = 5 \times 10^3 \text{ m}^2 \text{ h}^{-1} \quad EL^2 = 1 \text{ h}^{-1}$$

The effect of surface heat flux is neglected by taking

$$S(T) = 0$$

The initial temperature distribution and the numerical results for $t = 4 \times 10^{-5}$ and $10^{-4}$ hours are illustrated in Figure 17.3. Two time steps $k = 10^{-6}$ and $2 \times 10^{-5}$ hours are employed. Initially, the interval $0 < x/L < 1$ is discretized by 100 equally spaced nodes. Numerical results illustrated in Figure 17.3 are in

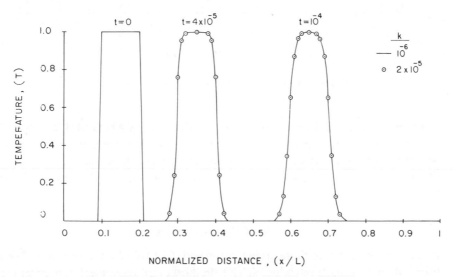

**Figure 17.3** Problem 2, numerical solution from FEMIC

good agreement for both values of the time step and do not exhibit any oscillation, overshoot or distortion of symmetry in this severely convection dominated transport problem.

Problem 3 is a moderately convection dominated transport problem with a variable cross-sectional area. The parameters of the problem are

$$A = (1 + x/L) \times 10^3 \text{ m}^2 \quad Q/L = 200 \text{ m}^2 \text{ h}^{-1} \quad E/L^2 = 3 \times 10^{-3} \text{ h}^{-1}$$

$$S(T) = 0 \quad T_u = 1 \quad T_d = T_0 = 0$$

The results from FEMIC for $t = 1, 3, 5$ hours are illustrated in Figure 17.4. The time step employed is $k = 0.25$ hour and the interval $0 < x/L < 1$ is discretized by 50 equally spaced nodes.

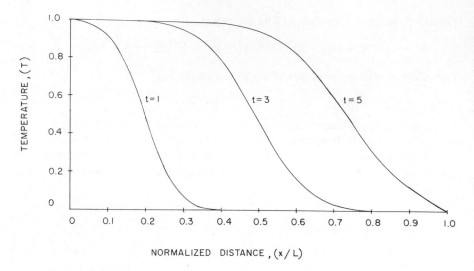

**Figure 17.4**　Problem 3, numerical solution from FEMIC

The numerical results presented were obtained on an AMDAHL 470 V/6-II computer at the University of British Columbia. The total processing time necessary for the generation of the mesh, and the construction and solution of the resulting system of linear equations at each time step was less than 0.05 s for all of the examples.

## 17.6　CONCLUSIONS

A characteristics based finite element method (FEMIC) for the solution of heat transport problems is introduced. The method employs spatial–temporal elements and incorporates the characteristics of the hyperbolic partial differential equation obtained by substituting $E = 0$ into the diffusion convection equation. These characteristics form the sides of the elements joining nodes at consecutive time levels.

The utility and accuracy of the method have been demonstrated by three numerical examples. Numerical results obtained are satisfactory and do not exhibit any oscillations, overshoots or numerical diffusion even for large time steps. The associated computer program has an algorithm to generate automatically a spatially variable mesh at each time step.

### Acknowledgements

Financial assistance under strategic grant No. 67-0215 of the Natural Sciences and Engineering Research Council of Canada is gratefully acknowledged.

# REFERENCES

1. D. L. Book, J. P. Boris and K. Hain (1975), 'Flux corrected transport II—generalization of the method', *J. Comp. Phys.*, **18**, 248–283.
2. P. P. Paily and O. E. Macagno (1976). 'Numerical prediction of thermal-regime of rivers', *ASCE, J. Hydr. Div.*, **102**, 255–274.
3. R. A. Adey and C. A. Brebbia (1974). 'Finite element solution for effluent dispersion', in *Numerical Methods in Fluid Dynamics*, C. A. Brebbia and J. J. Connor (Eds), pp. 325–354, Pentech Press, London.
4. N. D. Brocard and D. R. F. Harleman (1976). 'One-dimensional temperature predictions in unsteady flows', *ASCE, J. Hydr. Div.*, **102**, 227–240.
5. D. C. L. Lam (1977). 'Comparison of finite-element and finite-difference methods for nearshore advection–diffusion transport models', in *Finite Elements in Water Resources*, W. G. Gray, G. F. Pinder and C. A. Brebbia (Eds), pp. 1.115–1.129, Pentech Press, London.
6. C. Ehlig (1972). 'Comparison of numerical methods for solution of the diffusion–convection equation in one or two dimensions', in *Finite Elements in Water Resources*, W. G. Gray, G. F. Pinder and C. A. Brebbia (Eds), pp. 1.91–1.102, Pentech Press, London.
7. J. N. Lillington and I. M. Shepherd (1978). 'Central difference approximations to the heat transport equation', *Int. J. Num. Meth. Engng*, **12**, 1692–1704.
8. I. Christie and A. R. Mitchell (1978). 'Upwinding of high order Galerkin methods in conduction–convection problems', *Int. J. Num. Meth. Engng*, **12**, 1635–1650.
9. D. Spalding (1972). 'A novel finite difference formulation for differential expressions involving both first and second derivatives', *Int. J. Num. Meth. Engng*, **4**, 551–559.
10. A. K. Runchal (1972). 'Convergence and accuracy of three finite-difference schemes for a two-dimensional conduction and convection problem', *Int. J. Num. Meth. Engng*, **4**, 541–550.
11. E. Varoğlu and W. D. L. Finn (1978). 'A finite element method for diffusion–convection equation with constant coefficients', *J. Adv. Water Resources*, **1**, 337–343.
12. E. Varoğlu and W. D. L. Finn (1979). 'Finite elements incorporating characteristics for one-dimensional diffusion–convection equation', *J. Comp. Phys.*, to be published.
13. R. Bonnerot and P. Jamet (1974). 'A second order finite element method for the one-dimensional Stefan problem', *Int. J. Num. Meth. Engng*, **8**, 811–820.

*Numerical Methods in Heat Transfer*
Edited by R. W. Lewis, K. Morgan, and O. C. Zienkiewicz
© 1981 John Wiley & Sons Ltd

## Chapter 18

# Convective Flows in Closed Cavities with Variable Fluid Properties

*E. Leonardi and J. A. Reizes*

### SUMMARY

A compressible vector potential–vorticity formulation of the steady-state equations of motion with variable properties is developed for a Newtonian fluid. These equations include an equation for calculating the pressure. The two-dimensional form of the equations together with the appropriate boundary conditions is solved for vertical and inclined rectangular cavities differentially heated on the sides. It is shown that the conventional parameters, the Rayleigh and Prandtl numbers, aspect ratio and angle of inclination do not fully specify a problem and that an additional six parameters are necessary.

Solutions are presented in the form of flow and temperature fields for air at various Rayleigh numbers, aspect ratios, and angles of inclination. Although the streamline and isotherm patterns are substantially different from those obtained using the Boussinesq approximation, the Nusselt number is not significantly affected.

### 18.1 INTRODUCTION

Natural convection in cavities similar to Figure 18.1 has been extensively studied experimentally by Eckert and Carlson,[1] Elder,[2] and Mynett and Duxbury[3] and numerically, for example, by de Vahl Davis,[4] MacGregor and Emery,[5] Rubel and Landis,[6] Polezhaev,[7] and Mallinson and de Vahl Davis.[8,9] In general there is reasonable agreement between the experimental and theoretical results, but only three of the above authors[5,6,7] have studied numerically the effect of variable properties whilst the remainder have used the Boussinesq approximation. MacGregor and Emery[5] used the Boussinesq approximation and variable viscosity, while Rubel and Landis[6] used a linearized approach and reported results for moderate Rayleigh numbers and temperature differences. Polezhaev[7] solved the complete equations, including the continuity equation, for a square cavity and for one value of the temperature difference between the hot and cold wall. A thorough study of the effect of

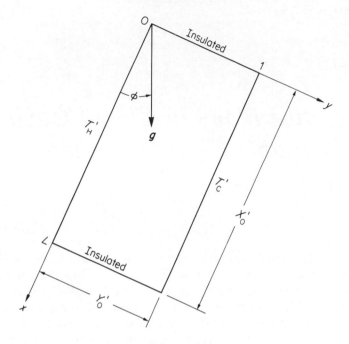

**Figure 18.1** Definition sketch

the various properties was therefore necessary. Since in general only the steady-state results are of interest, only steady flows are considered.

The technique proposed in this chapter avoids the necessity of solving the continuity equation in the case of steady flow and thus can take full advantage of the false transient method[8] for speeding up convergence. Further, the scheme proposed here permits an accurate determination of the very small changes in pressure which occur within the cavity.

The above mentioned method is used in a systematic analysis of the effects of fluid property variations on both the flow and thermal fields for the enclosed cavity, illustrated in Figure 18.1. In this chapter results are presented for vertical cavities and cavities inclined at −60° for aspect ratios of 1 and 2.

## 18.2  MATHEMATICAL FORMULATION

Since non-dimensional parameters are often used to generalize the scope of experimental and numerical results, a non-dimensional approach is used here. But, as mentioned by MacGregor and Emery,[5] and shown later in this section in the case of fluids with variable properties such a generalization does not occur. Further, since the results of the present work are to be compared with numerical and experimental results obtained by others, the conventional parameter, the Rayleigh number, is retained although simpler non-dimensional schemes are possible. With this in mind, the equations describing the motion of a Newtonian fluid may be manipulated from the standard equations[10] to yield,

$$\frac{\partial \rho}{\partial t} + \nabla \cdot (\rho \mathbf{V}) = 0 \qquad (18.1)$$

$$\rho\left(\frac{\partial \mathbf{V}}{\partial t} + \mathbf{V} \cdot \nabla \mathbf{V}\right) = \left(\frac{RaPr}{\varepsilon \eta_R}\right) \rho \hat{\mathbf{g}} - Pr\nabla \times (\mu \nabla \times \mathbf{V}) + 2Pr(\nabla \mathbf{V}) \cdot (\nabla \mu)$$

$$- 2Pr(\nabla \mu)(\nabla \cdot \mathbf{V}) + \nabla\left\{Pr(\lambda + 2\mu)\nabla \cdot \mathbf{V} - \left\{\frac{Pn}{Gn}\left(\frac{RaPr^2}{\varepsilon \eta_R}\right)^{2/3}\right\}p\right\}$$

$$(18.2)$$

and

$$\rho C_P\left(\frac{\partial \theta}{\partial t} + \mathbf{V} \cdot \nabla \theta\right) = \nabla \cdot (k\nabla \theta) + \eta\left(\frac{Pn}{\varepsilon}\right)\left(\frac{\partial p}{\partial t} + \mathbf{V} \cdot \nabla p\right) + \left[Gn\left(\frac{\eta_R^2}{Ra^2 Pr\varepsilon}\right)^{1/3}\right]\Phi$$

$$(18.3)$$

where

$$\rho = \frac{\rho'}{\rho'_R} \qquad \hat{\mathbf{g}} = \frac{\mathbf{g}'}{g'} \qquad t = \frac{t'\alpha'_R}{Y_0'^2} \qquad \alpha'_R = \frac{k'_R}{\rho'_R C'_{PR}},$$

$$V = \frac{V'Y_0'}{\alpha'_R} \qquad \mu = \frac{\mu'}{\mu'_R} \qquad \lambda = \frac{\lambda'}{\mu'_R} \qquad k = \frac{k'}{k'_R}$$

$$Ra = \frac{g'\beta'_R\rho'_R(T'_H - T'_C)Y_0'^3}{\alpha'_R\mu'_R} = \text{Rayleigh number}$$

$$Pr = \frac{\mu'_R}{\rho'_R\alpha'_R} = \text{Prandtl number}$$

$$\varepsilon = \frac{T'_H - T'_C}{T'_R} \qquad \eta = \beta'T' \qquad C_P = \frac{C'_P}{C'_{PR}}$$

$$Pn = \frac{p'_R}{\rho'_R C'_{PR}T'_R} \qquad Gn = \frac{(\mu'_R g'/\rho'_R)^{2/3}}{C'_{PR}T'_R}$$

$$\theta = \frac{T' - T'_R}{T'_H - T'_C}$$

and $\rho$ is the density, $\mathbf{V}$ the velocity vector, $\mathbf{g}$ the gravitational acceleration vector, $\mu$ and $\lambda$ the first and second coefficients of viscosity, $k$ the coefficient of thermal conductivity, $T$ the temperature, $C_P$ the specific heat capacity at constant pressure, $\beta$ the coefficient of volumetric expansion, $p$ the thermo-dynamic pressure, $t$ the time, $\Phi$ is the dissipation function and $\nabla$ is the gradient operator. The prime is used throughout this chapter to denote a dimensional quantity and the subscripts R, C and H refer to the reference state, cold wall and hot wall respectively.

For a given fluid $Pn$, $Gn$, $Pr$, and $\eta_R$ are only functions of the reference state. Further, it may be easily shown that, for a perfect gas $\eta \equiv 1$ and $Pn = (\gamma - 1)/\gamma$, where $\gamma = C'_{PR}/C'_{VR}$, is the ratio of the specific heat capacities.

The curl of Equation (18.2) is taken so that the last term on the right-hand side of Equation (18.2) is eliminated, because

$$\nabla \times \nabla \left\{ Pr(\lambda + 2\mu)\nabla \cdot \mathbf{V} - \left[ \frac{Pn}{Gn}\left( \frac{RaPr^2}{\varepsilon \eta_R} \right)^{2/3} \right]p \right\} \equiv 0$$

The pressure and the bulk viscosity have thus been eliminated from the momentum equation. After some manipulation Equation (18.2) then becomes the vorticity transport equation:

$$\rho\frac{\partial \boldsymbol{\zeta}}{\partial t} = \nabla \times (\rho \mathbf{V} \times \boldsymbol{\zeta}) + \left( \frac{RaPr}{\varepsilon \eta_R} \right)(\nabla \rho) \times \hat{\mathbf{g}} + Pr\, \nabla^2(\mu \boldsymbol{\zeta})$$

$$- Pr\, \nabla(\boldsymbol{\zeta} \cdot \nabla \mu) + 2Pr\, \nabla \times [(\nabla V) \cdot (\nabla \mu)]$$

$$+ 2Pr(\nabla \mu) \times [\nabla(\nabla \cdot \mathbf{V})] - \tfrac{1}{2}(\nabla \rho) \times (\nabla V^2) - (\nabla \rho) \times \frac{\partial \mathbf{V}}{\partial t} \qquad (18.4)$$

where as usual the vorticity, $\boldsymbol{\zeta}$, is defined as,

$$\boldsymbol{\zeta} = \nabla \times \mathbf{V} \qquad (18.5)$$

Since, as mentioned in the introduction, only steady flow is to be investigated in this chapter, $\partial \rho / \partial t \equiv 0$, with the result that Equation (18.1) becomes, $\nabla \cdot (\rho \mathbf{V}) = 0$. Hence $\rho \mathbf{V}$ is a solenoidal vector field, and it is possible to represent it by another vector field such that,

$$\rho \mathbf{V} = \nabla \times \boldsymbol{\xi} \qquad (18.6)$$

The vector field, $\boldsymbol{\xi}$, is not uniquely defined for any solenoidal field since,

$$\boldsymbol{\xi}_1 = \boldsymbol{\xi} + \nabla \psi$$

also satisfies Equation (18.6), where $\psi$ is any scalar field. It is possible to show[11] that this leads to selecting a vector potential which is solenoidal itself, that is,

$$\nabla \cdot \boldsymbol{\xi} = 0 \qquad (18.7)$$

To obtain a relationship between the vorticity, $\boldsymbol{\zeta}$, and the compressible vector potential, $\boldsymbol{\xi}$, the curl of Equation (18.6) is taken, which together with the vector identity,

$$\nabla \times (\nabla \times \boldsymbol{\xi}) \equiv \nabla(\nabla \cdot \boldsymbol{\xi}) - \nabla^2 \boldsymbol{\xi}$$

yields

$$-\rho\zeta + \mathbf{V} \times (\nabla\rho) = \nabla^2 \boldsymbol{\xi} \tag{18.8}$$

if $\boldsymbol{\xi}$ is chosen to satisfy Equation (18.7).

The compressible vector potential, $\boldsymbol{\xi}$, is similar to the vector potential used in solving incompressible three-dimensional flows by Mallinson and de Vahl Davis,[9] and to the stream function used in incompressible two-dimensional flow. The result of such a formulation is to remove the necessity for solving the continuity equation.

Hence for steady compressible flow, Equations (18.1), (18.2), and (18.3) may be replaced by the set of equations, written in conservative form,[12] as

$$0 = \nabla \times (\rho \mathbf{V} \times \boldsymbol{\zeta}) + \left(\frac{RaPr}{\varepsilon\eta_R}\right)(\nabla\rho) \times \hat{\mathbf{g}} + Pr\,\nabla^2(\mu\boldsymbol{\zeta})$$

$$- Pr\,\nabla(\boldsymbol{\zeta} \cdot \nabla\mu) + 2Pr\,\nabla \times [(\nabla\mathbf{V}) \cdot (\nabla\mu)] + 2Pr(\nabla\mu) \times [\nabla(\nabla \cdot \mathbf{V})]$$

$$- \tfrac{1}{2}(\nabla\rho) \times (\nabla V^2) \tag{18.9}$$

$$\nabla^2\boldsymbol{\xi} = -\rho\boldsymbol{\zeta} + \mathbf{V} + \nabla\rho \tag{18.10}$$

and

$$0 = -C_P\nabla \cdot (\rho\mathbf{V}\theta) + \nabla \cdot (k\nabla\theta) + \eta\left(\frac{Pn}{\varepsilon}\right)\mathbf{V} \cdot \nabla p + \left[Gn\left(\frac{\eta_R^2}{Ra^2 Pr\varepsilon}\right)^{1/3}\right]\Phi \tag{18.11}$$

Equations (18.9), (18.10) and (18.11) represent the vorticity–compressible vector potential formulation. For the case of the Boussinesq approximation, that is $\rho$ is constant except in the second term of Equation (18.9) where it becomes $[1 - \beta(T' - T'_R)]$, and $\mu$, $k$, $C_P$, and $\beta$ are assigned constant values, Equations (18.9), (18.10), and (18.11) reduce to the vorticity–vector potential formulation found to be convenient for numerical computations by Mallinson and de Vahl Davis.[9]

Although the above set of equations is sufficient to describe the motion for incompressible flows,[9] it is necessary to introduce additional equations relating the properties, for the compressible variable property case. The particular equations specifying fluid properties depend on the fluid being considered. It is always possible to define a property as a function of two other independent properties[13] so that the density of the fluid may be written as,

$$\rho = \rho'(p', T')/\rho'(p'_R, T'_R) \tag{18.12a}$$

The remaining properties are only weak functions of the pressure with the result that the temperature dominates[14] and the non-dimensional properties become,

$$k = k'(T')/k'(T'_R) \tag{18.12b}$$

$$\mu = \mu'(T')/\mu'(T'_R) \tag{18.12c}$$

$$\lambda = \lambda'(T')/\lambda'(T'_R) \tag{18.12d}$$

and
$$C_P = C'_P(T')/C'_P(T'_R) \tag{18.12e}$$

It follows that unless Equations (18.12a)–(18.12e) are linear $\rho$, $k$, $\mu$, $\lambda$, and $C_P$ can only be determined if the actual values of $p'$, $p'_R$, $T'$ and $T'_R$ are known. For the particular problem under consideration it is therefore necessary to specify the fluid, that is Equations (18.12a)–(18.12e), the temperatures $T'_R$, $T'_H$, $T'_C$, and the pressure $p'_R$. Since,

$$\beta' = -\frac{1}{\rho'}\left(\frac{\partial \rho'}{\partial T'}\right)_{p'} \tag{18.12f}$$

the coefficient of volumetric expansion, $\beta'$, can be derived immediately from Equation (18.12a). It follows that, if $g$ is constant, the only parameter not defined in the Rayleigh number is the length, $Y'_0$; and the parameters $Pn$ and $Gn$ are fully defined. Thus, a particular dimensional problem has been specified for given values of $Ra$, $T'_R$, $T'_C$, and $T'_H$ with the result that the advantage of the non-dimensionalization as a means of generalizing has been lost.

The density is a function of pressure (Equation (18.12a)) so that the pressure field must be known before the density can be evaluated. One possible method would be to integrate the momentum equation (18.2) for the pressure, but unfortunately this leads to results which depend upon the path of integration.[15] If a unique solution is sought the divergence of the momentum equation must be taken[15] leading to,

$$\left[\frac{Pn}{Gn}\left(\frac{RaPr^{1/2}}{\varepsilon\eta_R}\right)^{2/3}\right]\nabla^2 p = \left(\frac{Ra}{\varepsilon\eta_R}\right)\hat{\mathbf{g}}\cdot\nabla\rho - Pr^{-1}\nabla\cdot(\rho\mathbf{V}\cdot\nabla\mathbf{V})$$

$$+ (\nabla\times\zeta)\cdot\nabla(\lambda+\mu)+\nabla^2\lambda(\nabla\cdot\mathbf{V})+\nabla(\lambda+\mu)\cdot\nabla(\nabla\cdot\mathbf{V})$$

$$+ (\lambda+2\mu)\nabla\cdot(\nabla^2\mathbf{V})+\nabla^2\mathbf{V}\cdot\nabla(\lambda+2\mu)+\nabla\cdot(\mathbf{V}\cdot\nabla\nabla\mu)$$

$$+ \nabla\cdot[(\nabla\mathbf{V})\cdot(\nabla\mu)]-\mathbf{V}\cdot\nabla(\nabla^2\mu) \tag{18.13}$$

The set of Equations (18.9), (18.10), (18.11), (18.12a)–(18.12f), and (18.13) with proper boundary conditions are sufficient to obtain solutions, at least in principle. However, three-dimensional numerical solutions are expensive both in terms of computer time and the amount of storage necessary so that only the solution of two-dimensional problems has been attempted.

## 18.3   EQUATIONS FOR TWO-DIMENSIONAL FLOW

The equations describing the two-dimensional convective motion in an enclosure, the boundary conditions and the property relations for air are given in this section.

### 18.3.1 Equations of motion

For two-dimensional flow, Equations (18.9), (18.10), (18.11), and (18.13) become,

$$\frac{\partial \rho u \zeta}{\partial x} + \frac{\partial \rho v \zeta}{\partial y} = Pr\left(\frac{\partial^2 \mu \zeta}{\partial x^2} + \frac{\partial^2 \mu \zeta}{\partial y^2}\right) + \frac{RaPr}{\eta \varepsilon}\left(\frac{\partial \rho}{\partial x}\sin\phi - \frac{\partial \rho}{\partial y}\cos\phi\right)$$

$$+ \left(u\frac{\partial u}{\partial x} + v\frac{\partial v}{\partial x}\right)\frac{\partial \rho}{\partial y} - \left(u\frac{\partial u}{\partial y} + v\frac{\partial v}{\partial y}\right)\frac{\partial \rho}{\partial x}$$

$$+ 2Pr\left[\frac{\partial u}{\partial y}\frac{\partial^2 \mu}{\partial x^2} + \frac{\partial^2 \mu}{\partial x \partial y}\left(\frac{\partial v}{\partial y} - \frac{\partial u}{\partial x}\right) - \frac{\partial v}{\partial x}\frac{\partial^2 \mu}{\partial y^2}\right]$$

$$+ Pr\left[\frac{\partial \mu}{\partial x}\left(\frac{\partial^2 u}{\partial x \partial y} + \frac{\partial^2 v}{\partial y^2}\right) - \frac{\partial \mu}{\partial y}\left(\frac{\partial^2 u}{\partial x^2} + \frac{\partial^2 v}{\partial x \partial y}\right)\right] \qquad (18.14)$$

$$\frac{\partial^2 \xi}{\partial x^2} + \frac{\partial^2 \xi}{\partial y^2} = -\rho\zeta + \frac{1}{\rho}\left(\frac{\partial \xi}{\partial x}\frac{\partial \rho}{\partial x} + \frac{\partial \xi}{\partial y}\frac{\partial \rho}{\partial y}\right) \qquad (18.15)$$

$$\frac{\partial \rho u \theta}{\partial x} + \frac{\partial \rho v \theta}{\partial y} = k\left(\frac{\partial^2 \theta}{\partial x^2} + \frac{\partial^2 \theta}{\partial y^2}\right) + \frac{\partial k}{\partial x}\frac{\partial \theta}{\partial x} + \frac{\partial k}{\partial y}\frac{\partial \theta}{\partial y}$$

$$+ \frac{Pn}{\varepsilon}\eta\left(u\frac{\partial p}{\partial x} + v\frac{\partial p}{\partial y}\right) + Gn\left(\frac{\eta_R^2}{Ra^2 Pr\varepsilon}\right)^{1/3}\Phi \qquad (18.16)$$

and

$$\left[\frac{Pn}{Gn}\left(\frac{RaPr^{1/2}}{\varepsilon \eta_R}\right)^{2/3}\right]\left(\frac{\partial^2 p}{\partial x^2} + \frac{\partial^2 p}{\partial y^2}\right)$$

$$= \frac{Ra}{\varepsilon \eta_R}\left(\frac{\partial \rho}{\partial x}\cos\phi + \frac{\partial \rho}{\partial y}\sin\phi\right)$$

$$- Pr^{-1}\left[\left(u\frac{\partial u}{\partial x} + v\frac{\partial u}{\partial y}\right)\frac{\partial \rho}{\partial x} + \left(u\frac{\partial v}{\partial x} + v\frac{\partial v}{\partial y}\right)\frac{\partial \rho}{\partial y}\right]$$

$$- Pr^{-1}\rho\left[u\left(\frac{\partial^2 u}{\partial x^2} + \frac{\partial^2 v}{\partial x \partial y}\right) + v\left(\frac{\partial^2 u}{\partial x \partial y} + \frac{\partial^2 v}{\partial y^2}\right) + \left(\frac{\partial u}{\partial x}\right)^2 + 2\left(\frac{\partial u}{\partial y}\frac{\partial v}{\partial x}\right) + \left(\frac{\partial v}{\partial y}\right)^2\right]$$

$$+ 2\frac{\partial \lambda}{\partial x}\left(\frac{\partial^2 u}{\partial x^2} + \frac{\partial^2 v}{\partial x \partial y}\right) + 2\frac{\partial \lambda}{\partial y}\left(\frac{\partial^2 v}{\partial y^2} + \frac{\partial^2 u}{\partial x \partial y}\right)$$

$$+ 2\frac{\partial \mu}{\partial x}\left(2\frac{\partial^2 u}{\partial x^2} + \frac{\partial^2 v}{\partial x \partial y} + \frac{\partial^2 u}{\partial y^2}\right) + 2\frac{\partial \mu}{\partial y}\left(2\frac{\partial^2 v}{\partial y^2} + \frac{\partial^2 u}{\partial x \partial y} + \frac{\partial^2 v}{\partial x^2}\right)$$

$$\left(+\frac{\partial^2 \lambda}{\partial x^2} + \frac{\partial^2 \lambda}{\partial y^2}\right)\left(\frac{\partial u}{\partial x} + \frac{\partial v}{\partial y}\right) + (\lambda + 2\mu)\left(\frac{\partial^3 u}{\partial x^3} + \frac{\partial^3 u}{\partial x \partial y^2} + \frac{\partial^3 v}{\partial x^2 \partial y} + \frac{\partial^3 v}{\partial y^3}\right)$$

$$+ 2\left[\frac{\partial^2 \mu}{\partial x^2}\frac{\partial u}{\partial x} + \frac{\partial^2 \mu}{\partial y^2}\frac{\partial v}{\partial y} + \frac{\partial^2 \mu}{\partial x \partial y}\left(\frac{\partial v}{\partial x} + \frac{\partial u}{\partial y}\right)\right] \qquad (18.17)$$

where

$$u = \frac{1}{\rho}\frac{\partial \xi}{\partial y} \quad \text{and} \quad v = -\frac{1}{\rho}\frac{\partial \xi}{\partial x} \tag{18.18}$$

$$\Phi = \lambda\left(\frac{\partial u}{\partial x} + \frac{\partial v}{\partial y}\right)^2 + 2\mu\left[\left(\frac{\partial u}{\partial x}\right)^2 + \left(\frac{\partial v}{\partial y}\right)^2\right] + \mu\left(\frac{\partial u}{\partial y} + \frac{\partial v}{\partial x}\right)^2 \tag{18.19}$$

and

$$p = p_0 + \Delta p \tag{18.20}$$

where $p_0$ is the pressure at the left top corner ($x = 0$, $y = 0$) and $\phi$ is the angle of inclination (Figure 18.1).

In deriving Equation (18.16) it has been assumed that the variation in the values of the specific heat capacity at constant pressure, $C_P$, is small in comparison with other property variations[16] and that little error is therefore introduced for moderate temperature and pressure changes. This however is not true for the other properties.

### 18.3.2  Fluid properties

The only fluid which will be considered in this chapter is air. The equation of state for air at moderate temperatures and pressures is adequately represented by the perfect gas equation which when non-dimensionalized becomes,

$$\rho = \frac{p}{\varepsilon\theta + 1} \tag{18.21}$$

From Equation (18.17) it is possible to calculate the value of $\Delta p$. Since the actual pressure has to be known in Equation (18.21), $p_0$ has to be evaluated and substituted in Equation (18.20).

Now, if it is assumed that the initial conditions in the cavity are $p'_R$ and $T'_R$ and $T'_R$ and that the mass of gas in the cavity remains constant it follows that,

$$\frac{1}{L}\int_A \rho \, dA = 1 \tag{18.22}$$

where $L$ is the aspect ratio defined in Figure 18.1 and $A$ the area of the cavity. The substitution of Equations (18.21) and (18.20) into Equation (18.22) leads to,

$$\frac{1}{L}\int_A \frac{p_0 + \Delta p}{\varepsilon\theta + 1} \, dA = 1 \tag{18.23}$$

Since $p_0$ is a point value, Equation (18.23) can be rewritten as,

$$p_0 = \frac{L - \int_A [\Delta p/(\varepsilon\theta + 1)] \, dA}{\int_A dA/(\varepsilon\theta + 1)} \tag{18.24}$$

So that $p$ can be evaluated once the $\Delta p$ field is known.

Because for gases far from the critical state the first coefficient of viscosity and the coefficient of thermal conductivity differ only by a constant multiplier,[14] one equation will represent both properties, when written in non-dimensional form. The Sutherland formula has been extensively used for evaluating the first coefficient of viscosity and the coefficient of thermal conductivity[14] and may be written in non-dimensional form as,

$$k = \mu = \left(\frac{s}{\varepsilon\theta + s}\right)(\varepsilon\theta + 1)^{1.5} \tag{18.25}$$

where $s = 1 + 1.47 T'_{bp}/T'_R$ ; $T_{bp}$ is the temperature at boiling point.

There are considerable variations in the measured values of $\lambda$, the second coefficient of viscosity[17] and no empirical relationship has been found in the literature. Numerical studies carried out by the present authors show no detectable change in any of the parameters when $\lambda$ was changed from 0 to 50, so that the second coefficient of viscosity has been neglected in this study. There remains the need to specify the boundary conditions in order to obtain a solution.

### 18.3.3 Boundary conditions

The boundaries at $y = 0$ and $y = 1$ (Figure 18.1) are at constant temperatures $\theta = 0$ and $\theta = 1$ respectively. Since the other boundaries are adiabatic, $\partial\theta/\partial x = 0$ on $x = 0$ and $x = L$.

All the walls of the cavity are stationary, non-slip, and impermeable so that,

$$u_b = v_b = 0 \tag{18.26a}$$

and

$$\xi_b = \text{constant} \tag{18.26b}$$

where the subscript b indicates the boundary. Since the value of the constant $\xi_b$ is arbitrary, $\xi_b = 0$ was chosen.

The substitution of Equations (18.26) and (18.18) into Equation (18.15) yields the vorticity boundary condition:

$$\zeta_b = -\frac{1}{\rho}\left(\frac{\partial^2 \xi}{\partial n^2}\right)_b \tag{18.27}$$

where $n$ is the coordinate normal to the boundary.

It is not possible to write the boundary conditions for the pressure explicitly. Only the gradients can be calculated from Equation (18.2). Substitute Equation (18.26) into the two-dimensional form of Equation (18.2) to give the boundary conditions for the pressure:

$$\left[\frac{Pn}{Gn}\left(\frac{RaPr^2}{\varepsilon\eta_R}\right)^{2/3}\right]\left[\frac{\partial\Delta p}{\partial x}\right]_{x=0;L} = \rho\left(\frac{RaPr}{\varepsilon\eta_R}\right)\cos\phi + Pr\left[\frac{\partial u}{\partial x}\left(\frac{\partial\lambda}{\partial x}+2\frac{\partial\mu}{\partial x}\right)\right.$$
$$\left.+(\lambda+2\mu)\left(\frac{\partial^2 u}{\partial x^2}+\frac{\partial^2 v}{\partial x\,\partial y}\right)+\frac{\partial\mu}{\partial y}\frac{\partial v}{\partial x}-\mu\frac{\partial\zeta}{\partial y}\right]$$

(18.28a)

and

$$\left[\frac{Pn}{Gn}\left(\frac{RaPr^2}{\varepsilon\eta_R}\right)^{2/3}\right]\left[\frac{\partial\Delta p}{\partial y}\right]_{y=0;1} = \rho\left(\frac{RaPr}{\varepsilon\eta_R}\right)\sin\phi + Pr\left[\frac{\partial v}{\partial y}\left(\frac{\partial\lambda}{\partial y}+2\frac{\partial\mu}{\partial y}\right)\right.$$
$$\left.+(\lambda+2\mu)\left(\frac{\partial^2 u}{\partial x\,\partial y}+\frac{\partial^2 v}{\partial y^2}\right)+\frac{\partial\mu}{\partial x}\frac{\partial u}{\partial y}+\mu\frac{\partial\zeta}{\partial x}\right]$$

(18.28b)

There only remains the need to find a suitable numerical scheme for the solution of the problem.

## 18.4  NUMERICAL SOLUTION OF THE EQUATIONS

The finite difference approximations (FDA) for Equations (18.15) and (18.17) are solved using a combined Fourier analysis–fast Fourier transform (FA–FFT) direct method.[18,19] However, in the present authors' experience, solutions of the FDA for Equations (18.14) to (18.17) tend to diverge if the FA–FFT method is used. Successive over-relaxation or a time derivative has to be introduced in Equations (18.14) and (18.16) to make the solution converge. The fast convergence properties of the method of false transients[8] have therefore been used. The essential feature of the method is the addition of the false transient terms,

$$\frac{Pr}{\alpha_\zeta}\rho\frac{\partial\zeta}{\partial t} \quad \text{and} \quad \frac{1}{\alpha_\theta}\rho\frac{\partial\theta}{\partial t}$$

to the left-hand side of the vorticity transport equation (18.14) and energy equation (18.16), respectively. The coefficients $\alpha_\zeta$ and $\alpha_\theta$ are false transient factors which are used to enhance the rate of convergence. The FDA of the 'false transient' form of Equations (18.14) and (18.16) are solved using the Samarskii–Andreyev ADI scheme.[20]

The finite difference approximations were obtained by replacing the 'false transient' terms with a forward difference in time and the spatial derivatives with second-order central difference approximations. The cross derivative terms in Equations (18.14) and (18.16) were approximated using[15]

$$(\delta^2_{xy}\psi)_{i,j} = (\psi_{i+1,j+1} - \psi_{i+1,j-1} - \psi_{i-1,j+1} + \psi_{i-1,j-1})/4\,\Delta x\,\Delta y$$

$$(\delta^3_{x^2y}\psi)_{i,j} = [\psi_{i+1,j+1} - \psi_{i+1,j-1} - 2(\psi_{i,j+1} - \psi_{i,j-1}) + \psi_{i-1,j+1} - \psi_{i-1,j-1}]/2\,\Delta x^2\,\Delta y$$

where $i, j$ are the mesh point subscripts and $\Delta x$, $\Delta y$ are the mesh intervals, for the $x$- and $y$-directions respectively, and $\psi$ is any transport property. The derivative boundary conditions for the pressure are evaluated from the finite difference approximations of Equation (18.28), where second-order accurate forward difference approximations have been used for the normal derivatives and central difference approximations for the tangential derivatives, and the cross derivatives are given by,

$$(\delta^2_{xy}\psi)_{1,j} = (-\psi_{3,j+1} + 4\psi_{2,j+1} - 3\psi_{1,j+1} + \psi_{3,j-1} - 4\psi_{2,j-1} + 3\psi_{1,j-1})/4\,\Delta x\,\Delta y$$

on the $x = 0$ boundary with similar results on the other boundaries.

Boundary conditions on $\xi$ and $\theta$ are applied in the usual manner, using central differences and image points for derivative conditions. Vorticity boundary conditions can be obtained from either a Taylor's series expansion of Equation (18.27),[21] or from the assumption that the variations of vorticity[22] and density are linear between the boundary and the first mesh point. Either procedure leads to the same result in which the boundary condition, Equation (18.27), on say $x = 0$, becomes

$$\zeta_{1,j} = -\left(\frac{12}{\Delta x^2}\xi_{2,j} + \zeta_{2,j}(\rho_{1,j} + \rho_{2,j})\right)(3\rho_{1,j} + \rho_{2,j})^{-1} \qquad (18.29)$$

Similar equations can easily be written for the other boundaries.

## 18.5   RESULTS AND DISCUSSION

The specification of the Rayleigh number, $Ra$, Prandtl number, $Pr$, aspect ratio, $L$, and angle of inclination, $\phi$, fully defines the problem for the case in which the Boussinesq approximation is invoked. Unfortunately with variable properties, the use of these four parameters is not sufficient to fully specify the problem; in fact, ten parameters are required, namely the fluid, $Ra$, $Pr$, $L$, $\phi$, $Gn$, $Pn$, $T'_R$, $T'_H$, and $T'_C$. Once the fluid and the reference temperature have been selected $Gn$, $Pn$, and $Pr$ are fixed. Therefore the fluid, $Ra$, $L$, $T'_R$, $T'_H$, and $T'_C$ need to be specified. The three parameters, $T'_H$, $T'_C$, $T'_R$, can be reduced to two if the reference temperature, $T'_R$, is specified in terms of the wall temperatures, $T'_H$ and $T'_C$. The definition of the reference density and the assumptions used in deriving Equation (18.24) in which $p_0$ is evaluated, lead to the conclusion that initially the whole cavity is at a temperature $T'_R$ and pressure $p'_R$ ($p'_R$ need not be specified). It would seem obvious to define the reference temperature as $T'_R = (T'_H + T'_C)/2$. However, it was shown[23] that for air and this definition of

$T'_R$ the pressure in the cavity is always less than $p'_R$. Although not important for gases, a drop in pressure in the case of liquids could lead to the formation of vapour bubbles which would eventually aglomerate at the top of the cavity. Since in the long term, it is intended to study liquids as well as gases, the reference temperature has to be so defined as to avoid boiling. $T'_R = T'_C$ was chosen since other workers[1,4,7,24] have used it and because no pressure drop occurs, as may be seen in Figure 18.2.

There are few references in the literature to pressure fields in natural convection. Polezhaev[7] mentions that the pressure varies approximately linearly from the top to the bottom of the cavity, but does not specify the magnitude of the variations. In this study the maximum $\Delta p$ found in any problem examined was $\Delta p = 7 \times 10^{-5}$; a pressure change which would be difficult to detect using Polezhaev's technique. Hence the pressure variations in the cavity can be neglected, although $p_0$ changes with $\varepsilon$ (Figure 18.2).

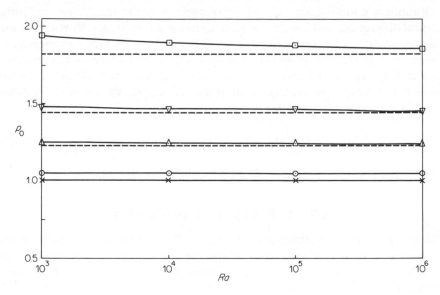

**Figure 18.2**   Corner pressure, $p_0$, as a function of Rayleigh number, $Ra$, for various values of $\varepsilon$: $\times$, $\varepsilon = 0.01$; $\odot$, $\varepsilon = 0.1$; $\triangle$, $\varepsilon = 0.5$; $\triangledown$, $\varepsilon = 1$; $\square$, $\varepsilon = 2$. The chain curves represent values calculated from Equation (18.30)

Suppose that a linear temperature distribution which is only a function of $y$ is imposed and that the pressure is constant within the cavity. It may be shown from Equation (18.24) that,

$$p_0 = \left( \int_0^1 \frac{dy}{\varepsilon + 1 - \varepsilon y} \right)^{-1} = \frac{\varepsilon}{\ln(\varepsilon + 1)} \tag{18.30}$$

so that $p_0 > 1$. As may be seen in Figure 18.2, the numerical results are in excellent agreement with those obtained from Equation (18.30).

For air, any particular solution can therefore be obtained for a given Rayleigh number, cold wall temperature, and $\varepsilon$, which because of the definition of the reference temperature is now $\varepsilon = (T'_H - T'_C)/T'_C$. Note that the Prandtl number, $Pr$, is no longer an independent parameter and this is a significant difference from the Boussinesq approximation approach.

Variations in reference temperature, $T'_R$, result in only small changes to the solution. This is explained by the fact that the non-dimensional density, $\rho$, is independent of $T'_R$, Equation (18.21), and that the non-dimensional coefficients of viscosity and conductivity are weakly dependent on $T'_R$, Equation (18.25). For air ($T'_{bp} = 78$ K), it can easily be shown that for a change in $T'_R$ from 288 K to 576 K (a variation of 100%) the variation in the coefficients of viscosity and conductivity at the hot wall, for $\varepsilon = 2$, is only 8%, so that one value of $T'_R$ need be used.

Results are presented for $T'_C = T'_R = 288$ K, $10^3 \leqslant Ra \leqslant 10^6$, and $0 \leqslant \varepsilon \leqslant 2$. For air at $T'_R = 288$ K, $Gn = 9.34 \times 10^{-9}$, $Pn = 0.28$, and $Pr = 0.72$. When $Ra \leqslant 10^4$, a $17 \times 17$ rectangular grid was used for $L = 1$ and a $33 \times 17$ rectangular grid for $L = 2$. For $Ra > 10^4$, a $33 \times 33$ grid, and $65 \times 33$ grid was used for $L = 1$ and $L = 2$, respectively.

From Figure 18.3 it can be seen that the use of the Boussinesq approximation (BA) instead of the perfect gas equation of state always leads to an overestimate of the density change. Hence stronger convective flows are predicted with the use of BA, as may be seen in Figures 18.4(a) and (b), in which the

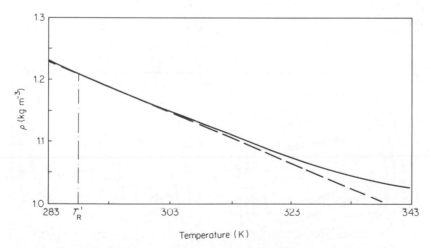

**Figure 18.3** Density, $\rho'$, as a function of temperature, $T'$, at constant pressure of one atmosphere. The full curve is obtained from the perfect gas equation of state. The chain curve is that used in the Boussinesq approximation

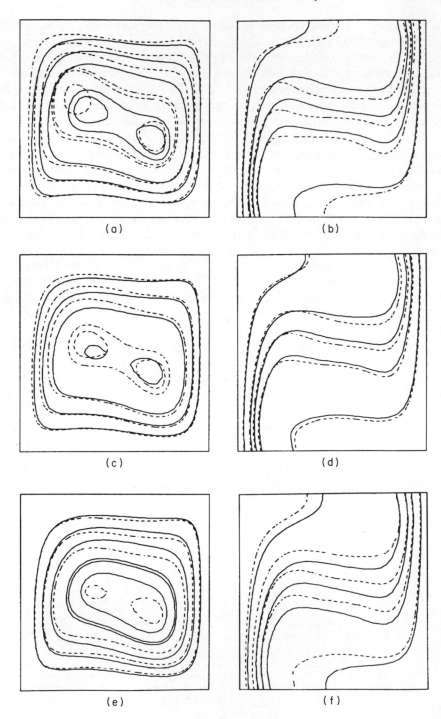

solution with the density as the only variable property is compared with the solution using BA.

Two secondary cells appear in both the solutions, but the symmetry apparent in the solution using BA is not present when the variable density is introduced. The motion has shifted to the lower right corner as noted by other workers.[6,7] This may be explained by the fact that the density in the hot region is lower than the density in the cold region, whilst the velocities do not change proportionately. Therefore the streamlines must move closer together in the cold region, resulting in the shift. The isotherms have been shifted towards the centre of the cavity (Figure 18.4(b)) due to the lower rate of the heat transfer.

When variable viscosity is introduced as well as density, Figures 18.4(c) and (d), the flow is further reduced since the viscosity is then always greater than in the case of constant viscosity. The secondary cells are more distorted, with the left cell considerably reduced in size due to the large increase in viscosity in this high temperature region. The isotherms have been shifted further towards the centre of the cavity (Figure 18.4(d)).

With all properties variable (i.e. variable conductivity has also been introduced) the motion is increased due to larger heat transfer into the cavity (Figure 18.4(e)). The two secondary cells are smeared into one cell; a considerable difference from the BA solution in Figure 18.4(a). The isotherms now shift towards the cold wall, due to the increased heat transfer. The increased rate of heat transfer means that a higher temperature gradient must exist at the cold wall near which the coefficient of thermal conductivity is not changed. Whereas near the hot wall the increase in the coefficient of thermal conductivity is greater than the decrease in the temperature gradient with a resultant increase in the rate of heat transfer. This reduces the effects of variable viscosity and density on the thermal field. Thus *all* properties must be variable; if only some are variable misleading results may be obtained.[6] However, as may be expected, at low $Ra$, where conduction dominates, it is the coefficient of conductivity that has the greatest effect.

It may be seen in Figure 18.5, even at very low Rayleigh numbers ($Ra = 10^3$), $\varepsilon$ has a marked effect on the flow and thermal fields. As $\varepsilon$ is increased the motion is reduced (Figures 18.5(a), (c), (e), and (g)) since the viscosity increases as $\varepsilon$ increases. As mentioned previously the streamlines are shifted towards the cold wall. The isotherms are straightened, as $\varepsilon$ increases, due to the reduced rate of convection, but are moved towards the cold wall because of the increased rate of heat conduction (Figures 18.5(b), (d), (f), and (h)).

---

**Figure 18.4** Streamlines ((a), (c), (e)) and isotherms ((b), (d), (f)) for a vertical cavity $L = 1$, $Ra = 10^5$, $\varepsilon = 1$ indicating the effect of the progressive introduction of variable properties. (a) and (b): chain lines are for the Boussinesq approximation, full lines are for variable density only; (c) and (d): chain lines are for variable density only, full lines are for variable density and viscosity; (e) and (f): chain lines are for variable density and viscosity, full lines are for variable density, viscosity, and conductivity

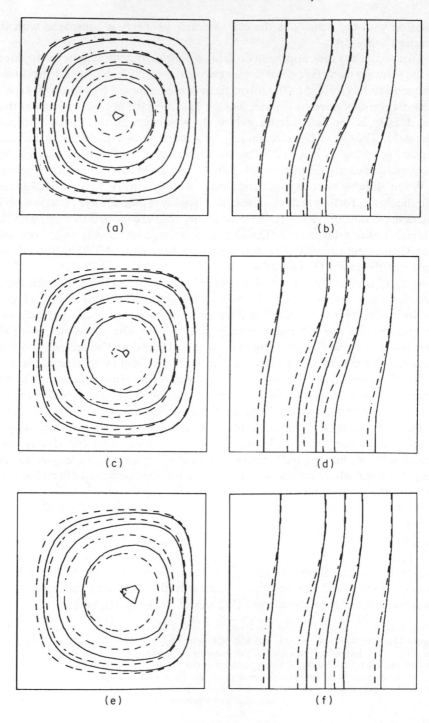

(a)

(b)

(c)

(d)

(e)

(f)

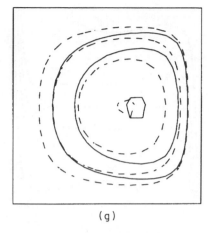
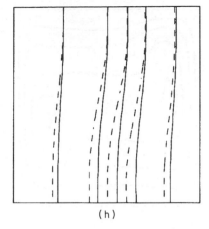

(g)            (h)

**Figure 18.5** Streamlines ((a), (c), (e), (g)) and isotherms ((b), (d), (f), (h)) for a vertical cavity $L = 1$, $Ra = 10^3$ at various values of $\varepsilon$; (a) and (b): chain lines are for the Boussinesq approximation, full lines, $\varepsilon = 0.1$; (c) and (d): chain lines, $\varepsilon = 0.1$, full lines, $\varepsilon = 0.5$; (e) and (f): chain lines, $\varepsilon = 0.5$, full lines, $\varepsilon = 1$; (g) and (h): chain lines, $\varepsilon = 1$, full lines, $\varepsilon = 2$

At high Rayleigh numbers ($Ra = 10^6$), where convection dominates, the cells are moved to the right and the small secondary cell which appears in the centre of the cavity for the BA and low $\varepsilon$ solutions, disappears (Figures 18.6(a), (c), (e), and (g)). The isotherms, on the other hand, are mainly affected near the boundaries (Figures 18.6(b), (d), (f), and (h)).

Similar effects are obtained for the flow and thermal fields for the other aspect ratio studied so that only $L = 1$ is presented for the vertical cavity and $L = 2$ is presented for the case of an inclined cavity.

The effect of $\varepsilon$ on the solution for $Ra = 10^5$, $L = 2$, and inclined at $-60°$ is illustrated in Figure 18.7. As mentioned above, the effect of increasing $\varepsilon$ is to reduce the rate of convection within the cavity. The motion is shifted to the right and up, to a region of lower viscosity (Figures 18.7(a), (c), (e) and (g)) and the core is reduced. Since convection is reduced, the isotherms near the centre of the cavity are less and less distorted as $\varepsilon$ is increased, but are shifted towards the centre near the boundaries due to the increase in heat conduction.

The average Nusselt number is a measure of the heat transferred through the cavity, since the average Nusselt number based on the width, $Nu_w$, is defined as,

$$Nu_w = \dot{q}'/L(T'_H - T'_C)k' \tag{18.31}$$

where $\dot{q}'$ is the heat transferred per unit depth of the cavity. Equation (18.31) may be written in non-dimensional form as,

$$Nu_w = \dot{q}/\varepsilon L \tag{18.32}$$

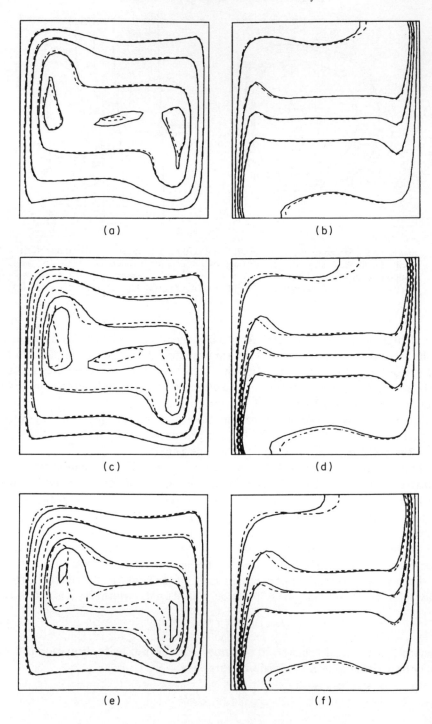

(a)  (b)

(c)  (d)

(e)  (f)

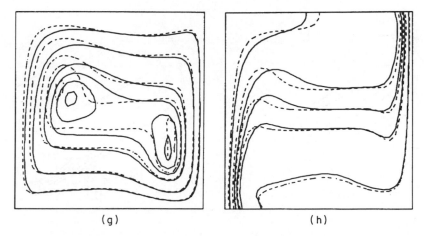

(g)                                         (h)

**Figure 18.6** Streamlines ((a), (c), (e), (g)) and isotherms ((b), (d), (f), (h)) for a vertical cavity $L = 1$, $Ra = 10^6$ at various values of $\varepsilon$; (a) and (b): chain lines are for the Boussinesq approximation, full lines, $\varepsilon = 0.1$; (c) and (d): chain lines, $\varepsilon = 0.1$, full lines, $\varepsilon = 0.5$; (e) and (f): chain lines, $\varepsilon = 0.5$, full lines, $\varepsilon = 1$; (g) and (h): chain lines, $\varepsilon = 1$, full lines, $\varepsilon = 2$

$\dot{q}$ in Equation (18.32) may be calculated from either the temperature gradient on the heated and cooled boundaries or from the rate of energy transport (i.e. the algebraic sum of the convection and conduction of heat) across any plane parallel to the hot and cold wall. Equation (18.32) becomes

$$Nu_w = -\frac{1}{L} \int_0^L k \left( \frac{\partial \theta}{\partial x} \right)_{y=0;1} dx \qquad (18.33a)$$

when the temperature gradients at the boundaries are used, and

$$Nu_w = \frac{1}{L} \int_0^L \left( \rho v \theta - k \frac{\partial \theta}{\partial y} \right) dx, \qquad (18.33b)$$

for a section through the cavity.

In the case of the BA,

$$\theta(x, y) = 2\theta_m - \theta(L - x, 1 - y) \quad \text{and} \quad v(x, y) = -v(L - x, 1 - y) \qquad (18.34)$$

where $2\theta_m = \theta_H + \theta_C$. Hence the average Nusselt number must be the same on the hot and cold boundaries, but its value depends on the approximation used for $\partial \theta / \partial y$ in Equation (18.33a).[23,25] Further, even for the case of BA the values of $Nu_w$ calculated at different sections in the cavity with the use of Equation (18.33b) *differ*,[26] and are not those obtained from Equation (18.33a).[27] This variation in 'sectional' Nusselt number is illustrated in Figure 18.8. The average Nusselt number increases to a peak near the hot wall, dropping to an almost constant value in the central region and then rising to another smaller

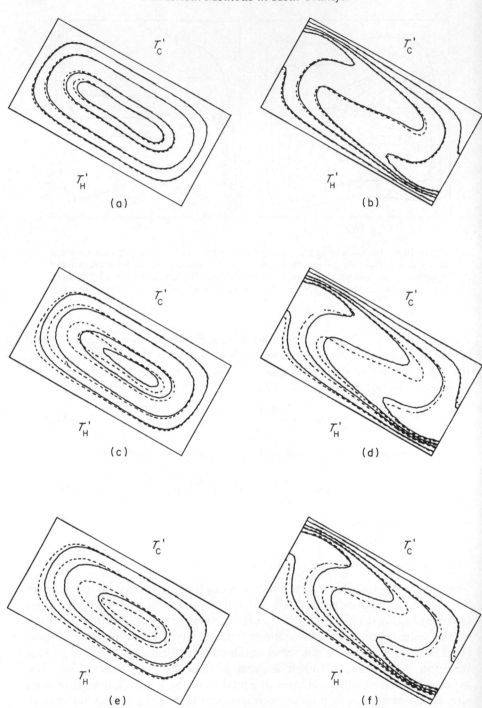

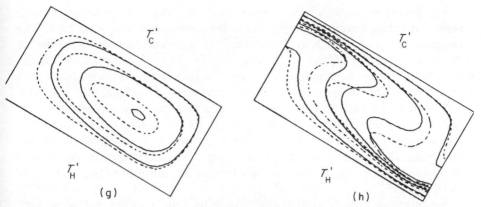

**Figure 18.7** Streamlines ((a), (c), (e), (g)) and isotherms ((b), (d), (f), (h)) for a cavity inclined at 60° to the vertical, $L = 2$, $Ra = 10^5$ at various values of $\varepsilon$; (a) and (b): chain lines are for the Boussinesq approximation, full lines, $\varepsilon = 0.1$; (c) and (d): chain lines, $\varepsilon = 0.1$, full lines, $\varepsilon = 0.5$; (e) and (f): chain lines, $\varepsilon = 0.5$, full lines, $\varepsilon = 1$; (g) and (h): chain lines, $\varepsilon = 1$, full lines, $\varepsilon = 2$

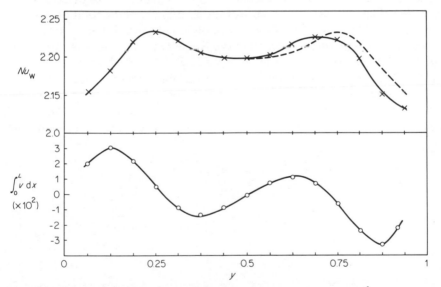

**Figure 18.8** Distribution of Nusselt number, $Nu_w$, and net volumetric flux, $\int_0^L v \, dx$, across a vertical cavity, $L = 1$, $Ra = 10^4$. Solution obtained using the Boussinesq approximation. The chain curve is obtained from Equation (18.35)

peak near the cold wall. Thus the distribution of average Nusselt number is not symmetric even when the BA is used. Using Equations (18.33b) and (18.34), it can be shown that for the BA,

$$Nu_w(y) = Nu_w(1-t) - \frac{2\theta_m}{L} \int_0^L v(L-x, 1-y) \, dx \qquad (18.35)$$

The second term on the right-hand side of Equation (18.35) is not equal to zero, but varies across the cavity and is the lower curve in Figure 18.8. The inclusion of this term leads to symmetrical results as may be seen on the dotted curve in Figure 18.8. Although Thomas[27] had a similar distribution of $Nu_w$, he did not explain the asymmetry. Roux *et al.*[26] show symmetric results because in their work the non-dimensional temperature is defined as $\theta(x, y) = -\theta(L - x, 1 - y)$, so that $\theta_m$ is zero, hence in their case the integral does not appear. The present authors, Thomas, and Roux *et al.* found that a reduction in mesh size and $Ra$ reduced the variation in Nusselt number across the cavity. Hence, in addition to the use of the variation in $Nu_w$ recommended by Roux *et al.*, the value of $\int_0^L v \, dx$ can be used as a qualitative indicator of the accuracy of the mesh used.

The symmetry mentioned above (Equation (18.34)) does not occur with variable properties so that the Nusselt numbers are no longer the same even on the hot and cold boundaries. It has been found that the Nusselt number on the hot wall is always larger than the Nusselt number on the cold wall.[3,7,23] The variation in sectional Nusselt number follows a similar trend to those observed in the BA (Figure 18.8), but because of the asymmetry of the solution, a similar expression to Equation (18.35) cannot be obtained. A choice has to be made as to which section to use for evaluating the Nusselt number. Because for the BA the flow field is skew symmetric, at $y = 0.5$, $\int_0^L v \, dx \equiv 0$, the mid-section was chosen. This section was also used for the variable property solutions.

The present authors,[23] found that when the reference temperature was defined as $T'_R = (T'_H + T'_C)/2$, no satisfactory correlation between $Nu_w$ and $Ra$ was obtained, since the $Nu_w$ was also a function of $\varepsilon$. On the other hand, when

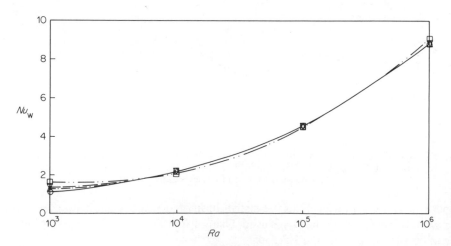

**Figure 18.9** Nusselt number, $Nu_w$, as a function of Rayleigh number, $Ra$, for a vertical cavity, $L = 1$, at various values of $\varepsilon$: —— Boussinesq approximation, $\odot$—— $\varepsilon = 0.1$, $\triangle$— · — $\varepsilon = 0.5$, $\triangledown$——· $\varepsilon = 1$, $\square$—·· $\varepsilon = 2$

$T'_R = T'_C$ is used, a good correlation between $Nu_w$ and $Ra$ is obtained, except at low Rayleigh numbers where conduction dominates. This was also observed by Eckert and Carlson[1] and Ostrach,[24] who found that 'if the fluid properties in the Nusselt number and Rayleigh number are evaluated at wall conditions, then these parameters correlate the data well'.[24] Further, it is presumed that, Polezhaev[7] in correlating the Nusselt and Rayleigh numbers, used fluid properties evaluated at both the cold and hot walls to obtain the best fit.

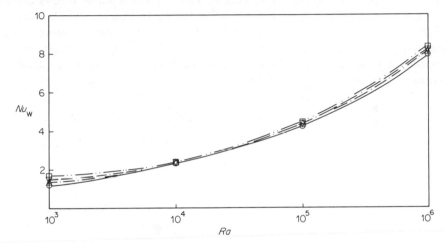

**Figure 18.10** Nusselt number, $Nu_w$, as a function of Rayleigh number, $Ra$, for a vertical cavity, $L = 2$, at various values of $\varepsilon$: —— Boussinesq approximation, $\odot$—— $\varepsilon = 0.1$, $\triangle$—·— $\varepsilon = 0.5$, $\triangledown$——· $\varepsilon = 1$, $\square$—·· $\varepsilon = 2$

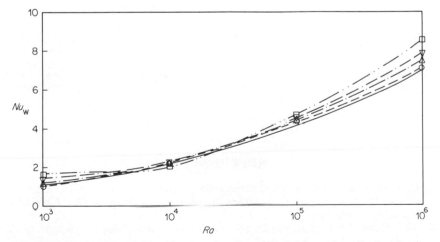

**Figure 18.11** Nusselt number, $Nu_w$, as a function of Rayleigh number, $Ra$, for a cavity inclined at 60° to the vertical, $L = 1$, at various values of $\varepsilon$: —— Boussinesq approximation, $\odot$—— $\varepsilon = 0.1$; $\triangle$—·— $\varepsilon = 0.5$, $\triangledown$——· $\varepsilon = 1$, $\square$—·· $\varepsilon = 2$

The average Nusselt number, evaluated at the mid-plane, is plotted against Rayleigh number for varying $\varepsilon$, two aspect ratios and angles of inclination in Figures 18.9, 18.10, 18.11, and 18.12. It is surprising to note that although the flow and thermal fields for the case of the BA and variable properties differ considerably (Figures 18.5, 18.6, 18.7) the Nusselt numbers are not significantly affected except for low $Ra$. For a vertical cavity with $Ra \geqslant 10^4$ the BA results for $Nu_w$ are in very good agreement with those obtained using variable properties (Figures 18.9, 18.10). The Nusselt number results using BA, for an inclined cavity with $Ra \geqslant 10^4$ differ by approximately 10% from those obtained using variable properties. If $Nu_w$ is the only parameter of interest, then for air and $10^4 \leqslant Ra \leqslant 10^6$, the Boussinesq approximation can be considered valid and the substantial complications of introducing variable properties avoided.

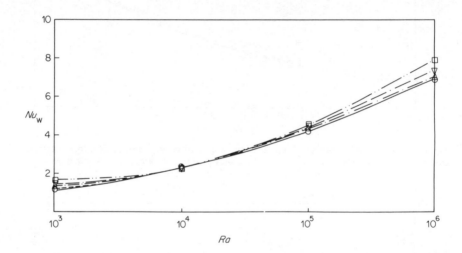

**Figure 18.12** Nusselt number, $Nu_w$, as a function of Rayleigh number, $Ra$, for a cavity inclined at 60° to the vertical, $L = 2$, at various values of $\varepsilon$: —— Boussinesq approximation, $\odot$—— $\varepsilon = 0.1$, $\triangle$—·— $\varepsilon = 0.5$, $\triangledown$——· $\varepsilon = 1$, $\square$—·· $\varepsilon = 2$

## REFERENCES

1. E. R. G. Eckert and W. O. Carlson (1961). 'Natural convection in an air layer enclosed between two vertical plates with different temperatures', *Int. J: Heat Mass Transfer*, **2**, 106–120.
2. W. J. Elder (1965). 'Laminar free convection in a vertical slot', *J. Fluid Mech.*, **23**, 77–98.
3. J. A. Mynett and D. Duxbury (1974). 'Temperature distributions within enclosed plane air cells associated with heat transfer by natural convection', *Proc. 5th Int. Heat Transfer Conf., Tokyo*, NC3.8, pp. 119–123.

4. G. de Vahl Davis (1968). 'Laminar natural convection in an enclosed rectangular cavity', *Int. J. Heat Mass Transfer*, **11**, 1675–1693.
5. R. K. MacGregor and A. F. Emery (1969). 'Free convection through vertical plane layers—moderate and high Prandtl number fluids', *J. Heat Transfer*, 91, *Series C*, **3**, 391–403.
6. A. R. Rubel and F. Landis (1970). 'Laminar natural convection in a rectangular enclosure with moderately large temperature differences', in *Heat Transfer 1970*, NC2.10, pp. 1–11, Publishers VDI, Dusseldorf.
7. V. I. Polezhaev (1967). 'Numerical solution of a system of two dimensional unsteady Navier–Stokes equations for a compressible gas in a closed region', *Fluid Dynamics*, **2**, 70–74.
8. G. D. Mallinson and G. de Vahl Davis (1973). 'The method of the false transient for the solution of coupled elliptic equations', *J. Comp. Phys.*, **12**, 435–461.
9. G. D. Mallinson and G. de Vahl Davis (1977). 'Three-dimensional natural convection in a box: a numerical study', *J. Fluid Mech.*, **83**, 1–31.
10. L. Howarth (1959). 'Laminar boundary layers', in *Handbuch der Physik VIII/1*, S. Flügge (Ed.), pp. 261–350, Springer-Verlag, Berlin.
11. L. I. Sedov (1972). *A Course in Continuum Mechanics*, Vol. 3, Wolters-Noordhoff Publishing, Groningen.
12. G. D. Mallinson (1973). 'Natural convection in enclosed cavities', *Ph.D. Thesis*, University of New South Wales, Sydney, Australia.
13. L. C. Woods (1975). *The Thermodynamics of Fluid Systems*, Clarendon Press, Oxford.
14. S. Bretsznajder (1971). *Prediction of Transport and other Physical Properties of Fluids*, Vol. 2, Pergamon Press, Oxford.
15. P. J. Roache (1972). *Computational Fluid Dynamics*, Hermosa Publishers, Albuquerque, New Mexico.
16. Y. R. Mayhew and G. F. C. Rogers (1973). *Thermodynamic and Transport Properties of Fluids*, 2nd edn, Basil Blackwell, Oxford.
17. L. Rosenhead (1954). 'A discussion of the first and second viscosities of fluids', *Proc. R. Soc. Lond. A*, *266*, 1–69.
18. R. C. Le Bail (1972). 'Use of fast Fourier transforms for solving partial differential equations in physics', *J. Comp. Phys.*, **9**, 440–465.
19. J. W. Cooley, A. A. Lewis and P. D. Welch (1970). 'The fast Fourier transform algorithm: programming considerations in the calculation of sine, cosine and laplace transforms', *J. Sound Vib.*, **12**, 315–337.
20. A. A. Samarskii and V. B. Andreyev (1963). 'On a high-accuracy difference scheme for an elliptic equation with several space variables', *USSR Comp. Math. & Math. Phys.*, **3**, 1373–1382.
21. L. C. Woods (1954). 'A note on the numerical solution of fourth order differential equations', *Aero. Q.*, **5**, 176–184.
22. A. D. Gosman, W. M. Pun, A. K. Runchal, D. B. Spalding and M. Wolfshtein (1969). *Heat and Mass Transfer in Recirculating Flows*, Academic Press, London.
23. E. Leonardi and J. A. Reizes (1979). 'Natural convection in compressible fluids with variable properties', in *Numerical Methods in Thermal Problems*, R. W. Lewis and K. Morgan (Eds), pp. 297–306, Pineridge Press, Swansea.
24. S. Ostrach (1972). 'Natural convection in enclosures', in *Advances in Heat Transfer*, J. P. Harnett and T. F. Irvine Jr (Eds), Vol. 8, pp. 161–226, Academic Press, London.
25. S. S. Leong and G. de Vahl Davis (1979). 'Natural convection in a horizontal cylinder', in *Numerical Methods in Thermal Problems*, R. W. Lewis and K. Morgan (Eds), pp. 287–296, Pineridge Press, Swansea.

26. B. Roux, J. C. Grandin, P. Bontoux and B. Gilly (1978). 'On a high-order accurate method for the numerical study of natural convection in a vertical square cavity', *Num. Heat Transfer*, **1**, 331–349.
27. R. W. Thomas (1970). 'Finite difference computation of heat transfer by natural convection', *Ph.D. Thesis*, University of New South Wales, Sydney, Australia.

*Numerical Methods in Heat Transfer*
Edited by R. W. Lewis, K. Morgan, and O. C. Zienkiewicz
© 1981 John Wiley & Sons Ltd

## Chapter 19

# A Program for Calculating Boundary Layers and Heat Transfer along Compressor and Turbine Blades

*Denis Dutoya and Pierre Michard*

### SUMMARY

This chapter presents a program calculating the development of a boundary layer along a blade profile, starting from the stagnation point, through its laminar, transitional, and turbulent states. The boundary layer equations are integrated with an implicit finite difference method. Eddy viscosity and turbulent diffusion are related to two characteristic quantities, the mean kinetic energy $\bar{K}$ and the molecular rate of dissipation of the turbulence $\bar{\varepsilon}$, whose values at each point of the flow are calculated with the classical $\bar{K}$ and $\bar{\varepsilon}$ transport equations. A few examples of application and of comparison between calculated and experimental results arc presented.

### NOMENCLATURE

| | | | |
|---|---|---|---|
| $C_f$ | skin friction coefficient | $r_c$ | radius of curvature of profile |
| $C_p$ | specific heat at constant pressure | $T$ | static temperature |
| $f$ | flow quantity | $T_u$ | turbulence rate of external |
| $F_i$ | value of flow quantity at grid point $i$ | | flow, $(2\bar{K}_e)^{1/2}/u_e$ |
| $H^*$ | stagnation enthalpy | $\bar{u}$ | $x$-component of average velocity |
| $\bar{K}$ | mean turbulent kinetic energy, $\frac{1}{2}\bar{u}_i'^2$ | $\bar{v}$ | $y$-component of average velocity |
| $L$ | integral scale of turbulence | $u', v', w'$ | component of velocity fluctuation, $u_i'$ |
| $M$ | molecular mass | $u_s$ | shear velocity, $(\sigma_w/\bar{\rho}_w)^{1/2}$ |
| $N$ | number of transverse mesh intervals | $x$ | abscissa along profile |
| $p$ | static pressure | $y$ | normal distance to wall |
| $q$ | heat flux | $R_x$ | $x$-Reynolds number, |
| $R$ | gas constant | | $\rho_e u_e x/\mu_e$ |

$R_{\delta_1}$  displacement thickness
Reynolds number,
$\rho_e u_e \delta_1 / \mu_e$

$R_{\delta_2}$  momentum thickness
Reynolds number

$R_T$  Reynolds number of
turbulence, $\bar{\rho}\bar{K}^2/\mu\bar{\varepsilon}$

$\alpha$  exchange coefficient

$\delta$  boundary layer thickness

$\delta_1$  displacement thickness,
$\int_0^\delta (1 - \bar{\rho}\bar{u}/\rho_e u_e)\,dy$

$\delta_2$  momentum thickness,
$\int_0^\delta [(1 - \bar{u}/u_e)\bar{\rho}\bar{u}/\rho_e u_e]\,dy$

$\Delta x$  $x$-wise progression

$\lambda$  Taylor's microscale

$\mu$  laminar viscosity

$\mu_t$  eddy viscosity

$\nu$  kinematic viscosity

$\rho$  density

$\bar{\varepsilon}$  mean dissipation rate of
turbulence

$\bar{\varphi}$  $\bar{\varepsilon} - 2\nu\left(\dfrac{\partial\sqrt{\bar{K}}}{\partial y}\right)^2$

$\omega$  normalized stream
function

## 19.1  INTRODUCTION

The prediction of boundary layer development along a compressor or a turbine blade is of great importance for engineering applications. During the last decade, flow procedures as well as turbulence modelling have seen constant developments. While these fields are being investigated, it may be interesting to apply such methods and modelling, now widely recognized among specialists, from a more practical point of view, and to build programs which, at the same time, are sufficiently elaborate to take into account wide classes of complex phenomena (such as generation and development of turbulence), and simple enough to be extensively used—at low cost—by applied research engineers.

An example of a problem of practical interest is the development of boundary layer characteristics and heat exchange along relatively low Reynolds number blades: in this case, the laminar, transitional, and turbulent states are present, each extending along a non-negligible part of the flow.

This chapter describes the physical and numerical basis of a program for calculating the development of a boundary layer in a compressible gas, through its laminar and turbulent states, with eventually pressure gradients and rapidly varying thermal conditions along the wall, and presents a few examples of numerical results and a comparison with experimental data.

## 19.2  PHYSICAL ANALYSIS

### 19.2.1  Flow quantities and statistical analysis

The system of curvilinear coordinates used in the following is presented in Figure 19.1: the abscissa $x$ is measured along the two-dimensional profile, with the origin $x = 0$ chosen as the stagnation point, and $y$ is the normal distance to

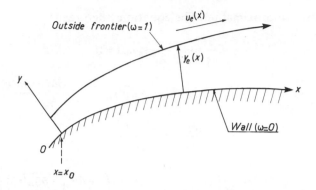

**Figure 19.1** Calculation domain in physical plane $(x, y)$

the wall. The longitudinal curvature radius of the profile is denoted $r_c(x)$, chosen as positive when the wall is convex.

At each point $(x, y)$, the dynamical state of the fluid is defined by a limited number of time varying quantities $g$, namely, density $\rho$, static pressure $p$, momentum per unit volume $\rho u_i$ ($\rho u$ along $x$ and $\rho v$ normal to the profile), and stagnation enthalpy per unit volume $\rho h^*$. In a turbulent flow, these quantities are more or less randomly varying with time, and a statistical study is necessary at this stage. Defining a statistical average operator (denoted by an 'overbar') and a random fluctuation (prime), one gets:

$$g = \bar{g} + g'$$

Other flow characteristics $f$ may be deduced from the above: for instance, the velocity vector $u_i$, the static temperature $T$, the energy per unit mass, etc. For convenience, we use a different averaging process for these quantities, the so called Favre or mass weighted average:[1]

$$u_i = \bar{u}_i + u_i' \quad \text{with} \quad \bar{u}_i = \overline{\rho u_i}/\bar{\rho}$$

$$T = \bar{T} + T' \quad \text{with} \quad \bar{T} = \frac{\overline{\rho T}}{\bar{\rho}} = \frac{M}{R} \frac{\bar{p}}{\bar{\rho}}$$

### 19.2.2 Transport equations for average flow quantities

The general conservation equations for average mass, momentum, and energy are derived in their tensorial form in references 1 and 2. In the following we shall consider a perfect gas, with laminar viscosity given as a function of temperature, and with constant Prandtl number. The flow is assumed to be stationary, two-dimensional (as far as average quantities are concerned), and confined within a thin layer $\delta(x)$ near the wall, so that we may use the classical boundary layer approximation. The equations then read:

$$\frac{\partial}{\partial x}(\bar{\rho}\bar{u}) + \frac{\partial}{\partial y}(\chi\bar{\rho}\bar{v}) = 0 \tag{19.1a}$$

$$\frac{\partial}{\partial x}(\bar{\rho}\bar{u}^2) + \frac{\partial}{\partial y}(\chi\bar{\rho}\bar{v}\bar{u}) = -\frac{\partial\bar{p}}{\partial x} + \frac{\partial}{\partial y}[\chi(\bar{\sigma}_{xy} - \bar{\rho}\overline{u'v'})] + \frac{\bar{\sigma}_{xy} - \bar{\rho}\bar{u}\bar{v} - \bar{\rho}\overline{u'v'}}{r_c} \tag{19.1b}$$

$$\frac{\partial\bar{p}}{\partial y} = \bar{\rho}\frac{\bar{u}^2 + \overline{u'^2}}{\chi r_c} \tag{19.1c}$$

$$\frac{\partial}{\partial x}(\bar{\rho}\bar{u}\bar{h}^*) + \frac{\partial}{\partial y}(\chi\bar{\rho}\bar{v}\bar{h}^*) = \frac{\partial}{\partial y}\left[\chi\left(\frac{\bar{\mu}C_p}{Pr}\frac{\partial\bar{T}}{\partial y} - \bar{\rho}C_p\overline{v'T'} + \overline{u\sigma_{xy}} - \bar{\rho}\overline{u}\overline{u'v'}\right)\right] \tag{19.1d}$$

with

$$\chi = 1 + y/r_c \qquad \overline{\sigma_{xy}} = \bar{\mu}\left(\frac{\partial\bar{u}}{\partial y} - \frac{\bar{u}}{\chi r_c}\right) \qquad \bar{p} = \bar{\rho}\frac{R}{M}\bar{T}$$

and

$$\bar{h}^* = C_p\bar{T} + \tfrac{1}{2}\bar{u}^2 + \tfrac{1}{2}\overline{u_i'^2}$$

When $\delta \ll r_c$, one does introduce a significant error by dropping all the terms containing the curvature radius $r_c$. Indeed, as mentioned by many authors, the main effect of streamline curvature is to modify the local turbulent structure, and hence the mean flow through the turbulent stress $\overline{u'v'}$.[3]

### 19.2.3   Closure assumptions and turbulent model

The averaging process performed on the local fluctuating conservation laws implies that some information is lost, namely, information about how average quantities $\bar{f}$ are transported from one point to another through the non-linear fluctuating motion. From a mathematical point of view, the problem is to express the second-order correlations $\overline{u'v'}$ and $\overline{v'T'}$ appearing in the above equations in terms of known quantities $\bar{f}$. The problem of turbulent modelling is still in a state of constant change, and many theories have been developed during the past few years, which are more or less complex and more or less satisfactory.[4] One of them, the so called 'two-equations' or '$K, \varepsilon$' model, has been extensively studied (see, for instance, Reynolds and Cebeci,[5] Lumley and Tennekes,[6] and Lumley and Khajeh-Nouri[7]), and, although some of its physical bases are not very clear, it is fairly convenient for engineering applications: it has the advantage of being both relatively simple, but complex enough to take roughly into account many aspects of non-homogeneous and non-equilibrium turbulence phenomena.

The model that we use here is similar to the one which has been developed by Jones and Launder[8] for low Reynolds number turbulence. The turbulent fluxes are expressed in terms of the gradients of associated average quantities:

$$\overline{\rho u' v'} = -\mu_t \left( \frac{\partial \bar{u}}{\partial y} - \frac{\bar{u}}{\chi r_c} \right) \tag{19.2a}$$

$$\overline{\rho v' T'} = -\frac{\mu_t}{Pr_t} \frac{\partial \bar{T}}{\partial y} \tag{19.2b}$$

The turbulent Prandtl number $Pr_t$ is taken as constant. The eddy viscosity $\mu_t$ is related to the turbulent kinetic energy $\bar{K} = \overline{u_i' u_i'}/2$ (mass weighted average) and to the rate of viscous dissipation $\bar{\varepsilon}$ of $\bar{K}$:

$$\bar{\rho} v_t = \mu_t = \bar{\rho} C_\mu C_r \bar{K}^2 / \bar{\varepsilon} \tag{19.3}$$

The term $C_\mu$ attempts to take into account the influence of laminar viscosity $\mu$ on turbulent diffusion, as the turbulent Reynolds number:

$$R_T = \bar{\rho} \bar{K}^2 / \mu \bar{\varepsilon}$$

tends to zero. Gathering many experimental data in the near wall region of turbulent boundary layers led us to the following choice (quite arbitrary, for measurements of $\bar{\varepsilon}$ near a wall are not precise enough):

$$C_\mu = C_{\mu 0} \left[ 1 - C_{\mu 1} \exp \left( -\left( \frac{R_T}{R_0} \right)^2 \right) \right] \tag{19.4}$$

with

$$C_{\mu 0} = 0.09 \quad C_{\mu 1} = 0.86 \quad R_0 = 600$$

The correction factor $C_r$ of (19.3) expresses the influence of curvature on the eddy viscosity itself (see next section). The turbulent energy $\bar{K}$, and the rate of viscous dissipation $\bar{\varepsilon}$ may be calculated at each point with the aid of the following transport equations:

$$\frac{\partial}{\partial x} (\bar{\rho} \bar{u} \bar{K}) + \frac{\partial}{\partial y} (\chi \bar{\rho} \bar{v} \bar{K}) = \frac{\partial}{\partial y} \left[ \chi \left( \mu + \frac{\mu_t}{\sigma_1} \right) \frac{\partial \bar{K}}{\partial y} \right] + \bar{\rho} (\bar{\Pi} - \bar{\varepsilon}) \tag{19.5a}$$

$$\frac{\partial}{\partial x} (\bar{\rho} \bar{u} \bar{\varphi}) + \frac{\partial}{\partial y} (\chi \bar{\rho} \bar{v} \bar{\varphi}) = \frac{\partial}{\partial y} \left[ \chi \left( \mu + \frac{\mu_t}{\sigma_2} \right) \frac{\partial \bar{\varphi}}{\partial y} \right] + \bar{\rho} \frac{\bar{\varphi}}{\bar{K}} (C_1 \bar{\Pi} - C_2 \bar{\varepsilon}) \tag{19.5b}$$

with

$$\bar{\varphi} = \bar{\varepsilon} - 2\nu \left( \frac{\partial \sqrt{\bar{K}}}{\partial y} \right)^2 = \text{isotropic dissipation}$$

and

$$\bar{\Pi} = -\overline{u_i' u_j'} \frac{\partial \bar{u}_i}{\partial x_j} \approx \nu_t \left( \frac{\partial \bar{u}}{\partial y} - \frac{\bar{u}}{r_c} \right)^2 = \text{production of turbulence}$$

*Numerical Methods in Heat Transfer*

According to Jones and Launder,[8] the correction factor $C_1$ and $C_2$ are taken as functions of the turbulent Reynolds number $R_T$. In the near-wall viscous layer, one may expect that the statistically isotropic character of the dissipative eddies is affected by the vicinity of the wall. A good measure of the scale of these eddies at low and moderate $R_T$ being Taylor's microscale $\lambda$:

$$\lambda^2 = 10\nu\bar{K}/\bar{\varepsilon}$$

We set:

$$C_1 = C_{10}\left[1 - C_{11}\exp-\left(\frac{R_T}{R_1}\right)^2\right] + C_{12}\left(\frac{\lambda}{y}\right)^2 \tag{19.6a}$$

$$C_2 = C_{20}\left[1 - C_{21}\exp-\left(\frac{R_T}{R_2}\right)^2\right] - C_{22}\left(\frac{\lambda}{y}\right)^2 \tag{19.6b}$$

The $\lambda/y$ correction terms allow a correct behaviour of $\bar{\varphi}$ in the near-wall region, and are negligible in the logarithmic and outer domains. In fact, they play the same role as the more popular 'second derivative' term,[8] and the main reason for their use is that they are easier to handle numerically.

The values of the constants $C_{10}, C_{11}, C_{12}, \ldots$ were chosen by numerical optimization in order to fit the calculated and experimental results for three test cases: decay of isotropic turbulence, equilibrium turbulent boundary layer along a flat plate, and onset of transition on a laminar boundary layer. The set of constants chosen, as reported in Table 19.1, is certainly not unique. At this level, many other choices might work as well, and are justified as far as one keeps using the same constants throughout any further engineering parameter study.

Table 19.1

| $C_{10}$ | $C_{11}$ | $C_{12}$ | $C_{20}$ | $C_{21}$ | $C_{22}$ | $R_1$ | $R_2$ | $\sigma_1$ | $\sigma_2$ |
|---|---|---|---|---|---|---|---|---|---|
| 1.35 | 0.04 | 0.25 | 2 | 0.3 | 0.08 | 50 | 50 | 0.90 | 0.95 |

Equations (19.5a) and (19.5b), along with relations (19.3), (19.4), and (19.6) constitute a closure scheme for the turbulence problem, and allow the resolution of the transport equations (19.1).

### 19.2.4 Influence of wall curvature on turbulent correlations $\overline{u_i' u_j'}$

It is possible to obtain an evaluation of this effect (namely, to express the correction factor $C_r$ of relation (19.3)) by considering what happens in homogeneous turbulence. The first step is to choose a model for the transport equations of the turbulent Reynolds stress $\overline{u_i' u_j'}$ among the more or less complex theories proposed in the literature (see, for instance, references 5, 9,

10, 11, 12, etc.). The model developed by Launder and coworkers[13] has the advantage of being simple. Following the procedure presented in reference 13, we set the $\overline{u_i'u_j'}$ equations according to the closure scheme proposed, then drop the differential terms by assuming homogeneity of turbulence quantities, which leads to an analytical expression for the stress tensor:

$$\overline{u_i'u_j'} = \tfrac{2}{3}\bar{K}\delta_{ij} + \frac{0.4}{C_1'}(\bar{\Pi}_{ij} - \tfrac{2}{3}\bar{\Pi}\delta_{ij})\frac{\bar{K}}{\bar{\varepsilon}}$$

where $\bar{\Pi}_{ij}$ is the generation tensor (rate of generation of $\overline{u_i'u_j'}$ by interaction with the mean rate of strain), and $\bar{\Pi}$ its trace. After some tensor analysis applied to the curvilinear system $(x, y)$, use of the boundary layer approximations, and of the equilibrium condition $\bar{\Pi} = \bar{\varepsilon}$, one gets:

$$\overline{u'v'} = -C_\mu C_r \frac{\bar{K}^2}{\bar{\varepsilon}}\left(\frac{\partial \bar{u}}{\partial y} - \frac{\bar{u}}{\chi r_c}\right) \tag{19.7}$$

with

and

$$C_\mu = \text{constant} \tag{19.8}$$

$$C_r = \left[1 + C_{\mu r}\frac{\bar{K}^2}{\bar{\varepsilon}^2}\frac{\bar{u}}{r_c}\left(\frac{\partial \bar{u}}{\partial y} + \frac{\bar{u}}{\chi r_c}\right)\right]^{-1}$$

The constant $C_{\mu r}$ depends on the value of $C_1'$ which, according to different authors, may vary from 1.5 to 3.5, so that $C_{\mu r} \sim 0.1$ to 0.5.

Equations (19.7) and (19.8) show that a positive curvature (convex wall) has a stabilizing effect on the turbulent flow—as might be expected from experimental evidence—through a reduction of the eddy viscosity proportional to the Richardson number:

$$Ri = \frac{\bar{u}}{r_c}\frac{\partial \bar{u}}{\partial y} \sim C_\mu^{1/2}\frac{\bar{u}}{r_c}\frac{\bar{K}}{\bar{\varepsilon}}$$

## 19.2.5  Initial and boundary conditions

The set of differential equations (19.1) and (19.5) is of the parabolic type. Therefore, it can be integrated numerically from any known starting section $x_0$, if one provides:

(a)  The initial profiles of velocity, temperature, and turbulent parameters at $x = x_0$.

(b)  The no-slip condition at the wall: $\bar{u} = \bar{v} = \bar{K} = \bar{\varphi} = 0$.

(c)  A thermal condition at the wall, namely, the wall temperature or wall heat flux distribution, or any linear wall flux–temperature relationship.

(d)  The wall pressure distribution $p_w(x)$ and wall curvature $r_c(x)$.

The 'outside' flow conditions at the edge of the boundary layer $u_e(x)$, $K_e(x)$, $\varepsilon_e(x)$ can then be calculated at any station $x$ by integration of the asymptotic forms of Equations (19.1) and (19.5) as the transverse gradients vanish, once their initial value $u_{e0}$, $K_{e0}$, $\varepsilon_{e0}$ are given.

## 19.3 NUMERICAL ANALYSIS

The numerical method used here is similar to the one developed by Patankar and Spalding.[14]

### 19.3.1 New variables

For convenience, the calculations are performed in a rectangular $(x, \omega)$ grid, the new independent variable $\omega$ being the normalized stream function:

$$\omega = \frac{\int_0^y \bar{\rho}\bar{u}\chi \, dy}{\int_0^{y_e} \bar{\rho}\bar{u}\chi \, dy} = \frac{\psi}{\psi_e(x)} \tag{19.9}$$

In this plane, the transport equation for any quantity $f$ reads:

$$\frac{\partial f}{\partial x} = b\omega \frac{\partial f}{\partial \omega} + \frac{\partial}{\partial \omega} c \frac{\partial f}{\partial \omega} + s_1 f + s_2 \tag{19.10}$$

where $b$, $c$, $s_1$, and $s_2$ are non-linear coefficients, depending on the entrainment rate

$$\frac{1}{\psi_e} \frac{d\psi_e}{dx}$$

on the diffusion coefficients ($\mu_t, \ldots$), and on the quantities $f$ themselves.

### 19.3.2 Transverse discretization

Each function $f(x_l, \omega)$ at a station $x_l$ is replaced by a vector $\mathbf{F}^{(l)}$ whose components $F_i^{(l)}$ are the $N+1$ values of $f$ at the nodes of the $N$-interval lateral mesh, and two 'slip values' for treatment of singular behaviour at $\omega = 0$ and $\omega = 1$ (Figure 19.2).

A second-order finite difference approximation of (19.10) can be written as:

$$\frac{\partial \mathbf{F}}{\partial x} = -L_\omega \mathbf{F} + \mathbf{S} \tag{19.11}$$

where $L_\omega$ is a tridiagonal operator, depending on local flow quantities, and $\mathbf{S}$ a vector function of boundary conditions at $\omega = 0$ and $\omega = 1$ and of local source terms $s_2$ (for instance, the pressure gradient).

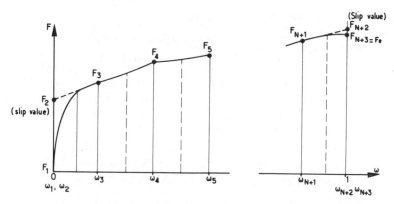

**Figure 19.2** Discretization of quantity $F(x, \omega)$ in $(x, \omega)$ plane

The three $i$-components of the non-linear operator $L_\omega$ are obtained (see reference 14) by integration of Equation (19.10) over the interval $[\omega_{i-\frac{1}{2}}, \omega_{i+\frac{1}{2}}]$, which guarantees the conservative character of the procedure.

### 19.3.3 $x$-wise integration procedure

The parabolic set of equations (19.11) is integrated by an implicit difference scheme, which has proved[14] to be very stable, and accurate enough for engineering calculations:

$$\mathbf{F}^{(l+1)} = \mathbf{F}^{(l)} + \Delta x(-L_\omega \mathbf{F}^{(l+1)} + \mathbf{S}) \qquad (19.12)$$

where $\mathbf{F}^{(l+1)}$ and $\mathbf{F}^{(l)}$ are the values of $\mathbf{F}$ at $x^{(l)}$ and $x^{(l+1)} = x^{(l)} + \Delta x$. The truncature error due to $x$-discretization is then of order $\Delta x^2$, the error due to the $\omega$-finite difference approximations being of the order $\Delta \omega^3$.

In fact, using a Taylor's development of $\mathbf{F}^{(l)}$ in terms of quantities at $x^{(l+1)}$, one can show that Equation (19.12) is a second-order approximation of the following differential equation:

$$\frac{\partial \mathbf{F}}{\partial x} = -L_\omega \mathbf{F} + \mathbf{S} + \frac{\Delta x}{2} \frac{\partial^2 \mathbf{F}}{\partial x^2} + O(\Delta x^2, \Delta \omega^2) \qquad (19.13)$$

Equation (19.13) shows that numerical diffusion along the streamlines will not exceed the neglected physical $x$-diffusion (cf. boundary layer approximations) as far as the numerical viscosity $\frac{1}{2}\bar{\rho}\bar{u}\,\Delta x$ is of the same order of magnitude as the eddy viscosity $\mu + \mu_t$. This is why we set:

$$\Delta x = \zeta\,\delta(x) \qquad (19.14)$$

$\delta(x)$ being the boundary layer thickness, and $\zeta$ a constant. Calculations have shown that the numerical solution differs by less than 5% from known exact or approximate (similarity) solutions, when:

$$\zeta < \begin{cases} 0.2 & \text{for low Reynolds number flow} \\ 0.05 & \text{for high Reynolds number flow} \end{cases}$$

It can be shown that the algorithm (19.12) is unconditionally stable about any existing similarity solution $\tilde{\mathbf{F}}$ when the operator $L_\omega$ is positive, which is the case for the boundary layer equations (19.10) as far as the linearized source term $(s_1)$ is negative (sink or dissipation).

The $N+1$ components of the vector Equation (19.12), which correspond, as we have seen, to the integration of the differential Equation (19.10) over the $N+1$ intervals $[\omega_1, \omega_{2.5}]$, $[\omega_{2.5}, \omega_{3.5}]$, etc., take the following form:

$$F_i^{(l+1)} = A_i F_{i+1}^{(l+1)} + B_i F_{i-1}^{(l+1)} + C_i(F_i^{(l)}) \quad i = 2, \ldots, N+2$$

At each step $x_l$, the components of the tridiagonal matrices $A_i$, $B_i$, and of the vector $C_i$ must be calculated in terms of known quantities $F_K^{(l)}$. Then, the unknowns $F_i^{(l+1)}$ are obtained by matrix inversion and use of the boundary conditions at $\omega = 0$ and $\omega = 1$. An iteration process may be necessary for readjustment of the coefficients $A_i$ and $B_i$.

### 19.3.4 Program performances

A detailed description of the program 'RENAULD' is given in reference 16. It was written with a view to being easy to handle for non-specialist engineers, and the calculation cost is reasonable: a typical calculation of the development of a boundary layer along a $10^6$ Reynolds number profile requires five seconds on a CDC 7600 computer, with limited memory storage.

### 19.3.5 Initial profiles

In general, the user must provide experimentally measured or guessed initial profiles at $\bar{u}$, $\bar{T}$, $\bar{K}$, and $\bar{\varepsilon}$ at the starting section $x_0$. However, if the user wants to start the calculations from a stable laminar region $(R_{x_0} < 10^4)$, for instance, near the stagnation point, the program automatically calculates the starting profiles as:

$$\bar{u} = u_{e0}f(y/y_{e0})$$

$$h^* = C_p T_{w0} + (h_e^* - C_p T_{w0})f$$

$$\bar{K} = K_{e0}f^2$$

$$\bar{\varphi} = C_\mu^{1/2} \bar{K}\left(\frac{\partial \bar{u}}{\partial y} - \frac{\bar{u}}{r_c}\right) \quad (\text{i.e. } \bar{\Pi} \sim \bar{\varepsilon})$$

where $f$ is the Blasius function and the user has only to provide the $\omega$-grid specifications, the outer flow stagnation conditions and turbulent characteristics, the wall pressure distribution, the wall thermal condition, and the initial boundary layer thickness $y_{e0}$ (see Schlichting[15] for estimation of $y_{e0}$).

## 19.4   A FEW EXAMPLES OF CALCULATION RESULTS

### 19.4.1   Test case

Extensive calculations have been done for the case of a boundary layer along an adiabatic flat plate, with different turbulence levels of the non-disturbed flow. The calculations are started in the stable laminar boundary layer region $(R_{x_0} = 10^3)$, with initial profiles chosen as presented in Section 19.3.5. The particular choice of $\bar{K}$ initial profile (here, constant turbulence rate throughout the laminar layer) appears to have no incidence on the downstream calculations, for any distortion of the $\bar{K}$ profile is wiped out after a few $x$-steps, as far as $R_{x_0} < 10^4$.

Figure 19.3 shows the evolution of the velocity profiles along the plate for an external turbulence rate $T_u = (u_e'^2 + v_e'^2 + w_e'^2)^{1/2}/\bar{u}_e$ of 5%. The $\bar{u}$ profiles

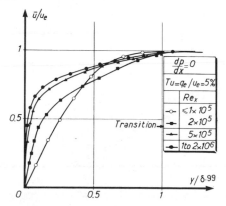

**Figure 19.3**   Calculated velocity profiles (flat plate)

remain of the laminar type up to $R_{x,C_r} = 1.5 \times 10^5$, then change quite rapidly to a turbulent type law, while the shape factor moves from 2.59 toward 1.4. At the same time, the turbulent energy is suddenly increased, as shown in Figure 19.4. The 'transitional' Reynolds number $R_{x,C_2}$ depends mainly on the turbulence rate $T_u$ of the external flow. Figure 19.5 represents the evolution of the skin friction $C_f$ in terms of the $x$-Reynolds number for various $T_u$.

Figure 19.6 shows that the calculated position of the beginning of transition fits with a set of experimental results gathered by McDonald and Fish.[17] Downstream of the transitional region, the skin friction coefficient (Figure

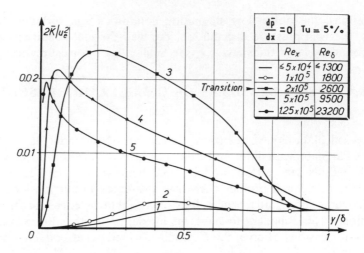

**Figure 19.4**   Turbulent energy profiles

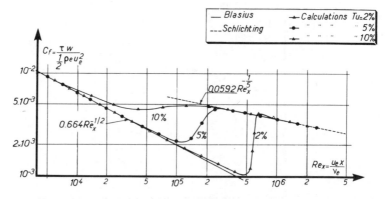

**Figure 19.5**   Calculated skin friction coefficient along a flat plate

19.5), velocity, and turbulent energy profiles (Figure 19.7) calculated by the program are in good agreement with the classical experimental results.

Of course, we do not intend here to say that the program takes into account the complexity of real transition phenomena,[18] but rather that it provides a convenient way of calculating a boundary layer starting from a stagnation point and of 'going through the transition' at low cost.

Other calculations have been performed with non-zero pressure gradient and finite curvature: Figure 19.8 shows the stabilizing effect of a convex curvature ($r_c > 0$) on a boundary layer, as obtained through the simple theory of Section 19.2.4. The same effect is obtained with an accelerated external flow ($dp/dx < 0$).

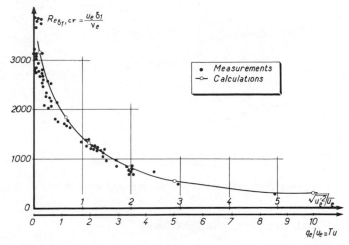

**Figure 19.6** Displacement thickness $\delta_1$ of a laminar boundary layer on a flat plate at the onset of transition in terms of free stream turbulence $T_u$

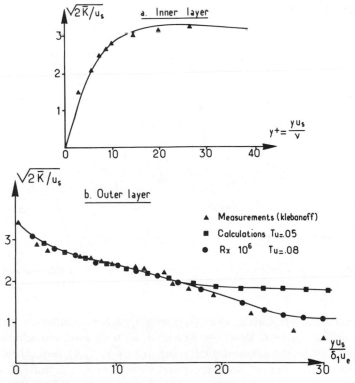

**Figure 19.7** Velocity fluctuation in an equilibrium boundary layer: (a) inner layer; (b) outer layer

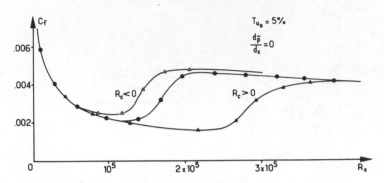

**Figure 19.8**   Influence of curvature on wall shear stress

### 19.4.2   Heat exchange along a turbine blade

Figure 19.9 shows the comparison between experimental and calculated exchange coefficient distributions along the pressure and suction sides of a cooled turbine inducer blade. The measurements were taken from a semi-industrial test set-up for turbine performance studies.[19]

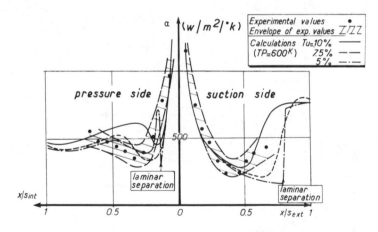

**Figure 19.9**   Exchange coefficient on cooled turbine blade calculations and measurements

Three series of calculations were performed, each with different turbulence rates $T_u$ of the upstream flow: the first one, with an assumed adiabatic wall condition, provides the adiabatic wall temperature $T_{w,ad}(x)$, the second and the third, with two different wall temperatures $T_w = 600$ K and $T_w = 900$ K respectively. The calculated exchange coefficients $\alpha = q_w/(T_{w,ad} - T_w)$ do not differ by more than 10%. Both measurements and calculations show that, on the suction side, the transition from laminar to turbulent state takes place at $x/s = 0.5$, in a zone where the external flow begins to decelerate. Good agreement between experiment and calculations occurs for $T_u = 0.07$ to $0.10$. On the pressure side, calculated results indicate a tendency for relaminarization near the leading edge, a trend which does not seem to be confirmed by measurements.

The main problem for the designer is to predict the skin temperature of the blade, which is the result of the interaction of three phenomena: heat exchange between the hot gas and the blade through the boundary layer, conduction within the material, according to the well known Fourier law, and internal cooling (Figure 19.10). This is why the boundary layer calculations must be

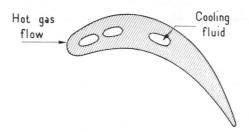

**Figure 19.10** Cooled blade

coupled with a finite element conduction program which calculates the temperature field within the blade material, given the internal cooling conditions (mass flow and temperature of cooling air), and the skin temperature $T_w(x)$ along the outer wall.[20] The boundary layer program calculates the temperature distribution $T_w(x)$, given an approximation of the wall flux distribution $q_w(x)$, and the conduction program uses the new $T_w$ distribution as an input to recalculate the heat flux $q_w(x)$, and so on. In fact, a few iterations are necessary to obtain the equilibrium $T_w(x)$ and $q_w(x)$ distributions. Figure 19.11 shows a comparison between the results of such calculations and measured temperature distribution.

## 19.5  CONCLUSION

A program calculating two-dimensional compressible laminar and turbulent boundary layers was written with a view to extensive and low cost industrial utilization for estimation of aerodynamical losses and heat exchange in turbomachinery. The relatively simple turbulent model on which it is based has proven to provide fairly good results for a wide range of Reynolds numbers. The calculations may be started in a stable laminar region near the leading edge, and the $\bar{K}, \bar{\varepsilon}$ turbulence model induces a 'transition' from laminar to

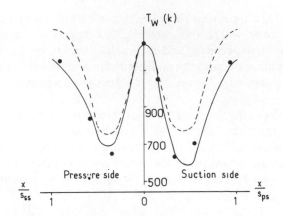

**Figure 19.11** Skin temperature distribution along a turbine blade

turbulent state whose position and extension is roughly the same as in a real situation, at least for the flat plate case. The modelling attempts to take into account the effect of streamline curvature on the eddy viscosity, but this theory needs to be further examined in the light of available experimental analysis.

The boundary layer program was used to calculate heat exchange characteristics along a cooled turbine blade, and was also coupled with a finite element heat propagation program, with a view to predicting the resulting skin temperature of the blade. The calculated temperature distribution has proven to agree with experimental data.

## REFERENCES

1. A. Favre *et al.* (1976). *La Turbulence en Mécanique des Fluides*, Gauthier Villars, Paris.
2. A. Favre (1965). 'Equations des gaz turbulents compressibles', *J. Mécanique*, **4**, Nos 3, 4.
3. J. P. Johnston (1976). 'Internal flows', in *Topics in Applied Physics*, Vol. 12 (*Turbulence*), P. Bradshaw (Ed.), Springer-Verlag, Berlin.
4. B. E. Launder and D. B. Spalding (1972). *Mathematical Models of Turbulence*, Academic Press, New York.
5. W. C. Reynolds and T. Cebeci (1976). 'Calculation of turbulent flows', in *Topics in Applied Physics*, Vol. 12 (*Turbulence*), P. Bradshaw (Ed.), Springer-Verlag, Berlin.
6. J. L. Lumley and M. Tennekes (1972). *A First Course in Turbulence*, MIT Press, Mass., USA.
7. J. L. Lumley and B. Khajeh-Nouri (1974). In *Advances in Geophysics*, Vol. 18A, F. N. Frenkiel and R. M. Munn (Eds). Academic Press, New York.
8. W. P. Jones and B. E. Launder (1973). 'The calculation of low Reynolds number phenomena with a two-equations model of turbulence', *J. Heat Mass Transfer*, **16**, No. 6.
9. J. O. Hinze (1959). *Turbulence*, McGraw-Hill, New York.

10. J. L. Lumley (1975). *Lecture Series No. 76*, Von Karman Institute, Belgium.
11. B. E. Launder *et al.* (1975). 'Progress in the development of Reynolds stress turbulence closure', *J. Fluid Mech.*, **68**, 537.
12. K. Hanjalic and B. E. Launder (1972). 'A Reynolds stress model for turbulence and its application to thin shear flow', *J. Fluid Mech.*, **52**, 609.
13. B. E. Launder (1976). 'Heat and mass transport', in *Topics in Applied Physics*, Vol. 12 (*Turbulence*), P. Bradshaw (Ed.), Springer-Verlag, Berlin.
14. S. V. Patankar and D. B. Spalding (1970). *Heat and Mass Transfer in Boundary Layers*, 2nd edn., Interest Books, London.
15. H. Schlichting (1963). *Boundary Layer Theory*, McGraw-Hill, New York.
16. D. Dutoya (1978). 'Description du programme de calcul de couche limite RENAULD', *ONERA Report* No. 01/3257 EN, ONERA, Paris.
17. H. McDonald and R. W. Fish (1973). 'Practical calculations of transitional boundary layers', *Int. J. Heat Mass Transfer*, **16**, No. 9.
18. D. Arnal and C. Juillen (1977). 'Etude expérimentale et théorique de la transition de couche limite', *La Recherche Aérospatiale* (1977–2).
19. Y. Le Bot, M. Charpenel and P. J. Michard (1977). 'Techniques de mesure dans les turbines à hautes températures', *High Temperature Problems in Gas Turbine Engines, AGARD Report* CP-229.
20. A. E. Cornut and P. J. Michard (1979). 'Calcul de la température de paroi d'aubes de turbine par couplage de la conduction interne et de la convection externe', *La Recherche Aérospatiale* (1979–4).

Numerical Methods in Heat Transfer
Edited by R. W. Lewis, K. Morgan, and O. C. Zienkiewicz
© 1981 John Wiley & Sons Ltd

Chapter 20

# Finite Element Thermal Analysis of Convectively-Cooled Aircraft Structures

*Allan R. Wieting and Earl A. Thornton*

## SUMMARY

The thermal environment in which convectively-cooled aircraft operate has a significant impact on the structural design. Strong interaction between the thermal and structural designs makes an integrated analysis based on a common method desirable. This chapter discusses some recent developments and applications in conduction/forced-convection finite element analysis. The applications show the finite element method to be fully competitive with the well established finite difference lumped-parameter method for combined conduction/forced-convection analysis.

## 20.1 INTRODUCTION

Recent design studies[1] of convectively-cooled engine and airframe structures for hypersonic transports have indicated significant thermal–structural design interactions. The design complexity and size of these structures necessitate the use of large general purpose computer programs for both thermal and structural analyses. Generally thermal analyses are based on the lumped-parameter finite difference technique, and structural analyses are based on the finite element technique. Because of differences in these techniques an efficient interface is difficult to achieve as illustrated in Figure 20.1. Thus, an integrated analysis based on a common technique is desirable. The finite element method has been universally accepted for structural analysis, but it has not yet received widespread acceptance for thermal analysis. One reason for the lack of complete acceptance for thermal analysis is that the finite element approach has not attained the full capabilities and efficiency of the lumped-parameter finite difference approach. However, with continuing development[2] in thermal analysis methodology, the finite element method offers high potential for an integrated thermal–structural analysis capability.

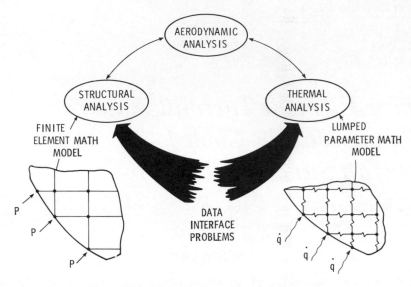

**Figure 20.1**   Thermal structural analysis interaction

One of the most recent developments in finite element thermal methodology is coupled conduction/forced-convection analysis. A simplified finite element solution procedure has been developed,[3,4] and new convective finite elements were derived by representing the flow passage with fluid bulk temperature nodes and fluid/structure interface nodes. The finite elements have been applied in the thermal analysis of a variety of problems of varying difficulty with excellent results.[3-6]

The purpose of this chapter is to summarize some of the efforts by the NASA Langley Research Center focused on the development of an integrated thermal structural analysis capability using the finite element method. In particular, the development of conduction/forced-convection finite element methodology is reviewed and applications to highlight capabilities are presented.

## 20.2   THERMAL–STRUCTURAL ANALYSIS PROGRAM

Although the design of high speed vehicles involves the interaction of many technical disciplines in addition to those indicated in Figure 20.1, this chapter addresses only the interaction between the thermal and structural analyses. Thermal–structural analysis capability at Langley Research Center is being developed within the framework of a computer program called 'SPAR'. SPAR[7] is a general purpose finite element program developed for structural analysis and recently extended to thermal analysis. The program consists of a number of technical processors which communicate with each other through a data base to perform basic analysis tasks. This architecture is extremely flexible and efficient. The communication link via the data base enhances thermal–structural analysis capability by automating the transfer of temperatures from a thermal analysis processor to a structural analysis processor.

## 20.3   CONDUCTION/FORCED-CONVECTION ANALYSIS

The finite element methodology for steady-state[3] and transient[4,5] conduction/forced-convection analysis has been developed by employing a number of assumptions customarily used in practical heat transfer analysis. The assumptions, governing equations, and finite element formulations are reviewed in the next sections.

### 20.3.1   Assumptions and governing equations

A general formulation of the finite element equations for practical conduction/forced-convection can be derived from the differential equations which describe energy balances on the solid and fluid in a typical flow passage (Figure 20.2). The flow passage shown consists of a thin tube of constant thickness and

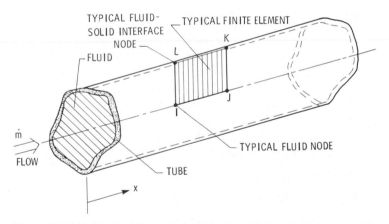

**Figure 20.2**   Finite element representation of coupled conduction–convection in a fluid passage

arbitrary cross section containing a fluid with specified mass flow rate, $\dot{m}$. The formulation is based on assuming that the thermal energy state of the fluid is characterized by the fluid bulk temperature $T_f$ which varies only in the flow direction, i.e. $T_f = T_f(x, t)$ where $t$ denotes time, and the flow is represented by a mean velocity $u$ which varies only in the flow direction. The mass flow rate in a fluid passage is then given by $\dot{m} = \rho_f A_f u$ where $\rho_f$ is the fluid density and $A_f$ is the flow area. A convection coefficient, $h$, is defined such that the heat transfer between the wall and the fluid is expressed by

$$q = h(T_w - T_f) = -k_f \frac{\partial T}{\partial n}\bigg|_{wall} \tag{20.1}$$

where $T_w(x, t)$ denotes the wall temperature and $k_f$ denotes the fluid thermal conductivity. In practical applications the convection coefficient may be expressed in terms of the fluid temperature alone by utilizing analytical–empirical equations for the Nusselt number. Utilizing these assumptions, energy balances on the fluid and tube give

$$-\frac{\partial}{\partial x}\left(k_f A_f \frac{\partial T_f}{\partial x}\right) + \dot{m}c_f \frac{\partial T_f}{\partial x} - hp(T_w - T_f) + \rho_f c_f A_f \frac{\partial T_f}{\partial t} = 0 \quad \text{(fluid)}$$
$$\tag{20.2}$$
$$-\frac{\partial}{\partial x}\left(k_w A_w \frac{\partial T_w}{\partial x}\right) + hp(T_w - T_f) + \rho_c c_w A_w \frac{\partial T_w}{\partial t} = 0 \quad \text{(tube)}$$

where $k_w$, $\rho_w$, $c_w$ are the tube thermal conductivity, density, and specific heat. The tube conduction area is $A_w$, and the perimeter of the fluid–tube interface is $p$.

### 20.3.2  Finite element formulation

A finite element representing these equations is characterized by fluid nodes and fluid–wall nodes such as the typical tube–fluid element shown in Figure 20.2. Within an element the fluid and wall temperatures are expressed in terms of element interpolation functions $(N)$. The Galerkin method with weighting functions $W_i$ is used to transform Equation (20.2) into discretized form. In the conventional finite element formulation $W_i = N_i$, but for the upwind formulation alternative weighting functions $W_i$ are utilized.[4,6] The discretized equations for a typical element take the form

$$\begin{bmatrix} C_f & 0 \\ 0 & C_w \end{bmatrix}\begin{pmatrix} dT_f/dt \\ dT_w/dt \end{pmatrix} + \begin{bmatrix} K_v + K_h + K_f & -K_h \\ -K_h & K_h + K_w \end{bmatrix}\begin{pmatrix} T_f \\ T_w \end{pmatrix} = 0 \tag{20.3}$$

where $[C_f]$ and $[C_w]$ are the capacitance matrices of the fluid and tube wall respectively; $[K_v]$, $[K_h]$, $[K_f]$, and $[K_w]$ are conductance matrices representing mass transport fluid convection, fluid–tube wall convection, fluid conduction, and tube wall conduction, respectively. Further details of these matrices are given in Thornton and Wieting[4] and Thornton.[6]

The capacitance matrix produced by the Galerkin method is known as a consistent formulation because it is derived in a manner consistent with conductance matrices. The consistent capacitance formulation requires an implicit time-integration algorithm because the time derivatives are coupled through off-diagonal terms in the matrices. As an approximation, a consistent capacitance matrix can be converted to a lumped or diagonal matrix by various procedures such as addition of rows of the consistent capacitance matrix. The advantage of the lumped formulation is that an explicit time-integration scheme can be used. If, however, a conditionally stable explicit algorithm is used with a lumped-capacitance matrix an extremely small time step can be required for stability when a nodal capacitance is small compared to other system nodal capacitances. Small fluid nodal capacitances are frequently encountered in convectively-cooled structures for low density coolants such as gases. A procedure, designated 'zero' capacitance nodes, for deletion of negligibly small capacitance nodes during the time integration was presented in Thornton and Wieting.[5] The procedure is applicable for both consistent and lumped-capacitance formulations. In this scheme the temperatures at negligible capacitance nodes are 'statically' related to the temperatures at the remaining nodes. This static relationship is used to eliminate the negligible capacitance nodes from each integration time step of the transient solution. The temperatures at the eliminated nodes are statically updated at the end of each integration.

## 20.4  APPLICATIONS

The finite element methodology has been applied to a variety of thermal problems ranging from simple problems with closed form solutions to practical convectively-cooled aircraft structures. Comparisons of finite element results with finite difference lumped-parameter and closed form analytical results have indicated the conventional finite element approach with a consistent capacitance matrix yields accuracy superior to all other methods. Use of the upwind formulation seriously degrades the accuracy of predicted temperatures due to artificial diffusion. Use of a lumped-capacitance matrix also degrades accuracy due to artificial dispersion/diffusion but to a lesser degree than upwinding. The conventional finite element formulation with a lumped-capacitance matrix gave accuracy superior or equal to the finite difference lumped-parameter method in all fluid convection applications. Selected problems are presented to highlight these conclusions.

### 20.4.1  Coolant passage

The results from one of the problems,[5] which is characterized in Figure 20.3, are shown in Figure 20.4. The problem consists of mass transport convection and conduction in the fluid and convection heat transfer between the fluid and passage wall. The partial differential equation and analytical solution of this problem are given in Thornton and Wieting.[5] The thermal parameters which characterize this problem are the thermal diffusivity $\alpha$, the Peclet number $Pe$, defined here as $Pe = \rho_f c_f u D / k_f$, and the Nusselt number, $Nu = hD/k_f$. The

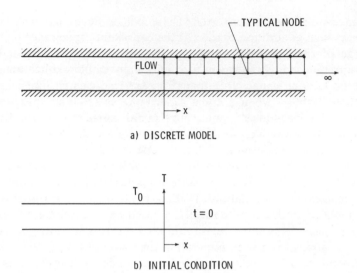

**Figure 20.3**  Schematic of discrete model and initial conditions for
coolant passage analysis: (a) discrete model; (b) initial condition

hydraulic diameter $D$ is defined by $D = 4A/p$. Finite element and finite
difference lumped-parameter fluid temperature distributions at $t = 6$ s are
compared in Figure 20.4 for $\alpha = 10^{-4}\,\mathrm{m^2\,s^{-1}}$, $D = 0.025$ m, $Pe = 10$, and
Nusselt numbers of 0.01 and 1.0.

The conventional finite element formulation with a consistent capacitance
matrix (Figure 20.4) gives excellent agreement with the analytical solution for
both Nusselt numbers. Using a lumped-capacitance matrix (Figure 20.4(b))
introduces dispersive effects at $Nu = 0.01$, but the lumped-capacitance matrix

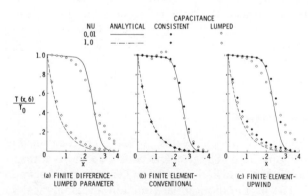

**Figure 20.4**  Comparison of finite difference and finite element predicted temperatures for coolant passage analysis. (a)
Finite difference—lumped parameter; (b) finite element—
conventional; (c) finite element—upwind

gives excellent agreement for $Nu = 1.0$. For $Nu = 0.01$ the upwind finite element (Figure 20.4(c)) and finite difference results (Figure 20.4(a)) show diffusive characteristics displayed in other applications.[4,5] However, for $Nu = 1.0$, the upwind finite element formulation with a lumped-capacitance matrix and the finite difference approach show better agreement with analytical results.

Overall, the results of this application show that as the convection heat transfer increases, the adverse dispersion and artificial diffusion effects of upwinding and lumping tend to decrease. However, the upwind finite element formulation remains inferior to the conventional formulation with either lumped- or consistent-capacitance matrices, and thus is not recommended for this type of application. The results suggest that for practical problems with mixed mode heat transfer the conventional finite element method and the finite difference method will yield acceptable results with the finite element showing superior accuracy through the use of the consistent-capacitance matrix.

### 20.4.2   Convectively-cooled panel

A convectively-cooled airframe structure for hypersonic flight is shown schematically in Figure 20.5. A finite element model of a convectively-cooled panel is shown in Figure 20.6. The panel is cooled by fluid flow through 'D' tubes

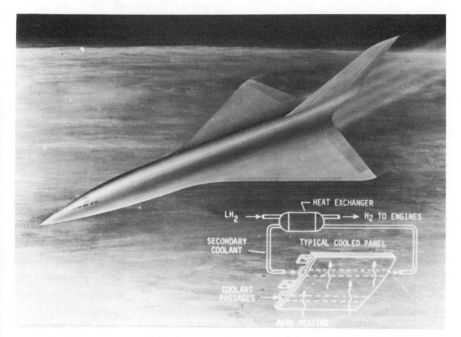

**Figure 20.5**   Convectively-cooled hypersonic aircraft and coolant system

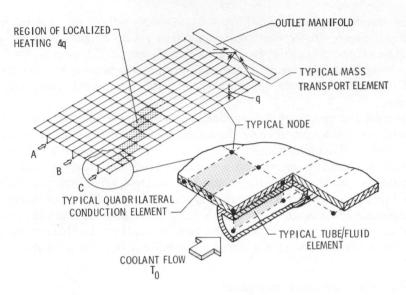

REGION OF LOCALIZED
HEATING 4q

OUTLET MANIFOLD

TYPICAL MASS
TRANSPORT ELEMENT

q

TYPICAL NODE

A

B

C

TYPICAL QUADRILATERAL
CONDUCTION ELEMENT

TYPICAL TUBE/FLUID
ELEMENT

COOLANT FLOW
$T_0$

**Figure 20.6**    Finite element model of a convectively-cooled panel

bonded to the panel. Flow from the coolant passages (denoted as A, B, C) is
collected in an outlet manifold. The panel is modelled with conduction
elements, the 'D' tubes with tube–fluid elements, and flow into the outlet
manifold by mass transport elements.

The cooled panel is subjected to aerodynamic heating of $q$ (in $W\,m^{-2}$)
uniformly distributed over the majority of the panel surface. The remainder of
the panel surface, as indicated by the cross-hatched area shown in Figure 20.6,
is subjected to heating of $4q$.

Finite element calculated coolant temperature distributions normalized to
the inlet coolant temperature $T_0$ are presented in Figure 20.7. The temperatures shown were calculated with consistent-capacitance matrices; however,
temperatures (not shown) were also computed with lumped-capacitance
matrices with a maximum difference of less than 0.3%. Flow in the 'D' tubes
was based upon the conventional conductance formulation, however, both the
conventional and upwind formulations were used for the mass transport
elements linking the outlets of the coolant passages to the manifold (Figure
20.7). With the localized asymmetric heating the coolant temperature distribution along each passage is different. The temperature in the outlet
manifold is an average of the coolant passage exit temperatures, and
consequently different thermal gradients develop between each passage exit
and manifold. The coolant temperatures in passages A (not shown) and C,
when computed with the conventional mass transport element (dotted lines),
show substantial spurious oscillations, but coolant temperatures (full lines)

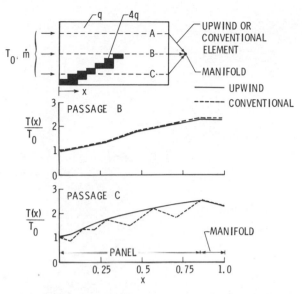

**Figure 20.7**   Panel coolant temperature distributions for conventional and upwind mass transport elements

show a realistic smooth variation when the upwind mass transport elements were used between the passage exits and manifold. Computed temperatures in the centre passage B were essentially the same with both conventional and upwind formulations because the passage B exit temperature was very close to the manifold temperature. This phenomenon did not occur under uniform heating because the passage exit temperatures were all equal.

The panel application illustrates an important practical situation for which the upwind finite element formulation can be beneficial. Fluid temperature oscillations due to downstream temperature gradients were eliminated by using the upwind mass transport elements and no artificial diffusion was introduced. Note that upwind elements were required at the coolant passage outlets only; the conventional finite element formulation was used for the tube–fluid elements.

### 20.4.3   Scramjet engine strut

A scramjet engine concept for hypersonic flight is shown in Figure 20.8. Two of the engines are shown with one of the side walls removed to reveal the internal engine surfaces. The three fuel-injection struts are exposed to a hostile environment characterized by a highly non-uniform aerodynamic heating distribution. The struts are cooled by hydrogen flowing through coolant passages on the external surface of the primary structure.

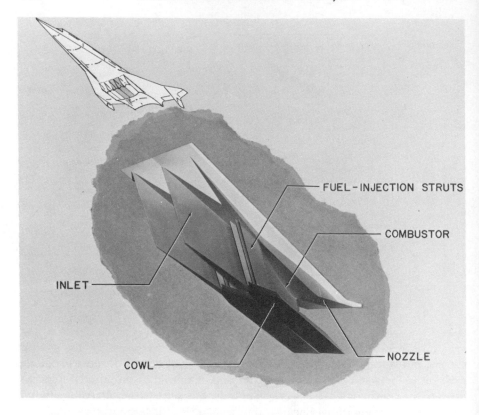

FUEL-INJECTION STRUTS

COMBUSTOR

INLET

NOZZLE

COWL

**Figure 20.8** Scramjet engine concept for hypersonic flight

A portion of the mathematical model for the strut thermal analysis is shown in Figure 20.9. Mass transport, surface convection elements, and an integrated plate-fin/fluid element were combined with conduction elements for both steady and transient analyses.[3,4,8] The finite element model and an identical lumped-parameter finite difference model used for comparison had 121 unknown nodal temperatures. The transient response of the strut to an instantaneous reduction in the heat load (flame-out simulation) on one side of the aft portion of the strut (Figure 20.10) was analysed.

Excellent agreement between finite element (FE) and lumped-parameter finite difference (FD) calculated coolant temperature distributions for one of the coolant passages was obtained as shown in Figure 20.10 for $t = 0$ s and $t = 2$ s. As in the previous example, upwind mass transport elements were required between exits of the two coolant passages and the outlet manifold to eliminate coolant temperature oscillations.[4]

A criticism of some finite element programs for non-linear transient thermal analysis has been their relatively low computer efficiency compared to lumped-

ELEMENTS

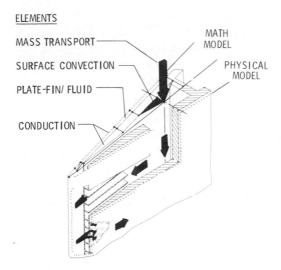

**Figure 20.9**  Forward portion of the finite element
thermal mathematical model of the strut

parameter finite difference programs. A comparison of relative computer times
for the strut problem is also indicated in Figure 20.10. The Crank–Nicolson
implicit algorithm used in one of the finite element programs took 63 s. The
explicit algorithms used in the finite element and finite difference lumped-
parameter programs took over 2300 s. The execution times of both of these
programs were significantly reduced to 57 s and 29 s respectively by the 'zero'
capacitance nodal elimination scheme.

## 20.5  CONCLUDING REMARKS

Recent advances in finite element thermal analysis methodology for conduc-
tion/forced-convection heat transfer have been summarized. Comparisons of
finite element and finite difference lumped-parameter calculated temperatures
with closed-form analytical solutions for fluid convection problems of increasing complexity have shown the conventional finite element approach with a
consistent-capacitance matrix to be more accurate than other methods. Use
of a lumped-capacitance matrix degrades accuracy due to artificial dis-
persion/diffusion. Use of the upwind finite element formulation can seriously
degrade the accuracy of predicted temperatures due to artificial diffusion.
However, upwind elements are beneficial for solution of problems where
multiple flow passages with different outlet temperatures culminate in a
common manifold.

The applications have shown the finite element method to be fully competitive with the well established finite difference lumped-parameter method for

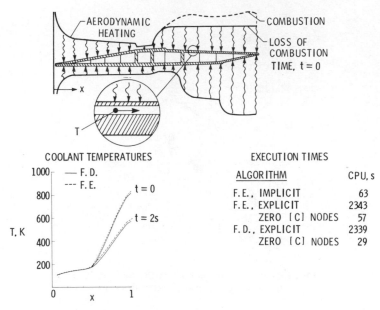

**Figure 20.10**  Comparison of FE and FD results for strut thermal transient
problem

combined conduction/forced-convection analysis. The applications indicate
that the methods currently have about the same conduction/forced-convection
analysis capabilities, and the finite element method is more accurate. Although
computer time requirements are highly program dependent and only limited
comparisons have been made, the results indicate that the finite difference
lumped-parameter method maintains an edge in efficiency. The studies also
indicate, however, that as additional finite element experience is gained,
improvements in efficiency can be achieved.

## REFERENCES

1.  H. N. Kelly, A. R. Wieting, C. P. Shore and R. A. Nowak (1978). 'Recent advances in
    convectively cooled engine and airframe structures for hypersonic flight', presented
    at *11th Congr. Int. Counc. Aeronautical Sci. (ICAS)*, *ICAS Proc.*, **1**, 137–151.
2.  A. R. Wieting (1979). 'Application of numerical methods to heat transfer and
    thermal stress analysis of aerospace vehicles', in *Numerical Methods in Thermal
    Problems, Proc. 1st Int. Conf., University College, Swansea, July 1979*, R. W. Lewis
    and K. Morgan (Eds), pp. 634–643, Pineridge Press, Swansea.
3.  E. A. Thornton and A. R. Wieting (1978). 'Finite element methodology for thermal
    analysis of convectively cooled structures', *AIAA 15th Aerospace Sciences Meeting,
    Los Angeles, CA, January 1977, AIAA Paper* 77-187, in *Progress in Astronautics
    Aeronautics*, Vol. 60, *Heat Transfer and Thermal Control Systems*, L. S. Fletcher
    (Ed.), pp. 171–189.

4. E. A. Thornton and A. R. Wieting (1979). 'Finite element methodology for transient conduction/forced-convection thermal analysis', presented at *AIAA 14th Thermophysics Conf., Orlando, Florida, June 1979, AIAA Paper* 79-1100.
5. E. A. Thornton and A. R. Wieting (1980). 'Evaluation of finite element formulations for transient conduction/forced-convection analysis', presented at *Natl Conf. on Numerical Methods in Heat Transfer, September 1979, College Park, Maryland. Numerical Heat Transfer*, accepted for publication.
6. E. A. Thornton (1979). 'Application of upwind convective finite elements to practical conduction/forced convection thermal analysis', in *Numerical Methods in Thermal Problems, Proc. 1st Int. Conf., University College, Swansea, July 1979*, R. W. Lewis and K. Morgan (Eds), pp. 402–411. Pineridge Press, Swansea.
7. W. D. Whetstone (1978). 'SPAR structural system reference manual', *Engineering Information Systems, Inc., San Jose, California, NASA Report* CR-145098.
8. A. R. Wieting and E. A. Thornton (1978). 'Thermostructural analysis of a scramjet fuel-injection strut', *NASA Report* CP-2065, pp. 119–144.

*Numerical Methods in Heat Transfer*
Edited by R. W. Lewis, K. Morgan, and O. C. Zienkiewicz
© 1981 John Wiley & Sons Ltd

## Chapter 21

# A Computational Procedure for Three-Dimensional Recirculating Flows Inside Can Combustors

*M. A. Serag-Eldin and D. B. Spalding*

## SUMMARY

This chapter describes a numerical prediction procedure for the computation of three-dimensional recirculating and swirling flows inside can combustion chambers. The predictions result from the simultaneous solution of differential equations for the main dependent variables, by means of a given finite-difference solution algorithm. This algorithm has been developed so as to handle efficiently the cyclic boundary conditions and axisymmetric properties of can-combustor problems.

This chapter also compares with measurements some predictions for turbulent non-reacting flow in a model of a gas-turbine combustion chamber.

## NOMENCLATURE

| | | | |
|---|---|---|---|
| $A$ | control surface area | $p$ | pressure |
| $C$ | FDE coefficient | $r$ | radial coordinate in the cylindrical system |
| $C_1, C_2, C_\mu$ | $k, \varepsilon$ model constants | | |
| $G_k$ | rate of generation of $k$ per unit volume | $u$ | axial velocity |
| | | $\mathbf{V}$ | velocity vector |
| $I$ | refers to $x = $ constant surfaces | $v$ | radial velocity |
| | | $w$ | swirl velocity |
| $J$ | refers to $r = $ constant surfaces | $x$ | axial coordinate in the cylindrical system |
| $K$ | refers to $\theta = $ constant surfaces | $\mu$ | viscosity |
| | | $\rho$ | density |
| $k$ | turbulence energy | $\theta$ | circumferential coordinate in the cylindrical system |
| $L$ | $I$ of last solution surface | | |
| $M$ | $J$ of last solution surface | | |
| $N$ | $K$ of last solution surface | $\sigma_k, \sigma_\varepsilon$ | $k, \varepsilon$ model constants |

$\varepsilon$ turbulence/energy dissipation rate

$\phi$ flow variable

*Subscripts*

l laminar

t turbulent

P at point 'P'

$x+, x-$ at larger and smaller $x$, respectively

$r+, r-$ at larger and smaller $r$, respectively

$\theta+, \theta-$ at larger and smaller $\theta$, respectively

$\phi$ pertaining to $\phi$

*Superscripts*

$'$ perturbed values

$*$ preliminary values

## 21.1 INTRODUCTION

This chapter is concerned with a numerical prediction procedure for steady, three-dimensional, recirculating, and swirling flows inside can-shaped combustion chambers. The flows investigated are turbulent and reacting, the actual physical processes considered being simulated mathematically by means of differential equations for the relevant dependent variables. Since these equations are partial differential and non-linear, their simultaneous solution requires a special solution algorithm; the presentation of which is the main objective of this chapter.

This algorithm is based on that of Patankar and Spalding;[1,2] it goes beyond those references in handling 'cyclic conditions', as will be described.

### 21.1.1 The problem considered

Figure 21.1 displays a sketch of a representative can-combustor geometry. Fuel and primary air enter the chamber through one end, while dilution air enters through discrete side-ports. For most applications, the fuel and primary air may be assumed to enter the chamber axisymmetrically; moreover, the side wall may be considered axisymmetric. However, the positioning of the discrete

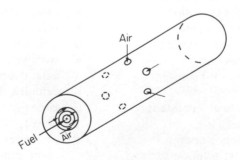

**Figure 21.1** Representative can combustor

dilution air ports causes the flow to be three dimensional. The spaces between the dilution air ports determine the degree of penetration of the dilution air jets;[3,4] hence, if *axisymmetric* computations were performed, with the dilution air jets approximated by a continuous peripheral slot, the flow field would not be predicted accurately. In order to stabilize the flame, a recirculation zone is usually created in the upstream region of the combustion chamber. Thus, the problem considered is that of predicting a three-dimensional, recirculating flow, with two special numerical characteristics. The first of these is the nearly axisymmetric shape of the combustor boundaries, a property which may be utilized advantageously to reduce the computation time. The other property is the appearance of 'cyclic boundary conditions', associated with the repetitive nature of the combustor-wall perforations.

## 21.2   THE GOVERNING DIFFERENTIAL EQUATIONS

A cylindrical system of coordinates is adopted as shown in Figure 21.2. The mass conservation equation states that:

$$\text{div}\,(\rho \mathbf{V}) = 0 \qquad (21.1)$$

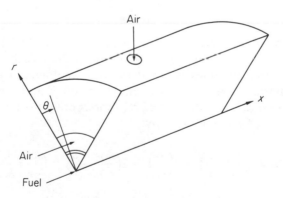

**Figure 21.2**   Integration domain

The three momentum equations may all be written in the following general form:

$$\text{div}\,(\rho \mathbf{V}\phi) = \text{div}\,(\Gamma_\phi\,\text{grad}\,\phi) + S_\phi \qquad (21.2)$$

where $\phi$ stands for any of the three variables, $v$, $wr$ or $u$, and the corresponding values of $\Gamma_\phi$ and $S_\phi$ are indicated in Table 21.1, where $\mu$ stands for the molecular (laminar) viscosity of the flow $\mu_l$.

**Table 21.1** The $\Gamma_\phi$ and $S_\phi$ expressions for the momentum equations in laminar flow

| $\phi$ | $\Gamma_\phi$ | $S_\phi$ |
|---|---|---|
| $v$ | $\mu$ | $\dfrac{\rho w^2}{r} - \dfrac{\partial p}{\partial r} + \dfrac{1}{r}\dfrac{\partial}{\partial r}\left(r\mu\dfrac{\partial v}{\partial r}\right) + \dfrac{1}{r}\dfrac{\partial}{\partial\theta}\left(\mu r\dfrac{\partial(w/r)}{\partial r}\right) - 2\dfrac{\mu}{r}\left(\dfrac{\partial w}{r\,\partial\theta} + \dfrac{v}{r}\right) + \dfrac{\partial}{\partial x}\left(\mu\dfrac{\partial u}{\partial r}\right)$ |
| $wr$ | $\mu$ | $-\dfrac{\partial p}{\partial\theta} + \dfrac{1}{r}\dfrac{\partial}{\partial r}\left(r\mu\dfrac{\partial v}{\partial\theta}\right) + \dfrac{2}{r}\left(\dfrac{\partial}{\partial\theta}(\mu v) - \dfrac{\partial}{\partial r}(\mu wr)\right) + \dfrac{1}{r}\dfrac{\partial}{\partial\theta}\left(\dfrac{\mu}{r}\dfrac{\partial wr}{\partial\theta}\right) + \dfrac{\partial}{\partial x}\left(\mu\dfrac{\partial u}{\partial\theta}\right)$ |
| $u$ | $\mu$ | $-\dfrac{\partial p}{\partial x} + \dfrac{1}{r}\dfrac{\partial}{\partial r}\left(r\mu\dfrac{\partial v}{\partial x}\right) + \dfrac{1}{r}\dfrac{\partial}{\partial\theta}\left(\mu\dfrac{\partial w}{\partial x}\right) + \dfrac{\partial}{\partial x}\left(\mu\dfrac{\partial u}{\partial x}\right)$ |

If the flow is turbulent, the same expressions are employed, but with $\mu$ standing for an effective viscosity given by:

$$\mu = \mu_l + \mu_t \qquad (21.3)$$

where $\mu_t$ is an estimated turbulent viscosity of the flow, derived from the $k$, $\varepsilon$ model of turbulence.[5] Numerous investigations with this model have confirmed its plausibility for many turbulent reacting and non-reacting flows (e.g. references 6 and 7).

The value of $\mu_t$ at any point in the field is calculated from the local values of $k$ and $\varepsilon$ according to:

$$\mu_t = \rho C_\mu k^2 / \varepsilon \qquad (21.4)$$

where $C_\mu$ is one of the constants whose values are reproduced in Table. 21.3.

The field values of $k$ and $\varepsilon$ are derived from the solution of differential equations for these quantities, which are solved simultaneously with the other governing differential equations. The equations for $k$ and $\varepsilon$ are of the same form as Equation (21.2), with $\Gamma_\phi$ and $S_\phi$ representing the terms displayed in Table 21.2. Here, $G_k$ is expressed by:

$$G_k \equiv \mu_t\left[2\left(\frac{\partial u}{\partial x}\right)^2 + 2\left(\frac{\partial v}{\partial r}\right)^2 + 2\left(\frac{\partial w}{r\,\partial\theta} + \frac{v}{r}\right)^2 + \left(\frac{\partial u}{\partial r} + \frac{\partial v}{\partial x}\right)^2\right.$$

$$\left. + \left(\frac{\partial w}{\partial x} + \frac{1}{r}\frac{\partial u}{\partial\theta}\right)^2 + \left(\frac{1}{r}\frac{\partial v}{\partial\theta} + r\frac{\partial(w/r)}{\partial r}\right)^2\right] \qquad (21.5)$$

**Table 21.2** The $\Gamma_\phi$ and $S_\phi$ expressions, for the $k$ and $\varepsilon$ equations

| $\phi$ | $\Gamma_\phi$ | $S_\phi$ |
|---|---|---|
| $k$ | $\mu/\sigma_k$ | $G_k - \rho\varepsilon$ |
| $\varepsilon$ | $\mu/\sigma_\varepsilon$ | $C_1\dfrac{\varepsilon}{k}G_k - C_2\rho\dfrac{\varepsilon^2}{k}$ |

and $C_1$, $C_2$, $\sigma_k$, and $\sigma_\varepsilon$ are all universal model constants, whose values are reproduced in Table 21.3.

Table 21.3 The $k$, $\varepsilon$ model constants

| $C_\mu$ | $C_1$ | $C_2$ | $\sigma_k$ | $\sigma_\varepsilon$ |
|---------|-------|-------|------------|----------------------|
| 0.09 | 1.44 | 1.92 | 1.0 | 1.3 |

Combustion and heat transfer models also exist which lead to transport equations of the form of Equation (21.2) (e.g. references 8–10). The simultaneous solution of the transport equations for the combustion and heat transfer variables, together with the rest of the flow equations, allows the evaluation of the field values of these variables. The combustion and heat transfer processes are assumed to influence the equations for conservation of mass momentum and turbulence, solely through the density gradients introduced.

## 21.3 INTEGRATION DOMAIN AND BOUNDARY CONDITIONS

### 21.3.1 Boundaries of the integration domain

If the dilution air ports of Figure 21.1 are identical and equally spaced, then the cross section of the integration domain needs only to cover a 60° sector of the combustor cross section, as indicated in Figure 21.2. This is because the combustor geometry may then be divided into six identical domains, each exhibiting a 60° sector cross section.

### 21.3.2 Types of boundary conditions encountered

The problem investigated exhibits the following types of boundaries:

(1) Wall boundaries where the boundary values or their fluxes are prescribed.
(2) Entry boundaries where the boundary values or their fluxes are prescribed.
(3) A downstream exit boundary where the gradients are assumed to be zero.
(4) Flow boundaries which are 'cyclic' and where only 'cyclic boundary conditions' are prescribed. Reference should be made to Figure 21.3, which displays a cross section of the integration domain sketched in Figure 21.2; the $\theta$-boundaries, B1 and B2, are 'cyclic planes', i.e. both planes exhibit identical profiles of the flow variables. This is a consequence of the identical position of the B1 and B2 planes with respect to the combustor geometry (air ports, walls, etc.). It also follows that all $\theta$-planes separated

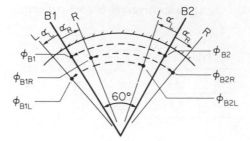

**Figure 21.3**  Cyclic boundary conditions for 60°
section

by an angle equal to the one subtended between B1 and B2 are cyclic
planes. Thus the 'cyclic boundary conditions' state that:

$$\phi_{B1} = \phi_{B2} \tag{21.6a}$$

$$\phi_{B1L} = \phi_{B2L} \tag{21.6b}$$

$$\phi_{B1R} = \phi_{B2R} \tag{21.6c}$$

where $\phi$ represents any of the variables, and the subscripts B1, B1L and
B1R refer to locations on the boundary B1, locations on the left-hand side
of B1 and locations on the right-hand side of B1, respectively; whereas
B2, B2L, and B2R indicate locations corresponding to those of B1, B1L,
and B1R on their respective cyclic planes, as displayed in Figure 21.3.
Alternatively, when the integration domain covers the whole cross section
of the combustor in Figure 21.1, then the two $\theta$-boundary planes B1 and
B2 physically overlap as sketched in Figure 21.4. However, as far as the

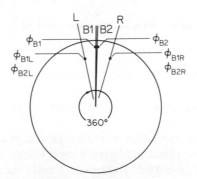

**Figure 21.4**  Cyclic boundary conditions for full cross section

solution of the finite difference equations (FDE's) is concerned, the two
planes are distinct and separated by 360°. Moreover, the $\theta$-boundary

values and fluxes are unknown prior to solution, and Equations (21.6) still apply; hence this case also imposes cyclic boundary conditions.

## 21.4 THE SOLUTION ALGORITHM

The solution algorithm employed here is an iterative finite difference one, which is based upon the well known 'SIMPLE' algorithm,[11] and which has been extended to permit efficient handling of the cyclic boundary conditions.

### 21.4.1 Grid and notation

*Grid arrangement*

Figures 21.5 and 21.6 illustrate the grid-node arrangements and control volume shapes employed in the $(r, x)$ and $(r, \theta)$ planes, respectively. The dots

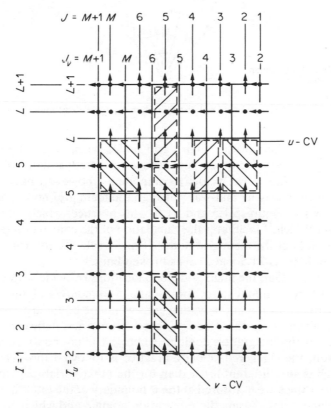

CV: control volume

**Figure 21.5** Grid distribution in *r*- and *x*-directions

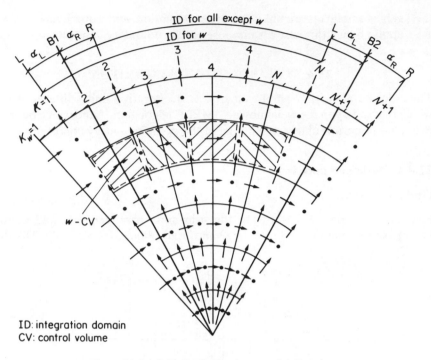

ID: integration domain
CV: control volume

**Figure 21.6** Grid distribution in $r_{\phi_p}$ and $\theta$-directions

indicate the locations of the 'main' grid nodes where the values of all the variables except the velocity components are evaluated; and the full lines show their control surfaces. The velocity components, however, have 'staggered' grid-node locations lying mid-way between the main grid nodes, as indicated by the arrows in Figures 21.5 and 21.6. This practice, which follows that of Harlow and Welch,[12] facilitates the calculation of the convection terms on the control surfaces of the main grid nodes. Typical velocity control volumes are displayed in Figures 21.5 and 21.6 as broken lines.

It is remarked that, the main grid-node locations coincide with those of the normal velocity components at the $x$- and $r$-boundaries of the integration domain. This arrangement is advantageous when the boundary values or their fluxes are given, as is the case for these boundaries. However, at the $\theta$-boundaries of the integration domain, where cyclic boundary conditions are encountered, the staggered-velocity practice is followed through the boundary; which is also different for $w$ than for the other variables. Moreover the $\theta$-boundary values are not stored at the $\theta$-boundary of the integration domain, but at $\theta$-planes lying outside the integration domain and which are located at such an angle as to form 'cyclic planes' with internal solution planes ($K = 1$ corresponding to $K = N$, and $K = 2$ to $K = N + 1$). These innovations over the original solution procedure[2] (which employs the practice adopted for the $r$- and $x$-boundaries, for the $\theta$-boundaries as well) greatly aid the accurate expression of the cyclic $\theta$ boundary conditions.

### Notation for the grid surfaces

Three indices, $I$, $J$, and $K$, are employed to refer to the different grid surfaces and grid-node locations, as shown in Figures 21.5 and 21.6. The index $I$ pertains to the $x =$ constant surfaces, $J$ to the $r =$ constant surfaces, and $K$ to the $\theta =$ constant surfaces. The location of any grid node is therefore specified by the values of $I$, $J$, and $K$.

Since each of the velocity components grid-node locations is staggered in one direction, the $I$, $J$, and $K$ indices may indicate different surfaces for the velocity components, from those which they indicate for the main grid surfaces, as apparent from Figures 21.5 and 21.6. Therefore, in order to avoid confusion, in Figures 21.5 and 21.6 the pertinent velocity symbol is subscripted on the index of the special velocity grid surface, i.e. $I_u$ for the special $u$ surfaces, $J_v$ for the special $v$ surfaces and $K_w$ for the special $w$ surfaces.

### 21.4.2 The finite difference equations

The FDE for any dependent variable $\phi$, is derived at any grid node P by integrating the pertinent differential Equation (21.2) over the control volume surrounding P. The resulting equation is of the form:

$$\phi_P = (C_{\phi,x+}\phi_{x+} + C_{\phi,x-}\phi_{x-} + C_{\phi,r+}\phi_{r+} + C_{\phi,r-}\phi_{r-} + C_{\phi,\theta+}\phi_{\theta+}$$
$$+ C_{\phi,\theta-}\phi_{\theta-} + S_{\phi,P})/C_{\phi,P} \tag{21.7}$$

where

$$C_{\phi,P} \equiv C_{\phi,x+} + C_{\phi,x-} + C_{\phi,r+} + C_{\phi,r-} + C_{\phi,\theta+} + C_{\phi,\theta-} \tag{21.8}$$

$\phi_P$ is the value of $\phi$ at the grid node 'P', the $C$'s are the finite difference coefficients evaluated at the control surfaces, and the subscripts $(x+, x-, \ldots, \theta-)$ to the $\phi$'s refer to the locations of the immediate neighbouring grid nodes relative to the considered node P; whereas the same subscripts to the $C$'s refer to the locations of the coefficient evaluation points with respect to P, as displayed in Figure 21.7. The coefficients $C$ express the contributions of the convection and diffusion crossing the control surfaces.[13]

The finite difference counterpart of Equation (21.1), however, is not directly employed in the solution procedure. Instead, mass conservation is ensured through the solution of a 'pressure correction' equation of the same form as Equation (21.7); as will be explained in Section 21.4.3.

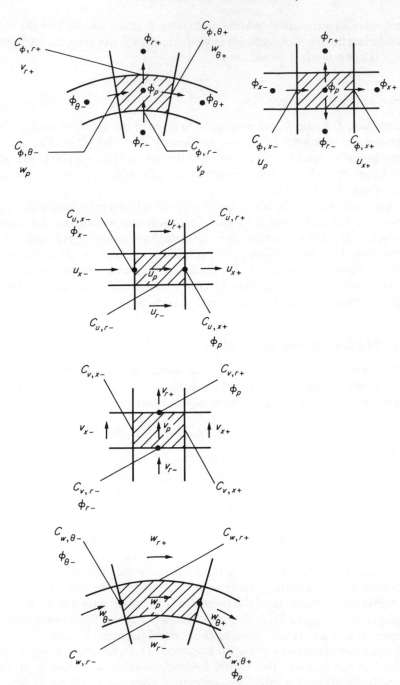

**Figure 21.7**  Convention for neighbouring nodes and coefficients

### 21.4.3 The pressure correction method

The central idea behind the pressure correction method may be summarized as follows.

First, the momentum equations are solved for a guessed pressure field. The solution of these will result in preliminary velocity components $u^*$, $v^*$, and $w^*$, which in general will not satisfy the continuity equation, because the initially assumed pressure field is incorrect.

Then, in the next step, simultaneous corrections are made to both the velocity and pressure fields, such that the newly corrected velocity components $u$, $v$, and $w$, satisfy mass conservation and are connected to the pressure corrections through the following linearized expressions of the momentum equations:

$$u_P = u_P^* + \frac{A_{x,P}}{C_{u,P}}(p'_{x-} - p'_P) \qquad (21.9a)$$

$$v_P = v_P^* + \frac{A_{r,P}}{C_{v,P}}(p'_{r-} - p'_P) \qquad (21.9b)$$

$$w_P = w_P^* + \frac{A_{\theta,P}}{C_{w,P}}(p'_{\theta-} - p'_P) \qquad (21.9c)$$

where $p'$ represent local pressure corrections and their subscripts denote the position of the pressure grid nodes relative to the pertinent velocity component, according to the notation displayed in Figure 21.7. $A_{x,P}$, $A_{r,P}$, and $A_{\theta,P}$ are the main-grid-node control surface areas, which are normal to $u_P$, $v_P$, and $w_P$ respectively; whereas $C_{u,P}$, $C_{v,P}$, and $C_{w,P}$ are the coefficients given by Equation (21.8) for $u$, $v$, and $w$, respectively.

The values of $p'$ are calculated from the solution of FDE's for $p'$, which are of the same form as Equation (21.7). The FDE for $p'$ at any main grid node P, is derived from the substitution of the right-hand side of Equation (21.9) for the velocity components appearing in the expression for the conservation of mass over the control volume surrounding P. Consequently, the values of $u_P$, $v_P$, and $w_P$ derived from Equation (21.9), for the calculated values of $p'$, satisfy mass conservation over each of the main grid-node control volumes.

### 21.4.4 The sequence of solution

Within each iteration, the sequence of solution is as follows:

(1) The FDE's for $u$, $v$, and $wr$ are solved successively, in conformity with the previous-iteration pressure field. For each of the variables, the FDE's at the different internal grid nodes constitute a set of algebraic equations which is solved by a line-iteration method described in Section 21.4.5.

From Figures 21.5 and 21.6 it is noticed that the first $I_u$ internal surface is $I_u = 3$ and the first $J_v$ internal surface is $J_v = 3$; whereas, the first $K_w$ internal surface is $K_w = 2$, again as a consequence of the special treatment for cyclic boundary conditions on the $\theta$-boundaries.

(2)  The FDE's for $p'$ are solved and the pressure field is updated.
(3)  The velocity components are updated according to Equation (21.9).
(4)  If any other dependent variables are present, their sets of FDE's are solved successively. The sequence of solution among these variables is chosen according to their relative influence on each other.
(5)  When present, the auxiliary quantities such as $\mu_t$, $T$, and $\rho$, are calculated from the dependent variables.

At the end of each iteration, a check is made on the degree of convergence of the solution, and if this is not satisfactory, a new iteration is initiated and the procedure is repeated.

### 21.4.5  The solution of the FDE's

From Equation (21.7), it is apparent that, for any of the variables, the FDE at a given grid node P links the value of the variable at P to its values at the neighbouring grid nodes, $x+$, $x-$, $r+$, .... Thus, the set of FDE's at the different grid nodes constitutes a system of simultaneous algebraic equations; the solution of which is obtained here by performing successive line-iterations in the three coordinate directions. In this procedure, the FDE's at all the grid nodes along a given grid line are solved simultaneously for the values along that line, whilst the neighbouring values along the other grid lines are fixed at their previous iteration values.

Along the grid lines in the $x$- and $r$-directions, the set of FDE's yields a tridiagonal matrix of coefficients whose solution is obtained rapidly by means of the tridiagonal matrix algorithm (see, e.g. reference 14).

Along the $\theta$-direction however, the resulting matrix of coefficients is not a true tridiagonal one. This is because, as a consequence of the cyclic $\theta$-boundary conditions, the FDE for $\phi_2$ contains values of $\phi_N$ (the subscripts referring to the pertinent $K$-planes in Figure 21.6), and vice versa, so that the resulting matrix of coefficients contains corner values as displayed in Figure 21.8. This matrix cannot be solved by an ordinary tridiagonal matrix algorithm; instead, it requires a more complex algorithm termed 'cyclic tridiagonal matrix algorithm', which is presented in an appendix to this chapter.

It may be mentioned here that the tridiagonal matrix algorithm could also be employed in the $\theta$-direction, provided fixed $\theta$-boundary values were assumed. Since .he whole solution procedure is iterative, the $\theta$-boundary values may be fixed from the previous iteration solution and updated at each iteration. However, investigations with this practice[15] revealed that the solution con-

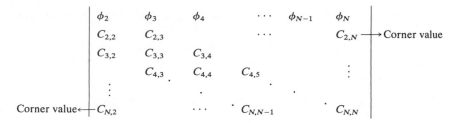

**Figure 21.8** Matrix of coefficients for the $\theta$-direction solution

verged much more slowly than when the $\theta$-boundary values are introduced implicitly by way of the algorithm.

### 21.4.6 Expression of the boundary conditions

The given boundary values are stored at the boundary grid nodes before the solution commences, and retained there throughout the solution.

The given boundary fluxes are implemented through the modification of the coefficients of the FDE's at the near-boundary grid nodes.

The cyclic boundary conditions, which are expressed by Equation (21.6), are satisfied inherently by allocating common computer storage to the corresponding array elements in the $K = 1$ and $K = N$ planes, and the $K = 2$ and $K = N + 1$ planes, displayed in Figure 21.6.

## 21.5 THE STARTING PROCEDURE

To improve the convergence rate of the solution, the following starting procedure is adopted.[15]

(1) An axisymmetric solution is obtained for an axisymmetric geometry similar to the three-dimensional one, except that the dilution air holes, which are the cause of the asymmetries, are replaced by a continuous axisymmetric slot of the same flow area. This solution employs the same $(r, x)$ grid distribution as the three-dimensional solution, but uses only a single $\theta$-plane for the computations. Typically, the axisymmetric solution requires only about 2% of the computation time required for the three-dimensional solution.

(2) Once the axisymmetric solution is obtained, this solution and its boundary values are ascribed to all the $\theta$-planes composing the grid of the three-dimensional solution. Subsequently, the solution proceeds in three dimensions, the given three-dimensional boundary conditions being introduced gradually over a number of specified iterations (typically twenty). In this stage, the dilution air flow areas are incremented gradually over the specified number of iterations, the flow area increasing at the $\theta$-planes where the actual dilution air ports are positioned and decreasing in the other $\theta$-planes; the sum

of all the dilution air flow areas is kept constant, so that overall mass conservation is satisfied.

(3) Finally, when the correct three-dimensional boundary conditions are reached, the solution progresses iteratively with fixed boundary conditions.

## 21.6 THE TURBULENT COLD-FLOW TEST

### 21.6.1 The case investigated

In order to demonstrate the capability of the prediction procedure, and to verify the use of the $(k, \varepsilon)$ model of turbulence to predict three-dimensional flows in can-combustor geometries, the prediction procedure comprising the equations for mass conservation, momentum, $k$, and $\varepsilon$, was applied to predict the three-dimensional flow inside a model of an industrial gas-turbine combustion chamber for which experimental cold-flow measurements were available.[16] A sketch of this combustion chamber geometry is reproduced in Figure 21.9. For the sake of computational simplicity, the slightly non-uniform cross section of this geometry is approximated to a uniform one (Figure 21.10).

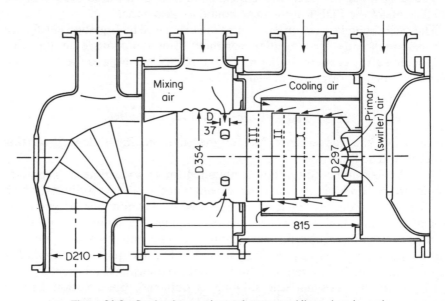

**Figure 21.9** Section in experimental geometry (dimensions in mm)

*Main parameters*

| | |
|---|---|
| Mean chamber pressure | atmospheric |
| Flow rates | |
|    primary air | $0.215 \text{ kg s}^{-1}$ |

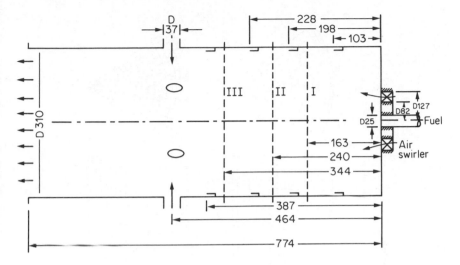

**Figure 21.10** Section in simplified geometry (dimensions in mm)

| | |
|---|---|
| dilution air | $0.258 \text{ kg s}^{-1}$ |
| first film cooling air | $0.083 \text{ kg s}^{-1}$ |
| second film cooling air | $0.0571 \text{ kg s}^{-1}$ |
| third film cooling air | $0.0649 \text{ kg s}^{-1}$ |
| fourth film cooling air | $0.0435 \text{ kg s}^{-1}$ |
| Swirler vane angles | 56° (flow deflection) |

Further details of this particular investigation may be found in Serag-Eldin.[15]

## 21.6.2 Some computational details

*The integration domain and grid distribution*

The integration domain for this case covers a longitudinal slice of the combustor whose cross section is a 60° sector of the combustor cross section, as in Figure 21.2. In order to investigate the effect of grid refinement on the solution obtained, The computations were performed with two grids of different sizes. The coarser of the two grids employs $24 \times 17 \times 7$ grid nodes in the $x$-, $r$- and $\theta$-directions respectively, whereas the finer one employs $30 \times 22 \times 12$ grid nodes in the same directions. The grid-node distributions for both grids are deducible from Figure 21.11 where the main grid-node surfaces are displayed.

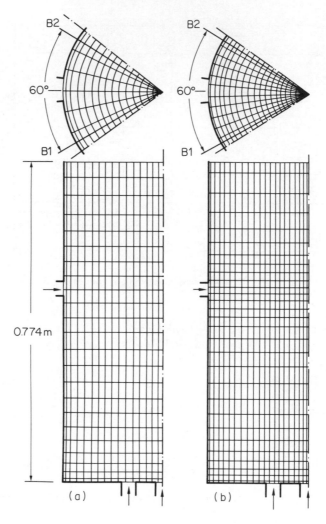

**Figure 21.11** (a) The $24 \times 17 \times 7$ grid; (b) the $30 \times 22 \times 12$ grid

*Computational time and storage requirements*

The following were the computational requirements for a well-converged solution on an IBM 360/195 computer:

Storage requirement            60k words for coarse grid
                                      145k words for fine grid
Computation time              6 min for coarse grid
                                   32 min for fine grid

### 21.6.3   The corresponding measured and predicted velocities

Figures 12.12–12.16 display the corresponding measured and predicted velocity profiles, at the three measurement sections denoted by I, II and III in Figure 21.9. It should be noted that grid refinement exerted little influence on the computed results. Moreover, agreement between the measured and computed results is generally good.

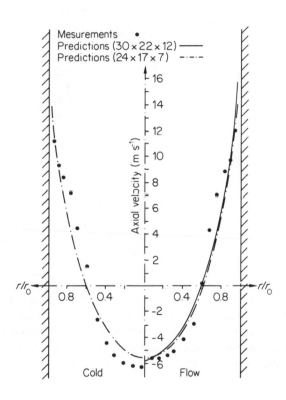

**Figure 21.12**   Axial velocity profile at I

## 21.7   SUMMARY AND CONCLUSION

This chapter describes a numerical prediction procedure for three-dimensional recirculating and swirling flows, inside can-combustor geometries. At the heart of this procedure is a computational algorithm which features special refinements for the expression of cyclic boundary conditions, as well as a starting procedure which exploits the axisymmetric properties of some can-combustor problems.

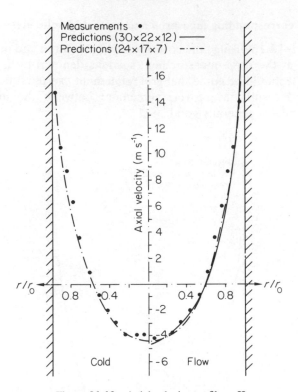

**Figure 21.13**   Axial velocity profile at II

The computational algorithm, coupled with the $(k, \varepsilon)$ model of turbulence, is shown to predict successfully the turbulent cold flow inside a model of an industrial gas-turbine combustion chamber. Further incorporation of available models of combustion, should allow the computation of turbulent reacting flows.

### Acknowledgements

The authors are grateful to Combustion, Heat and Momentum Ltd for lending them the basic code, and to the Science Research Council for sponsoring the work that led to this publication.

## APPENDIX   THE CYCLIC TRIDIAGONAL MATRIX ALGORITHM

The cyclic tridiagonal matrix algorithm is employed to solve systems of algebraic equations of the form:

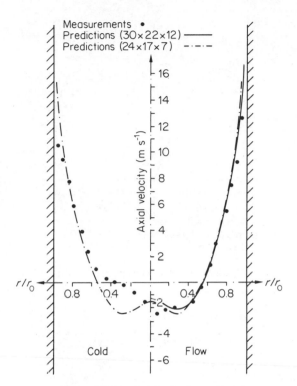

**Figure 21.14** Axial velocity profile III

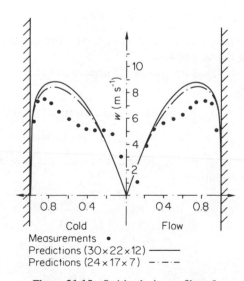

**Figure 21.15** Swirl velocity profile at I

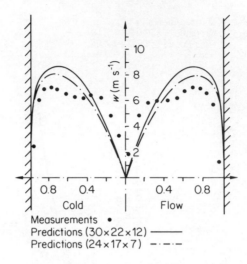

**Figure 21.16** Swirl velocity profile at II

$$d_1 x_1 = a_1 x_2 + b_1 x_n + c_1$$

$$d_2 x_2 = a_2 x_3 + b_2 x_1 + c_2$$

$$\vdots$$

$$d_i x_i = a_i x_{i+1} + b_i x_{i-1} + c_i \qquad (21.A1)$$

$$\vdots$$

$$d_{n-1} x_{n-1} = a_{n-1} x_n + b_{n-1} x_{n-2} + c_{n-1}$$

$$d_n x_n = a_n x_1 + b_n x_{n-1} + c_n$$

where $a_i$, $b_i$, $c_i$, and $d_i$, for $i = 1, 2, \ldots, n$, are the coefficients, and $x_i$, for $i = 1, 2, \ldots, n$, are the unknowns.

The solution is obtained by means of successive substitutions, according to the following steps:

(1)   New coefficients $a'_1$, $b'_1$, $c'_1$, and $d'_1$ are ascribed the following values:

$$a'_1 = a_1 \quad b'_1 = b_1 \quad c'_1 = c_1 \quad d'_1 = d_1 \qquad (21.A2)$$

(2)   New coefficients $a'_i$, $b'_i$, $c'_i$, and $d'_i$, for $i = 2, 3, \ldots, n$, are obtained from the following equations:

$$a'_i = a_i \quad b'_i = \frac{b_i b'_{i-1}}{d'_{i-1}}$$

$$\qquad (21.A3)$$

$$c'_i = \frac{c'_{i-1} b_i}{d'_{i-1}} + c_i \quad d'_i = d_i - \frac{b_i a'_{i-1}}{d'_{i-1}}$$

by working upwards from $i = 2$ to $i = n$.

(3)  New coefficients $a''_{n-1}$ and $b''_{n-1}$ are calculated from

$$a''_{n-1} = \frac{a'_{n-1} + b'_{n-1}}{d'_{n-1}} \quad b''_{n-1} = \frac{c'_{n-1}}{d'_{n-1}} \tag{21.A4}$$

(4)  New coefficients $a''_i$ and $b''_i$, for $i = 1, 2, \ldots, n-2$, are evaluated from the following equations:

$$a''_i = \frac{a'_i a''_{i+1} + b'_i}{d'_i} \quad b''_i = \frac{a'_i b''_{i+1} + c'_i}{d'_i} \tag{21.A5}$$

by working down and from $i = n-2$ to $i = 1$.

(5)  Finally, the value of $x_n$ is calculated from:

$$x_n = \frac{b''_1 a'_n + c'_n}{d'_n - b'_n - a''_1 a'_n} \tag{21.A6}$$

and the values of $x_i$, for $i = 1, 2, \ldots, n-1$, are derived from

$$x_i = a''_i x_n + b''_i \tag{21.A7}$$

The derivation of this algorithm may be found in (21.15).

## REFERENCES

1.  S. V. Patankar and D. B. Spalding (1972). 'A calculation procedure for heat, mass and momentum transfer in three-dimensional parabolic flows', *Int. J. Heat Mass Transfer*, **15**, 1787–1806.
2.  S. V. Patankar and D. B. Spalding (1973). 'A computer model for three-dimensional flow in furnaces', *Proc. 14th Int. Symp. on Combustion*, pp. 605–614, The Combustion Institute.
3.  Y. U. A. Spiridonov and I. G. Maschenko (1971). 'Influence of the area of the holes in a flame tube on the non-uniformity of gas temperature distribution', *Teploenergetika*, **18**, No. 5, 24–27.
4.  J. D. Holdeman, R. E. Walker and D. L. Kors (1973). 'Mixing of multiple dilution jets with a hot primary air-stream for gas turbine combustors', *NASA Report* TM X-714 26.
5.  B. E. Launder and D. B. Spalding (1974). 'The numerical computation of turbulent flows', *Comp. Meth. Appl. Mech. Engng*, **3**, 269–289.
6.  P. Hutchinson, E. E. Khalil, J. H. Whitelaw and G. Wigley (1976). 'The calculation of furnace flow properties and their experimental verification', *J. Heat Transfer*, **98**, 276.
7.  E. E. Khalil, D. B. Spalding and J. Whitelaw (1975). 'The calculation of local flow properties in two-dimensional furnaces', *Int. J. Heat Mass Transfer*, **18**, 775–791.
8.  M. A. Serag-Eldin and D. B. Spalding (1976). 'Prediction of the flow and combustion processes in a three-dimensional combustion chamber', *Imperial College, Mechanical Engineering Department Report* HTS/76/3.
9.  D. B. Spalding (1971). 'Turbulent, physically-controlled combustion processes', *Imperial College, Mechanical Engineering Department Report* RF/TN/A/4.

10. D. B. Spalding (1976). 'Development of the eddy-break-up model of turbulent combustion', *Imperial College, Mechanical Engineering Department Report* HTS/76/1.
11. L. S. Caretto, A. D. Gosman, S. V. Patankar and D. B. Spalding (1973). 'Two calculation procedures for steady, three-dimensional flows with recirculation', *Proc. 3rd Int. Conf. on Numerical Methods in Fluid Mechanics*, Vol. II. Springer-Verlag.
12. F. H. Harlow and J. E. Welch (1965). 'Numerical calculation of time-dependent viscous incompressible flow of fluid with free surface', *Phys. Fluids*, **8**, 2182–2189.
13. D. B. Spalding (1972). 'A novel finite-difference formulation for differential expressions involving both first and second derivatives', *Int. J. Num. Meth. Engng*, **4**, 551–559.
14. P. J. Roache (1976). *Computational Fluid Dynamics*, Hermosa Publications, Albuquerque, NM.
15. M. A. Serag-Eldin (1977). 'The numerical prediction of the flow and combustion processes in a three-dimensional can combustor', *Ph.D. Thesis*, London University. Also *Imperial College, Mechanical Engineering Department Report* HTS/77/21.
16. T. E. A. Youseff (1968). 'Experimental determination of the velocity and pressure profiles in the combustion zone in a model of a gas turbine combustion chamber', *Trans. SAE, Sect. 2*, 1068–1082.

*Numerical Methods in Heat Transfer*
Edited by R. W. Lewis, K. Morgan, and O. C. Zienkiewicz
© 1981 John Wiley & Sons Ltd

*Chapter 22*

# Enclosed Radiation and Turbulent Natural Convection Induced by a Fire*

*David W. Larson*

## SUMMARY

A mathematical model of a fire in a room size enclosure is developed. The resulting mass, momentum, and energy conservation equations are solved by numerical methods. The effects of radiation, turbulent natural convection, and wall heat conduction are included. The flame is assumed to be radiatively participating and the radiative transport equation within the non-grey, non-uniform, non-isothermal flame is solved. External to the flame, the integral radiosity equations within the enclosure are solved. A stream function/vorticity formulation of the conservation equations is solved by an alternating-direction implicit (ADI) finite difference method. The resulting velocity and temperature fields in the enclosure indicate that radiative heat transfer is predominant and that it must be included, along with turbulent flow, to achieve realistic modelling of enclosed fires.

## 22.1 INTRODUCTION

In the last two decades increased efforts have been made to gain an understanding of the free-burning fire in an attempt to achieve more effective methods of detection and control. One of the many individual problems which must be addressed to obtain a more complete knowledge of the problem of fires and fire spreading is that of understanding and predicting the processes by which heat is transferred from a fire to the surroundings. In an effort to gain some of the needed insight into this problem, a finite difference numerical analysis was conducted of the mass, momentum, and energy transfer which occurs in a two-dimensional rectangular enclosure due to the presence of a fire. This chapter will present the mathematical formulation of the problem, describe the finite difference solution method that was used, and show

* United States Government Work.

representative results that were obtained. The effects of radiation, wall heat conduction, and turbulent natural convection are included.

A large body of literature exists on various numerical methods of solving the conservation equations. Specifically, the text by Roache[1] has a very complete discussion of the various finite difference methods (FDM). The more recent finite element methods are discussed in Gartling's survey paper.[2] A review of the early experimental and analytical work on enclosed natural convection is given by Elder[3] and Ostrach.[4] Although the related literature is too extensive for a detailed discussion here, some of the notable developments of applying numerical techniques to natural convection problems include the work of Hellums and Churchill[5] who developed an explicit FDM for solving the problem of natural convection on an isothermal vertical surface. Wilkes and Churchill[6] extended that solution method to analyse a rectangular enclosure. Aziz and Hellums[7] reported the first numerical results on three-dimensional natural convection. Torrance[8] determined that most of the early numerical techniques did not conserve energy or vorticity and Torrance and Rockett[9] reported on the first successful numerical solutions at Grashof numbers ($Gr$) greater than $10^6$. The effects of coupling radiation with laminar natural convection were reported by Larson and Viskanta.[10] With the exception of this work,[10] all of the work to date on natural convection includes temperature-specified and/or adiabatic-wall boundary conditions. In addition, certain of the solution techniques determine only the steady-state results. Neither of these conditions is appropriate for certain classes of problems, in particular the enclosed-fire or heat-source problems. For an enclosed fire the transient development of the flow patterns and the transient heat flow is of great interest. In addition, the thermal response of finite-thickness and finite-heat-capacity conducting walls is of interest since it would, for example, determine the rate of fire spread. Further, laminar flow has been assumed in the studies reported to date. This is, of course, unrealistic for the actual conditions likely to exist in a flaming room, where the Grashof number will vary from $10^9$ to $10^{13}$, well beyond the laminar/turbulent transition.

The numerical analysis described here utilizes a fully transient solution and extends the earlier work on enclosed natural convection into the turbulent flow regime ($Gr \geqslant 10^9$) by incorporating a simple turbulence model. Finite-thickness heat-conducting walls are included in the analysis. In addition, the effects of a radiatively participating fire in the enclosure are also investigated.

## 22.2 MATHEMATICAL FORMULATION

### 22.2.1 Physical model

The model used for this study is illustrated in Figure 22.1. The two-dimensional rectangular enclosure is surrounded by walls of uniform thickness. Initially, the walls and enclosed air are at some uniform temperature and the air is stagnant.

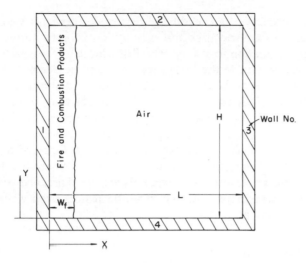

**Figure 22.1** Model of enclosure

At time $t > 0$, a specified temperature (non-isothermal) flame is presumed to exist at some arbitrary (specified) part of the enclosure. For this analysis the air and combustion productions are confined to the enclosure and are presumed to be radiatively participating within the flame and non-participating external to the flame. The basic assumptions made in the study are as follows:

(a)  the fluid motion and heat transfer processes are two dimensional;
(b)  the fluid is Newtonian and compressibility is negligible;
(c)  viscous heat dissipation is negligible;
(d)  all physical properties are constant except for the density, whose variation with temperature is allowed in the buoyancy;
(e)  the walls are grey diffuse emitters and reflectors of radiant energy;
(f)  the Boussinesq approximation is valid.

The first assumption can be questioned for a turbulent fire simulation because the size of the enclosures where fires occur is such that three-dimensional effects will be present and turbulence is inherently three dimensional. However, considerable insight into the flow and heat transfer phenomena can be gained initially without introducing the complexity and large computer storage and computational requirements of three-dimensional motion. The second and third assumptions are readily justifiable due to the relatively low velocities of natural convection flows. The assumption of constant (temperature-independent) physical properties is primarily a numerical convenience and is frequently adopted in generalized analyses in order to decrease the number of independent parameters. In dimensionless form the parameters of the governing conservation equations frequently have a much weaker temperature dependence than do the physical properties. If necessary, this restriction can readily be removed. Toor and Viskanta[11] have shown that for most enclosures the assumption of grey diffuse walls is a reasonable approximation to reality. The Boussinesq approximation accounts for the density variation in the buoyancy term, but neglects it in the inertial terms of the equations of motion, consistent with the assumptions of constant physical properties. This approximation also implies that, in the momentum equation, $\beta(T - T_m)$ is small relative to unity, where $\beta$ is the thermal expansion coefficient, $T$ is the local temperature, and $T_m$ is the mean temperature in the enclosure. The approximation is least correct for high-Grashof-number situations, early in the transient response particularly, when the temperature differences in the enclosure are large. However, Torrance and Rockett[9] obtained a good agreement with experimental results even when the Boussinesq condition was not satisfied.

### 22.2.2  Governing Equations

The governing equations include the conservation equations of mass, momentum, and energy, the radiative transport equation, and a single equation eddy turbulence model.

*Radiative transfer*

For the problem presented here, the conditions within the radiatively participating fire are specified, therefore the radiative transport equation can be solved 'once and for all' for each specified flame condition. This simplifies the solution of the governing equations considerably since the integral radiative transport equation and the differential energy equation can be solved separately as opposed to solving an integrodifferential equation, i.e. the radiative flux divergence does not appear in the energy equation. Physically, this implies an uncoupling of the chemical kinetics of combustion and the heat transfer processes. The radiation from the flame results from soot particle radiation and emission from combustion gases, primarily $H_2O$ and $CO_2$. The gaseous radiation contribution was determined using the Edwards and Menard[12] exponential wide-band model. The soot contribution was predicted using the small particle Mie theory and the optical properties determined by Dalzell and Sacrofim.[13] For the soot particle concentration expected for naturally occurring fires, the radiation contribution from soot was predominant; and therefore, for the purposes of this study the gaseous radiation was neglected.

The equation of radiative transfer for a non-scattering medium in local thermodynamic equilibrium can be written for a particular direction **s** as[14]

$$\frac{dI_\nu}{ds} = \kappa_\nu(I_{b\nu} - I_\nu) \tag{22.1}$$

where $I_\nu$ is the spectral radiative intensity, $\kappa_\nu$ is the spectral absorption coefficient, and $I_{b\nu}$ is Planck's function. Equation (22.1) can be solved using an integrating factor to yield

$$I_\nu(s) = I_\nu(0) \exp\left[-\tau_\nu(s)\right] + \int_0^s \kappa_\nu I_{b\nu} \exp\left\{-\left[\tau_\nu(s) - \tau_\nu(s')\right]\right\} s' \quad (22.2)$$

where the optical distance is defined as

$$\tau_\nu(s) = \int_0^s \kappa_\nu(s')\, ds' \quad (22.3)$$

and $I_\nu(0)$ is the intensity leaving the boundary. Once the geometry and the conditions within the flame (non-uniform and non-isothermal) are specified, Equation (22.2) can be used to determine the spectral intensity at any point in any direction. For simplicity in this study, the walls adjacent to the flame are assumed to be black bodies (Figure 22.2). The particular item of interest is the radiative flux leaving the free surface of the flame. The spectral flux normal to the flame is expressed as

$$q_{rx\nu} = \int_{4\pi} I_\nu \cos\theta_x\, d\omega \quad (22.4)$$

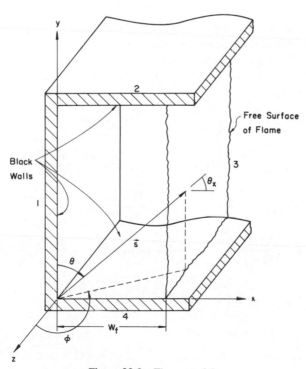

**Figure 22.2** Flame model

where $\omega$ is the solid angle and $I_\nu$ is determined from Equation (22.2). Using the geometry of Figure 22.2, Equation (22.4) becomes

$$q_{rx\nu} = \int_0^\pi \left[ \int_0^\pi I_\nu \sin^2 \theta \, d\theta \right] \sin \phi \, d\phi \qquad (22.5)$$

Integrating over wavelength, the total flux from the flame is given by

$$q_{rx} = \int_{\nu=0}^\infty q_{rx\nu} \, d\nu \qquad (22.6)$$

External to the flame, the equation for radiative flux along wall $i$ for grey diffuse walls with an arbitrary temperature distribution is given by

$$q_{ri}(s_i) = \frac{\varepsilon_i}{\rho_i} [\sigma T_i^4 (s_i) - J_i(s_i)] \qquad (22.7)$$

where $\sigma$ is the Stephan–Boltzmann constant, $\varepsilon_i$ and $\rho_i$ are the wall emissivity and reflectivity, respectively, $s_i$ represents a position on wall $i$, and the radiosity $J_i(s_i)$ is given by

$$J_i(s_i) = \varepsilon_i \sigma T_i^4 (s_i) + \rho_i \sum_{\substack{j=1 \\ j \neq i}}^m \int_{A_j} J_j(s_j) K(s_i, s_j) \, dA_j \qquad (22.8)$$

where $A_j$ is the area of wall $j$ and the kernel $K(s_i, s_j)$ is a function of the configuration. When the flame is on wall $i$, the first term on the right-hand side of Equation (22.8) must be replaced with the radiative flux from the flame (Equation (22.6)). For the geometry of Figure 22.1, Equation (22.8) can be evaluated to yield:

$$J_1(y_1) = \varepsilon_1 \sigma T_1^4 (y_1) + \frac{\rho_1}{2} \left( \int_0^L \frac{J_2(x_2)(H - y_1)(x_2)}{[(H - y_1)^2 + x_2^2]^{3/2}} \, dx \right.$$

$$\left. + \int_0^H \frac{J_3(y_3)L_2}{[L^2 + (y_3 - y_1)^2]^{3/2}} \, dy_3 + \int_0^L \frac{J_4(x_4)(y_1)(x_4)}{(y_1^2 + x_4^2)^{3/2}} \, dx_4 \right) \qquad (22.8a)$$

$$J_2(x_2) = \varepsilon_2 \sigma T_2^4 (x_2) + \frac{\rho_2}{2} \left( \int_0^H \frac{J_1(y_1)(x_2)(H - y_1)}{[x_2^2 + (H - y_1)^2]^{3/2}} \, dy_1 \right.$$

$$\left. + \int_0^H \frac{J_3(y_3)(H - y_3)(L - x_2)}{[(H - y_3)^2 + (L - x_2)^2]^{3/2}} \, dy_3 + \int_0^L \frac{J_4(x_4)H^2}{[H^2 + (x_4 - x_2)^2]^{3/2}} \, dx_4 \right) \qquad (22.8b)$$

$$J_3(y_3) = \varepsilon_3 \sigma T_3^4 (y_3) + \frac{\rho_3}{2} \left( \int_0^H \frac{J_1(y_1)L^2}{[L^2 + (y_1 - y_3)^2]^{3/2}} \, dy_1 \right.$$

$$\left. + \int_0^L \frac{J_2(x_2)(H - y_3)(L - x_2)}{[(H - y_3)^2 + (L - x_2)^2]^{3/2}} \, dx_2 + \int_0^L \frac{J_4(x_4)(Y_3)(L - x_4)}{[y_3^2 + (L - x_4)^2]^{3/2}} \, dx_4 \right) \qquad (22.8c)$$

$$J_4(x_4) = \varepsilon_4 \sigma T_4^4(x_4) + \frac{\rho_4}{2}\left(\int_0^H \frac{J_1(y_1)(x_4)(y_1)}{(x_4^2 + y_1^2)^{3/2}} \, dy_1\right.$$

$$\left. + \int_0^L \frac{J_2(x_2)H^2}{[H^2 + (x_2 - x_4)^2]^{3/2}} \, dx_2 + \int_0^H \frac{J_3(y_3)(L - x_4)(y_3)}{[y_3^2 + (L - x_4)^2]^{3/2}} \, dy_3\right) \qquad (22.8d)$$

Equations (22.6), (22.7), and (22.8) can readily be non-dimensionalized. As a practical consideration, however, it should be mentioned that in solving the radiosity equations the integral terms are evaluated in their dimensional form and then the result is non-dimensionalized. This procedure takes advantage of symmetry of the kernel which would not be possible for any $L/H \neq 1$ if the equations were non-dimensionalized before evaluation of the integrals. In order to prevent additional complication in the solution of the radiosity equations external to the flame (Equations (22.8)), the radiative flux leaving the free surface of the flame is assumed to be diffusely distributed. This assumption is valid for an isothermal or one-dimensional temperature flame, but is less correct for a two-dimensional flame. However, since the temperature variation normal to the flame is much larger than the variation parallel to the flame, and since also the radiation at every point on the free surface receives energy from every elementary volume within the flame, this assumption is expected to be reasonable even for two-dimensional temperature flames. Without this assumption the directional distribution of energy leaving the flame would have to be included in the solution of the radiosity equations.

*Turbulence model*

The approach followed in this study was to use the so called eddy diffusivity concept wherein the molecular kinematic viscosity and thermal diffusivity are replaced by an 'effective' or 'virtual' turbulent diffusivity. The procedure precludes examination of the detailed fluctuating motion in turbulent flow and only the 'average' values of the dependent variables are determined. The limitations of this procedure are well known, but a generalized more correct approach is not yet available. The assumption is made that the effective viscosity $\nu_e$ and effective thermal diffusivity $\alpha_e$ can be expressed as the sum of a molecular and a turbulent component:

$$\nu_e = \nu + \varepsilon_M \qquad (22.9)$$

and

$$\alpha_e = \alpha + \varepsilon_H \qquad (22.10)$$

If we define the ratio of effective to molecular viscosity, $\varepsilon^+ = \nu_e/\nu$, then numerous semi-empirical correlations are available for the prediction of $\varepsilon^+$ in forced convection flow. Spalding's correlation[15] is used here since it yields a single expression valid over all regions of the flow,

$$\varepsilon^+ = 1.0 + 0.04432[\exp(0.4u^+) - 1.0 - 0.4u^+ - 0.08(u^+)^2] \quad (22.11)$$

where $u^+ = u/\sqrt{(\tau_w/\rho)}$, $u$ is velocity, $\tau_w$ is wall shear stress, and $\rho$ is density. Since $u^+$ has been non-dimensionalized by the shear velocity, which is an unknown quantity in free convection, an additional assumption was required in order to use Equation (22.11) to evaluate turbulent viscosity. It was assumed that beyond the turbulent transition point, the following equation was valid:

$$u^+ = \left(\frac{(u^+)_{tr}}{(U)_{tr}}\right)U \quad (22.12)$$

where the subscript tr stands for transition and $U = u/\sqrt{(g\beta\Delta TL)}$, $g$ is gravity, $\beta$ is thermal expansion coefficient, $\Delta T$ is the maximum initial temperature difference, and $L$ is the characteristic length of the enclosure. This is equivalent to assuming that the wall shear velocity is proportional to the velocity potential created by the buoyancy forces. This can be seen by replacing the dimensionless transition terms in Equation (22.12) with dimensional and reference quantities resulting in

$$u^+ = \left(\frac{g\beta\Delta TL}{\tau_w/\rho}\right)^{1/2}U \quad (22.13)$$

Introducing the turbulent Prandtl number, $Pr_t = \varepsilon_M/\varepsilon_H$ yields the method for determining the eddy diffusivity from the eddy viscosity. Presently, there is considerable disagreement in the variation of $Pr_t$ across the boundary layer. In this study, the constant condition $Pr_t = 0.80$ was used as recommended by Kestin and Richardson.[16] The above procedure for predicting turbulence parameters requires both theoretical and experimental verification. At present, there is a serious lack of data on turbulent free convection. The method used is justified only to the extent that it requires no *a priori* assumptions on the velocity profile and it yields reasonable results for the eddy diffusivity distribution.

*Convection*

The conservation equations of mass, momentum, and energy are transformed to a stream function/vorticity formulation by cross-differentiating and combining the momentum equations to eliminate the pressure term and introducing a stream function which automatically satisfies continuity. The equations can be written in dimensionless form by introducing the variables $U = u/u_0$, $V = v/u_0$, $\theta = T/T_H$, $X = x/L$, $Y = y/L$, $\tau = tu_0/L$, $\Omega = \omega L/u_0$, $\Psi = \psi/u_0L$, where $u_0 = \sqrt{(g\beta\Delta TL)}$, $u$ is the horizontal velocity, $v$ is the vertical velocity, $T_H$ the maximum (source) temperature in the enclosure, $t$ is the time, $\omega$ the vorticity, and $\psi$ the stream function, and $Gr$ is the Grashof number

$g\beta\Delta TL^3/\nu^2$. As a result, we obtain the dimensionless turbulent energy equation

$$\frac{\partial\theta}{\partial\tau}+\frac{\partial(U\theta)}{\partial X}+\frac{\partial(V\theta)}{\partial Y}=\frac{1}{Pr_t\sqrt{Gr}}\left[\frac{\partial}{\partial X}\left(\varepsilon^+\frac{\partial\theta}{\partial X}\right)+\frac{\partial}{\partial Y}\left(\varepsilon^+\frac{\partial\theta}{\partial Y}\right)\right] \quad (22.14)$$

the dimensionless turbulent vorticity transport equation

$$\frac{\partial\Omega}{\partial\tau}+\frac{\partial(U\Omega)}{\partial X}+\frac{\partial(V\Omega)}{\partial Y}=\frac{1}{Pr_t\sqrt{Gr}}\left[\frac{\partial}{\partial X}\left(\varepsilon^+\frac{\partial\Omega}{\partial X}\right)+\frac{\partial}{\partial Y}\left(\varepsilon^+\frac{\partial\Omega}{\partial Y}\right)\right]$$

$$+\frac{T_H}{\Delta T}\frac{\partial\theta}{\partial X} \quad (22.15)$$

and the dimensionless stream function and velocity equations

$$\Omega=-\left(\frac{\partial^2\Psi}{\partial X^2}+\frac{\partial^2\Psi}{\partial Y^2}\right) \quad (22.16)$$

$$U=\partial\Psi/\partial Y \quad (22.17a)$$

$$V=-\partial\Psi/\partial X \quad (22.17b)$$

The initial conditions in the enclosure are a specified uniform temperature and a stagnant gas. The boundary conditions are a specified temperature source in some part of the enclosure, no slip at the walls, and a zero value and zero gradient of the stream function at the walls. The boundary condition on vorticity is not known explicitly, but this presents no particular difficulty as noted later.

The above dimensionless equations are in slightly different forms from those developed and used by previous investigators of natural convection in enclosures. There are two reasons for the differences. Since there is no characteristic velocity under natural convection conditions there is no unique way of defining a dimensionless velocity. Instead, a reference velocity must be selected from examination of the physical problem and a consideration of the controlling forces. Previous investigators have typically selected a characteristic velocity related to the viscous forces, $u_0=\nu/L$. It appears, however, that a more meaningful dimensionless parameter may result, in a buoyancy-dominated situation, by relating the characteristic velocity to the buoyancy forces, and thus the selection $u_0^2=g\beta\Delta TL$ was made. The second reason for the differences results from the definition of dimensionless temperature. In purely convective (or conductive) heat transfer situations, it is convenient to define the dimensionless temperature as the ratio of temperature differences, and all previous analyses have done so. However, when radiation is also present such a definition is inappropriate since the driving potential for heat transfer is no longer a linear temperature difference. Therefore, because of the radiative–convective coupling, the dimensionless temperature is defined here as a temperature ratio.

*Conduction*

The heat conduction in the wall was assumed to be one dimensional. This assumption is reasonable since the thermal conductivity of the non-metallic walls is generally quite low and the temperature gradient in the normal direction is considerably larger than along the plane of the wall. Therefore, the dimensionless energy equation in the wall can be written as

$$\frac{\partial \theta}{\partial \tau} = \frac{N_\alpha}{N_L^2 Pr \sqrt{Gr}} \frac{\partial^2 \theta}{\partial \zeta^2}$$ (22.18)

where $N_\alpha$ is $\alpha_w / \alpha_a$, the ratio of the wall thermal diffusivity to the air thermal diffusivity, and $N_L = L_w / L$, and $\zeta$ is the coordinate normal to the wall.

An energy balance at the wall yields the boundary condition

$$\frac{\partial \theta_w}{\partial \zeta} = \frac{1}{N_k} \frac{\partial \theta_a}{\partial \zeta} + N_r Q_r$$ (22.19)

where $N_k$ is the thermal conductivity ratio $k_w / k_a$, $N_r$ is the dimensionless radiation-to-conduction parameter $\sigma T_H^3 L / k_a$, and $Q_r = q_r / \sigma T_H^4$. It is this boundary condition whch couples the radiative, convective and conductive heat transfer in the problem. Any change in wall surface temperature simultaneously affects the convection to the wall, the conduction inside the wall, and the net radiant heat flux at the wall.

The external surface of the walls can either be insulated or have a specified convective-heat-transfer coefficient. Owing to the low conductivity of the walls the results for either condition are essentially the same; however, the results reported here assumed the horizontal walls to be externally insulated and the external vertical walls to have convective transfer to ambient temperature.

## 22.3  METHOD OF SOLUTION

Equations (22.2), (22.4)–(22.8), (22.11), and (22.14)–(22.18) and their appropriate initial and boundary conditions are sufficient to describe the problem completely. With the exception of the radiative transport equations internal to the flame, the equations are all coupled either through the variables, the boundary conditions, or both.

### 22.3.1  Integral equations

The radiative transport in the flame was determined by constructing a grid over the flame and specifying the flame temperature and composition at each grid nodal point and linearly interpolating between nodal points. The computer code is written so as to handle any arbitrary distribution of temperature and combustion products. A numerical integration is then performed at each nodal point on the free surface of the flame over the distance s for all $\theta$ and $\phi$, evaluating Planck's function and the spectral absorption coefficient at the local conditions. Numerically, the procedure is quite straightforward, but the geometrical considerations and 'book-keeping' are quite involved. By examining Equations (22.2)–(22.6), it can be seen that five sequential integrations are required. In general, Simpson's rule was used for the numerical integrations, although when uneven spacing on a variable was required the trapezoidal rule was employed. The integral radiosity equations (22.8) were solved by the method of successive approximation.

### 22.3.2 Conservation equations

The conservation equations (Equations (22.14)–(22.18)) were solved in conservation form by using an alternating-direction implicit (ADI) finite difference scheme. The non-linear advection terms were represented by a second-order-accurate conservative 'upstream' finite difference form as suggested by Torrance[8] for buoyancy-dominated flow.

Since the boundary conditions for the vorticity transport equation are not known, it is necessary to determine the wall vorticities from additional considerations. Following Wilkes and Churchill[6] this is accomplished by a Taylor series expansion of the stream function in the vicinity of the wall. By using the boundary conditions on $\Psi$ and noting from the stream function equation (22.16) that $\Omega_{0,j} = \partial^2 \Psi / \partial X^2$, an approximation for vorticity at the new time step at the wall is obtained from the values of the interior stream function,

$$\Omega_{0,j} = -\frac{8\Psi_{1,j} - \Psi_{2,j}}{2(\Delta X)^2} \tag{22.20}$$

where the subscripts refer to nodal point locations and zero is at the boundary.

The elliptic stream function equation was converted to a parabolic equation by the addition of a time-dependent term and was then solved to steady state at each time step for the vorticity distribution at that time by the ADI method. The velocity Equations (22.17a, b) were solved using four-point central difference representations at all nodal points more than one grid space from the wall. The velocities one grid space from the wall were obtained by a four-point non-central difference form.

### 22.3.3 Numerical accuracy

The solution procedure consisted of advancing the interior temperatures through a time step by solving Equation (22.14). The vorticity transport equation (22.15) was then solved for all interior nodal points. Next the stream function equation (22.16) was solved. From the solution of the stream function, the boundary vorticities were obtained and the velocity field determined. The radiosity equations (22.8) were then solved and finally the energy equation in the wall was evaluated. Allowance was made for iterating on the non-linear advection terms during each time step.

Space prevents a detailed discussion of the errors introduced by the truncation, round-off, finite difference approximations of the differential equations, and artificial (numerical) viscosity in the above solution procedure. A large number of numerical experiments were conducted in an effort to determine the time step, grid size, and convergence criteria which would lead to a reasonable compromise between accuracy and computing time. An internal check on the solution scheme was conducted by solving selected problems with both explicit and implicit finite difference techniques and also by comparing limiting cases (temperature specified boundaries, laminar flow, and no radiative transfer) with earlier studies.[6,17] The spatial discretization was varied from $50 \times 20$ to $100 \times 40$ (uniform grid) and, in general, indicated good convergence with this grid structure, although at $Gr = 10^{11}$ and above, the coarser grid led to some stability problems evident by physically impossible local temperature peaks. As discussed by Torrance[8] and Roache,[1] the artificial viscosity problem increases with increasing Grashof number, therefore the results should be viewed with caution until experimental comparisons are available. However, the large physical (eddy) viscosity in turbulent flow is expected to minimize the problem. The accuracy of the radiative transport solutions (22.2) was determined by comparison with a closed-form solution for a grey isothermal medium. The difference in predicted flux from the flame was less than 1.5%.

The solution of the radiosity equations (22.8) external to the flame was checked by reducing the enclosure to a special case of two radiating parallel walls and comparing the resulting radiosity distribution for various wall spacing and emissivity to those obtained in a numerical study by Sparrow *et al.*[18] The results were identical to five significant figures. The computer time required for the full transient solution to $\tau = 20.0$ with a $100 \times 40$ grid and a time step of 0.1 (i.e. 200 time steps) was approximately 200 seconds on a CDC 7600 computer.

## 22.4 RESULTS

Due to space limitations, only a very limited sample of the wide class of problems which can be solved will be presented here. The parameter values were selected to be representative of room-size enclosures. In a room-size enclosure with a fire, the Grashof number would typically be in the range of $10^9$ to $10^{13}$, depending upon the room dimensions, characteristic flame temperature, and the temperature at which the kinematic viscosity and thermal expansion coefficient are evaluated. In these simulations a non-grey, non-uniform, non-isothermal flame was placed in an arbitrary (but, specified) location in the enclosure. The temperature variation in the flame was from 660 K ($\theta = 0.3$) to 2200 K ($\theta = 1.0$). The volumetric soot concentration ranged from $10^{-8}$ to $5 \times 10^{-6}$ cm$^{-3}$/cm$^3$. Other parameter values used in the simulations were $Pr_t = 0.8$, $N_k = 4$, $N_r = 2000$, $N_L = 0.025$, $N_\alpha = 0.001$, and $\varepsilon = 0.9$. The results shown here are for a square enclosure.

The radiative flux from both one- and two-dimensional flames (non-grey, non-uniform, and non-isothermal) was determined in order to examine the effects of a two-dimensional temperature variation and soot concentration. The temperature distribution used for two such conditions is shown in Figure 22.3. For the two-dimensional flame, the temperature variation with height

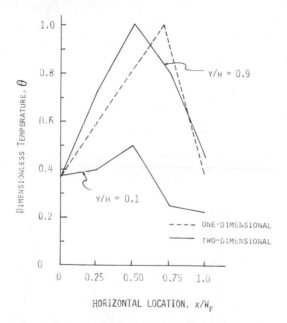

**Figure 22.3** Dimensionless temperature distribution within the flame for a one-dimensional (–––) and a two-dimensional (——) flame

was assumed to be linear between the two curves showing the lateral temperature variation at two different heights. The volumetric soot concentration, $C_v$ (cm$^3$/cm$^3$) is shown in Figure 22.4. The resulting total radiative flux distribution along the surface of the flame is shown in Figure 22.5. The effect of the cool pyrolysing walls near the floor and ceiling is apparent for both cases.

The transient isotherms and streamlines at two different points in the time are shown in Figure 22.6 for $Gr = 10^{13}$ with a flame on the left wall (flame thickness is 5% of the enclosure width). The effect of the fluid motion in

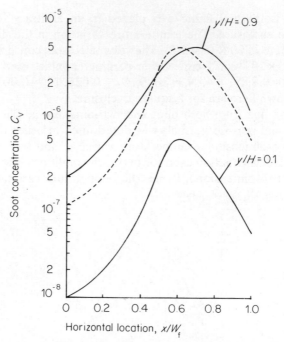

**Figure 22.4** Volumetric soot concentration, $C_v$, for a one-dimensional (---) and a two-dimensional (——) flame

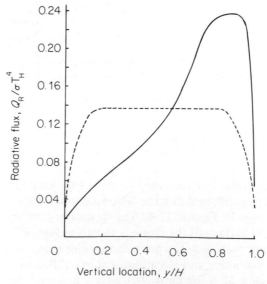

**Figure 22.5** Radiative flux distribution leaving the surface of the flame for a one-dimensional (---) and two-dimensional (——) flame

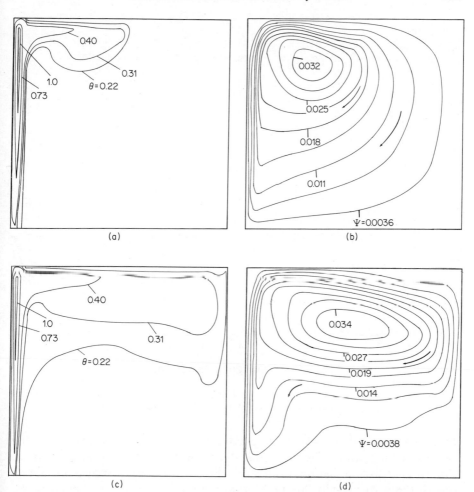

**Figure 22.6** Transient streamlines and isotherms in the enclosure with flame on left wall (wall No. 1) at $Gr = 10^{13}$: (a) isotherms at $\tau = 4$; (b) streamlines at $\tau = 4$; (c) isotherms at $\tau = 20$; and (d) streamlines at $\tau = 20$

deforming the isotherms is very apparent. The transient behaviour is of more interest than steady-state conditions when examining the enclosure response to the fire, since the steady-state condition would occur when the entire enclosure approached the temperature of the flame.

The effect of Grashof number variation and the turbulence model can be seen in Figure 22.7 where the vertical velocity profile across the mid-height of the enclosure is shown. Both the laminar and turbulent flow models were run at $Gr = 10^9$, the approximate transition condition. As shown, the laminar flow model results in a higher and narrower velocity spike adjacent to the flaming wall. The turbulent mixing created by the eddy viscosity in the turbulence

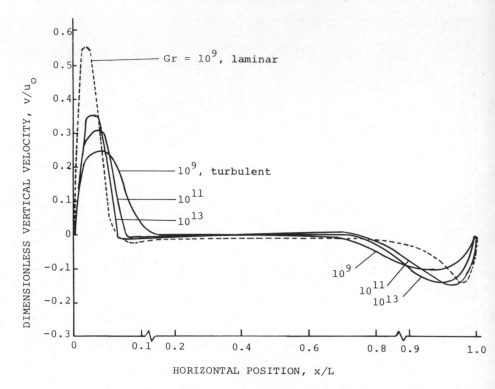

**Figure 22.7** Vertical velocity at enclosure mid-height ($y/H = 0.5$) for three different Grashof number conditions at $\tau = 10$ with a flame on the left wall

model causes the upward velocity to extend farther into the enclosure, i.e. the rapidly moving gases in the flame drag the air outside of the flame along with it. For every condition that was examined where a flame was assumed present on one of the vertical walls, the rapid fluid motion was generally limited to the regions near the walls, while the core of the enclosure remained relatively stagnant. The early temperature distribution across the enclosure at two different elevations and the progress of the thermal front across the ceiling for $Gr = 10^{11}$ can be seen in Figure 22.8. At the enclosure mid-height, the temperature remains unchanged at $\tau = 10$ since the hot gases from the flaming wall have yet to reach the opposite cool wall.

For the flaming left wall, the effect of radiative heat transfer can be seen in Figure 22.9. A comparison of the temperature profile outside the flame across the mid-height of the enclosure for $Gr = 10^9$ at $\tau = 20$ is shown when the radiative heat transfer is neglected. The air temperature is not significantly different at this point in time, however, the right wall temperature ($x/L = 1.0$) is considerably higher due to the radiation. This result is typical of all of the simulations that were run. The wall temperatures were controlled almost

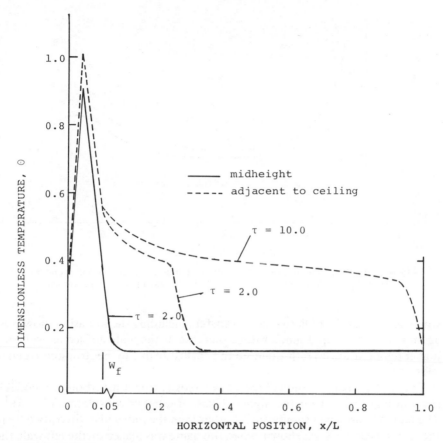

**Figure 22.8** Temperature profile at enclosure mid-height ($y/H = 0.5$) (——), and adjacent to ceiling ($y/H = 0.975$) (– – –) for $Gr = 10^{11}$ and a flame on the left wall

completely by the radiative heat transfer. The convective heat transfer rates to the walls (including the floor and ceiling) were typically an order of magnitude lower than the radiative transfer.

The influence of the radiative heat transfer is even more pronounced when the fire is located along the ceiling. Figure 22.10 shows the effect of neglecting radiation on both the velocity and temperature profiles. When radiative transfer is neglected, the walls have a cooling effect and, therefore, the convective air velocity moves downward adjacent to the wall. However, when it is included, the walls heat up rapidly and the vertical velocity near the wall is two orders of magnitude larger and opposite in direction. In addition, there is a strong upward plume in the centre of the enclosure, resulting in strong multiple convective cells. The temperature profiles show that at $\tau = 50$, the air and walls at the mid-height of the enclosure have yet to feel the effects of the ceiling flame

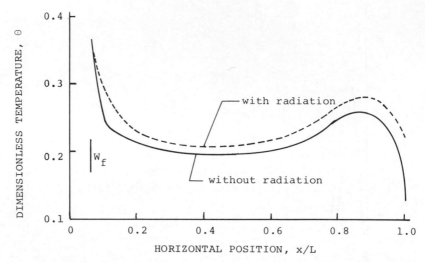

**Figure 22.9** Effect of radiative heat transfer on temperature distribution at enclosure mid-height ( $y/H = 0.5$ ) and $\tau = 20$ for $Gr = 10^9$ and a flame on the left wall

when neglecting the radiative heat transfer. Including the radiation, however, again results in a rapid temperature increase at the wall surface as well as a higher temperature level throughout the enclosure due to the strong convective cells.

The predicted distribution of the eddy viscosity from the turbulence model is show in Figure 22.11 for the highest Grashof number considered, $Gr = 10^{13}$. The eddy viscosity was significantly higher than the molecular viscosity only in the region near walls. As shown, when the flame was placed on the left wall, the turbulent mixing in this region created eddy viscosities as much as 2000 times greater than the molecular viscosity.

## 22.5  CONCLUSIONS

To obtain realistic results when modelling a fire in an enclosure, it appears to be necessary to include the effects of turbulence and radiative heat transfer. In fact, when portions of an enclosure are at temperatures associated with fires, radiation is the predominant mechanism of heat transfer. The numerical model discussed here yields an improved understanding of fire-induced gas flow and temperature fields. The results are useful, for example, in the design, place-ment, and evaluation of fire detectors or suppressors, whether they are actuated by temperature radiative flux, or smoke. The model is sufficiently general that parametric studies which examine the effects of enclosure geometry, conduction, convection, radiation parameters, flame composition and temperature, and turbulence can be considered.

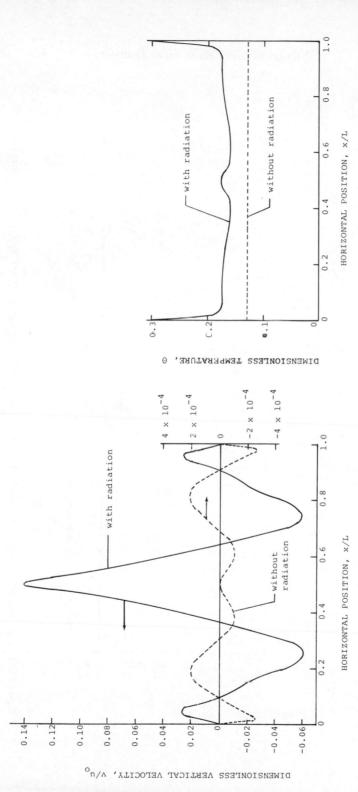

**Figure 22.10** Effect of radiative heat transfer from a ceiling flame on: (a) vertical velocity and (ɔ) temperature at enclosure mid-height ($y/H = 0.5$) for $Gr = 10^9$ at $\tau = 50$

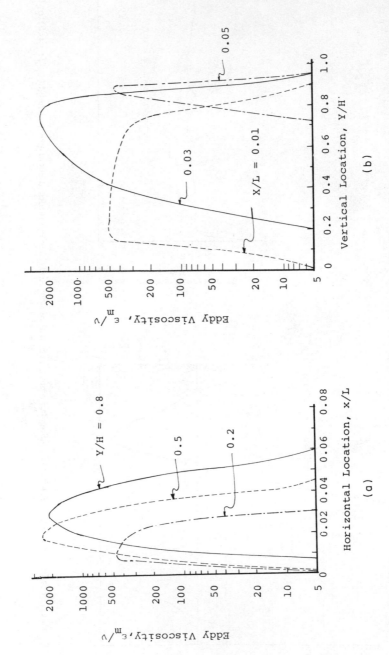

**Figure 22.11** Eddy viscosity distribution: (a) across the boundary layer and (b) along the boundary layer at $Gr = 10^{13}$ with a flame on the left wall

## Acknowledgement

This study was supported by the US Department of Energy (DOE) under Contract AT (29-1)-789.

## REFERENCES

1. P. J. Roache (1972). *Computational Fluid Dynamics*, Hermosa Publishers, Albuquerque, NM.
2. D. K. Gartling (1976). 'Recent developments in the use of finite element methods', *Computing in Applied Mechanics*, AMD-Vol. 18, R. F. Hartung (Ed.), ASME Winter Annual, New York.
3. J. W. Elder (1965). 'Laminar free convection in a vertical slot', *J. Fluid Mech.*, **23**, 77–98.
4. S. Ostrach (1972). 'Natural convection in enclosures', *Adv. Heat Transfer*, **8**, 161–227, Academic Press, New York.
5. J. D. Hellums and S. W. Churchill (1962). 'Transient and steady state free and natural convection, numerical solutions', *A.I.Ch.E. J.*, **8**, 690–695.
6. J. O. Wilkes and S. W. Churchill (1965). 'The finite-difference computation of natural convection in a rectangular enclosure', *A.I.Ch.E. J.*, **12**, 161–166.
7. K. Aziz and J. D. Hellums (1967). 'Numerical solution of the three-dimensional equations of motion for laminar natural convection', *Phys. Fluids*, **10**, 314–324.
8. K. E. Torrance (1968). 'Comparison of finite-difference computations of natural convection', *J. Res. Natl Bur. Stand.*, **72B**, 281–300.
9. K. E. Torrance and J. A. Rockett (1969). 'Numerical study of natural convection in an enclosure with localized heating from below: creeping flow to the onset of laminar instability', *J. Fluid Mech.*, **36**, 33–54.
10. D. W. Larson and R. Viskanta (1976). 'Transient combined laminar free convection and radiation in a rectangular enclosure', *J. Fluid Mech.*, **78**, 65–85.
11. J. S. Toor and R. Viskanta (1972). 'A critical examination of the validity of simplified models for radiant heat transfer analysis', *Int. J. Heat Mass Transfer*, **15**, 1553–1567.
12. D. K. Edwards and W. A. Menard (1964). 'Comparison of models for correlation of total band absorptance', *Appl. Opt.* **3**, 621–625.
13. W. H. Dalzell and A. F. Sarofim (1969). 'Optical constants of soot and their application to heat flux calculations', *J. Heat Transfer*, **91**, *Trans. ASME, Series C*, 100–104.
14. R. Viskanta (1966). 'Radiation transfer and interaction of convection with radiation heat transfer', *Adv. Heat Transfer*, **3**, 175–251, Academic Press, New York.
15. D. B. Spalding (1961). 'A single formula for the law of the wall', *J. Appl. Mech.*, **28**, 455–457.
16. J. Kestin and P. D. Richardson (1963). 'Heat transfer across turbulent incompressible boundary layers', *Int. J. Heat Mass Transfer*, **6**, 147–189.
17. T. S. Chen (1971). 'A three-dimensional steady state mathematical model of a glass furnace', *Ph.D. Thesis*, Purdue University.
18. E. M. Sparrow, J. L. Gregg, J. V. Szel and P. Manus (1961). 'Analysis results and interpretation for radiation between some simply arranged gray surfaces', *J. Heat Transfer*, **83**, *Trans. ASME, Series C*, 207–214.

*Numerical Methods in Heat Transfer*
Edited by R. W. Lewis, K. Morgan, and O. C. Zienkiewicz
© 1981 John Wiley & Sons Ltd

*Chapter 23*

# Numerical Computations of Turbulent Reacting Combustor Flows

*Essam Eldin Khalil*

## SUMMARY

This chapter was prepared to describe primarily a numerical procedure that can calculate the local flow properties and heat transfer characteristics of flow in furnaces and combustion chambers. The present procedure solves the governing conservation equations of mass, momentum, and energy expressed in a finite difference form. Several mathematical models were employed to represent the turbulent nature of the flow, the reaction rate, and the heat transfer by convection and radiation. It was necessary to identify the capabilities of and prospects for the procedure, together with the mathematical modelling assumptions embodied in it, through comparisons with available measured data. Qualitatively the procedure produced very good results in most flows, but quantitatively, to obtain better agreement more work is needed to improve and refine the mathematical models. However, the general agreement obtained by the procedure justifies the use of this method in the design of furnaces and combustion chambers.

## NOMENCLATURE

$A_n$   coefficients of the finite difference equations
$C_\mu, C_1, C_2$   constants of the turbulence model
$C_p$   specific heat at constant pressure
$d, D$   diameter
$e$   dissipation rate
$E$   constant in log law of the wall
$f$   mixture fraction $= (M_{fu} - M_{ox})/i$
$g$   square of the concentration fluctuations $g = \overline{f'^2}$
$H$   total enthalpy
$i$   stoichiometric oxygen to fuel mass ratio
$k$   kinetic energy of turbulence
$l, L$   length

$M_a$    mass fraction of species $a$
$m_a$    fluctuating component of mass fraction of species $a$
$p$      pressure
$q$      heat flux per unit area
$r$      radius, radial coordinate
$S_\phi$   source term of the entity $\phi$
$T$      temperature
$\bar{U}$    mean axial velocity component
$\bar{V}$    mean radial velocity component
$\bar{W}$    mean tangential velocity component
$x$      distance along the axis of the furnace from the burner exit
$y_a$    characteristic length
$\Gamma_\phi$   turbulent exchange coefficient
$\rho$      density
$\phi$      general dependent variable
$\mu$      viscosity
$\mu_t$    turbulent viscosity
$\delta x$     distance between adjacent grid nodes
$\chi$      constant in the wall function
$\sigma_\phi$   Prandtl/Schmidt number
$\psi_0$    function of pressure gradient
$\tau_W$    wall shear stress

*Subscripts*

$a$    species $a$
$f$    furnace
fu    fuel
ox    oxidant
$t$    turbulent
W    wall

*Superscripts*

$'$    fluctuating
$^-$    average

## 23.1  INTRODUCTORY REMARKS

### 23.1.1  The problem

In this day and age there is a continuing demand for increased performance in furnaces, gas turbine combustors and heat exchangers. This demand has led in turn to new design philosophies that require that these systems should operate at higher pressure levels, higher temperatures, and turbulence intensities than those known before, provided, of course, that an optimum rate of heat transfer is attained. Currently available analytical/empirical combustor design procedures, which are based on empirical correlations or prior combustor data, are becoming more inadequate tools in facing the task of extrapolating experience to low fuel consumption and low emission concepts. Designers were, and still are, aided foremost by lengthy and costly experimental programs which are performed on full scale models. However, the recent development in digital computers has encouraged designers to produce numerical procedures, utilizing various mathematical models, that can readily produce the required data to help the designer in his task and at a far lower cost than that of an experimental program.[1]

The flow in combustion chambers is governed by a set of conservation equations; without the recent advent in digital computers the solution of these equations would have been almost an impossible task. It is now possible to express these equations in a form amenable to solution by computers. In this process various mathematical models, which represent the turbulent nature of the flow, the reaction process, and the radiation heat transfer, were utilized to make the solutions of these equations possible. These models are increasing steadily in their physical realism and refinement.[2,3,4]

This chapter is devoted to the task of demonstrating exactly what a numerical procedure which solves the conservation equations can do. This procedure employs the primitive, pressure–velocity variables, and solves the conservation equations expressed in a finite difference form. The computer code, namely TEACH-T, used in the present chapter, was developed by Gosman and Pun,[5] tested and validated by Gosman *et al.*[6] in non-reacting flows, and was extended to include reacting flows as described by Hutchinson *et al.*[7,8] Through all the validation tests it was established that the predicted results always represented correctly the trends of the flow, i.e. good qualitative agreement was generally reported.[4,6,8,9,10] In reacting flow configuration quantitative agreement was obtained between measurements and calculations of the flow pattern and flame structure,[1,7] however this agreement was limited and extensive work is necessary to produce better results.

### 23.1.2 The way forward

It is the task of this chapter to describe a numerical procedure, which involves the numerical simulation of combustion processes, and to discuss its present capabilities and its prospects. This present procedure allows the solution of the partial differential equations of mass, momentum, and energy. These equations are of the form:

$$\frac{\partial}{\partial x}\left(\bar{\rho}\bar{U}\bar{\phi} - \Gamma_\phi \frac{\partial\bar{\phi}}{\partial x}\right) + \frac{1}{r}\frac{\partial}{\partial r}\left(\bar{\rho}r\bar{V}\bar{\phi} - \Gamma_\phi r \frac{\partial\bar{\phi}}{\partial r}\right) = S_\phi \tag{23.1}$$

where $\bar{\phi}$ stands for any dependent variable, namely the axial velocity $\bar{U}$, radial velocity $\bar{V}$, swirl velocity $\bar{W}$, stagnation enthalpy $\bar{H}$, fuel mass fraction $\bar{M}_{fu}$, mixture fraction $\bar{f}$, turbulent kinetic energy $k$ and its dissipation rate $e$, the mean square fluctuations of concentrations, $g = \overline{f'^2}$, $\overline{m_{fu}^2}$ and any other variable such as radiative flux in the coordinate directions. $\Gamma_\phi$ is the exchange coefficient and $S_\phi$ is the source term; details of these two terms are given in Table 23.1.

The closure of the governing equations is affected by a set of mathematical models that represents the various physical processes taking place inside the combustion chamber. Our limited understanding and knowledge of what goes on exactly in a combustion chamber, and the size limitation of the computers, both impose restrictions on the degree of refinement and sophistication of these models. The three main features that are modelled in this procedure are the turbulence characteristics of the flow, the chemical reaction, and finally the heat transfer by radiation. The calculation procedure employs the two-equation turbulence model,[11] the combustion models of Khalil,[12] and the radiation models of Khalil and Truelove.[13]

### 23.1.3 Outline of the chapter

The numerical procedure is first discussed through describing the governing equations, the solution procedure, and the major features of mathematical modelling. The capabilities and prospects of the procedure are then analysed through comparisons with measured data; this is followed by utilizing the procedure in design situations. Finally the chapter ends with a conclusion section which aims at aiding the designer, in his search for the best performance possible, by pointing out the advantages and limitations of the present procedure.

## 23.2  GOVERNING EQUATIONS AND SOLUTION PROCEDURE

### 23.2.1  Differential equations

The partial differential conservation equations governing the flows considered herein are compactly represented by Equation (23.1), and the accompanying Table 23.1 where the dependent variables and the associated definitions of $\Gamma_\phi$ and $S_\phi$ are listed. Substituting the contents of this table into Equation (23.1), gives rise to the more familiar continuity, three components of momentum, and thermal energy equations, together with the equations for the transport of turbulent kinetic energy and its dissipation rate. The energy equation is complemented by the transport equations for radiative heat fluxes, which are to be used in reacting flows with significant radiation. The task of the solution procedure is to solve the set of these equations with the appropriate boundary conditions (due to the elliptic nature of these equations, boundary conditions must be prescribed for $\phi$ or its normal gradient at all boundaries of the solution domain), and the algebraic equations for the effective viscosity and exchange coefficients.

Table 23.1 Conservation equations

| Variable $\varphi$ | $\Gamma_\varphi$ | $S_\varphi$ |
|---|---|---|
| $\bar{U}$ | $\mu_{\text{eff}}$ | $\dfrac{\partial}{\partial x}\left(\mu_{\text{eff}}\dfrac{\partial \bar{U}}{\partial x}\right)+\dfrac{1}{r}\dfrac{\partial}{\partial r}\left(\mu_{\text{eff}}r\dfrac{\partial \bar{V}}{\partial x}\right)-\dfrac{\partial P^*}{\partial x}$ |
| $\bar{V}$ | $\mu_{\text{eff}}$ | $\dfrac{\partial}{\partial x}\left(\mu_{\text{eff}}\dfrac{\partial \bar{U}}{\partial r}\right)+\dfrac{1}{r}\dfrac{\partial}{\partial r}\left(\mu_{\text{eff}}r\dfrac{\partial \bar{V}}{\partial r}\right)-\dfrac{\partial P^*}{\partial r}-\dfrac{2\mu_{\text{eff}}\bar{V}}{r^2}+\dfrac{\bar{\rho}\bar{W}^2}{r}$ |
| $\bar{W}$ | $\mu_{\text{eff}}$ | $-\left(\dfrac{\bar{\rho}\bar{V}}{r}+\dfrac{1}{r^2}\dfrac{\partial}{\partial r}(r\mu_{\text{eff}})\right)\bar{W}$ |
| $k$ | $\dfrac{\mu_{\text{eff}}}{\sigma_k}$ | $G_{k1}-\bar{\rho}e-\tfrac{2}{3}\bar{\rho}k\left(\dfrac{\partial \bar{U}}{\partial x}+\dfrac{1}{r}\dfrac{\partial}{\partial r}(r\bar{V})\right)$ |
| $e$ | $\dfrac{\mu_{\text{eff}}}{\sigma_e}$ | $\dfrac{e}{k}(C_1 G_{k1}-C_2\bar{\rho}e)$ |
| $\bar{H}$ | $\dfrac{\mu_{\text{eff}}}{\sigma_h}$ | $\bar{S}_{\text{H}}$ |
| $\bar{f}$ | $\dfrac{\mu_{\text{eff}}}{\sigma_f}$ | $0$ |
| $\bar{M}_{\text{fu}}$ | $\dfrac{\mu_{\text{eff}}}{\sigma_{\text{fu}}}$ | $-A_0\bar{\rho}^2(\bar{M}_{\text{fu}}\bar{M}_{\text{ox}}+\overline{m_{\text{fu}}m_{\text{ox}}})\exp(-E/R\bar{T})$ |
| $\overline{m_{\text{fu}}^2}$ | $\dfrac{\mu_{\text{eff}}}{\sigma_{\text{fu}}}$ | $C_{g1}G_{g1}-C_{g2}\bar{\rho}\dfrac{e}{k}\overline{m_{\text{fu}}^2}-2\bar{\rho}^2 A_0\bar{M}_{\text{ox}}\bar{M}_{\text{fu}}\exp(-E/R\bar{T})\left(\dfrac{\overline{m_{\text{fu}}^2}}{\bar{M}_{\text{fu}}}+\dfrac{\overline{m_{\text{fu}}m_{\text{ox}}}}{\bar{M}_{\text{ox}}}\right)$ |
| $\overline{m_{\text{ox}}^2}$ | $\dfrac{\mu_{\text{eff}}}{\sigma_{\text{ox}}}$ | $C_{g1}G_{g2}-C_{g2}\bar{\rho}\dfrac{e}{k}\overline{m_{\text{ox}}^2}-2\bar{\rho}^2 A_0\bar{M}_{\text{ox}}\bar{M}_{\text{fu}}\exp(-E/R\bar{T})\mathrm{i}\left(\dfrac{\overline{m_{\text{ox}}^2}}{\bar{M}_{\text{ox}}}+\dfrac{\overline{m_{\text{fu}}m_{\text{ox}}}}{\bar{M}_{\text{fu}}}\right)$ |
| $\overline{m_{\text{fu}}m_{\text{ox}}}$ | $\dfrac{\mu_{\text{eff}}}{\sigma_{\text{fu}}}$ | $C_{g1}G_{g3}-C_{g2}\bar{\rho}\dfrac{e}{k}\overline{m_{\text{fu}}m_{\text{ox}}}-\bar{\rho}^2 A_0\bar{M}_{\text{ox}}\bar{M}_{\text{fu}}\exp(-E/R\bar{T})\left[\dfrac{\overline{m_{\text{ox}}^2}}{\bar{M}_{\text{ox}}}+\mathrm{i}\dfrac{\overline{m_{\text{fu}}^2}}{\bar{M}_{\text{fu}}}\right.$ $\left.+\left(\dfrac{1}{\bar{M}_{\text{fu}}}+\dfrac{\mathrm{i}}{\bar{M}_{\text{ox}}}\right)\overline{m_{\text{fu}}m_{\text{ox}}}\right]$ |

| $C_1$ | $C_2$ | $C_\mu$ | $C_{g1}$ | $C_{g2}$ | $\sigma_k$ | $\sigma_e$ | $\sigma_f$ | $\sigma_h$ | $\sigma_{\text{fu}}$ | $\sigma_{\text{ox}}$ | $A_0$ | $E/R$ |
|---|---|---|---|---|---|---|---|---|---|---|---|---|
| 1.42 | 1.92 | 0.09 | 2.8 | 2.0 | 0.9 | 1.22 | 0.9 | 0.9 | 0.9 | 0.9 | $10^{10}$ | 19,000 |

$$\mu_{\text{eff}} = \mu + \mu_t$$

$$G_{k1} = \mu_{\text{eff}}\left\{2\left[\left(\dfrac{\partial \bar{U}}{\partial x}\right)^2+\left(\dfrac{\partial \bar{V}}{\partial r}\right)^2+\left(\dfrac{\bar{V}}{r}\right)^2\right]+\left(\dfrac{\partial \bar{W}}{\partial x}\right)^2+\left[r\dfrac{\partial}{\partial r}\left(\dfrac{\bar{W}}{r}\right)\right]^2+\left(\dfrac{\partial \bar{U}}{\partial r}+\dfrac{\partial \bar{V}}{\partial x}\right)^2\right.$$
$$\left.-\dfrac{2}{3}\left(\dfrac{\partial \bar{U}}{\partial x}+\dfrac{1}{r}\dfrac{\partial}{\partial r}(r\bar{V})\right)^2\right\}$$

$$G_{g1} = \mu_{\text{eff}}\left[\left(\dfrac{\partial \bar{M}_{\text{fu}}}{\partial x}\right)^2+\left(\dfrac{\partial \bar{M}_{\text{fu}}}{\partial r}\right)^2\right]$$

$$G_{g2} = \mu_{\text{eff}}\left[\left(\dfrac{\partial \bar{M}_{\text{ox}}}{\partial x}\right)^2+\left(\dfrac{\partial \bar{M}_{\text{ox}}}{\partial r}\right)^2\right]$$

$$G_{g3} = \mu_{\text{eff}}\left[\left(\dfrac{\partial \bar{M}_{\text{fu}}}{\partial x}\dfrac{\partial \bar{M}_{\text{ox}}}{\partial x}\right)+\left(\dfrac{\partial \bar{M}_{\text{fu}}}{\partial r}\dfrac{\partial \bar{M}_{\text{ox}}}{\partial r}\right)\right]$$

$$P^* = \bar{p}+\tfrac{2}{3}\bar{\rho}k-\dfrac{2}{3}\dfrac{\mu_{\text{eff}}}{\bar{\rho}}\left(\bar{U}\dfrac{\partial \bar{\rho}}{\partial x}+\bar{V}\dfrac{\partial \bar{\rho}}{\partial r}\right)$$

### 23.2.2 Finite difference equations

The conservation equations governing the flow field are difficult to solve analytically due to their complexity. With the advent of digital computers, however, it is now possible to solve them numerically. The steps that should be followed in the development of a numerical procedure for solving the governing differential equations are: first to superimpose a grid distribution on the flow domain in order to discretize the differential equations on all the grid points of the field; and second to obtain 'equivalent' algebraic expressions, known as the finite difference equations. The accuracy of the set of finite difference equations which approximates the partial differential equation set, is dependent on the method of formulation of the equations,[5] and on the number of grid nodes which represent the flow field.

All the dependent variables, except the axial and radial velocities, are calculated at the intersection of the grid lines, i.e. at the grid nodes. The axial and radial velocity components are calculated at locations mid-way between the locations where the pressure is calculated, as shown in Figure 23.1, this arrangement has the advantage of providing the pressure gradients at the locations of the calculation of the velocities. The finite difference counterpart of Equation (23.1), excluding the continuity which receives special treatment,

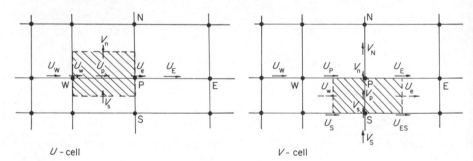

$U$ - cell                                    $V$ - cell

**Figure 23.1**   Grid arrangement

is derived by integrating the partial differential equations (PDE's) over a control volume surrounding each node as illustrated in Figure 23.1. Appropriate assumptions to describe the relation between the nodal values of $\phi$ and the rates of creation/destruction of this entity within the cell and its transport by convection and diffusion across the cell boundaries are made. The rate of creation/destruction is represented by a linearized form as:

$$S_\phi = S_u + S_p \phi_p \tag{23.2}$$

while the convection and diffusion across the cell boundaries are represented by the expression:

$$U_w(\phi_P + \phi_W)A/2 - \Gamma_{\phi_w}(\phi_P - \phi_W)A/(\delta x_{PW}) \tag{23.3}$$

when the quantity $Pe_{\mathrm{w}} = \rho U_{\mathrm{w}} \delta x_{\mathrm{PW}} / \Gamma_{\phi \mathrm{w}}$ is small; and by

$$\rho U_{\mathrm{w}} \phi_{\mathrm{w}}, \ U_{\mathrm{w}} > 0 \quad \text{and} \quad \rho U_{\mathrm{w}} \phi_{\mathrm{P}}, \ U_{\mathrm{w}} < 0 \qquad (23.4)$$

when $Pe_{\mathrm{w}}$ is large.

Here the subscripts P and W refer to the central and west nodes respectively, and w denotes the intervening cell boundary. Assembling the above expressions, and similar expressions for the remaining boundaries yields the finite difference equation in the form:

$$(A_{\mathrm{p}} - S_{\mathrm{p}}) \phi_{\mathrm{p}} = \sum_{n} A_{n} \phi_{n} + S_{\mathrm{u}}' \qquad (23.5)$$

where $\sum_{n}$ denotes summation over the neighbouring nodes, N, S, E, and W; $A_{\mathrm{p}} = \sum_{n} A_{n}$; and $S_{\mathrm{u}}$ and $S_{\mathrm{p}}$ are deduced from the $S_{\phi}$'s of Table 23.1. Equations of this type are written for each of the variables, $\bar{U}$, $\bar{V}$, $\bar{W}$, $k$, $e$, $\bar{M}_{\mathrm{fu}}$, $\bar{f}$, $m_{\mathrm{fu}}^{2}$, $m_{\mathrm{fu}} m_{\mathrm{ox}}$, and $\bar{H}$, at every cell, with appropriate modifications to the total flux expressions (23.3) and (23.4) at cells adjoining the boundaries of the solution domain, in order to account for the conditions imposed there.

An equation for the remaining unknown pressure is obtained by combining the continuity and momentum equations in the manner explained by Gosman *et al.*;[6] this entails connecting changes in pressures, denoted by $P''$, with changes in the velocities $U''$ and $V''$ using approximate formulae derived from the momentum finite difference equations (FDE's). Substitution of these formulae into the continuity equation yields a FDE for $P''$ similar to Equation (23.5), with $S_{\mathrm{u}}$ now representing the local mass imbalance in the prevailing velocity field.

### 23.2.3 Solution algorithm

The FDE's are solved iteratively, employing inner and outer sequences. The outer iteration sequence involves the cyclic application of the following steps: first, a field of intermediate axial and radial velocities, denoted by $U^{*}$ and $V^{*}$, is obtained by solving the associated momentum equations using prevailing pressure $P^{*}$. Continuity is then enforced, by solving the equations for $P''$ and thereby determining the required adjustments to the velocities and pressures. The equations for the remaining variables are then solved in turn. The whole process is repeated until a satisfactory solution is obtained. It is, of course, very expensive and time consuming to reduce the residuals to absolute zero; the value of $10^{-4}$ of the specified inlet value of the entity in question, in any of the FDE's has been proved to give adequate results.

The inner iteration sequence is employed to solve the equations set for the individual variables. Solution is carried out in the form of block iteration, where a simple recurrence formula, described by Gosman *et al.*,[6] is used to solve simultaneously for the $\phi$'s along each grid line, in the line-by-line counterpart of point Gauss–Seidel iteration. Complete convergence of the inner sequence is not necessary, and usually one to three applications of the block procedure suffices. The computer program size is given as $(20,000 + 4x$ number of grid nodes $\times$ number of equations solved) words.

### 23.2.4 The assumptions of initial and boundary conditions

Due to the elliptic nature of the flow, boundary conditions should be specified at the four boundaries of the solution domain. Special practice is employed to impose the boundary conditions at the impermeable walls: these are described in Section 23.3.1. Conditions at the inlet planes are seldom measured in complex geometries, therefore unless measurements are provided, it is reasonable to assume inlet profile distributions that comply with physical observations. Generally the following can be assumed, for various flow properties:

(1)  The mean axial velocity profile at inlet is expressed as a power law:

$$\bar{U} = \bar{U}_0(1 - 2r/d)^n \qquad (23.6)$$

where $0.2 \geqslant n \geqslant 0$ and $\bar{U}_0$ is the centre line velocity at inlet and $d$ is the diameter of the inlet port.
(2)  The mean radial velocity profile should satisfy the continuity.
(3)  The mean tangential velocity profile in a swirling flow is commonly expressed in a forced vortex form as $W = \lambda r$, $\lambda$ varies as a function of the swirl number. In order to maintain numerical stability and fast convergence, it is recommended that the introduction of the swirl intensity factor $\lambda$ should be gradual. In a typical swirling flow configuration the value of the inet swirl intensity may be introduced through fifty iterations; sudden introduction of large disturbances to the momentum equations results in divergence and unstable solution as reported by Khalil.[14]
(4)  The inlet profile of the kinetic energy of turbulence exhibits an increase towards the walls of the inlet pipe; the ratio of the kinetic energy near the wall to that at the centre line of the jet is of the order of 3, as discussed by Hinze[15] and measured by laser Doppler anemometry.[16] An assumed profile, based on the above observations, is expressed as:

$$k = k_0[1 + \delta(2r/d)^2] \qquad (23.7)$$

and $2 \geqslant \delta \geqslant 0$, while $k_0$ is the centre line value of kinetic energy of turbulence at inlet to the solution domain.
(5)  It is rather difficult to measure the dissipation rate, therefore, assumption of the inlet profile is made through a mixing length as:

$$e = C_\mu k^{3/2}/(0.03 y_\text{a}) \qquad (23.8)$$

where $y_a$ is the characteristic length, i.e. jet radius, width of burner annular gap, etc.

(6) The inlet profiles of gas temperature conform to measured values, or are assumed in the form of a turbulent pipe flow temperature distribution with a power law index $1/7$.

(7) The inlet profiles of species concentrations are commonly assumed uniform across the respective fuel and oxidant ports.

At the centre line of the flow and at the plane of symmetry, the gradient type boundary condition assumption, $\partial\phi/\partial r = 0$, is made. At the exit planes, where the conditions are seldom known, unless otherwise stated, it is common practice to locate (by trial and error) the outlet boundary condition in a region where the flow is strongly outwards-directed, and therefore insensitive to downstream conditions. It then suffices to estimate the streamline direction, which in the present examples is taken as normal to the exit boundary; the upwind differencing practice ensures that the remaining conditions are automatically applied. In the flow situations where recirculation zones extend till the exit section (strongly swirled flows), floating boundary conditions are used, this implies a successive change of the assumed exit conditions until momentum and continuity equations are satisfied.

## 23.2.5 Speed of convergence and accuracy

The simultaneous and non-linear characteristics of the FDE's necessitate the employment of special measures to procure numerical stability and convergence, these include:

(a) Under-relaxation of the solution of the momentum and turbulence equations. Typically the factors are 0.3 for momentum, 0.5 for the turbulence equations; however, for the species and energy conservation equations, these factors can be as high as 0.7. A common practice, to speed up the convergence, is to increase these factors to their limiting values, stated above, with the progress of iterations. In the tangential momentum equation the under-relaxation factor is 0.2; in reacting flows the density is under-relaxed heavily.

(b) Linearization of the non-linear source/sink terms in the conservation equations of the kinetic energy of turbulence $k$ and its dissipation rate '$e$'. The source term in the fuel conservation equation is also linearized.

(c) To use the log law of the wall properly, the first grid node near the wall is made to lie within the range in which this law can be used, i.e. $11.5 < y^+ < 100$.

The results obtained should be grid independent in the sense that a considerable increase in the grid size should not alter the computed results significantly.[14] It is the computer size that limits the grid size for the solution of a particular reacting flow problem. In the present calculation, a $22 \times 22$ non-uniform orthogonal grid was used, and the run time was 400 s for a swirling non-reacting flow, which increased to 1000 s when performing the swirling flame calculations.

## 23.3  PHYSICAL MODELLING

### 23.3.1  Turbulence model

As can be seen from Table 23.1, the equations for turbulence energy and dissipation rate involve four constants. In addition, the relation:

$$\mu_t = C_\mu \rho k^2 / e \tag{23.9}$$

is necessary to close the set of equations; this expression includes one additional constant. The five constants necessary for the turbulence model are $C_\mu = 0.09$, $C_1 = 1.42$, $C_2 = 1.92$, $\sigma_e = 1.22$, $\sigma_k = 0.9$ and are identical to those recommended by Gosman *et al.*[6]

The validity of the two-equation turbulence model has been tested previously for several recirculating isothermal flows, e.g. in sudden expansions and downstream of bluff bodies.[16] It was demonstrated that, although the model underpredicted the length of recirculation zone by 20%, it represented qualitatively the flow pattern in these complex flows. In reacting flows a few validation tests using this turbulence model were reported by, among others, Hutchinson *et al.*,[7,8] Khalil *et al.*,[17] and Lilley,[18] and these indicated that general agreement was obtained between measured and predicted flow pattern, flame structure, and heat flux. The discrepancies observed in reacting flows are likely to be due to combustion modelling assumptions.

In order to minimize computer storage and run times, the dependent variables at the wall were linked to those at the grid node next to the wall by equations which are consistent with the logarithmic law of the wall. Thus, the resultant velocity parallel to the wall in question and at a distance $y_p$ from it (corresponding to the first grid node), was assumed to be represented by the modified law of the wall,[19]

$$\tau_W = U_p \chi \mu y_p^+ / \{y_p \ln [E y_p^+ (1 + \psi_0)]\}$$
$$y_p^+ = \rho (k_p C_\mu^{1/2})^{1/2} y_p / \mu \tag{23.10}$$
$$e_p = (C_\mu^{1/2} k_p)^{3/2} / \chi y_p$$
$$\int_0^{y_p} e \, dy = (C_\mu^{1/2} k_p)^{3/2} \frac{1}{\chi} \ln [E y_p^+ (1 + \psi_0)]$$

where $\psi_0 = (dP/dx)(C_\mu^{1/2} y_p / \rho k)$. This modified log law of the wall was found suitable for flows with steep pressure gradients. If $\psi_0$ is put equal to zero, the resultant is the more common form of the wall function.[11]

The heat flux to the wall can be represented in a similar manner, with the wall temperature and heat flux per unit area denoted as $T_W$ and $q_W$:

$$\rho C_p (T_P - T_W) C_\mu^{1/4} k_p^{1/2} = q_W \left( \frac{\sigma_{h,t}}{\chi} \ln \left[ E y_p^+ (1 + \psi_0) \right] + p_j \right) \quad (23.11)$$

and $p_j$ is a function of the laminar and effective Prandtl numbers, details of such function can be found in Khalil *et al.*[17]

### 23.3.2 Combustion models

In turbulent reacting flows, the calculation of the rate of reaction, and consequently the heat release, is essential to allow the evaluation of the local gas temperatures, densities and hence velocities through the various conservation equations. Many previous attempts reported the modelling of diffusion, premixed, and arbitrary fuelled flames.[4,7,8,9,12,17,18,20] These three categories of combustion modelling differ in the assumptions of spatial and temporal fuel and oxidant concentrations and in the reaction progress. The model for diffusion reaction assumes infinitely fast one-step chemical reaction, with the turbulent mixing controlling the reaction, i.e. 'mix–burn' model.[7,8,17] Fuel and oxidant are allowed to coexist in the flame brush at the same place but at different times, through the assumption of a random variation of the mixture fraction $f = (M_{fu} - M_{ox})/i$ with time. The corresponding probability density function is Gaussian and is clipped at the physical limits of $f = 0$ and 1.[20] This model was further modified and the constraint of the clipped Gaussian distribution was removed by calculating the shape of the probability density function $P(f)$ from a corresponding transport equation as suggested by Pope[21] and tested by Khalil.[10]

In flame situations where the reaction is chemically influenced, i.e. the time scale of chemistry is much larger than that of turbulent mixing, as in premixed flames, the simple and rigorous eddy break-up assumptions of Spalding can be used. An equation for the concentration fluctuations, $\overline{m_{fu}^2}$, is solved for in the form (23.1), this equation accounts for the effect of the rate of reaction on the production of the entity $\overline{m_{fu}^2}$. This model does not, however, account for detailed reaction kinetics.

The third flame category, and the one most commonly used in practical applications, is the arbitrary fuelled flames where the effects of chemistry and turbulence are of the same magnitude, i.e. unity Damkohler number. In such a case, an averaged Arrhenius reaction rate, which incorporates the correlations between the fluctuating components of species concentration and temperatures, is used. The modelling of such terms requires the solution of the transport equations for $\overline{m_{fu}m_{ox}}$, $\overline{m_{fu}^2}$, $\overline{m_{ox}^2}$, $\overline{m_{fu}T'}$, $\overline{m_{ox}T'}$, $\overline{T'^2}$, etc., expressed in the form (23.1). Further details of this model and that for premixed flames can be found in Hutchinson *et al.*[8] and Khalil.[12]

### 23.3.3 Radiation model

The radiative heat flux is represented in the total energy equation by a source term. Two radiation models are commonly used with the numerical procedure of the present chapter, namely the four-flux model[22] and the discrete ordinate model.[13] In the former, the radiation intensity vector, expressed in a Taylor series expansion, when substituted in the radiative transfer equation for an absorbing, emitting, non-scattering, grey medium, yields the transport equations of the net flux in the coordinate directions. These equations are represented by Equation (23.1) with zero convection term; the value of $S_H$, the radiative source term in the energy equation is obtained from the solution of these equations. In the latter, the angular distribution of intensity is approximated by a finite number of intensities in discrete directions, spanning the solid angle at each point. The discrete ordinates equations are derived by evaluating the radiation transfer equation for the discrete direction; $S_H$ is obtained from the summation of the product of local intensities and weight factors. Both models were assessed in Hutchinson *et al.*[8] and Khalil and Truelove[13] and were found suitable for furnace and combustion chamber calculations.

## 23.4 COMPUTED RESULTS

After describing the solution procedure, the remaining task in this chapter is to demonstrate the capabilities of this procedure and the associated modelling assumptions. This is accomplished by calculating the flow pattern and local flow properties for several examples and comparing the results with available measurements. Furthermore, predictions are carried out where no measurements are available; this is done to give a clear idea of the wide limits of the procedure.

In furnace and combustor design, it is essential to be able to calculate the local flow pattern and heat transfer characteristics, as this leads to the correct estimation of the rate of heat release and the flame structure, hence the performance of the combustor can be properly evaluated. In most reacting flows, the flow field results from coaxial jets entering a confined sudden enlargement configuration. It must be noted, therefore, that the corner recirculation zone is provoked by this sudden expansion (Figure 23.2). For flame stability purposes, the combustion air normally enters the combustor with a certain degree of swirl; such a tangential momentum creates a reverse pressure gradient along the centre line of the furnace, hence, a central recirculation zone is formed downstream the burner.

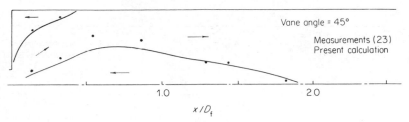

**Figure 23.2** Measured and predicted flow patterns in an axisymmetric furnace for a 45°
vane swirler

Previous computational investigations were carried out in this type of
furnace flows and reviewed in several places such as in the work done by
Hutchinson *et al.*[7,8] and Khalil;[10,12] the comparisons between the calculated
and measured data were displayed and it was established that there is a general
agreement between them.

In this chapter, however, attention is dedicated to comparisons between the
measured and calculated flow patterns such as the recirculation zones, flame
structure, and heat transfer, rather than the local flow properties.

### 23.4.1 Verification of analysis

The first case to be dealt with is the experimental furnace arrangement of
Beltagui and Maccallum;[23] the swirled air jet issues into a confined sudden
expansion. The flow fields at different expansion ratios, namely of 0.4 and 0.2,
were investigated for several different swirler vane angles. The burner was
0.092 m in diameter while the length of the furnace was 1.4 m. The inlet mean
axial velocity was 15.2 m s$^{-1}$ which corresponds to a Reynolds number equal to
$9 \times 10^4$.

The calculated and measured reversed flow regions, for an expansion ratio of
0.2 and swirler vane angle of 45° which in turn corresponds to a swirl number of
0.67, are shown in Figure 23.2. The high intensity of the swirl creates a large
central recirculation zone which persists up to $x/D_f = 1.9$ and, at the same time,
occupies a large portion of the flow. A much smaller recirculation zone is
created at the corner, and is well predicted. At low swirl intensities the central
recirculation zone may disappear while the corner one extends further downstream. Beltagui and Maccallum[23] reported that at a swirl number lower than
the critical swirl number of 0.46 the central recirculation zone disappears; this
swirl number corresponds to a vane angle of 35°. As illustrated in Figure 23.3,
no central recirculation zone was either measured or predicted for vane angles
of 15° and 30°, while the corner recirculation zone is present both in the
measurements and predictions.

It is worthwhile to consider another example that can demonstrate the
capabilities of the procedure. This second example is the flow of Hasenack;[24]

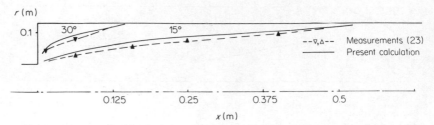

**Figure 23.3** Measured and predicted flow patterns for various swirl intensities in an axisymmetric furnace of Beltagui and Maccallum[23]

the burner arrangement is illustrated in Figure 23.4. It comprised a central excess air jet surrounded by an annular stream of fuel/air mixture; combustion air flows around this mixture with a swirl intensity of 1.7 corresponding to a vane angle of 70°. The burner, whose diameter was 0.87 m, was located at the upstream end of furnace whose height was 11 m and cross section was a 4 m × 4 m square. Attention is devoted in this case to the non-reacting flow in the vicinity of the burner model, the detailed predictions and comparisons of flame structure and heat transfer characteristics can be found in Khalil and Truelove.[13,25]

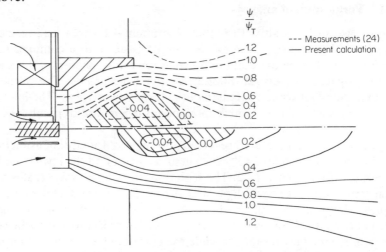

**Figure 23.4** Measured and predicted stream lines in the vicinity of the burner of Hasenack[24]

The measured and calculated streamlines for the swirl intensity of 1.7, under non-reacting conditions, are illustrated in Figure 23.4. The combined effect of the swirling motion and the burner quarl creates a central recirculation zone inside the burner quarl (mouth). There is a shift between the measured and calculated recirculation zone, but there is a general agreement between both which suffices engineering purposes. This difference between the measured and predicted values can be attributed to a number of factors; for example the curved quarl wall was approximated, in the numerical procedure, to the shape of a stepped wall. The finite accuracy of the measurements and the approximations associated with the turbulence model for highly swirling flows can be among the factors, especially for the complex flow investigated here.

It is necessary at this stage to focus the attention on the more complicated problem of reacting flows. To predict correctly the flame performance it is essential to be able to predict the local velocities and gas temperature with a reasonable degree of accuracy. The Delft furnace of Wu and Fricker[26] was chosen as an example of turbulent reacting flows. Wu and Fricker[26] obtained their measurements in this axisymmetric vertical furnace, whose diameter was 0.9 m and height was 5.0 m. The furnace was segmented into ten water-cooled segments. The fuel nozzle, in the measurements procedure, was of a pepper mill shape; it was replaced in the calculation procedure by an annular slot, but still the same rate of fuel issued from the latter burner at an angle of 35° to the burner axis. The fuel jet was surrounded by the combustion air issuing from a divergent water-cooled burner quarl. The fuel was the Dutch natural gas whose composition was 81.3% $CH_4$ + 14.4% $N_2$ + 3.2% $C_2H_6$ + traces. The comparisons which are discussed here are for the non-swirling flame at stoichiometric conditions.

To calculate the flow pattern and flame structure, the combustion model of Khalil,[10] which accounts for the turbulence–chemistry interaction and allows for the presence of the fuel and oxidant in the flame brush, was incorporated in the numerical procedure. The shape of the species probability density function was obtained from the solution of the corresponding transport equation. Furthermore, the discrete ordinate radiation model of Khalil and Truelove[13] was chosen to calculate the radiative heat transfer characteristics of the flame.

The radial distributions of the measured and calculated mean gas temperatures are shown in Figure 23.5, for various axial locations downstream of the burner exit. The general agreement is satisfactory as the maximum discrepancy does not exceed 150 °K. The steep gradients in the temperature profiles trace the reaction zone boundaries. The peak of the temperature, in the vicinity of the burner, is off-centre and is actually located in the layer formed between the air and the fuel streams. Further downstream this peak shifts its location towards the centre line, while the temperatures of the gases increase and become more uniform in the post-flame region. The agreement between the measurement and calculation improves further downstream of the reaction zone. However, the discrepancies observed can be created by the approximations associated with the combustion–turbulence interaction modelling assumptions. Nevertheless, the level of agreement obtained through this procedure clearly justifies its utilization in the design of furnaces and combustion chambers.

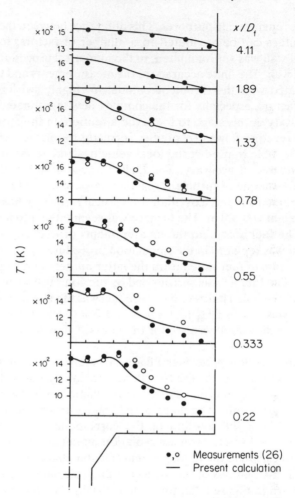

**Figure 23.5** Measured and predicted radial profiles of mean gas temperature in the furnace of Wu and Fricker[26] at zero swirl

Calculation of the total wall heat flux is well within the capabilities of the present procedure. This can be emphasized by calculating the total wall heat flux for the furnace of Wu and Fricker.[26] The convective part is estimated through the local heat transfer coefficient and the temperature difference in the vicinity of the wall. The radiative part is estimated using the discrete ordinate model,[13] the radiation contribution to the total wall heat flux amounts to more than 70% for the above furnace configuration. As shown in Figure 23.6, the calculated and measured total wall heat fluxes are in good agreement with a maximum discrepancy not exceeding 10%. On the same figure, the total heat

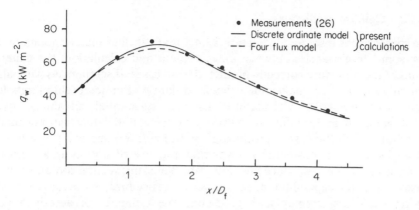

**Figure 23.6** Wall heat flux distribution in the furnace of Wu and Fricker[26]

flux calculated from the four-flux model,[13] is shown; it is clear that the two models perform equally well.

Another example is to calculate the wall heat flux distribution for the furnace of Hasenack.[24] The distribution, which is shown in Figure 23.7 for a high swirl

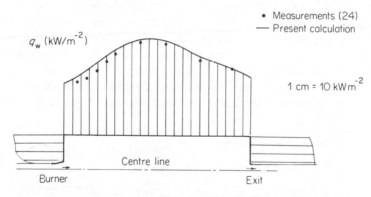

**Figure 23.7** Wall heat flux distribution in the furnace of Hasenack.[24] Swirl number, 5

number of 5, exhibits a peak which is located in the first furnace diameter and corresponds to a short intense flame. The predicted heat flux is in good agreement with the measured data, i.e. the differences were less than 15%. However, in the vicinity of the furnace exit the discrepancy was as high as 25%. The total radiant heat transfer to the cooling tubes, which is obtained from the area under the flux curve, was predicted with an accuracy of 4%.

### 23.4.2 General predictions

The comparisons discussed in Section 23.4.1 make it clear that, although there exist some discrepancies between measurement and calculation, the general trends of the flow were correctly predicted in all cases; discrepancies were also generally small which makes the calculated data within acceptable limits for engineering purposes. Calculations have a substantial advantage over measurement, namely in the cost of producing calculated data. It must be made clear that though the cost of producing and utilizing a numerical procedure is not negligible, it is considerably lower than the cost of setting up an experimental programme to produce the same data. Thus, this section contains a brief illustration of the capabilities of the numerical procedure, in the range outside the reported measurements, in predicting the influence of aerodynamics, and burner and combustor geometry on the local flow and heat transfer characteristics.

The first example which can support this point is provided by the furnace of Hasenack;[24] measurements of the wall heat flux were reported only along the furnace wall, while the two end walls at inlet and exit planes were ignored. Incorporating the calculation of the heat flux at those two walls was readily performed and the predictions, shown in Figure 23.7, agree with what is to be anticipated from the physical nature of such a flow.

The second example is a flame tube whose length to diameter ratio, $L/D_f$, is 3. The fuel was admitted through a central jet and was surrounded by non-swirling combustion air; the products of combustion were allowed to escape through an annular exit in order to be used in a second pass. This configuration is a simulation of a practical fire tube boiler. In this case the procedure was used to calculate the temperature contours, as shown in Figure 23.8, in order to investigate closely the performance of the flame. It is clear

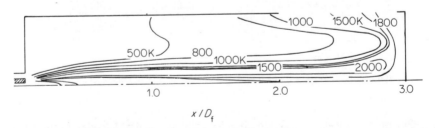

**Figure 23.8** Calculated temperature contours in a fire tube boiler furnace. Swirl number, 0.0

that, in this case, the flame is a long one and impinges on the downstream end wall, i.e. the metal at this end was exposed to very high temperatures. It is obvious that it is important to minimize the thermal stresses on the walls caused by the impingement of the flame. These stresses can be avoided by shortening the flame, which can be brought about in several ways. Introducing swirl to the combustion air, for example, is one of the ways normally used; changing the shape of the burner can be another affecting factor or maybe placing a bluff body in the air and fuel stream. All these methods, and their consequent effects on the flame structure can be investigated thoroughly and inexpensively by the present procedure.

## 23.5 SUMMARY OF CONCLUSIONS

To end this chapter some brief conclusions are drawn from the discussions presented so far:

(1) The elliptic form of the conservation equations is solved by the procedure presented, expressed in finite difference form. The equations were solved simultaneously and iteratively at each grid node of an orthogonal non-uniform grid superimposed on the flow domain.

(2) The two-equation turbulence model, which utilizes the kinetic energy of turbulence $k$ and its dissipation rate $e$, was found to be an adequate representation of the turbulence characteristics.

(3) Several combustion models and radiation models were utilized to provide closure to the species and energy equations, in order to yield the furnace flame properties.

(4) The ability of the present procedure and the associated modelling assumptions, to predict the flow pattern and flame characteristics in furnaces and combustion chambers, was demonstrated through comparisons with measured data. Considerable agreement was obtained between both the measured and predicted data; the discrepancy in the flow pattern amounted to 10%, while that in the gas temperature amounted to 150 K and in the heat flux amounted to 10%.

(5) The procedure was shown to be able to carry out predictions beyond the range of measurements. Parametric investigations, which are of utmost importance to the designer, can therefore be carried out thoroughly and inexpensively to produce the necessary data to improve the performance of furnaces and combustion chambers.

(6) The present work gives a clear picture of a recent development in modelling the various characteristics of flames, and indicates how successfully can these models be employed in this procedure to produce valuable information.

(7) It is important to stress that, although the program is operating satisfactorily, far more work is necessary to improve the various models. A deeper insight into the physical situation must be attained to reach a more ideal procedure.

## REFERENCES

1. E. E. Khalil, P. Hutchinson and J. H. Whitelaw (1979). 'The calculation of the flow and heat transfer characteristics of gas fired furnaces', *Imperial College Report* FS/79/15.
2. B. E. Launder and D. B. Spalding (1972). *Mathematical Models of Turbulence*, 1st edn, Academic Press, London.
3. D. B. Spalding (1979). 'Theories of turbulent reacting flows', *Imperial College Report* HTS/79/1.
4. E. E. Khalil and J. H. Whitelaw (1979). 'The calculation of turbulent reacting flows', *Acta Astronautica*, **6**, 1011–1015.
5. A. D. Gosman and W. M. Pun (1974). 'The calculation of recirculating flows', *Imperial College Report* HTS/74/2.
6. A. D. Gosman, E. E. Khalil and J. H. Whitelaw (1979). 'The calculation of two dimensional turbulent recirculating flows', in *Turbulent Shear Flows I*, F. Durst, B. E. Launder, F. W. Schmidt and J. H. Whitelaw (Eds), pp. 237–255. Springer-Verlag, Berlin.
7. P. Hutchinson, E. E. Khalil, J. H. Whitelaw and G. Wigley (1976). 'The calculation of furnace flow properties and their experimental verification', *J. Heat Transfer*, **98**, 276–283.
8. P. Hutchinson, E. E. Khalil and J. H. Whitelaw (1977). 'Measurement and calculation of furnace flow properties', *J. Energy*, **1**, 210–221.
9. S. ElGhobashi and W. M. Pun (1974). 'A theoretical and experimental study of turbulent diffusion flames in cylindrical furnaces', *Proc. 15th Int. Symp. on Combustion*, pp. 1353–1365, The Combustion Institute.
10. E. E. Khalil (1978). 'Numerical computations of turbulent swirling flames in axisymmetric combustors', in *Flow Mixing and Heat Transfer in Furnaces*, K. H. Khalil, F. M. ElMahallawy and E. E. Khalil (Eds), pp. 231–246. Pergamon Press, Oxford.
11. B. E. Launder and D. B. Spalding (1974). 'The numerical computation of turbulent flows', *Comput. Meth. Appl. Mech. Engng*, **3**, 269–289.
12. E. E. Khalil (1979). 'On the modelling of turbulent reacting flows in furnaces and combustion chambers', *Acta Astronautica*, **6**, 449–465.
13. E. E. Khalil and J. S. Truelove (1977). 'Calculation of radiative heat transfer in large gas fired furnace', *Lett. Heat Mass Transfer*, **4**, 353–365.
14. E. E. Khalil (1979). 'Initial and boundary conditions and their influence on numerical computations of confined elliptic flows', in *Numerical Methods in Thermal Problems*, R. W. Lewis and K. Morgan (Eds), pp. 458–467. Pineridge Press, Swansea.
15. J. O. Hinze (1966). *Turbulence*, McGraw-Hill, London.
16. M. A. Habib and J. H. Whitelaw (1978). 'Velocity characteristics of a confined coaxial jet', *Imperial College Report* FS/78/6.
17. E. E. Khalil, D. B. Spalding and J. H. Whitelaw (1975). 'The calculation of local flow properties in two dimensional furnaces', *Int. J. Heat Mass Transfer*, **18**, 775–791.
18. D. G. Lilley (1979). 'Computer modelling of turbulent reacting flows in practical combustion chamber design', *AIAA Paper* 79-0353.
19. T. Cebeci, E. E. Khalil and J. H. Whitelaw (1979). 'Calculation of separated boundary layer flows', *AIAA Paper* 79-0284. See also *AIAA Journal*, **17**, 1291–1292.
20. A. S. Naguib (1975). 'The prediction of axisymmetrical free jet turbulent reacting flows', *Ph.D. Thesis*, London University.

21. S. B. Pope (1976). 'The probability approach to the modelling of turbulent reacting flows', *Combustion and Flame*, **27**, 299–312.
22. A. G. Demarco and F. C. Lockwood (1975). 'A new flux model for the calculation of three dimensional radiative heat transfer', *Revista Combustibile*, No. 5, **29**, 184–196.
23. S. A. Beltagui and N. R. L. Maccallum (1976). 'Aerodynamic of vane swirled flames in furnaces', *J. Inst. Fuel*, **49**, 183–190.
24. J. Hasenack (1977). Private communication.
25. E. E. Khalil and J. S. Truelove (1977). 'Calculation of radiative heat transfer in furnaces', *UK Atomic Energy Research Establishment Report* AERE R/8747.
26. H. L. Wu and N. Fricker (1976). 'The behaviour of swirling jet flames in narrow cylindrical furnace', *J. Inst. Fuel*, **49**, 144–151.

*Numerical Methods in Heat Transfer*
Edited by R. W. Lewis, K. Morgan, and O. C. Zienkiewicz
© 1981 John Wiley & Sons Ltd

*Chapter 24*

# A Heat Transfer Analysis of Automotive Internal Combustion Gasoline Engines

*Hai Wu and Robert A. Knapp*

## SUMMARY

An analytical method developed to study thermal conditions in automotive internal combustion gasoline engines is presented in this chapter. Heat transfer from the combustion gases inside the engine cylinders to the coolant is considered. A two-dimensional, steady-state formulation subject to given sets of boundary conditions is derived and solved in the analysis using the finite element method. Calculated temperature distributions and heat rejection rates in sections of three different types of automotive gasoline engines are presented as examples to illustrate the procedure and the potential use of the technique. The analysis can be utilized to investigate the thermal effects of various hardware designs, to detect possible thermal problem spots in an engine, and to provide input data for general engine modelling and engine lubrication and wear studies.

## 24.1 INTRODUCTION

Increasingly stringent design criteria are required by automotive engine emissions, fuel economy, durability, weight, and overall performance considerations. Consequently more detailed information on engine thermal conditions would be useful to engineers and designers. In practice, extensive tests are conducted with thermocouples imbedded at key locations to determine the thermal conditions of an engine. Such tests are in general very expensive and time consuming to run, especially when experimental parts must be fabricated. Analytical methods would therefore be very useful in screening various initial design concepts, in providing directions for improvement and for testing, and in evaluating proposed final designs. An analytical method that has been developed to study the heat transfer in automotive internal combustion gasoline engines is presented in this chapter.

The analysis considers the heat transfer from the combustion gases inside the engine cylinders to the coolant. It is two dimensional and steady state. The finite element method is used to solve the heat transfer problem for given sets of boundary conditions. The temperatures and heat transfer coefficients used in describing the boundary conditions on both the combustion gas side and the coolant side are assumed time-averaged values corresponding to a typical engine operating condition.

This chapter first describes the model, its governing equations and boundary conditions, and numerical method, followed by a discussion of results from three examples.

The first example considers a typical production V-8 gasoline engine operating at a high speed and heavy load. The calculated temperature distributions and heat flux through various sections of the engine are presented and discussed.

The second example considers the heat transfer in a prototype piston of a developmental PROCO engine.[20] The effects of some design changes on piston temperature are discussed.

The third example presents a heat transfer analysis of an experimental engine with a 'unique' cooling design.[19] Its use as a design aid is also discussed.

## 24.2 ANALYSIS

### 24.2.1 Basic equations

Heat transfer in internal combustion engines is in general three dimensional and unsteady.[18] However, the periodic variation of the engine wall temperature is very small in comparison with the cyclic fluctuation[1–15,18,21,24] of the temperature difference between the combustion gas and the coolant. The depth of penetration of the temperature fluctuation inside the wall is also very limited.[18,21] Since the main interests in this analysis are the spatial temperature distribution and the heat rejection rate, only the steady-state case is considered. In addition, it is assumed that representative sections as shown in Figure 24.1 can be found such that one can treat the problem either two dimensionally or axisymmetrically.

From Wu and Knapp[18] one has the governing equation for the two-dimensional, steady-state case

$$\frac{\partial}{\partial y}\left(k\frac{\partial T}{\partial y}\right)+\frac{\partial}{\partial z}\left(k\frac{\partial T}{\partial z}\right)=0 \tag{24.1}$$

and for the axisymmetrical, steady-state case

$$\frac{1}{r}\frac{\partial}{\partial r}\left(kr\frac{\partial T}{\partial r}\right)+\frac{\partial}{\partial z}\left(k\frac{\partial T}{\partial x}\right)=0 \tag{24.2}$$

the boundary conditions are

$$k\frac{\partial T}{\partial n}\bigg|_{\text{gas side}} = h_g(T_g - T_{wg}) \qquad (24.3)$$

$$-k\frac{\partial T}{\partial n}\bigg|_{\text{coolant side}} = h_c(T_{wc} - T_c) = q_c/A_{wc} \qquad (24.4)$$

$$-k\frac{\partial T}{\partial n}\bigg|_{\text{general case}} = h(T_{ws} - T_s) = q_s/A_{ws} \qquad (24.5)$$

where $T_w$, $T_g$, $T_c$, and $T_s$ are respectively the wall temperature, the gas temperature, the coolant temperature, and the surrounding temperature. They can all be functions of space. The other notations are: solid density $\rho$, specific heat $C$, thermal conductivity $k$, heat transfer coefficient $h$, direction normal $n$, and heat flux $q/A$.

In order to predict the wall temperature distribution and heat flux accurately the exact boundary conditions on both the gas side and the coolant side as described in Equations (24.3)–(24.5) must be known. However, because of the complexity of the engine geometry and the large number of parameters involved, this is not possible. Simplifying assumptions are thus introduced.

In this analysis an empirically determined set of gas temperatures and gas side heat transfer coefficients are selected corresponding to engine operating conditions under consideration.[18] These are time-averaged values and are shown in Figures 24.3, 24.8, and 24.13. Since detailed spatial gas temperature variations inside the engine cylinder and exhaust port are not readily available, a single temperature value is used for each region studied. However, local variations in heat transfer are partially accounted for by variations in gas side heat transfer coefficient.

As shown in Figure 24.3, it is assumed in this study that localized coolant boiling may occur.[18] The 50–50 mixture of ethylene glycol and water has a boiling temperature of 128 °C at a pressure of 89.6 kPa. In this study it is considered that coolant boiling occurs whenever the coolant side surface temperature reaches 142 °C, 14 °C above the coolant boiling temperature.[17] In practice, this transition point depends mainly on the surface conditions, coolant properties, and flow characteristics.

Below the transition point, the heat transfer is assumed convective and an empirically determined heat transfer coefficient[18,21] is used. No spatial variation in convective heat transfer coefficient is considered. Above the transition point, a trial heat flux value is first assumed. This heat flux is then adjusted corresponding to the calculated surface temperature in an iterative procedure to comply with the experimentally determined boiling curve.[17,18,21,22]

For the piston heat transfer the sump oil temperature and the oil side heat transfer coefficient are assumed to be 115 °C and 2440 Wm$^{-2}$°C$^{-1}$ respectively.[5,8,18,24]

Special considerations are given to the contact areas such as valve face and seat. In the case of the exhaust valve seat, it receives heat from the exhaust gas when the valve is open, about one-third of the total time. The seat receives heat from the exhaust valve when the valve is closed, the remaining two-thirds of the total time. In this study the heat transfer into the exhaust valve seat is assumed to be the product of the temperature difference between the valve face and seat and a fictitious heat transfer coefficient, $h_{fh}$, as defined in the following equation:

$$h_{fh}(T_v - T_h) = \frac{1}{t_o + t_c}[t_o h_g(T_g - T_h) + t_c h_k(T_v - T_h)] \tag{24.6}$$

where $h_k$ is the contact conductance and $t_o$ and $t_c$ are respectively the time intervals for exhaust valve open and closed.

On the other hand, the exhaust valve face receives heat from the exhaust gas when the valve is open and rejects heat to the seat when the valve is closed. The steady-state heat flux out of the exhaust valve face is assumed to be

$$h_{fv}(T_v - T_h) = \frac{1}{t_o + t_c}[t_o h_g(T_g - T_v) - t_c h_k(T_v - T_h)] \tag{24.7}$$

where $h_{fv}$ is a second fictitious heat transfer coefficient. The values for $T_v$, $T_h$, $h_{fv}$, and $h_{fh}$ are first assumed and then iterated until they are self-consistent. The intake valve is treated in a similar fashion.

Heat transfer between the valve stem and valve guide is also considered in the study. The heat transfer coefficient is assumed to be $410 \text{ W m}^{-2} \, {}^\circ\text{C}^{-1}$.[6,8] However, in the example shown in the study, it is assumed that there is no heat transfer between the piston and the engine block. This yields more conservative temperature estimation in piston.

### 24.2.2  Computational method

The problem is to solve Equations (24.1) and (24.2) with boundary conditions (24.3)–(24.5). It is difficult to solve the governing equations, even in the simplified form, for the problems encountered in engines. The finite element method is used in this analysis because of its advantages in handling complex geometry, different material properties, and mixed boundary conditions.[16,18,23]

In solving the problem, the finite element representations of the various sections are constructed first and the corresponding boundary conditions are assigned. The values used in the boundary conditions are assumed time-averaged values corresponding to the engine operating condition under consideration. These values should be adjusted for different operating conditions and for different types of engines. The problem is then solved section-by-section by using a Honeywell 6000 computer. The CPU time for one typical calculation of the cylinder head valve bridge section, for instance, is around 20 seconds. On average, four iterations are required to make the conditions at contact surface self-consistent. The results will be discussed in the next section.

## 24.3 RESULTS AND DISCUSSION

Three examples are presented below to illustrate the technique of the analysis. The first example considers a typical production V-8 automotive gasoline engine operating at a high speed and heavy load.[18] The calculated temperature distributions and heat flux through various sections of the engine are presented and discussed.

The second example considers the heat transfer in a prototype piston of a developmental PROCO engine,[20] and the third example presents a heat transfer analysis of an experimental engine with a 'unique' cooling design.[19] The sensitivity of design changes on engine thermal conditions is discussed.

### 24.3.1 V-8 Engine

A 5.8 L V-8 automotive gasoline engine is considered in the first example. Figure 24.1 shows a typical section through the engine head and cylinder block. It also shows the relative locations of the sections that are considered in this example. The valves and piston are treated as axisymmetrical bodies and the rest as two-dimensional bodies.

Figure 24.2 shows the finite element representation of the cylinder head valve bridge section. The temperatures and heat transfer coefficients used in

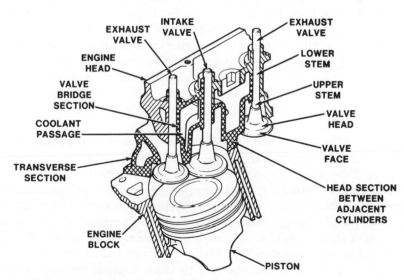

**Figure 24.1** Sketch of a conventional automotive gasoline engine section

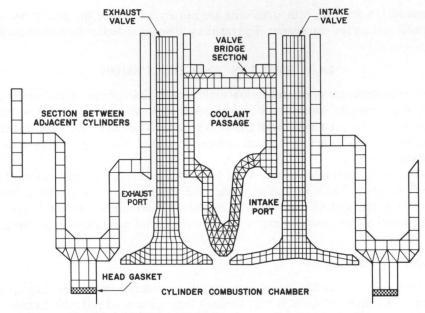

**Figure 24.2** Finite element representation of engine head

the boundary conditions are shown in Figure 24.3. Localized coolant boiling is assumed in the engine head coolant jacket which will be confirmed subsequently. The calculated temperature and heat flow distributions are shown in Figures 24.4 and 24.5. It is seen that the hot-spot in the cylinder head bridge section is at the exhaust valve seat. The hot-spots in the exhaust valve, however, are at the upper stem and around the centre of the valve head.

From Figure 24.5 it is seen that the largest heat flow goes into the head at the exhaust valve seat and then goes out into the coolant along the valley where localized coolant boiling takes place. On the intake valve side, a relatively small amount of heat flows into the coolant along the valley and a negligible amount of reverse heat goes from the coolant into the fresh air and fuel mixture in the intake port.

Figure 24.6 shows the temperature distribution of the transverse head section which also includes part of the exhaust port.

A three-dimensional composite temperature distribution of the exhaust valve is shown in Figure 24.7. It is interesting to see that the temperatures are not circumferentially uniform. This may generate hot-spots in the valve and may cause valve warpage. However, it should be noted here that the non-uniformity of the circumferential temperature distribution in an actual valve should be less than that shown in Figure 24.7 because of the heat transfer among the three sections and also due to valve rotation. Similar comments apply to temperatures at the valve seat.

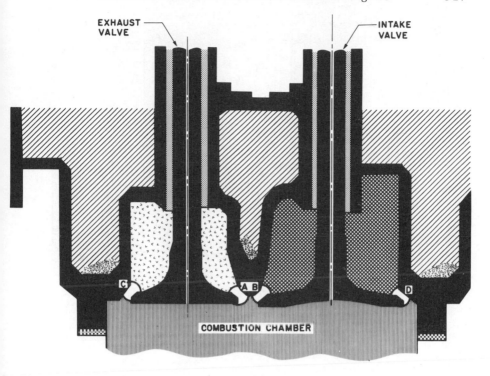

| | | |
|---|---|---|
| **LEGEND** | | |

| | | |
|---|---|---|
| | EXHAUST PORT | T - 815 °C   h - 570  W/m² °C |
| | INTAKE PORT | T - 65      h - 240 |
| | VALVE GUIDE | T - CALCULATED   h - 410 |
| | COMBUSTION GAS | T - 870      h - 440 |
| | COOLANT | T - 93      h - 5700 |
| | BOILING COOLANT | T - 93      h - 5700 - 18000 |
| A | EXHAUST VALVE | FACE  T - 298      h - 2100<br>SEAT  T - 587      h - 3400 |
| B | INTAKE VALVE | FACE  T - 269      h - 3800<br>SEAT  T - 381      h - 1900 |
| C | EXHAUST VALVE | FACE  T - 251      h - 2100<br>SEAT  T - 564      h - 3400 |
| D | INTAKE VALVE | FACE  T - 202      h - 3400<br>SEAT  T - 344      h - 2200 |

**Figure 24.3**   Boundary conditions for engine head

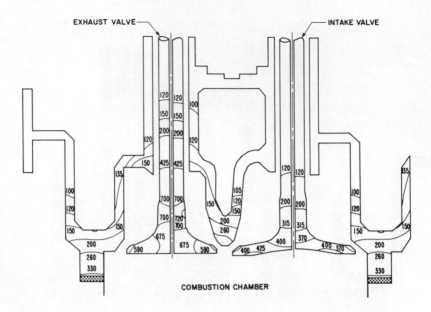

**Figure 24.4**  Temperature distributions of engine head

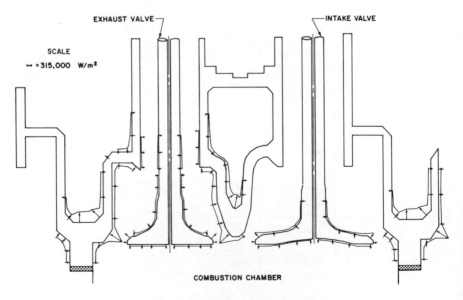

**Figure 24.5**  Heat flux distributions

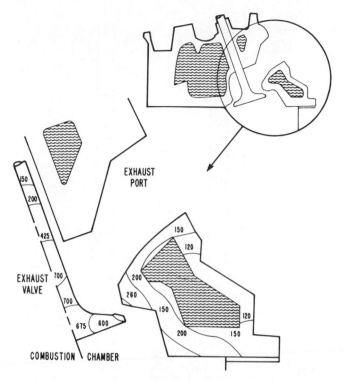

**Figure 24.6**  Temperature distribution of head transverse section
including part of exhaust port

Since detailed engine wall temperature measurements are not readily available, the results presented in this example were compared with a few published data at several discrete points in Wu and Knapp[18] and the general agreement was good.

This type of analysis can be used to investigate the thermal effects of different engine and cooling system designs and to detect any potential thermal trouble spots in the engine. The detailed temperature information may be used to provide input data for general engine modelling, engine knock and flame quenching analyses, and engine lubrication and wear studies.

### 24.3.2  PROCO piston

The second example considers the heat transfer in the piston of a prototype PROCO—a stratified engine.[20] The prototype PROCO piston has an internal bowl-shaped combustion chamber as shown in Figure 24.8. This configuration is considered the baseline design in the example. The presence of the bowl exposes more surface area of the piston to the hot combustion gases and,

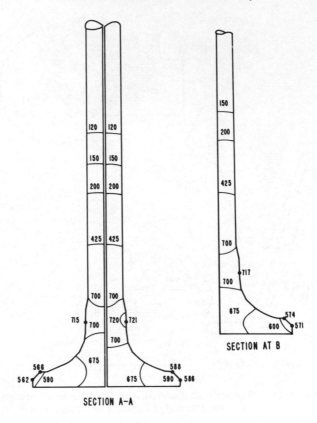

SECTION A-A

SECTION AT B

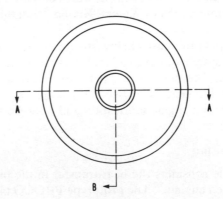

**Figure 24.7** Three-dimensional composite temperature distribution of exhaust valve

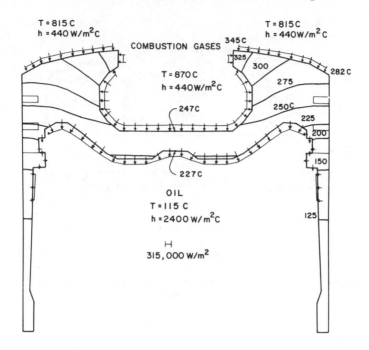

**Figure 24.8** Thermal condition in major axis section of prototype PROCO piston

consequently, may result in developing hot-spots around the upper edge of the bowl and the top piston ring area.

Figures 24.8 and 24.9 show the boundary conditions and the calculated temperature and heat flux distributions in the major-axis section and the pin-boss section of the piston, respectively. It can be seen, as suspected, that the maximum temperatures occur at the upper edge of the bowl and the top-ring area. It should be reiterated here, however, that the example assumes no heat transfer between the piston and the engine block. This, therefore, yields more conservative temperature distribution in the piston.

To alleviate these hot-spots in the piston, several design concepts, which were intended either to increase the cooling of the piston or to reduce the heat input into the piston, were evaluated. Figure 24.10 shows a design modification that improves the heat transfer from the piston to the sump oil by enlarging the undercut. It can be seen that this modification is effective in reducing temperatures around the piston rings.

Figure 24.11 illustrates a design concept which retards the heat transfer to the piston by coating the piston crown with a ceramic layer. This modification reduces the overall piston temperature. Figure 24.12, on the other hand, demonstrates the effects of a different bowl geometry which limits the

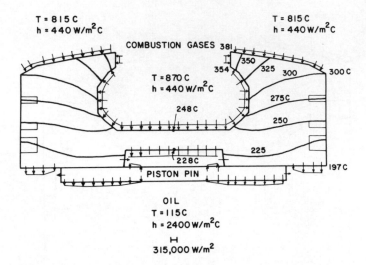

**Figure 24.9** Thermal condition in pin-boss section of prototype PROCO piston

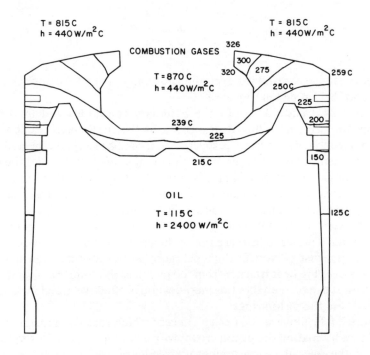

**Figure 24.10** Thermal condition in major axis section of PROCO piston with increased cooling

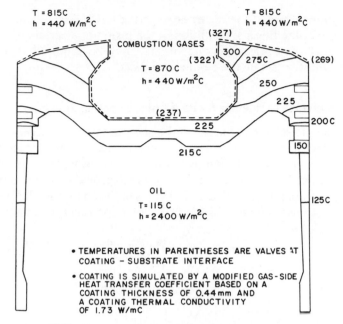

**Figure 24.11**   Thermal condition in major axis section of PROCO piston with ceramic coating

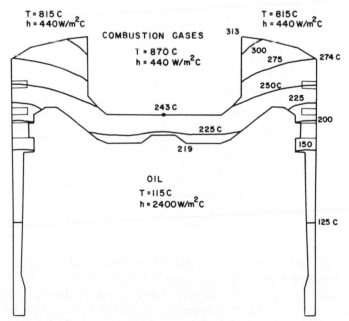

**Figure 24.12**   Thermal condition in major axis section of PROCO piston having different bowl geometry

maximum temperature at the upper edge of the bowl. Each of these modifications improves the thermal condition in the piston to a certain extent but a combination of these ideas, among others, may be required to meet the thermal specification set by the engine designers.

### 24.3.3 Aluminium engine head with a unique cooling design

The third example illustrates the use of the analysis technique presented in this chapter as a design aid to the engine engineers. In order to make the aluminium alloy engine cost effective, a 'unique' cooling concept was proposed.[19] This concept improves the coolant jacket design which provides better control over the operating temperature of all the vital areas of the engine head. The general concept is illustrated in Figure 24.13. The coolant passages in the head consist

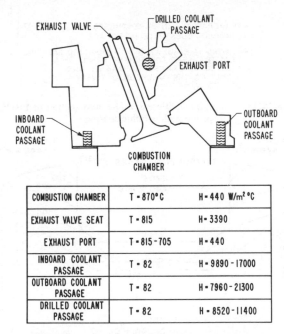

| COMBUSTION CHAMBER | T = 870° C | H = 440 W/m² °C |
|---|---|---|
| EXHAUST VALVE SEAT | T = 815 | H = 3390 |
| EXHAUST PORT | T = 815 - 705 | H = 440 |
| INBOARD COOLANT PASSAGE | T = 82 | H = 9890 - 17000 |
| OUTBOARD COOLANT PASSAGE | T = 82 | H = 7960 - 21300 |
| DRILLED COOLANT PASSAGE | T = 82 | H = 8520 - 11400 |

**Figure 24.13** Cross section and boundary conditions of unique cooling design aluminium engine head

of two troughs or channels cast around the combustion chambers with a single drilled hole adjacent to the valve guides. All the coolant enters from the front, flows through the engine block, and returns through the cylinder heads.[19] When the coolant enters each cylinder head, it is divided into the three parallel channels described above. The analysis technique is used to estimate thermal conditions of such a design and to size the three coolant passages.

Figure 24.13 shows the cross section through the exhaust valve of the engine head and the boundary conditions used in the analysis. Figure 24.14 shows the calculated temperatures.

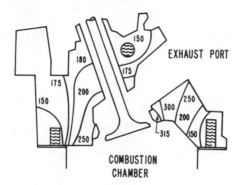

**Figure 24.14** Temperature distribution in unique cooling design aluminium engine head

In order to conserve exhaust heat, a design with a 2.5 mm ceramic exhaust port liner was also considered. Figure 24.15 shows the calculated temperatures. It can be seen that the overall temperature with the liner is much less than that without the liner.

The heat flux distributions for the cases with and without the liner are summarized in Table 24.1. The information of heat flux into the inboard channel, outboard channel, and drilled hole, as shown, is used to size these three passages to achieve a coolant flow distribution directly proportional to the flux distribution.

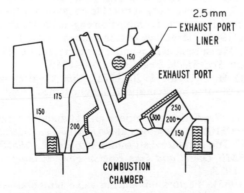

**Figure 24.15** Temperature distribution in unique cooling design aluminium engine head with ceramic exhaust port liner

Table 24.1   Heat flux distribution in unique cooling design aluminium engine head

|  | Without liner | With liner |
|---|---|---|
| Inboard channel | 21,100 W m$^{-1}$ 35.2% | 19,800 W m$^{-1}$ 37.1% |
| Outboard channel | 23,900 W m$^{-1}$ 39.9% | 22,000 W m$^{-1}$ 41.3% |
| Drilled hole | 14,900 W m$^{-1}$ 24.9% | 11,500 W m$^{-1}$ 21.6% |
| Section total | 59,900 W m$^{-1}$ 100% | 53,300 W m$^{-1}$ 100% |

## REFERENCES

1. G. Eichelberg (1939). 'Some new investigations on old combustion engine problems', *Engineering, Lond.*, **148**, 436–466, 547–549, 603–605.
2. V. D. Overbye, J. E. Bennethum, O. A. Uyehara and P. S. Myers (1961). 'Unsteady heat transfer in engines', *SAE Trans.*, **69**, 461–494.
3. W. J. D. Annand (1963). 'Heat transfer in the cylinders of reciprocating internal combustion engines', *Proc. Instn Mech. Engrs*, **177**, 973–996.
4. G. L. Borman (1964). 'Mathematical simulation of internal combustion engine process and performance including comparisons with experiments', *Ph.D. Thesis*, University of Wisconsin.
5. N. D. Whitehouse, A. Stotter and C. Gray (1964, 1965). 'Piston thermal loading', *Proc. Instn Mech. Engrs*, **179**, pt. 34, 158–167.
6. A. Stotter, K. S. Woolley and E. S. Ip (1965). 'Exhaust valve temperature—a theoretical and experimental investigation', *SAE Paper* 969A, presented at *SAE Automotive Engineering Congress and Exposition, Detroit 1965.*
7. G. Woschni (1967). 'A universally applicable equation for the instantaneous heat transfer coefficient in the internal combustion engine', *SAE Trans.*, **76**, 3065–3083.
8. S. J. Pachernegg (1967). 'Heat flow engine pistons', *SAE Trans.*, **76**, 2995–3030.
9. F. Nagao (1970). 'Measurement of cylinder gas temperature of internal combustion engines', *Bull. Jap. Soc. Mech. Engrs*, **64**, 1240–1246.
10. W. J. Seale and D. H. C. Taylor (1970, 1971). 'Spatial variation of heat transfer to pistons and liners of some medium speed diesel engines', *Proc. Instn Mech. Engrs*, **185**, 203–218.
11. R. Limpert (1971). 'Prediction of temperatures attained in a diesel engine cylinder head', *SAE Paper* 710617, presented at *SAE Mid-Year Meeting, Montreal 1971.*
12. M. Goyal (1973). 'Temperature distribution in pistons', *ASME Paper* 73-DGP-8, presented at *ASME Diesel and Gas Engine Power Conference and Exhibition, Washington DC 1973.*
13. J. M. Cherrie (1965). 'Factors influencing valve temperatures in passenger car engines', *SAE Paper* 650484, presented at *SAE Mid-Year Meeting, Chicago 1965.*
14. S. Furuhama and Y. Enomoto (1973). 'Piston temperature of automobile gasoline engine in driving on the road', *Bull. Jap. Soc. Mech. Engrs*, **16**, 1385–1400.

15. C. C. J. French and K. A. Atkins (1973). 'Thermal loading of a petrol engine', *Proc. Instn Mech. Engrs*, **187**, 561–573.
16. E. L. Wilson and R. E. Nickell (1966). 'Application of the finite element method to heat conduction analysis', *Nucl. Engng Des.* **4**, 276–286.
17. R. A. Knapp (1972). 'Limiting heat flow rates in engine coolants', *M.Sc. Thesis*, Michigan Technological University.
18. H. Wu and R. A. Knapp (1978). 'Thermal conditions in an internal combustion engine', in *Gas Turbine Heat Transfer, 1978*, V. L. Eriksen and H. L. Julien (Eds), ASME, New York.
19. R. P. Ernest (1977). 'A unique cooling approach makes aluminum alloy cylinder heads cost effective', *SAE Paper* 770832, presented at *SAE Passenger Car Meeting, Detroit 1977.*
20. A. J. Scussel, A. O. Simko and W. R. Wade (1978). 'The Ford PROCO engine update', *SAE Paper* 780699, presented at *SAE West Coast Meeting, San Diego 1978.*
21. E. R. G. Eckert and R. M. Drake (1959). *Heat and Mass Transfer*, McGraw-Hill, New York.
22. W. H. McAdams (1954). *Heat Transmission*, 3rd edn, McGraw-Hill, New York.
23. O. C. Zienkiewicz (1972). *The Finite Element Method in Structural and Continuum Mechanics*, McGraw-Hill, New York.
24. G. Woschni and J. Fieger (1979). 'Determination of local heat transfer coefficients at the piston of a high speed diesel engine by evaluation of measured temperature distribution', *SAE Paper* 790834, presented at *SAE Off-Highway Vehicle Meeting and Exposition, Milwaukee, 1979.*

# Author Index

The page numbers in ordinary type indicate where the author's work is mentioned in the text. The numbers in bold type refer to the pages where the journal or book reference is given.

# Subject Index